15X1$31.95

Essentials of Geography and Development

Charles E. Merrill Publishing Company
A Bell & Howell Company
Columbus   Toronto   London   Sydney

Contributors
Leonard Berry
Louis De Vorsey, Jr.
James S. Fisher
Don R. Hoy
Douglas L. Johnson
Clifton W. Pannell
Roger L. Thiede
Jack F. Williams

Essentials of Geography and Development

Concepts and Processes

Edited by

Don R. Hoy

Second Edition

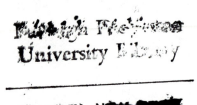
Published by Charles E. Merrill Publishing Company
A Bell & Howell Company
Columbus, Ohio 43216

This book was set in Univers
Text Designer: Cynthia Brunk
Cover Designer: Cathy Watterson
Production Coordinator: Rebecca Money
Cover photograph by Volker Hinz/STERN

Photo Credits: iii—Volker Hinz/STERN; 1, 239, 283, 309,
341—United Nations; 71—United Nations/Toronto Transit
System; 125—Danish Ministry of Foreign Affairs; 175—TASS
from Sovfoto; 211—Consulate General of Japan, New York

Library of Congress Catalog Card Number: 83–63014
International Standard Book Number: 0–675–20142–X
Printed in the United States of America
    2 3 4 5 6 7 8 9—89 88 87 86 85

PREFACE

The world is getting smaller, more crowded, and more interdependent. It is smaller because improved transport technology has shortened travel time, and better electronic communications bring almost any place on earth to our home. Faraway places with strange-sounding names seem neither so far away nor so unusual to our ears. The world is more crowded because a veritable explosion in population numbers has occurred: a hundred years ago about 1.5 billion people inhabited the earth, but by 1960 there were 3 billion. At the end of the twentieth century, the world will hold more than 6 billion people. We have become more interdependent as complex economic systems have evolved with labor and regional specialization. Nations with increasing populations and higher levels of living need more supplies of diverse origin and in return need markets for their surplus production.

What happens in other parts of the world does affect us directly. For example, the United States has long been a major producer and exporter of soybeans; the crop is raised principally in the Midwest and the Southeast. In recent years, however, Brazil has challenged traditional United States soybean markets. Brazil is attempting to occupy effectively the tropical upland region to the north and west of Rio de Janeiro. Much of the eastern part of the uplands is already settled, and the agricultural frontier is being pushed inland (Figure 17–4). One of the cash crops grown on these uplands is soybeans, and today Brazil is a leading exporter of this crop. The increased competition created by Brazil has led to a significant drop in bean prices received by the United States farmer. In turn, depressed soybean prices mean fewer tax dollars. For college students, reduced governmental funds for university support could lead to higher tuitions. Thus increased soybean production in Brazil could contribute to higher tuitions in some United States universities.

Most Americans are geographic illiterates—a strange circumstance, considering the high level of educational attainment and the status of the United States as a world power. We need to develop an awareness of the world around us, for events both near and far have direct and indirect effects on our daily lives. It is imperative that we become more knowledgeable about the world, for ignorance can hurt, problems can magnify, and opportunities can be lost.

This book is the second edition of *Essentials of Geography and Development.* This edition preserves those aspects found useful in its predecessor and contains some modifications that make the book still more serviceable. Preserved is the basic thrust of development processes, broadly defined, and several aids to assist the reader. For example, the text is generously illustrated with maps, graphs, tables, and photographs. There is also a glossary, which gives definitions or explanations of many terms, and a table of selected national data in the Appendix provides a current ready reference for comparing nations and world regions. At the end of each part is a short annotated list of readings for those who are interested in exploring a particular idea a bit further or who wish more background information. Finally, in this age of interdependence and standardization, the move toward the metric system is recognized by presenting material in both the English and metric systems of measurement. Conversion factors for some of the more widely used measures are given in the Appendix.

Several modifications and new features have been introduced. First, throughout the book, key words, concepts, and points have been italicized. Second, a chapter on geography has been added to introduce the reader to this exciting field. Third, the area of Africa and the Middle East has been divided into two major regions for discussion. As part of this reorganization, the coverage of North Africa and the Middle East and Subsaharan Africa has been substantially increased. Fourth, the coverage of Western Europe has also been increased so that all nations of the region are examined. Last, the data used in all sections of the book have been updated. Not only were statistical data revised, but also maps and charts were redone to incorporate new information.

D. R. H.

CONTENTS

FIGURES/TABLES

FIGURES

TABLES

I

Some Basic Concepts and Ideas

James S. Fisher
and Don R. Hoy

1

Geography: An Exciting Discipline

James A. Michener, the noted American novelist, is a staunch supporter of geography. He believes that his novels *(Hawaii, Caravans, The Covenant, Centennial* and *Chesapeake)* enjoy much success because he fixes them firmly within a *regional geographic context.* In researching materials for his novels Michener relies heavily on geographic works. Modern corporations employ geographers and use *geographic principles* to aid in locating new manufacturing plants and retail stores. An *optimal location* can increase profits and lower costs. A stock brokerage employs a geographer to monitor weather conditions in the Midwest in order to anticipate the volume of the harvest. Based on predicted supply figures, the company's brokers can advise their clients on buying or selling animal and grain futures and stocks related to agriculture.

DEVELOPMENT OF GEOGRAPHY

These examples of geography and geographers illustrate only a sample of *geography's character and utility.* Geography is an old discipline with a rich and varied heritage. Its solid foundation rests on the works of ancient scholars who recorded the *physical and cultural characteristics* of lands near and far. Although the study of geography apparently arose in many civilizations, the *Greeks* made the most enduring contributions. In fact, the term *geography* comes to us from the Greek *geo* (the earth) and *graphos* (to write about or describe).

Greek Contributions

The *early Greeks* faced formidable geographic problems. They were concerned about the size and shape of the earth and its relation to the rest of the cosmos; how does one show where places are and what people do in the various parts of the world? If the Greeks did not solve these problems, they pointed the way to their solution. For example, consider the case of *Eratosthenes* (276?–196? B.C.), who, among other things, measured the

James A. Michener, the noted American novelist, often uses geographic concepts and principles. In his novels *Hawaii, Centennial, The Covenant,* and *Chesapeake* he used the concept of sequential occupancy as the principal method of organizing and presenting his story. (John Kings)

Eratosthenes and other Greeks recognized the *need for maps* in order to show the *relationships* of one place or region with another. Moreover, they needed a means of locating themselves on the earth and of describing their location to others. (To visualize their difficulty, place a mark on a smooth, uniform ball and try to describe the mark's position.) Fortunately the earth has an axis around which it rotates, and thus we have two known points where the axis intersects the surface (north and south poles). With these points the Greek geographers established a reference line (equator) halfway between the poles and another line from pole to pole. From these geographic reference points and lines a grid of *latitude* and *longitude* lines was drawn, and any point on earth could be located by just two numbers. (For example, Washington, D.C.'s location is 38°50′ N, 77°00′ W.) The next step was to use the *geographic grid* to construct a map whereby all or a part of the earth could be reduced to a two-dimensional plane, and the relative positions of places and regions could be ascertained. The Greek geographers (indeed all other geographers down through the ages) were particularly fond of maps, and maps became the hallmark of geography.

During Greek and Roman times, *geography thrived,* for new lands were discovered, and it was

earth's circumference. Eratosthenes lived in Alexandria, Egypt, and learned that only on one day of the year (northern hemisphere summer solstice) the noon sun shone directly down a well near the modern city of Aswan. He measured the sun's angle at Alexandria, some 500 miles (800 kilometers) north of Aswan, at the same time and date and found that the sun was not vertical but cast a shadow of 7.2 degrees. He reasoned that the sun's rays were parallel to each other and applied the geometric rule that when a straight line intersects two parallel lines, the alternate interior angles are equal (Figure 1–1). Therefore, the distance between Aswan and Alexandria intercepts a 7.2 degree arc of the earth's surface covering a surface distance of 500 miles. He then computed the value of 360 degrees and arrived at an estimate of 25,200 miles (40,554 kilometers) for the earth's circumference. (The earth's circumference at the equator is actually 24,901.5 miles.)

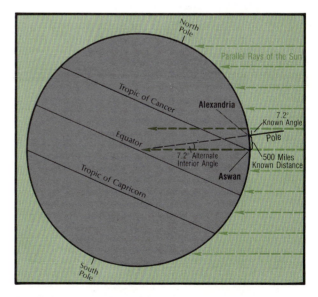

FIGURE 1–1 Eratosthenes' Calculation of the Earth's Circumference.
About 200 B.C. Eratosthenes measured the earth's circumference with an amazing degree of accuracy. By careful observation and application of simple geometry, he computed the earth's circumference to be 25,200 miles (40,554 kilometers).

Ptolemy (A.D. 100?–165?) was both a geographer and an astronomer. He helped develop the geographic grid of latitude and longitude and constructed one of the first maps of the world. Ptolemy and other ancient Greeks laid the foundation for modern geography. (Courtesy of the Newberry Library, Chicago)

of great practical importance to inventory the resources and characteristics of these new areas. The accomplishments of geographers such as Herodotus, Ptolemy, Strabo, and Hipparchus established them as leading figures in their society. Geography's position was an acclaimed one until the Roman Empire began to decline and the Dark Ages approached.

Dark Ages

Geography, indeed all the sciences, stagnated and fell into disrepute with the coming of the Dark Ages. Concurrent with the European Dark Ages, however, was a *golden period of Arab civilization,* and it was the Arabs who preserved much of the Greek and Roman geographic legacy. The Arabs themselves were formidable geographers. *Arab geographers* continued the tradition of mapmaking and recording data on maps. They also traveled widely and left copious notes of their journeys. So too did they continue the interest in the physical earth and the processes that created differences from one place to another. Avicenna (980–1037), for example, was one of the first to understand some of the processes by which mountains are built and destroyed. Ibn Batuta recorded his fourteenth-century travels throughout the Arab world in northern Africa and southern Asia and beyond into northern China.

The Renaissance and Age of Discovery

Beginning in the fifteenth century, Western Europe gradually emerged from the Dark Ages, and a new period of the *Renaissance and Age of Discovery* brought a *resurgence of geography* and other sciences. New routes to the Orient were needed and a new age of the geographer-explorer began. These Renaissance geographers were much like their forebears. They devised *better maps,* described the *physical and cultural characteristics of foreign lands,* and tried to understand the *processes that created differences and similarities* from one place to another. Explorers such as those under the aegis of the world's first geographic institute, founded in 1418 by Prince Henry of Portugal, and others such as Christopher Columbus and Ferdinand Magellan, led in developing new routes to the outside world and new information about the world.

The revelations from these *geographic voyages of discovery* had a profound impact in Europe. Scholars began to question age-old concepts and found some of them wanting in the light of discoveries in other parts of the world. An age of *scientific reasoning* began with experimentation and hypothesis testing. In the natural sciences new explanations were presented that challenged old ideas of the origin of continents and oceans, the formation of landforms, and the evolution of plants and animals. Following the early explorers were the scientific travelers (natural historians) who sought evidence and explanation for the varied world in which we live. Among them was the great German geographer Alexander von Humboldt (1769–1859). Von Humboldt traveled widely in Europe and Latin America. His curiosity, ability to observe carefully, and broad background of study (botany, physics, chemistry, Greek, archeology, and geology) allowed him to synthesize information from a variety of fields into a *coherent geographic analysis.*

By the mid-nineteenth century geography was a respected discipline in *European universities,* and

Modern Geography

Modern geography has grown beyond the stricture of the Greek earth description. Today, geographers not only describe by way of words, maps, and statistics but also search to *explain why things are distributed* as they are over the earth. Modern geography might best be characterized as the *study of distributions and relationships* among different distributions (for example, the distribution of economic activities and per capita income) and the resultant regional features. Yet at the same time modern geography remains true to its heritage. *Four traditions* persist in geography: (1) how things are organized in area (spatial distribution), (2) the relationship between people and the land that supports them, (3) the study of regions including analysis and explanations of why things are dis-

Alexander von Humboldt (1769–1859) was one of the founders of modern geography. He traveled widely in Eurasia and Latin America and carefully recorded his observations and conclusions. Although primarily a physical geographer, von Humboldt also formulated the outline of systematic regional geography. (The Geoffrey J. Martin Collection)

geographical societies served as important meeting places for scholars of all disciplines interested in the world around them. From Europe the age of geography spread to the far corners of the world.

In the *United States,* geography found a fertile environment in which to grow. People were eager for knowledge about their country and especially the frontier zones. Geographical literature was particularly popular. In 1852 the American Geographical Society was formed, followed by the National Geographic Society (1888) and the Association of American Geographers (1904). Geography as an academic field of study began to flower in the latter part of the century and in the subsequent one hundred years spread from a few centers to nearly every college and university.

William Morris Davis (1850–1934) was the principal founder of the Association of American Geographers and served as its first president. Davis was an ardent disciple of geography and a prolific writer. He is most famous for his work describing geomorphic cycles. (The Geoffrey J. Martin Collection)

tributed as they are, and (4) the study of the physical earth, perhaps the oldest of all geographic traditions.

KINDS OF GEOGRAPHY

There are many *kinds of geography*. Geography's principal subdivisions are twofold: (1) physical and human and (2) systematic and regional. *Physical geographers* study the environment from the viewpoint of distribution and process. *Landform geography* (geomorphology), for example, is concerned with the location of terrain features and why those features are located where they are. The geomorphologist may study the impact of stream deposition in a floodplain, the features created by wind erosion in a dry land, or the formation of coral reefs around a tropical island. A *biogeographer* is interested in the distribution of plants and animals, how they live together, the processes that affect the biologic earth, both natural and people-induced, and the effect of the process-induced change on human life. A *climatologist* studies the long-term characteristics of the atmosphere and how temperature or energy and moisture conditions create differences in climate from one part of the earth to another. Physical geography involves the study of the physical earth as a stage for human activities.

The *human geographer* studies various aspects of humanity's occupancy of the earth. *Urban geographers,* for example, examine the location and structure of cities in an attempt not only to explain why urban areas are distributed as they are but also to account for the pattern of distributions within the cities. Urban geographers are interested in the process of urban growth and decline, the types of activities carried out in cities, and the movement of goods and people within urban places. *Cultural geography* examines the way in which a group of people organizes itself and studies cultural institutions such as language, religion, and social and political structure. *Economic geography* involves the study of people's systems of livelihood, especially the distribution of these activities and explanations for the distribution. The economic geographer is concerned with the analysis of natural and cultural resources and how these resources are utilized for economic gain. *Historical geography* is the study of past landscapes. How, for example, did the population of the Great Plains organize itself in 1870 in comparison to 1920? What characteristics existing at one time have per-

sisted and what effect do these characteristics have on present-day distributional patterns? Michener used the geographic method of *sequential occupancy* as the organizing theme of some of his novels. He described the geography of an area at different periods, gradually building an image of how an area is organized today. Historical geography thus provides us with a depth perception in time to explain present patterns and their reasons for being.

The various kinds of geography cited above are examples of *systematic (topical) geography.* Systematic geographers normally study one aspect of the field—landforms, economic activities, or urban places. Another method of geographic study is *regional geography.* Regional geography involves the analysis of environmental and human patterns within a single area and organizes a wide variety of facts into a coherent form in order to explain how a region is organized and functions. Regional geographers, in essence, are experts on a particular area of the world and apply systematic approaches to the understanding of that region.

All fields of geography, although studying diverse sets of phenomena, are linked together by the *geographic viewpoint;* that is, geographers analyze the spatial arrangement *(distributions)* of whatever it is they are studying and search for an explanation for the patterns and *interrelationships* between these patterns and other phenomena. All geographers rely on the *map as an analytical tool,* and many have added the use of the computer and remote sensing techniques to aid in recording and analyzing data.

CAREERS IN GEOGRAPHY

Modern geographers differ from their forebears in the sense that of the two fundamental questions of geography—where is it and why is it there?—emphasis has turned more to explanation than description. This *change in emphasis* has increased the geographer's utility in solving many problems of our contemporary world.

Education

Traditionally and even today most geographers are employed as *teachers* at the high school and university level. Yet most American students are *geographic illiterates.* A recent public poll of subjects essential for high school students indicated that 81 percent of the respondents believe geography is an essential subject. No comparable survey exists for geography's role at the university level, but numerous educators and laymen have decried the geographic ignorance of college students. Professor Marion J. Levy, Jr., a sociologist at Princeton University, states that "almost any test of geographical literacy would separate American students from students trained abroad . . . and the same may be said of the vast majority of American faculty members."[1] Bob Wedrich, a columnist for the *Chicago Tribune,* recently lamented the lack of geographic training at the college level. Somewhat tongue-in-cheek he wrote: "Based on the disquieting reports of geographic shortcomings, it is not unreasonable to suspect that some people still think the world is flat. But without classroom motivation, it is unlikely they ever will question what happens when the oceans splash over the edge."[2]

Aside from the obvious need for geographic education for its own sake, the need for geographic instruction is increasingly critical for several other reasons. First, *economic interdependence:* There is a growing need for products and materials not sufficiently available within our national borders. We are increasingly reliant on foreign sources. For example, much of the petroleum, nickel, bauxite, tin, and various ferroalloys we use comes to us from other countries.

Second, *environmental concerns:* As our population grows and our levels of living increase, we are placing mounting pressure on our natural resource base. The interrelations between mankind and the environment have been a traditional part of geography. Now more than ever, it is critical for us to understand the environmental results of our actions and to find ways to reduce and avoid degrading the environment. Third, *political and social interdependence:* Not only are we increasingly interdependent for products and materials, but we are also affected by actions in other countries. For example, in recent years a large number of immigrants (legal and illegal) have entered the United States from Cuba, Haiti, Mexico, and Southeast Asia. Many of these immigrants come to the United States because of adverse conditions in their home countries and perceived opportunities in this nation. The impact of this migration is manifold. Without a knowledge of the geography of our own country and other countries that affect us, we have little hope of establishing policies that mit-

igate potentially dangerous situations and capitalize on advantageous opportunities. The opportunities for geographers as teachers must expand greatly as we become culturally, politically, and economically more interdependent with the rest of the world.

Business

Geographers are also applying *geographic skills in the business world* in many ways. Many firms use geographers for *location analysis.* Every business firm has special problems of location selection; for example, if a grocery store chain decides to expand its retail outlets in a city, where should new stores be located? Should they be opened in the central city or the suburbs? Selecting the location may involve a *market analysis* of the potential customers, an analysis of *traffic flow,* and any competitor's position for attracting the firm's potential customers. The directors of the grocery chain might wish to establish an outlet not at the present point of maximum profit but at another point whose location will prove better at some future date. The *applied geographer* must then study trends in neighborhood characteristics including patterns of growth and decline, projected road construction, and a host of other variables. Similar techniques are used for the location of factories and services.

The work of Professor Joseph A. Russell of the University of Illinois illustrates another aspect of *applied geography's contribution to business.* Russell was employed by the Ford Motor Company as a consultant and was given an essentially open charge to demonstrate how a geographer could provide a valuable service to the company. He decided to map the relative sales positions of Ford, General Motors, and Chrysler cars of comparable price. He placed these data (by county) on a United States map and found that the *relative sales position varied regionally;* that is, in some parts of the country Chevrolet sold more than Ford and Ford sold more than Chrysler, but in other regions the relationship was different. Ford officials were surprised that Russell's results for sales data had never before been mapped and a perusal of his map led to many questions. Why, for instance, did the relative sales position regionalize? What was Ford doing right to outsell its competitors? What was it doing wrong where the competitors outsold Ford? Russell helped Ford answer these questions that are truly geographic—*where is it and*

[1]Marion J. Levy, Jr., *New York Times,* 12 November 1961.
[2]Bob Wedrich, *Chicago Tribune,* 6 November 1980.

why is it there? Later Russell worked with Ford to assess the *geographic influences that affect dealerships.* He analyzed dealership location in relation to traffic flow, pointing out that in cities service and repair business is increased if the dealership is located on the right side of the street for customers going to work. Such a location is further enhanced if public transport facilities are nearby and if access to the garage is easy.

Other geographers concentrate on *area analysis.* Using area analysis techniques they study the growth potential of a market area and assist in preparing programs for industrial development, resource use opportunities, and transport lines. Some geographers, using the skills of both natural and social sciences, assess the impact of new construction to avoid possible detrimental repercussions. Such studies are especially vital in determining the *environmental impact.* Many geographers specialize in the *study of foreign areas,* and their knowledge of the cultural, human, and physical characteristics of these places is invaluable in providing information on market opportunities, resources, location sites, and problems in conducting business in such locations. Still other geographers do specialized studies for businesses such as weather forecasting for specific agricultural crops or analysis of the supply of specific labor skills, or they may serve as cartographers, travel agents, and consumer behaviorists, among a myriad of possibilities.

Government

Geographers hold a number of *governmental positions* at all levels. At the local level many are *urban and regional planners* charged with facilitating orderly residential, business, and industrial growth. At the national level geographers are employed in literally dozens of different types of positions. Many use their *cartographic skills* and *remote sensing techniques* to map and analyze many types of activities. A few serve as science and geographic attachés in the State Department Office of the Geographer and as State Department foreign service officers. Many others work as *intelligence specialists* or *research analysts* for various agencies, departments, and the Congress. Some are employed by the various scientific agencies such as the National Science Foundation and the United States Geological Survey. Geographers also work for the Agency for International Development, where they assist foreign countries in initiating and

carrying out development programs such as *resource analysis, regional development, urban reconstruction,* and *economic growth.* At the international level many geographers work on inventory analysis for socioeconomic development purposes. Various agencies of the United Nations, the World Bank, and the Inter-American Bank employ geographers to measure *natural and human resources* in various parts of the world.

GEOGRAPHY AND OTHER DISCIPLINES

Geography is a *bridge discipline* uniting the social and physical sciences. As such, geography possesses many characteristics similar to its sister fields. Yet at the same time geography is unique in that its *primary focus is spatial.* This spatial orientation means that geographers look at the world from a distinct point of view. For example, a *botanist* is concerned primarily with how plants grow, plant structure, propagation, and taxonomy. A *geographer* views plants from the point of view of plant distribution and analyzes why plants, or more properly groups of plants, are distributed as they are. As part of this analysis, the geographer will be concerned with the processes of plant growth and the distribution and interrelationships that exist between plant groups and the things that help or inhibit their survival.

Economists are interested in the production, distribution, and consumption of goods and services. They study how people choose to use resources to gain a livelihood by investigating topics such as the various costs and benefits of how resources are allocated, the causes of changes in the economy, the impact of monetary policy, how different economic systems work, supply and demand problems, and business cycles and forecasting. The *geographer* also is interested in how a group of people gain their livelihood but views economic activity from the point of view of where the activity takes place and the reasons for its location. In economic geography, one is concerned with the characteristics of an activity, how the activity is similar and different from one area to another, and how an activity is related to other phenomena, for example, the dietary preferences of a society.

In both examples, geography and botany and geography and economics, there is an obvious relationship between geography and its sister disciplines. The same type of relationship exists between geography and other related social and physical sciences, some of which are illustrated in Figure 1–2. What must be kept in mind is that each discipline views the world from a different perspective. There are *linkages* among all fields of learning, and *geography's linkages are especially extensive.*

This linkage with other fields permits geography to be a synthesizing and integrating subject. Geography is a *synthesizing field* that draws on knowledge in many disciplines and puts this knowledge in a *geographical context.* For example, a geographer who studies wheat growing on the Great Plains or the Argentine Pampas needs information on climate, soils, and landforms (physical science) of the region as well as knowledge of the farmers' cultural characteristics, the transport network to supply the farmers and send the wheat to market, the costs of wheat farming especially in relation to other economic opportunities, and a host of other socioeconomic factors (social science). Geography is an *integrating discipline.* As history is an integrator of time, so too is geography an *integrator of space* (regions). Neither history nor geography can be studied effectively without a knowlege of the other. Isaiah Bowman, former president of Johns

Hopkins University, said: "A man (or woman) is not educated who lacks a sense of time (history) and place (geography)."[3] By integrating information in a regional context the geographer pulls together knowledge from a variety of disciplines into a coherent, encompassing single picture.

GEOGRAPHY AND DEVELOPMENT

This book is organized on a *regional basis* and is designed to provide an introduction to geography by highlighting various geographic concepts. Our purpose is to acquaint students with the world in which we live and to organize this knowledge within a framework of *economic development,* broadly defined.

Development is a word that is used in a variety of ways. Basically it means a progressive change for the better. *Economic development* signifies a process of continual improvement in our livelihood. As we shall see in Chapter 4, economic develop-

[3]Quoted in Alfred H. Meyer and John H. Strietelmeyer, *Geography in World Society* (Philadelphia: J. B. Lippincott, 1963), p. 31.

FIGURE 1–2 The Scope of Geography.
Geography is an integrating and synthesizing discipline. This diagram shows that geography interrelates with many fields of study, including the sciences, engineering, social sciences, and humanities.

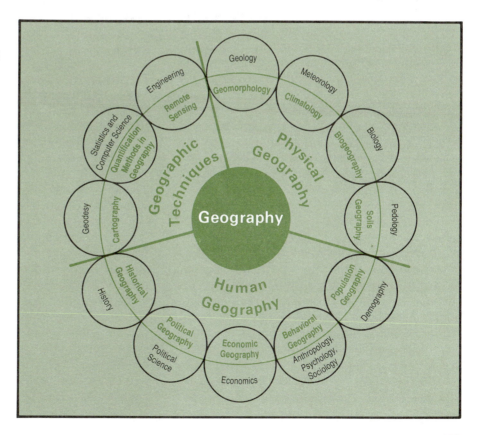

ment is measured in several ways such as income, energy use, and employment. Such measures tell us nothing about other aspects of an area, nation, or people. For example, differences in the measures from one area to another may reflect differences in cultural goals and values.

It is an increasingly trite but accurate truism that the *world is getting smaller.* Not only are we constantly bombarded with news of places such as Kuwait, India, Israel, Switzerland, and Japan, but events in these and other nations also influence our daily lives. The quintupling of crude oil prices by the Organization of Petroleum Exporting Countries (OPEC) awakened the public to the degree we rely on other nations to supply our needs. A coffee crop failure in Brazil, the development of a new high-yielding variety of wheat, the discovery of a chemical process for making plastic, an outbreak of insect pests, and 'a myriad of other events all materially affect our life-styles. As world population grows and standards of living improve, the level of *international interdependence* also increases. Consequently, we are required to know something about the world in which we live, for without knowledge we can exert little influence on our fate.

In this book our concern is knowledge about economic achievement. There is a *great disparity* in material well-being among the world's societies. With a more intimate world brought about by better communication and transportation, knowledge of how others live is at our finger tips. Those economically less fortunate often wish to emulate their materially richer neighbors but may be frustrated in their attempts by cultural, economic, and political constraints. The disparity in economic achievement is widening, and the *social and political ramifications* are manifold. The reasons for this disparity provide the focus and theme of this book.

Economic development was until recently basically a *Western societal concept.* As we shall see, Western ideas of *materialism* are spreading widely throughout the world. To support development, modern technology has been accepted by other cultures. Penetration of Western ideas and technology into other cultures often has led to disruption and conflict. Not only are traditional economic patterns altered, but also new modes of behavior and interpersonal relations are established. Few nations have "Westernized" completely, and *cultural and economic pluralism* has resulted.

People

To appreciate economic development, we must examine the process from *four principal points of view.* The first point of view is the *people:* their number, growth rates, and distribution. Improved sanitation and medical science have greatly lowered the death rate and have caused a rapid rate of population growth—so great as to constitute a *"population explosion."* Where birth rates have not decreased to a comparable level, advances in economic development and improved human well-being may be difficult to achieve.

Environment

The second point of view is the *natural environment,* which provides the stage and materials used in economic activities. Some environments are rich in resources that can be used for economic gain. A well-watered alluvial plain with a long growing season provides many opportunities for productive agriculture. Similarly, highly mineralized areas with easily extractable ores can support other means of livelihood. Conversely, areas with steeply sloping land, thin soils, or moisture deficiency are relatively poor for cropping and present obstacles to development.

Culture

The third point of view is *culture.* The way in which a society organizes itself in terms of beliefs, customs, and life-styles greatly influences the direction and degree of economic development. One aspect of Western culture is materialism. The Puritan work ethic is an example of our material bent with wealth as an index of success. Although not Western in tradition or culture, Japan has a similar work goal. Some other cultures do not place so high a priority on the material advantages of life. The wandering nomad cannot accumulate many possessions and still practice a migratory existence. Of more direct influence is the character of economic organization. Some cultures (including some Western cultures) have an economic structure ill-designed to use modern technology effectively, or they have such a rigid social and economic system that the human resource base is constrained. Other more flexible systems are able to adapt to new ideas and accept technology relatively easily.

History

The fourth point of view is that of *history*. That the past is a key to the present and a guidepost to the future is well demonstrated through a study of the evolution of economic activity and the world's various cultures. Economic development is not a short-term process. The necessary foundations or prerequisite conditions for development of most nations that are now undergoing rapid change or that have attained a high level of economic well-being were laid decades, even centuries ago. The cornerstones of Europe's Industrial Revolution, which began in the mid-eighteenth century, were formed during the preceding Renaissance period with roots going back to ancient Roman and Greek times and even earlier. A more recent example is Taiwan, where many of the foundations for the island's recent remarkable growth were laid in the early part of this century.

We begin our discussion with a unit of three chapters that provide a *background of ideas and knowledge* developed more fully in succeeding chapters. Chapter 2 presents an overview of population and the resource concept. Chapter 3 views the natural environment primarily from the resource standpoint. Also in this chapter is an appraisal of the elements of culture focused mainly on characteristics influencing development. Chapter 4 presents the rich and poor regions of the world based on measures of economic well-being along with characteristics of rich and poor regions and some theories of development.

TWO WORLDS

The development theme permits a broad *twofold division of the world* into regions with relatively rich populations and those inhabited by relatively poor people.[4] The rich regions are considered first and comprise Anglo-America, Western Europe, East-

ern Europe and the Soviet Union, Japan, and Australia–New Zealand. The poor regions are Latin America, North Africa and the Middle East, Subsaharan Africa, South Asia, China and its neighbors, and Southeast Asia.

Rich World

Each regional unit is organized along the same lines to include a historical perspective, an examination of the physical basis for development, culture, economic structure, and present patterns, trends, and prospects of development. Emphasis, however, varies from region to region. The high standard of living in Anglo-America is viewed in the light of a varied and bountiful physical resource base. Poverty pockets and cultural conflicts are also indicated. Western Europe has a greater historical stress to explain the region's multiplicity of nation-states and present advanced technological attainment. The Eastern European and the Soviet Union chapters contain an analysis of Communist development theory and appraisal of the Soviet Union's drive for rapid economic growth. Japan's status in the rich world is unique, and we must view Japanese growth, in spite of a poor natural endowment, from the perspective of a blend of Western and local culture traits.

Poor World

In the poor world, Latin America's status is viewed from the perspective of cultural pluralism, societal attitudes to the resource base, and rapid population growth. The African and Middle Eastern regions, truly diverse, have a recent history of colonialism, and the many newly independent nations struggle for self-identity that is expressed in different ways. The Arab-Israeli conflict and the oil-rich states present another aspect of the development process. Finally, in Monsoon Asia (South Asia, China and its neighbors, and Southeast Asia) emphasis is on the origin of different cultures and the relationship of economic organization to other aspects of culture. Attention is paid especially to the roles of the Monsoon wind system and religion in South Asia and the contrasts between traditional and Communist China.

---

[4]Another system based on development levels divides the world into the First World, composed mainly of the rich Western nations of Anglo-America, Western Europe, and Japan; the Second World, comprising principally the more wealthy Communist nations such as the USSR and those of Eastern Europe; and the Third World, which includes the poor or less-developed countries.

2

People and Resources

For most of mankind's existence, the world's population has been relatively small, but the dramatic increase over the last hundred years has been so great that the phrase *"population explosion"* has been applied. A rapid and sustained increase in population places a strain on the capacity of a society to fulfill the material needs and aspirations of its members. To counterbalance the increased demand, society's productive capacity must be expanded by development of new resources and accelerated exploitation of present resources. Even if production capacity matches population growth, previously minor problems like pollution, interpersonal and group rivalries, and occasional scarcity of goods may become serious. In addition, new situations may challenge basic and traditional societal values and ways of life. The society itself may collapse or be drastically altered.

PEOPLE

At the dawn of the *Agricultural Revolution* some 7000 to 10,000 years ago, the world's population probably numbered about 5 million (Figure 2–1). Plant cultivation and animal domestication, however, heralded a long and sustained period of population growth. Population clusters, at first confined to areas of agricultural innovation, later spread worldwide with the diffusion of agriculture. In a few places, such as Australia, the *diffusion process* was delayed until the coming of European colonists. Today only in polar zones, remote dry lands, and other harsh environments a small and dwindling number of people still live by the age-old hunting, fishing and gathering occupations.

Impact of Agricultural Revolution

The Agricultural Revolution led to many fundamental changes. No longer were people dependent solely upon nature, for domestic crops and animals far surpassed their wild cousins in utility. Many more people could live closely together at a

FIGURE 2–1 World Population Growth (in millions). For most of human history the world held relatively few people. In the last 400 years, however, the number of people has increased greatly and at an increasingly rapid rate. The term *population explosion* is often used to describe this rapid growth.

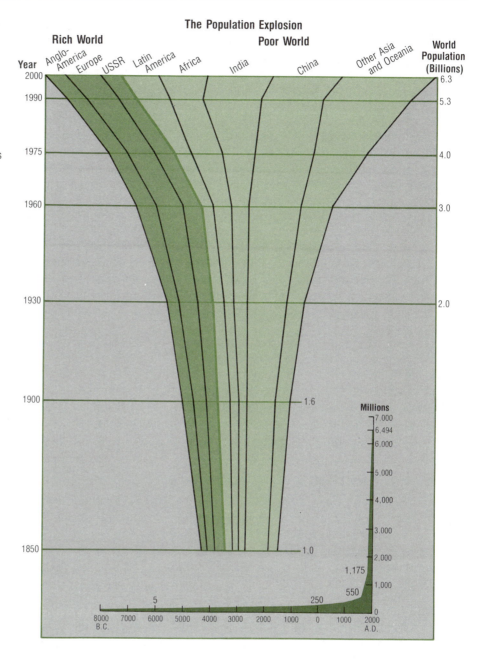

The Population Explosion

Rich World

Poor World

higher level of living. Villages grew, and a new social order was created to resolve the increasing conflicts brought about by more people living together.

Life was easier and more secure. Permanent homes, even substantial houses, replaced the crude huts or caves that had served as temporary lodgings. Many tools and other large and small luxuries—a chair, a table, and a bed—were acquired that previously were impractical because of the migratory nature of life. Indeed, *materialism* had its true beginnings with the development of agricul-

ture. Possessions could be accumulated and passed on to new generations. A sense of security increased, too, as production increased. Surpluses could be stored for an emergency. Protection from hostile groups was also an attribute of village life. Permanent fortifications and many defenders discouraged outside attempts to capture and control the agriculturist. Defense was not always successful; for example, the hunting and gathering Plains Indians of the United States, after obtaining the horse from the Spanish, were more than a match for the sedentary agricultural Indians along the eastern margin of the Plains. Another example is the Romans' loss of their agricultural tributary regions to more primitive northern and eastern tribes.

Increased population, production and interpersonal contact created a need for group action and led to numerous organizations. A political organization was established to settle disputes, govern, and provide leadership for collective action in warfare and such public works as irrigation, drainage, and roads. The formation of a priestly class helped formalize religion. Religious leaders frequently were the holders of both philosophical and practical knowledge, often playing a role in village life as medical men and weather forecasters. In the Mayan civilization of southern Mexico and Guatemala, for example, the priests developed an *agricultural calendar*. This calendar, based on the progression of the sun, planets, and stars, predicted the beginning of wet and dry seasons and alerted the farmer when to prepare the land for planting to take full advantage of the seasonal rains. Increased production per worker provided more than the family unit needed. A portion of the labor force was freed not only for government and religious activities but also for other nonagricultural activities such as pottery making, metallurgy, and weaving.

From A.D. 1 to 1650

By the beginning of the Christian era, the Old World's population had grown to about 250 million. Most of these people lived within three great empires: the Greco-Roman around the fringe of the Mediterranean Sea and northward into Europe; the Han dynasty of China extending into Southeast Asia; and the Mauryan of northern India. In these empires the simpler political, economic, and social organizations of the agricultural village vied with the more complex, integrating structure of the empire and the newly created city. *Urbanism* in all its facets became a way of life.

Urban life meant *specialization of labor*. The city dweller became dependent on the farmer for food and fiber. The city was the focal point of national life. The riches of the countryside and tribute from afar were concentrated in the city. Like a magnet, the city drew people to its core. Arts flourished, and education was available to those who could pay. The best lawyers and doctors practiced there. Government employed increasing numbers to construct, maintain, and supply sewage lines, roads and streets, and drinking water. In all, the activities of these empire cities were not much different from those of a modern metropolis.

The empire and the city fostered *regional specialization*. Each part of the empire, rather than producing all of its own needs, traded what it produced in surplus and received in turn what it did not process or could only produce with difficulty. To be sure, regional specialization was not fully developed, but the concept was recognized and used. Rome, for example, exported Arretine pottery, wines and oils, metalware, glass, and perfume; gold, too was an important export. In return Rome imported wheat from Egypt and North Africa, cattle and hides from Sicily, metals and livestock from Spain, slaves and fur from Germany, and even rare Indian spices and gems.

In contrast to the Old World, the New World's population probably numbered only about 10 million at the beginning of the Christian era. The Agricultural Revolution, which was concentrated in Mesoamerica (Central Mexico to Honduras) and the Peruvian-Bolivian Andes, did not affect the area as a whole.

By the end of the first 1650 years of the Christian era, the world's population had grown to more than 500 million. Most of the growth was in and around the preexisting centers with gradual expansion of the populations into areas that had been sparsely settled. Productive capacity expanded with improved technology and new resources. *Urbanization* became more pronounced, although agriculture remained the base of livelihood.

From 1650 to the Present

From A.D. 1650 to the present, the world's population has increased more and more rapidly. It took 1650 years for the population to double from 250 million to 500 million. By 1850 the population had doubled again and was estimated at 1175 million. Within the next hundred years the population doubled a third time to 2 billion. Sometime in 1975 the population again doubled, to 4 billion. During this relatively short period a second revolution, the Industrial Revolution, produced an impact on mankind the results of which are still incomplete. The revolution is continuing and not yet applied everywhere.

There are now nearly 5 billion people in the world, and prospects are for a continued rapid increase in the foreseeable future. As we shall see, however, a controversy exists whether the "population explosion" poses a threat to mankind.

Industrial Revolution

The *Industrial Revolution* is characterized by countless innovations. These range in complexity from the paper clip to interplanetary flight. By the beginning of the nineteenth century the age of invention had arrived, and each new idea seemed to spawn many others. Muscle power from people and animals was replaced by inanimate power: the steam engine, water turbine, and internal combustion engine. In the agricultural sector the use of the tractor and its attachments has made the farmer so productive in some parts of the world that only a small part of the labor force is needed to supply an abundance of food. To be sure, other scientific advances, such as improved higher yielding seed and the application of fertilizers, herbicides, and insecticides, have also contributed substantially to this productivity.

Inanimate energy greatly facilitated the growth of cities. More raw materials from agriculture, mining, and forestry, combined with new energy sources, spurred industrialization. Especially since the eighteenth century, energy and manufacturing innovations have made the factory worker many times more productive. Manufactured products have become relatively cheaper and more readily available. Craftsmen and small guilds have gradually given way to the modern factory where the worker is primarily a machine tender. The Industrial Revolution brought about large-scale factory production, which drew larger numbers of people to the cities to work in the factories. These cities, small and large, became service, financial, educational, governmental, wholesale, and retail centers. Many cities grew several times over (Table 2–1).

As the Industrial Revolution continues and its effects spread, urbanization increases. The revolution, which began in Western Europe, moved quickly to Anglo-America and other areas where European colonists settled. It moved more slowly into Eastern Europe, the Soviet Union, southern Europe, and Japan. Since the end of World War II, however, industrialization has become a major force nearly everywhere.

Distribution and Density of Population

A map of the distribution and density (number of people per unit area) of the world's population shows a strong tie with the past (Figure 2–2).

TABLE 2–1 Growth of Selected Cities

City	Population (millions)				
	1800	1850	1900	1950	1980
New York	0.1	0.7	3.4	9.6	20.4
London	0.8	?	1.8	8.3	10.3
Tokyo	1.0	2.3	4.5	8.4	20.0
Paris	0.5	1.1	2.7	6.3	9.9
Moscow	0.3	0.4	1.0	5.4	8.2
Buenos Aires	0.04	0.1	0.8	5.2	10.1
Shanghai	0.3	0.4	0.9	5.0	13.4
Chicago	—	0.03	1.7	5.1	7.8
Calcutta	0.6	?	0.8	4.6	9.1
Berlin	0.2	0.4	1.9	3.4	3.4
Mexico City	0.1	0.3	0.3	3.0	15.0
Rio de Janeiro	0.04	0.3	0.8	3.0	10.7
Detroit	—	0.02	0.3	3.0	4.4
Cairo	0.3	0.4	0.6	2.3	10.0
Washington, D.C.	0.003	0.04	0.3	1.5	3.0

SOURCES: W. S. Woytinsky and E. S. Woytinsky, *World Population and Production: Trends and Outlook* (New York: Twentieth Century Fund, 1953); *World Population Situation in 1970* (New York: United Nations, 1970); *Demographic Yearbook* (New York: United Nations, 1982); and estimates of national governments.

Three principal *centers of dense population* are readily apparent: the Indian subcontinent, eastern China and adjacent areas, and Europe. China and India represent old areas of large populations based on an early start in the Agricultural Revolution and in empire building. In these areas agriculture and village life are important facets of society. At the same time, however, the modern city with its service-manufacturing functions is also present. Population density in India and China varies considerably, usually in association with the relative productivity of the land. On the coastal and river plains of rich alluvial soils and abundant water, rural densities of 3000–4000 per square mile (1158–1544 per square kilometer) are not uncommon. Away from the well-watered lowlands, densities diminish but still may be in the range of 250–750 people per square mile (97–290 per square kilometer). At least one-half of the world's population lives in southern and eastern Asia.

High population density in Europe is traced from technological developments in the Middle East that were adopted by the Greeks and Romans and further expanded by the Industrial Revolution. The increase of the European population in the north and west is readily associated with the shift in the center of the technological revolution and political strength. Europe's population density is high but significantly less than in the Indian and Chinese areas. Moreover, the agricultural village and agriculture itself are overshadowed by the modern metropolis and manufacturing.

Secondary centers of high population density

FIGURE 2–2 World Population Distribution. The world's population is unevenly distributed. Three large areas of dense population are China, Indian subcontinent, and Europe. Most of the sparsely populated areas have environmental problems such as aridity or mountainous terrain.

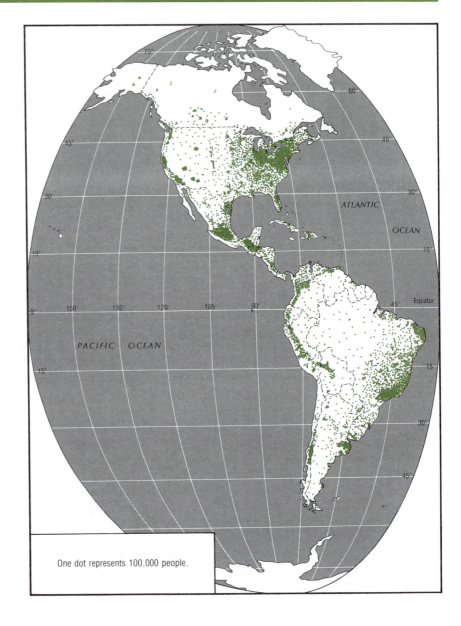

One dot represents 100,000 people.

are more numerous. The northeast quarter of the United States and adjacent Canada are considered by some as a principal cluster, although total population numbers and density are lower than Europe's. In many respects, for example rate of urbanization and employment, this region closely resembles the European pattern. In Africa, along the Guinea Coast and the Nile River and in the eastern highlands, high densities occur, but the total number of people involved in each cluster is relatively small. Similarly, around major urban centers of Latin America and in the old Aztec, Mayan, and Inca realms, small but locally dense population centers are common. Other pockets of high density are found in Asia, Java, the Malay Peninsula, and parts of the Middle East.

Most of the rest of the world's land surface (80%) is more sparsely inhabited. Many of these areas have serious *environmental problems*—cold, dryness, and rugged terrain—that have kept them from being made productive. The sparse populations of other areas, such as some of the humid tropics of South America and Africa, are not so easy to explain, especially since similar environments in Asia are densely settled. It is tempting to relate population density to the broad physical patterns described in Chapter 3, and indeed some writers have done so. Yet a population-physical environment correlation is an oversimplification. As we have already seen, technology and political organization are cultural factors to be considered. Additionally, other aspects of culture, such as de-

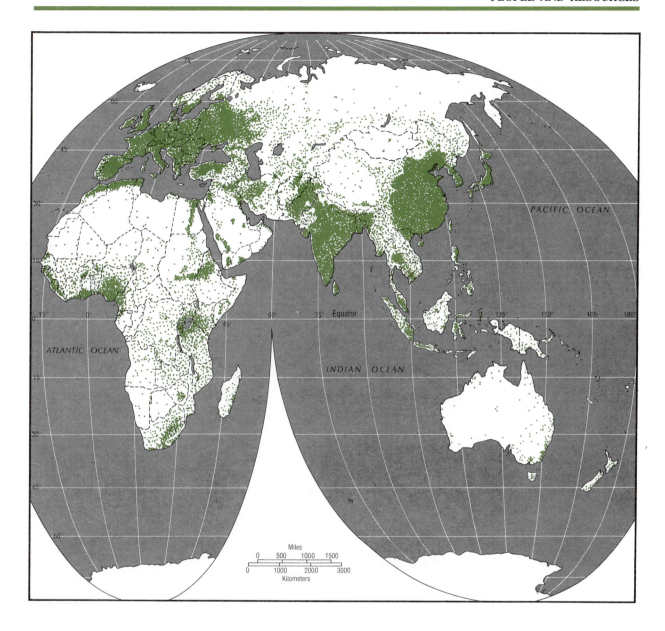

Miles
0   500   1000   1500
0   1000   2000   3000
Kilometers

sired family size and economic organization, have an important effect on population numbers.

Theories of Population Change

Overall, the world's population is increasing at a rate approaching 2% yearly, but this growth is by no means uniform. One explanation of the varied growth pattern is the theory of *demographic transformation,* which is based on four population stages (Figure 2–3). Stage I postulates an agrarian society where birth and death rates are high, creating a stable or very slowly growing population. Population density is low to moderate. Productivity is limited, and children, even at a young age, can contribute to the family's wealth. Large families are an asset, particularly since life expectancy is low and family security is dependent upon its members. Employment opportunities aside from agriculturally related ones are few, and technology is stagnant or nearly so.

In stage II the cultural custom of large families persists, and the birth rate remains high. The death rate, however, drops dramatically because of better sanitation and medical treatment and greater productivity. Productivity may be increased in the agricultural sector, but more important is the advent of alternative economic activity resulting from industrialization. With industry come urbanization and labor specialization. The principal aspect of stage II is rapid population increase.

Stage III is characterized by continued urbaniza-

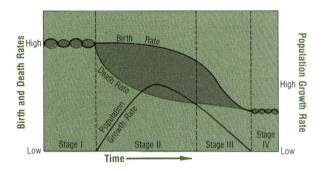

FIGURE 2–3 The Model of Demographic Transformation.

The demographic transformation model is based on the European experience. It may not represent what will happen elsewhere, especially in non-European culture areas. The model assumes an initial period of traditional rural life with high birth and death rates and no population growth. Then, with increased production and improved sanitation, the death rate declines, creating a condition of rapid population growth. As the population becomes more urbanized, the birth rate also declines until finally low birth and death rates prevail and population no longer grows.

FIGURE 2–4 World Birth and Death Rates.

There are great differences in population characteristics from one part of the world to another. Areas of the world that are highly urbanized and have advanced economies have low birth and death rates and consequently low rates of population growth. Areas that are less urbanized and less economically advanced have high birth rates, declining death rates, and moderately to very rapidly growing populations.

SOURCE: *Population Data Sheet* (Washington, D.C.: Population Reference Bureau, 1983).

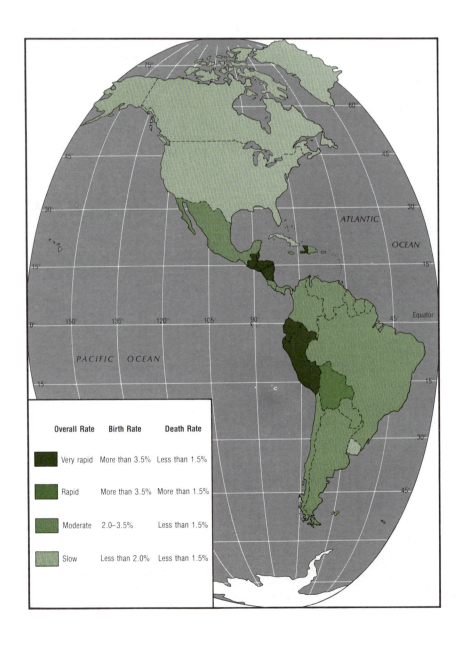

Overall Rate	Birth Rate	Death Rate
Very rapid	More than 3.5%	Less than 1.5%
Rapid	More than 3.5%	More than 1.5%
Moderate	2.0–3.5%	Less than 1.5%
Slow	Less than 2.0%	Less than 1.5%

tion, industrialization, and other economic trends begun in previous stages. Demographic conditions, however, evidence a significant change. The birth rate begins to drop rapidly as small families become more prevalent. The development of small families may be related to the fact that children in an urban environment are generally an economic liability rather than an asset. Still, in stage III the population continues to grow but at an ever slowing pace.

Stage IV finds the population growth rate stable or increasing only very slowly. Birth and death rates are low. The population is now urbanized, and birth control is in general practice. Population density may be quite high.

We must remember that demographic transfor-

mation is but a theory based on analysis of the European experience. It cannot be assumed that other areas with different culture bases will follow the European example. If the theory is valid, however, population growth will slow down in many parts of the world as stages III and IV are reached. If and when population growth will be reduced or stabilized is, of course, unknown (Figure 2–4).

Another theory that has received widespread attention and has numerous advocates is the *Malthusian theory*. Thomas Malthus, an Englishman, first presented his theory in 1798. He based his theory on two premises:

1 Humans tend to reproduce prolifically, that is, geometrically: 2, 4, 8, 16, 32.

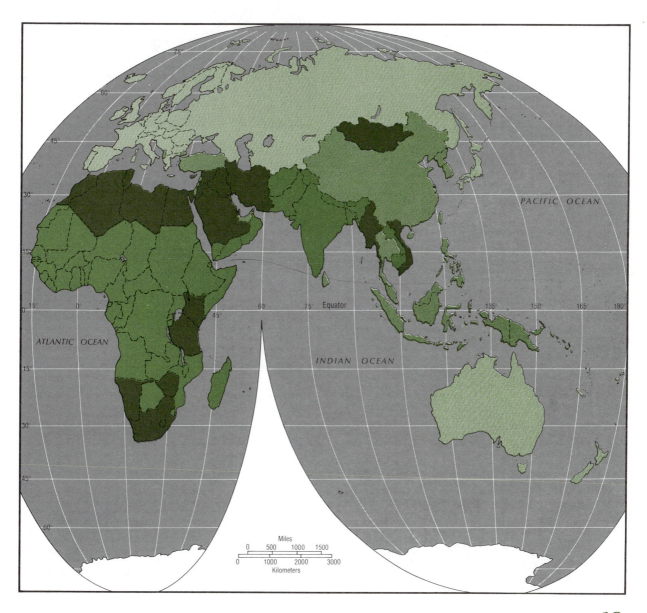

2 The capacity to produce food and fiber expands more slowly, that is, arithmetically: 2, 3, 4, 5, 6.

If population growth is not checked by society, nature will reduce surplus populations by war, disease, and famine.

If we plot Malthus's idea graphically, three stages of the population-production relation are apparent (Figure 2–5). In stage I human needs are less than productive capacity. By stage II productive capacity and human needs are about equal. In stage III population has grown to the point where these needs can no longer be met. When stage III occurs, there is a die-off of the population, and we cannot be sure what follows. One idea is that stage III perpetuates itself with alternating periods of growth and die-off, represented by curve (a) in Figure 2–5. Another idea is that the die-off is so great that stage I is reproduced (b).

Karl Marx viewed the population question differently. According to *Marxist philosophy,* economic and social benefits can only be enhanced by an increase in the labor force. So-called overpopulation and population pressure are seen as imperfections of the economic system, not as an actual excess of people. Communist nations today still espouse Marx's view from an ideological standpoint but

Thomas Malthus (1766–1834), an English clergyman, wrote his famous *Essay on the Principles of Population* in 1798. He predicted that population growth would outpace the means of supporting the population unless cultural constraints could check growth in some way. (The Bettmann Archive, Inc.)

some (China, for example) from a pragmatic point of view actively encourage family planning.

The Marxist idea that increasing population can lead to greater prosperity is in conflict with Malthusian thought. Malthus painted a dark future for mankind without population controls. Marxists believe population growth can lead to greater productivity if resources are adequately developed and distributed. Both theories suggest that the balance between population and resources (productive capacity) is central to the development process. The demographic transformation idea likewise relates population growth to resources but resources as measured by economic activity.

THE RESOURCE CONCEPT

Some Definitions

The resource concept contains three interacting components: resources, obstacles, and inert ele-

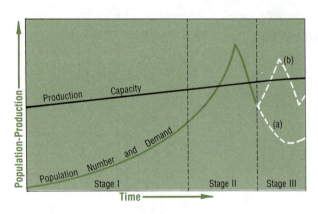

FIGURE 2–5 Malthusian Theory.
The Malthusian theory can be illustrated by a three-part diagram. In stage I the needs of the population are less than production capacity. In stage II, however, the increase in population is so great that need soon exceeds production capacity. For a while the population continues to grow by using surpluses accumulated from the past and overexploiting the soil resources. Eventually, the pressure on the resource base is too great and the population begins to die off. Stage III may be a continual repetition of stage II (b) or a return to stage I (a).

These simple examples illustrate several points. First, resources are not just material objects. The farmer's knowledge of when and how to tend crops is absolutely necessary. The mining of a mineral deposit requires skill and organization. Other examples of nonmaterial resources are as diverse as inventiveness, good government, useful education, cooperation, and adequate social order.

Second, material resources are not just natural resources. The farmer's and miner's tools are culturally derived. Even the farmer's seed and plants are culturally modified. In fact, most material items in the surroundings of an urban dweller are man-made—for example, books, chairs, cars, bridges, and buildings.

Third, elements in the environment may be resources, obstacles, or inert at the same time. The river is a barrier to the walker, inert to the miner, but a resource to the farmer from which he obtains water to irrigate his fields. The hills are inert to the farmer, a resource to the miner (because they contain useful minerals), and an obstacle to the hiker.

Finally, resources are not static or finite. The farmer's tools were improved with the development of draft animals and later the tractor. His seed has been improved by selective breeding. The miner's pick may be discarded for power-driven equipment. Resources are created.

Resources and Culture

Just as the hiker, miner, and farmer use different resources, so diverse cultures have different resources. Each culture group has developed a set of customs, laws, and organizations that effectively structure the lives and attitudes of its members. These *cultural controls* are evident in how resources are viewed. Those few groups who still live by hunting, gathering, and fishing view their surroundings from the standpoint of nature's productivity of useful plants and animals. They are not concerned about soil quality, growing season, and precipitation amount and distribution, even though these factors bear on natural production. These hunters and gatherers probably do not recognize differences in clay deposits, some of which other groups use as raw materials for brick and pottery manufacture. Hunters and gatherers do view as resources obsidian and other rock deposits from which they can fashion cutting and hunting tools; other cultures that have metal instruments would consider these deposits of little significance. Simi-

Karl Marx (1818–1883), the German founder of modern communistic and socialistic doctrine, believed that wealth is created by increasing the labor force. Marxist philosophy is based partly on the idea that population growth is desirable.

ments. A *resource* is anything that can be used to satisfy a need or a desire; it is a means to an end. An *obstacle* is anything that inhibits the attainment of a need or desire—the opposite of a resource.

Suppose a man needs to walk from point A to point B. His legs are a resource that can be used to satisfy this need. If he must cross a river and low hills to reach his goal (B), these inhibit his walking and in this case are obstacles. If he were driving a car, the river would be a formidable obstacle. *Inert elements* in our surroundings neither help nor hinder. In the example, neither the hiker's doctorate in astrophysics nor the arable land across which he passes helps or hinders his walking.

Let us now suppose that the man's brother is a local farmer. His resources may include seed and fertilizer, but the river is also a resource for irrigation. The hills may be obstacles to plowing for a farmer; however, minerals located in the same hills could be the basic resource of a thriving mining community.

21

larly, various plants used for making mats and containers are also resources, but to those of "Western" culture these same plants may be weeds.

Perhaps more important is the role that culture plays in directing economic activity. Resources are basically an economic idea. Some cultures, such as the Bedouin of the Middle East and several Asian cultures, are less materialistic than ours. To them the accumulation of wealth is not a prime goal, and economic organization is structured to provide little more than basic needs. In other cultures, social, political, and economic organization is such that large portions of the population have little or no opportunity to develop resources. Cultural controls inhibit the application of technology and limit individual opportunity.

In nearly all cultures certain economic activities are socially more acceptable than others. These attitudes tend to direct individuals to use different sets of resources. The Bedouin of the Arabian Desert is a nomadic herder because in his culture that is the highest occupation one can hold. The existence of large quantities of petroleum under the land is of minor import. Neither is he overawed by the possibility of becoming an oasis agriculturist, even though water and good soil may be available and the life of a farmer more secure. In the United States urban-oriented, white-collar employment is the goal of most Americans. Many students come to college to prepare themselves for these jobs because of parental and peer pressure, a form of culture control. Our population is so large and diverse, however, that there is a wide range of economic activities.

Resource Development

One of the major fallacies surrounding the resource concept is that resources are static or fixed in amount. The Agricultural Revolution and the Industrial Revolution are excellent examples of multiple-resource innovation. A simple stick becomes a resource when used to poke, pry, and shake trees and bushes for food. When someone discovers that the stick can also be used as a planting tool, its function as a resource is increased. In a like manner, the rich coal deposits of western Pennsylvania were not resources prior to the coming of the Europeans because the Indians made no use of coal. In more recent times the widespread use of aluminum has been made possible by generation of cheap electrical power.

It is not difficult to see that nonmaterial resources can be constantly created. New ideas and

better organization have no limit. We can also see that material resources that can be replenished—trees, crops, animals, soils, rivers, and so on, which are known as *flow resources*—are used repeatedly and improved upon. What may be more difficult to see is that *fund resources,* often called nonrenewable resources, are also created. Examples of fund resources are nonreusable minerals such as coal and petroleum.

Technological improvement can result in the creation of fund resources in two major ways. First, a need may be found for a formerly unused mineral or element. Uranium had no use until the development of atomic power. Many rare earths were inert until high-temperature technology created a need for them. It was only in this century that techniques for extracting nitrogen from the atmosphere were perfected, giving rise to an important segment of the fertilizer industry. Going back in time, we find that the early phases of the Industrial Revolution led to the use of aluminum, numerous ferroalloys, and natural gas.

Second, perhaps more important in the last hundred years, has been the addition to the fund-resource base of low-quality mineral deposits. Prior to the advent of mass transport and suitable processing equipment, only the richest mineral deposits could be mined economically. These deposits represent only a small portion of the total amount of any mineral in the earth's crust (Figure 2–6). Once technology became available to mine low-quality deposits, they became resources. For example, most of the world's copper is produced from ores containing less than 3% copper. The rich iron ores of the famous Mesabi Range in north-

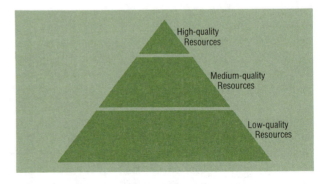

FIGURE 2–6 Resource Pyramid.
The pyramid concept applies to all types of resources: land, minerals, fish, people, crops. Most of the resources we use are high-quality. As technology improves, demand increases, or depletion occurs, we use more medium- and low-quality resources.

eastern Minnesota have been largely worked out, but *taconite,* a low-quality ore previously thought worthless, has become the basis of continued exploitation. Another example of expanded mineral-fund resources is petroleum, much of which could not be obtained without numerous advances in drilling technology.

Technology has also expanded the agricultural-resource base in several ways, one of which is examined here. The amount of *arable land* has been increased through improvement of seeds and mechanization of cropping. By seed selection and crossing (hybridization) new varieties of crops have been developed that are more drought tolerant or can mature in a shorter time than formerly. In the United States corn can now be raised farther west in the Great Plains and farther north than was feasible fifty years ago, and wheat is now grown in areas of only 15 inches (38 centimeters) of annual rainfall. Mechanization of agriculture permits the farmer to cultivate what was formerly nonarable land. By using power-driven equipment, one farmer can cultivate several hundred acres, so that although each acre may yield a small return, the total area farmed provides an adequate income. In the United States and Canada, the expansion of arable land by means of mechanization is well developed. In other areas such as China similar land is little used despite great need.

Fund resources can also be expanded by an increase in demand. If the demand for a particular item grows, creating higher prices, new resources are developed. There are large quantities of petroleum in oil shales and tar sands that have not been exploited because cheaper oil is available from other sources. If petroleum prices continue to rise, these reserves of oil will become profitable to extract.

Loss of Resources

Although demand can create resources, a decrease in demand may reduce resources. Steel cutting instruments have largely replaced the need for stone and obsidian tools. Synthetic rubber has decreased the need for natural rubber. Artificial heating and the fact that we live and work mainly indoors have decreased our need for heavy woolen clothing. Productivity increases in a crop without a significant change in demand may cause marginal producers to withdraw from farming. Similarly, the development of a new product (aniline dye for indigo) or discovery of a new mineral deposit may lead to the disuse of other products or mines.

Resources can be lost through use. Each drop of oil or gasoline burned and each morsel of wheat or corn eaten represent a decrease in resources. On the other hand, some resources are lost if not used. A tree allowed to decay means lost pulp or lumber, and unharvested crops are a lost resource, although these may help fertilize the soil and feed wild animals.

Unfortunately, resources also are destroyed by improper use. Poor farming techniques lead to erosion and soil depletion. Factories and cities may discharge harmful chemicals and sewage into rivers, destroying aquatic life and ruining water quality for those downstream. War destroys not only people—the most important resource—but also buildings, bridges, and other resources.

Technocratic Theory

A number of authorities have observed the great technological advances made during the ongoing Industrial Revolution and have derived what is often called the *technocratic theory.* This optimistic theory is directly counter to that formulated by Malthus (compare Figures 2–5 and 2–7). The technocrats accept the Malthusian population-growth curve but assume an increase in productive capacity (resources) at a rate greater than population growth. To support their contention, the

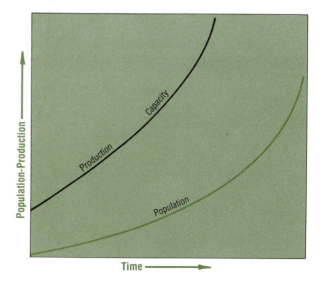

FIGURE 2–7 Technocratic Theory.
The technocratic theory assumes the same population curve as the Malthusian theory. What is different is the rate at which productive capacity is assumed to grow. If productive capacity increases faster than population, higher standards of living are possible.

technocrats point to the expansion of the resource base during the past four hundred years and the increase in the standard of living of Western cultures. The technocrats believe that technology will continue to expand, supporting still greater populations.

The present-day advocates of Malthus *(neo-Malthusians)* admit that the Industrial Revolution has postponed the day of disaster (Figure 2–8).

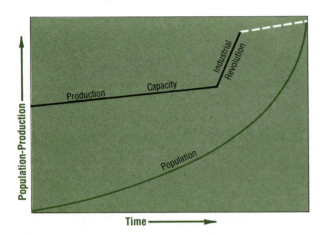

FIGURE 2–8 Neo-Malthusian Theory.
The neo-Malthusian theory is a recent refinement of classical Malthusian thought. It takes into account the great technological advances of the Industrial Revolution. Neo-Malthusians argue, however, that these rapid advances in productive capacity cannot be maintained.

They maintain, however, that the expansion of productive capacity cannot continue indefinitely; when population growth surpasses capacity, the prediction of Malthus will be fulfilled. The neo-Malthusians suggest that in such areas as India, Bangladesh, and parts of Africa the prediction of Malthus may soon be fulfilled.

PROSPECTS

The prospects are for continued population growth for the near future. By the year 2000 more than 6 billion people are expected to be living on the earth. By the year 2020 more than 7 billion will inhabit the earth. A few daring forecasters estimate a world population of 51 billion by the year 2100. Such long-range predictions, however, must be viewed with great caution.

Predictions of resource development are not easy to come by because there is no way to determine what resources will be created. The Food and Agriculture Organization of the United Nations believes that technology now exists to feed and clothe the world's population "adequately" for many decades. The rub is that the technology may not be available where most needed. As noted in the next chapters, physical and cultural systems create patterns over the world that directly affect resources, development, and population growth.

3

The Physical and Cultural Components of the Human Environment

There are great differences among nations in economic conditions and, therefore, also in state of biological health. In some parts of the world people live at high levels, whereas in other parts there is a constant struggle for enough food. Three principal factors affect the level of living in an area: the physical environment; the political, economic, and social systems of society; and the number of people and their rate of increase. These factors provide a framework within which material demands and expectations are rising rapidly, and so also is the need for earth resources. In Chapter 2 we briefly examined population. This chapter focuses on the physical environment and human cultural systems.

Professor Carl Sauer has provided us with the following:

> Every human population, at all times has needed to evaluate the economic potential of its inhabited area, to organize its life about its natural environment in terms of the skills available to it and the values which it accepted. In the cultural *mise en valeur* (exploitation) of the environment, a deformation of the pristine, or prehuman, landscape has been initiated that has increased with the length of occupation, growth in population, and addition of skills. [Wherever] men live, they have operated to alter the aspect of the earth, both animate and inanimate, be it to their boon or bane.[1]

Sauer made several important points. Society has an overriding economic concern; survival is based on some form of production. Humanity is continuously concerned with a utilitarian relationship with the environment. The nature of this relationship, however, depends upon the skills a society accumulates and the value system that motivates it. For example, the way we in the United States use a desert environment differs substantially from that of the Bushman of Africa.

---

[1]Carl O. Sauer, "The Agency of Man on the Earth," in *Readings in Cultural Geography,* ed. Philip L. Wagner and Marvin W. Mikesell (Chicago: University of Chicago Press, 1962), p. 539.

Environmental modification begins with the arrival of human beings. Our recent experience with water and air pollution and power problems, as important as they are, sometimes encourages us to think that landscape modification and environmental problems are new phenomena. This attitude is fallacious. Ancient Middle Eastern inhabitants had culturally induced environmental problems. Increased salinity of soils, a result of irrigation practices, ultimately rendered many agricultural lands useless and contributed to the decline of some Mesopotamian civilizations. *Landscape modification* is not new; it is as old as mankind. Most, if not all, landscapes are not "natural" but "cultural," shaped and formed as societies occupy and use the surface of the earth. Culture is part of the operational environment, just as is the physical world. Culture is, therefore, an important consideration in our effort to understand the condition of the human race in various regions of the world.

ENVIRONMENTAL ELEMENTS

The numerous components of the physical environment—rocks, soils, landforms, climate, vegetation, animal life, minerals, and water—are interrelated. Climate is partially responsible for variations in vegetative patterns, soil formations, and landforms. Organic matter from vegetation affects soil development. The environmental elements are all part of a large *ecosystem* that often exists in a delicate balance. As important as rec-

FIGURE 3–1 World Mean Annual Precipitation. Precipitation varies greatly from one part of the world to another. Moreover, there is considerable variability in precipitation from one year to the next. Variability is usually greatest in areas of limited precipitation.

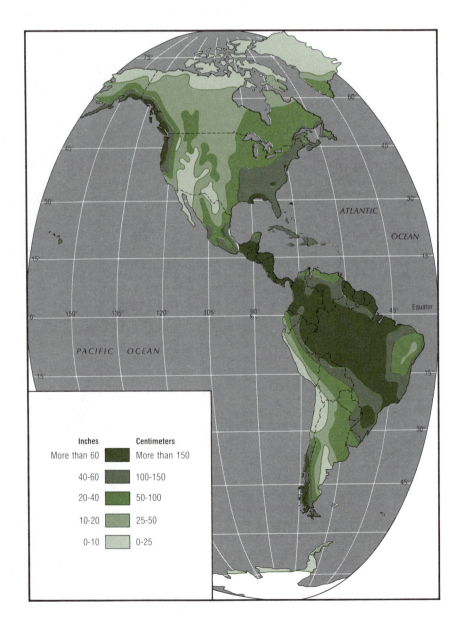

ognizing the relationship of natural processes is the recognition that humans interfere with natural processes. The human race is the most active agent for environmental change on the surface of the earth.

Climate

Climate is of direct importance in our effort to produce food and industrial crops. Agricultural production in great quantities is basic to our civilization. Each plant has specific requirements for optimum growth. Some plants require substantial amounts of moisture (rice); others are more drought tolerant (wheat). Coffee requires a year-round growing season yet does best in the tropics at elevations that provide cooler temperatures. Humans have modi-

fied the character of many plants (*cultigens*); yet climatic differences ensure great geographic variation in crop cultivation.

The most important climatic elements are precipitation and temperature. *Precipitation* is moisture removed from the atmosphere and dropped on the surface of the earth. Rainfall is most common, but the solid forms—snow, sleet, and hail—are also important. Melting snow has value as a source of water for streams or soil moisture that can be used in later seasons. Sleet and hail normally are quite localized in occurrence but nevertheless can cause damage and significant economic losses.

The precipitation map (Figure 3–1) shows the rainy areas of the world to be the tropics and the

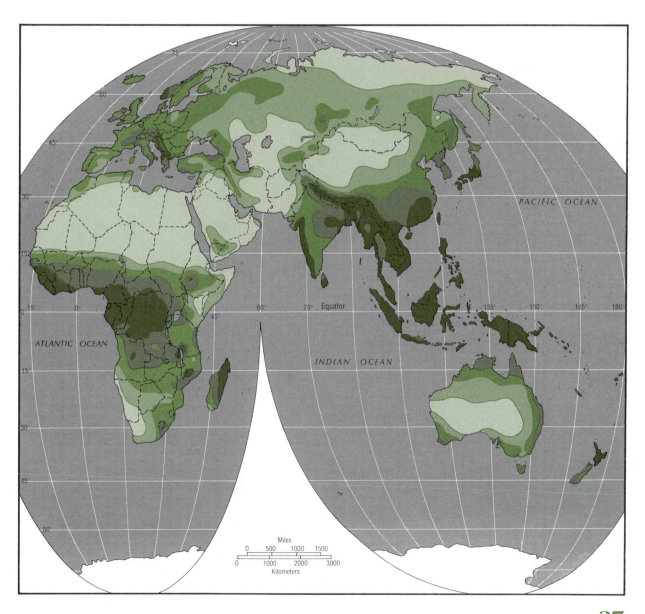

27

middle latitudes. In the latter areas it is particularly heavy on the western sides of continents. The eastern sides of middle-latitude continents may be less rainy but still receive sufficient precipitation to be considered humid regions, particularly if exposure to major water bodies (moisture sources) is favorable. For example, notice the eastern United States and Canada on Figure 3–1. Continental interiors, such as Asia's or North America's, or areas on the leeward side of mountains experience moisture deficiency. Some subtropical areas—northern Africa, Australia, northern Chile and Peru, northwest Mexico, and the southwestern United States—experience meager rainfall. Subsiding air and divergent wind patterns reduce the like-

lihood of precipitation, and cold waters offshore accentuate the aridity.

The map of *average annual precipitation* is useful for analyzing the distribution of mankind's activities. Areas of meager rainfall normally are not heavily populated. Exceptions require an exotic water supply (the Nile Valley and southern California). It is also apparent that some areas of very high rainfall and high temperatures contribute to other environmental conditions that are difficult to overcome, namely, infertile soils in tropical regions.

Seasonal rainfall patterns are as important as the yearly amount. Most equatorial areas receive a significant amount of rainfall during all seasons. The adjacent tropical wet and dry regions have a dis-

FIGURE 3–2 Climatic Regions of the World. Climate is the long-term condition of the atmosphere. Although there are many elements of climate, most classifications use only the two most important: temperature (amount and seasonality) and precipitation (amount and seasonality).

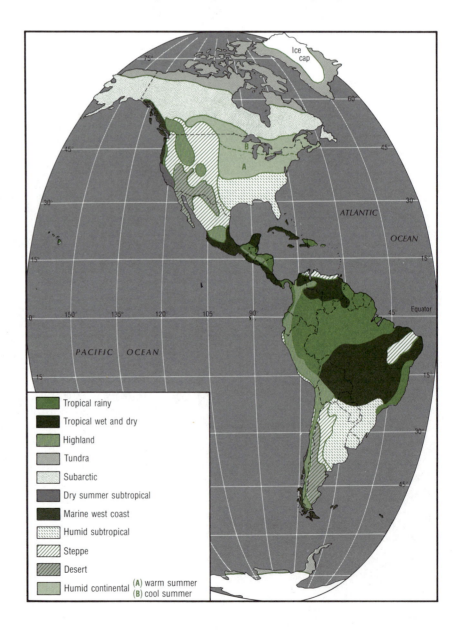

Tropical rainy
Tropical wet and dry
Highland
Tundra
Subarctic
Dry summer subtropical
Marine west coast
Humid subtropical
Steppe
Desert
Humid continental   (A) warm summer
                     (B) cool summer

tinct rainy summer season contrasting with an excessively dry season; the advantage of a yearlong growing season is therefore partly negated. The dry summer subtropical climates also experience distinct seasonal variations, except that summers are dry and winters have the maximum rainfall. In trying to understand land use in many areas of the world, one must consider seasonal rainfall patterns.

Variability of precipitation is the actual amount received in a given year expressed as a percentage of annual average. The greatest variability is experienced in areas of minimal rainfall. Unless a special source of water is available, settlement in dry areas is limited. Transitional areas (steppes) between humid and dry regions, however, frequently are important settlement zones in which production of staple foods (grains) is significant. Under normal conditions the United States Great Plains, the Black Earth Belt of the Soviet Union, the North China Plain, and the southern edge of the Sahara are significant food-producing regions. Unfortunately, these and other similar areas have experienced repeated droughts. A recent example is the *Sahel* (southern Sahara), where thousands have died of starvation. Drought and famine emphasize the danger of dependence on such areas. The human suffering associated with prolonged drought has encouraged the study of subhumid environments and the process by which they are ren-

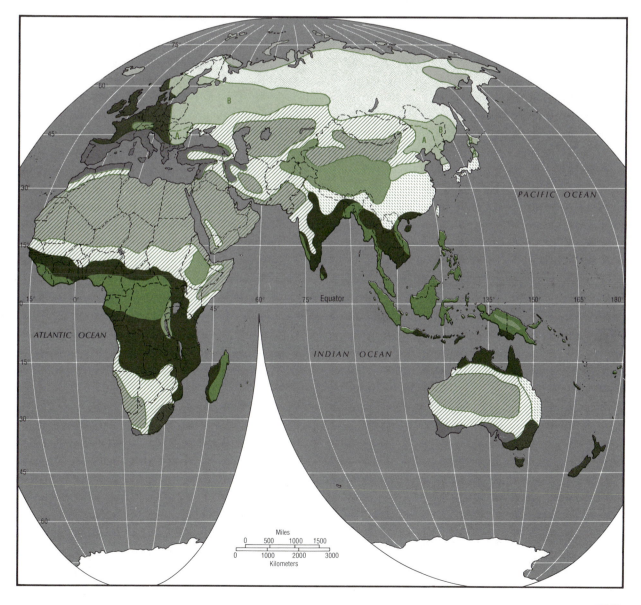

dered less productive (*desertification*). Ecosystem deterioration, especially of soils and vegetation, occurs as a result of both drought and the nature of human activities. Overpopulation may encourage either grazing or cultivation practices which have long-term detrimental effects on the environment—particularly during periods of recurring drought.

Data on annual amount, seasonality, and variability of rainfall reveal much about the utility of an environment; however, the moisture actually available also depends upon temperature conditions. High temperatures mean great potential for evaporation and plant transpiration—a high *evapotranspiration* rate. An unfavorable relationship between precipitation and evapotranspiration means moisture deficiency for plant growth.

Temperature and Plant Growth

Perhaps the most important aspect of temperature is the length of the *frost-free period*, which is related to latitude primarily but also to altitude and location of large bodies of water. The modifying influence of water on temperature is evident at maritime locations; growing seasons are longer than would normally be expected at those latitudes. Comparing northwest Europe with similar latitudes in the Soviet Union and eastern North America illustrates this point.

Winter temperatures are important, too. Many middle-latitude fruit trees require a specific dormancy period with temperatures below a certain level. Cold temperatures (*chill hours*) are necessary to assure vernalization, or activation of a new flowering cycle.

Photoperiod (the length of day or active period of photosynthesis) and daily temperature range during the growing season also affect plant growth. Some plants require a long daily period of photosynthesis to flower (barley), yet others (such as soybeans or rice) are favored by shorter day length. Each plant variety has a specific range of daily low and high temperatures, called *cardinal temperatures*, within which plant growth proceeds most rapidly. Cardinal temperatures for most plants are known. The numerous varieties of wheat, for example, have cardinal temperatures that range for the low between 32° and 41°F (0° and 4°C), and for the maximum between 87° and 98°F (31° and 37°C). Optimum temperatures for most varieties of wheat are between 77° and 88°F (25° and 31°C). Although less than optimum conditions do not preclude the cultivation of a particular crop, departures from the optimum reduce the efficiency with which a crop is produced. Efficiency, translated into cost per unit of production, is of major concern in commercial agriculture.

Climatic Classification

Figure 3–2 (pp. 28–29) presents the global distribution of climates. Climatic classification is based upon temperature and precipitation conditions (Table 3–1). Although conditions may vary within a climatic region, a generalized classification system is useful to compare and contrast characteristics among different areas. A review of the climatic map reveals a close relationship among latitude, continental position, and climate. The *global climatic pattern*, therefore, is not difficult to comprehend. Notice, for example, the locational similarity of humid subtropical climates in the United States, in China, and in Argentina.

A fundamental distinction in the classification is that between *humid* and *dry climates*. Areas are classified as *desert* or *steppe* if the potential for evaporation exceeds actual precipitation. A single specific precipitation limit cannot be used to separate the dry and humid climates; in the midlatitudes, however, deserts normally have less than 10 inches (25 centimeters) and steppes or semiarid regions between 10 and 20 inches (51 centimeters) of precipitation. Higher latitudes have lower limits and lower latitudes a higher limit before the dry classification is applied. Twenty-five inches (64 centimeters) of precipitation in higher middle latitudes may provide a humid climate and forest growth, but the same amount in tropical areas may result in a semiarid and treeless environment.

The *highland climates* shown on Figure 3–2 are a group with variable temperature and rainfall conditions. Along with the other controls of climate—such as latitude, marine exposure, prevailing winds, and atmospheric pressure systems—elevation is a major factor. Precisely what conditions may be expected depends upon elevation and position within mountains. Many mountainous areas are sparsely settled and provide a meager resource base for farmers; however, in Latin America, East Africa, and Indonesia, highland settlement is important. The variations of highland climate, and therefore vegetation and soils, are treated in the chapters dealing with those regions.

Vegetation

Vegetation patterns are closely associated with climate. This relationship is readily evident in the ma-

TABLE 3–1 Characteristics of World Climate Types

Climate	Location by Latitude [Continental Position If Distinctive]	Temperature	Precipitation (inches/year)
Tropical rainy	Equatorial	Warm, range[a] less than 5°F; no winter	>60; no distinct dry season
Tropical wet and dry	5°–20°	Warm, range[a] of 5–15°F; no winter	25–60; summer rainy, low sun period dry
Steppe	Subtropics and middle latitudes [sheltered and interior positions]	Hot and cold seasons, dependent on latitude	Normally less than 20, much less in middle latitudes
Desert	Subtropics and middle latitudes [sheltered and interior positions]	Hot and cold seasons, dependent on latitude	Normally less than 10, much less in middle latitudes
Dry summer subtropical	30°–40° [western and subtropical portions of continents]	Warm to hot summers; mild but distinct winters	20–30; dry summer, maximum precipitation in winter
Humid subtropical	0°–40° [eastern and southeastern subtropical portion of continents]	Hot summers; mild but distinct winters	30–65; rainy throughout the year, occasional dry winter (Asia)
Marine west coast	40°–60° [west coast of middle-latitude continent]	Mild summers and mild winters	Moisture throughout the year, some winter maximum tendency; total amount highly variable, from 20 to 100
Humid continental (long summer)	35°–45° [continental interior and east coast, northern hemisphere only]	Warm to hot summers; cold winters	20–45; summer concentration, no distinct dry season
Humid continental (short summer)	35°–45° [continental interior and east coast, northern hemisphere only]	Short, mild summers; severe winters	20–45; summer concentration, no distinct dry season
Subarctic	46°–70° [northern hemisphere only]	Short, mild summers; long, severe winters	20–45; summer concentration, no distinct dry season
Tundra	60° and poleward	Frost anytime; short growing season for limited vegetation	Limited moisture (5–10), except at exposed marine locations
Ice cap	Polar areas	Constant winter	Limited precipitation, but surface accumulation
Undifferentiated highland (see Fig. 3–2)			

[a]Range: temperature difference between warmest and coldest months.

jor distinction between grasslands (herbaceous plants) and forests (woody plants). *Grassland vegetation* on Figure 3–3 is associated with areas of moisture deficiency (steppes and deserts) or with areas of extreme seasonal variation in precipitation, but humans have extended grasslands in numerous areas by the use of fire and extensive grazing, so this kind of vegetation is now found in association with quite varied environmental realms.

The very cold areas (*tundra*) are also treeless but because of a short or nonexistent growing season. *Forest growth* usually requires a minimum rainfall between 15 and 20 inches (38 and 51 centimeters), depending on evapotranspiration rates, and is associated with humid climates.

The phrase "*natural vegetation*" should be used carefully. Climatic climax vegetation is that expected in an area if vegetation succession is al-

FIGURE 3–3 World Vegetation Regions.
The distribution of vegetation closely corresponds to climatic patterns. A map of vegetation can be used to study an area's agricultural potential since crops and "natural" vegetation use the same environmental elements for growth.

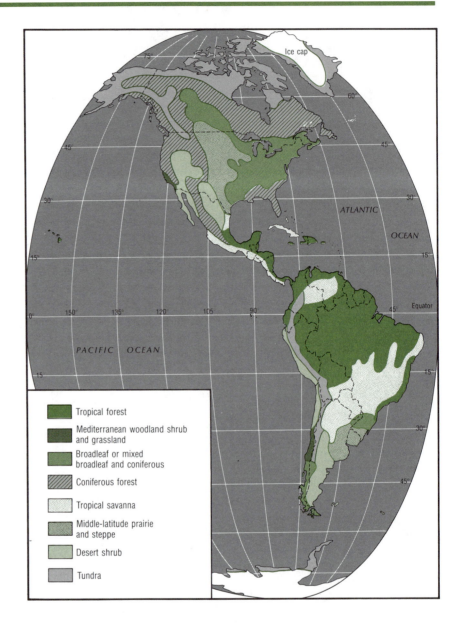

Legend:
- Tropical forest
- Mediterranean woodland shrub and grassland
- Broadleaf or mixed broadleaf and coniferous
- Coniferous forest
- Tropical savanna
- Middle-latitude prairie and steppe
- Desert shrub
- Tundra

lowed to proceed over a long period without human interference, but landscape modification has been going on for so long that few areas of natural vegetation remain. Most vegetative cover is really part of a *cultural landscape* and reflects human activity.

Vegetation provided mankind a direct source of food as well as an indirect food supply through its support of animal life. Wild plants were certainly the most significant biotic element with which humans dealt during the earlier millennia. Grass-covered plains facilitated movement and interaction with others. Dense forests functioned as barriers, isolating culture groups, and provided refuge for those trying to remain separate. For example, the Sahara was not the major barrier inhibiting Subsaharan African cultural exchange with the Arab world. In fact, interaction did occur; the grassland zone south of the Sahara is a cultural transition zone with features of both African and Arab cultures. The equatorial rain forest was a more effective barrier, not only as a forest that made travel difficult but also because of other environmental elements. The tsetse fly, for example, inhibited the adoption of cattle raising as a life-support system.

Forests have been removed from vast areas in middle-latitude Asia, North America, and Europe over the centuries (*broadleaf* and *mixed broadleaf* and *coniferous forests* on Figure 3–3). Increases in

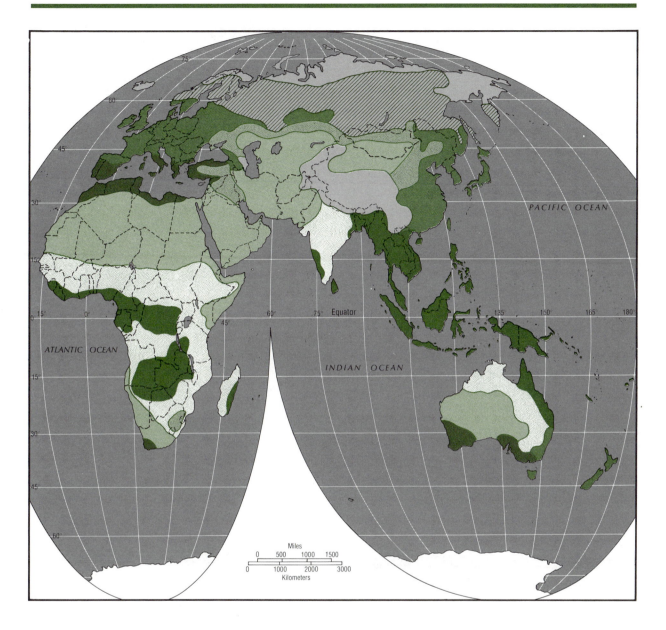

population and changes in agricultural technology placed great demands on such environments with favorable temperature, soil, and moisture conditions. Crop agriculture could only expand at the expense of forest. Forests were not only a barrier to travel but also had to be cleared and removed so that field agriculture could expand. The vast forested areas that remain, the high-latitude *coniferous forests* and *tropical rain forests,* do so less from the protective concern of people than because they are located in relatively less accessible places or contain other environmental features unfavorable for permanent settlement. Rapid population growth and the need for space may stimulate more clearing. The rapid expansion of Brazilians

into the Amazon Basin forests (tropical rain forest) for settlement purposes is one example. There are those who view this expansion into the Amazon with great expectation—that is, as a source of national pride and unity for Brazilians, as a means of easing population pressure, and as a source of new economic wealth. Yet, others view these as short-term gains which may exact a high price in damage to forest, soils, streams, and the atmospheric environment. History is replete with examples where short-term or immediate need has clouded our vision on matters of environmental stewardship.

Attitudes toward natural vegetation have recently begun to change dramatically. The significance of vegetation in numerous aspects of life is increasingly recognized, including its relationship to other components of our environment (soils and air). Forest vegetation becomes more critical as population increases and consumes ever greater amounts of lumber and paper. *Sustained-yield forestry* (harvesting only at the annual growth rate) is becoming more common as people recognize that wasteful usage can only result in greater resource problems.

Forests have many uses other than for wood. In the more affluent countries they are prized recreation areas. They help to prevent erosion, which, in addition to destroying land, contributes to silting problems in rivers, streams, and reservoirs. By reducing water runoff after precipitation, forests also contribute to flood prevention.

Soils

In addition to moisture and sun energy, nutrients are needed for plant growth. Nutrients are derived from minerals in the earth and from organic material added to the soil by vegetation (humus). Underlying parent material, slope of land, climate, vegetative cover, microorganisms, and time all contribute to soil formation.

Mankind has *modified soils* in many ways, the most common effects being soil depletion and erosion. Declining fertility resulting from prolonged use can sometimes be overcome by adding chemical or organic fertilizers. Lime is added to reduce acidity, a soil condition detrimental to many plants. Terraces are built to prevent erosion or to aid in the distribution of irrigation water. Crops that place a serious drain on soil nutrients may be rotated with less demanding plants or with leguminous plants that have the capability of adding rather than removing nitrogen. Farmers in many areas of the United States have placed drainage tile in the ground to remove excessive moisture from poorly drained areas. Other drainage techniques have been used in Asia and Europe to bring into use soils that otherwise would be unproductive. Primitive societies shift from one plot to another as soil fertility decreases. Leaving formerly cultivated land out of production for some years (*fallowing*) allows soil restoration by natural processes. Efforts to reduce soil destruction or to improve fertility involve expenditures of labor or capital, but the long-term needs of society require that we accept the costs of maintaining the environment. People, however, are not always willing or able to bear such costs, particularly when short-term profits or immediate survival are their main concerns. An *exploitative approach* leads to maximum returns with minimal inputs over a short period but often results in rapid destruction of one of our most basic resources.

An *American experience* illustrates the environmental damage and human hardship that result from less than cautious resource use. The years 1910 to 1914 are often referred to as the "golden years" of American agriculture because of the high prices received by farmers for agricultural commodities. The favorable prices continued through World War I and into the 1920s. The high prices and above-average rainfall encouraged farmers on the western margin of the Great Plains to convert grassland into wheat fields. Then followed several years of severe drought coinciding with the depression of the 1930s. The unprotected cultivated lands suffered great damage by wind erosion. The economic hardship wrought upon the farmers led to widespread migration, land abandonment, bank failure, and even the abandonment of towns.[2]

Soils share a generally close association with climate and vegetation. Climate is intricately related to the processes of vegetation evolution and soil formation. Figure 3–4 illustrates this correspondence. Although the associations are somewhat more complex than the diagram suggests, it provides an impression of various environmental realms available for human use. For example, using Figures 2–2 and 2–3, compare the environments of the southeastern United States with those of China.

Landforms

The surface of the earth is normally classified into four landform categories:

1 Plains have little slope or local relief (variations in elevation).
2 Plateaus are level land found at high elevations.
3 Hills have moderate to steep slopes and moderate local relief.
4 Mountains have steep slopes and great local relief.

---

[2]James A. Michener in his novel *Centennial* (New York: Random House, 1974) vividly describes the occupation of the western Great Plains and the problems faced by farmers in attempting to use the land. John Steinbeck's novel *The Grapes of Wrath* (New York: Viking Press, 1939) portrays what happened to the farmers of the western Great Plains during the depression and drought years of the 1930s.

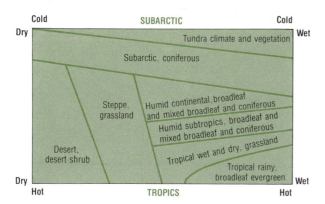

FIGURE 3–4 A Generalized Association of Climate and Vegetation.
As the diagram indicates, there are regional associations of climate with dominant vegetation. The boundaries are transition zones and the correspondence is not perfect, but the general scheme is useful for identification of environmental realms.

Plains are the most used landforms for settlement and production when other environmental features are available. Large areas of land with low slope and little relief are most suitable for agriculture. For thousands of years the human has been an agrarian being who produced food by crop cultivation and required large amounts of space. Ease of movement over plains, particularly those with grassland vegetation, has facilitated exchange with other societies. Rivers flowing through plains have long been routes of travel and transport.

The great densities of population occur in areas where intensive agriculture is practiced or in regions where industrial and commercial activity is concentrated. The early high-density agricultural populations were associated with plains. In mountainous regions the small basin and valley become the focus of settlement. Those plains that do exhibit limited settlement and utilization are less desirable for climatic reasons. The features of plains that contribute to their utility in peacetime can be handicaps in wartime; with few natural barriers to afford protection, they are easily traveled and overrun. As seen during World War II in Europe, a mechanized army can move very rapidly over a plains area.

Hills and mountains have a quite different meaning as a habitat. These less favorable areas are difficult to penetrate, and their isolation may lead to the formation of distinct cultures. While such areas provide security from attack, they may also contribute to economic disparity.

Major differences and even conflicts between highland and lowland inhabitants are common and are part of regional history in many parts of the world. Unifying a country that incorporates these contrasting environments and cultures remains difficult. Separatist movements are an expression of these differences, and isolated highland areas provide excellent guerrilla army bases. The mountain-dwelling Kurds of Iran and Iraq have been troublesome to the governments of both these states, and in Burma the Karen of the Shan plateau have

Hills and mountains provide a difficult habitat for people. In the mountains of Nepal intricate terracing systems, which require great effort in construction and maintenance, are necessary for intensive crop production. (Agency for International Development)

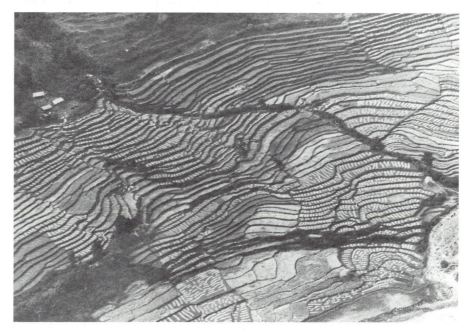

raised Communist and indigenous insurrections against a government that is controlled by the lowland river plain majority.

Minerals

The politically and economically powerful countries of today are those that have built an industrial structure by processing large quantities of minerals. These countries require huge quantities of *fossil fuels* (coal, petroleum, and natural gas). Any country striving for inclusion among the modern industrialized nations must either possess such resources or acquire them. Recently developed nuclear energy as yet contributes only minute quantities of power. Water is another source, but hydroelectric power generally has only local importance and provides a small proportion of a nation's total power needs.

The use of petroleum and natural gas has increased rapidly since World War II in all industrialized areas but particularly in the United States. This trend has given those less developed countries who share significantly in the world's oil wealth a special importance in international politics, far greater than their size or military strength could otherwise provide. Although huge quantities of coal are available, especially in the United States, the Soviet Union, and China, petroleum is now being used to meet expansion needs and as a substitute for coal. Petroleum and natural gas are cleaner and contribute less to air pollution. Whether the increasing cost of petroleum will generate a reverse trend—that it, a move back to coal for power—is not yet clear.

Iron, aluminum, and copper are the most important *metallic minerals* used in industry. Other important metallic minerals are chromium, copper, zinc, lead, gold, and silver. *Nonmetallic minerals* such as nitrogen, calcium, potash, and phosphate are used for fertilizer. Salt, building stones, lime, sulfur, and sand are other commonly used nonmetallic minerals.

It is possible that future discoveries of significant mineral deposits in parts of the world outside of today's producing regions will greatly affect our evaluation of the industrial potential of such areas. At present, however, a crude balance of power exists between the United States and the Soviet Union. Their power and position in world society are based upon effective use of industrial and power resources. Europe is a close third but somewhat more vulnerable because of limited petroleum supplies. Japan is clearly vulnerable because of the great dependence of its industrial structure on imported resources. But could the centers of power one day shift, based upon newly developed resources for industrial production? What is the potential of a country such as Brazil?

Environmental Use

We have reviewed a select few of the many components of our physical environment. Emphasis is on the highly varied character of the physical environment and on those aspects that have inhibited or supported mankind's activities. Two considerations are of utmost importance.

First, a basic dilemma is the need to resist environmental deterioration at the same time that increasing population and expectations are placing greater demands on environment. The problem will not be easily solved. Nonuse is not the solution, but neither is environmental destruction. What is necessary is the intelligent use of the environment for supplying the needs of mankind. This accomplishment necessarily will be costly. Moreover, such intelligent use will be elusive and complicated by the varied nature of the environment, of which there is yet so much to learn.

Second, the meaning of environment cannot be separated from the culture of the user society, and societies differ. The motivations, attitudes, accumulated technology, and ways of organizing economically and politically—all cultural considerations—greatly affect the use of the environment.

Third, there is need for long-range approaches to our use of the physical environment. Our attitudes and immediate real concerns seldom encourage us to formulate approaches to the use of the environment which incorporate the needs of mankind more than two or three generations into the future. This temporal perspective is not surprising but is ultimately unacceptable.

HUMANITY AND CULTURE

The human race is unique in being able to accumulate learned behavior and transmit it to successive generations. These behavior patterns have utility in assuring sustenance and preservation of social order. With this accumulation of learned behavior (*culture*), humanity makes decisions and executes life-styles. Some behavior is based upon earlier experiences of society (inherited culture), and other actions are based on the experiences of societies with which we have contact (*diffused culture*). The entire set of elements that identify a group's life-style comprises the culture of a society.

Another way of viewing culture is as a hierarchy of traits, complexes, and realms. A *trait* is the way a society deals with a single artifact, for example, how people plant seeds. A *complex* is a group of traits employed together in a more general activity, such as agricultural production. A *culture realm* is an area in which numerous activities or culture complexes are adhered to by most of the population. Not all of the world regions that we shall examine exhibit easily definable culture realms. Some are transition zones, in which numerous distinct cultures have met and clashed. *Cultural pluralism* is frequently evident even within political units, adding to the difficulty of achieving national unity.

A *culture hearth* is a source area where a culture complex has become so well established and advanced that its attributes are passed to future generations within and outside the immediate hearth area. It is no longer accepted that human progress stems from achievements spread from a single hearth. More likely, several different hearths contributed to culture and became the basis for advancement over much larger areas.

Societies have advanced at very uneven rates. For example, the Bantu culture of western Africa advanced more rapidly than those of the Bushmen or Hottentots of southern Africa. Furthermore, societies may progress culturally along different paths. The culture complexes utilized for food production in China and Europe during the nineteenth century had different traits. This cultural distinctiveness need not be construed as a difference in level of achievement but rather as a difference in kind. The cumulative nature of cultural evolution, its dynamic quality, the unevenness with which it occurs, and the different orientations all contribute to fundamental and intriguing variations among the more than 4 billion people of our contemporary world.

The Growth of Culture

Most of the major human accomplishments have occurred during the last few thousand years, a short time span when measured against the 2 million years or so that *hominids* (the family of humanlike beings) have occupied the earth. When the long *Paleolithic*, or Old Stone Age, period

ended approximately 10,000 years ago, people had spread over most of the habitable portions of the earth, although low population densities were prevalent. Cultural accomplishments included the manufacture of stone tools, the use of fire (perhaps the most significant achievement), and the construction of shelters. The *Neolithic*, or New Stone Age, period (which followed the Paleolithic) marks the beginning of one of humanity's great revolutions. It should not be thought of as an abrupt occurrence but rather as a period of several thousand years during which the human relationship to the environment changed slowly but fundamentally as new technologies were developed. People shifted from direct use of the environment by hunting and gathering to systems of agriculture.

Early Primary Culture Hearths

The Middle East contained one of the world's foremost culture hearths (the Fertile Crescent), or more accurately, several hearths in proximity (Figure 3–5). The earliest plant domestication apparently took place in adjacent hill lands, and by 10,000 years ago agricultural villages appeared in the Mesopotamian lowlands. From the fourth millenium B.C., there followed flourishing civilizations, city-states, and empires. The peoples of these civilizations codified laws, made use of metals, developed the functional use of the wheel, established mathematics, and contributed several of the world's great religions (Zoroastrianism, Judaism, Christianity, and Islam). These are only a few of their major achievements.

A second of the world's great culture hearths developed in the *Indus Valley*. A mature civilization existed there by 2500 B.C. The exchange of ideas and materials with Mesopotamia began early and continued over a long period of time (3000–1000 B.C.). Much of this exchange of culture was by way of ancient Persia (Iran). The two hearths may even have attracted migrant peoples from the same areas. The Indus Valley experienced repeated invasions and migrations of people from northwestern Asia and central Asia. Each new invasion meant an additional infusion of racial and cultural traits. The Indus and the adjacent Ganges Valley served as a source area for the Indian culture realm. Significant contributions were made through literature, architecture, the working of metals, and city planning. Philosophy evolved that later contributed to Hinduism, and the idea of living the sinless and good life may have been transferred to

37

FIGURE 3–5 Early Culture Hearths of the World. Three major culture hearths are recognized as the principal contributors to modern societies over the world. From these hearths have spread ideas, plants and animals, religions, and other cultural characteristics. Minor hearths, although locally important, have not had much impact outside their source areas.

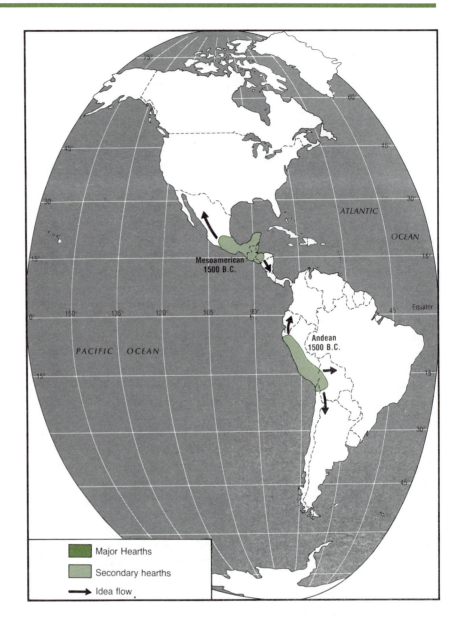

the Middle East, where it was incorporated into Judaism and later into Christianity.

The middle *Huang He Valley* (Yellow River Valley) and its tributaries served as a hearth area for the evolution of Chinese culture. China, however, undoubtedly benefited from contacts with other areas. Wheat and oxen, a part of the north China agricultural complex, may have Middle Eastern origins. Rice, pigs, poultry, and the water buffalo have Southeast Asian origins. Nevertheless, the cultural achievement accumulated over the long period since early Neolithic time, as well as the peculiarly Chinese traits maintained even after contact with other areas, clearly suggest that this area is a hearth from which an original culture emanated. Crop domestication, village settlement, distinctive architecture, incipient manufacturing, and metalworking were in evidence at the time of the Shang dynasty (1700 B.C.). Chinese culture spread into south China and northeast into Manchuria, Korea, and eventually Japan.

Each of the early primary culture hearths appears to have accomplishments acquired independently, yet each was also a recipient of ideas and commodities from other areas. Each hearth was able to maintain a distinct identity that was trans-

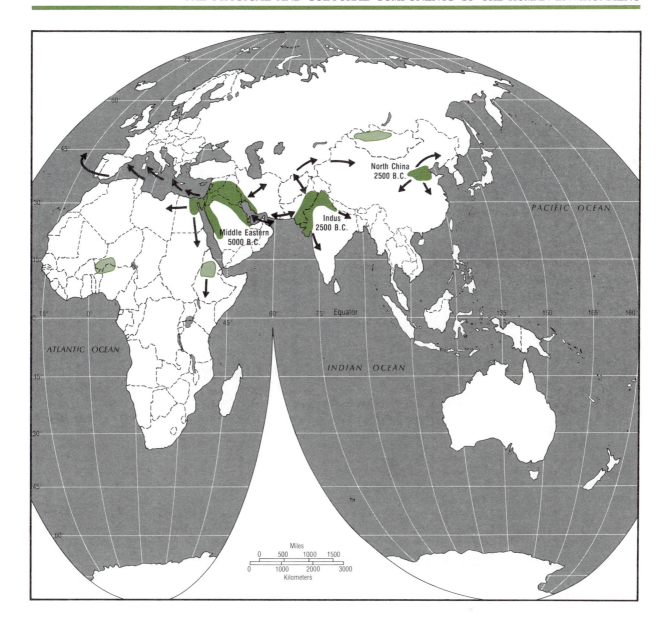

ferred to succeeding generations as population increased or to invading or conquered people eventually assimilated and acculturated.

Secondary Culture Hearths

Two areas in the Americas served as culture hearths for major civilizations. The first was in northern Central America and southern Mexico (Mayan civilization) and extended to central Mexico (Aztec civilization). This large *Mesoamerican region* supported a sizable sedentary population and was characterized by cities, political-religious hierar-

chies, numeric systems, and domesticated crops (maize, beans, and cotton).

The *Middle Andean area* of Peru and Bolivia contained the second American culture hearth. The Inca civilization advanced more slowly than its Middle American counterpart but by the sixteenth century had developed irrigation, worked metals, domesticated the white potato, established complex political and social systems, set up a transport network, and controlled an empire.

These culture hearths are considered secondary because the civilizations never evolved into lasting culture realms. The Latin American culture realm as we now know it is clearly of mixed origin—indigenous American and European.

Areas in *West Africa,* the *Ethiopian highlands,* and *Central Asia* were also secondary culture hearths. The Bantu language family seems to have its source area in West Africa. The Ethiopian highlands were a secondary domestication center for various grains (wheat, millets, sorghums). Central Asia was also a domestication center for grains and certainly a trade route along which ideas and commodities were exchanged among major civilizations.

Europe as a Culture Hearth

Europe is an additional primary culture hearth but must be viewed in a different perspective from other hearths. First, it flowered much later than the civilizations of the Middle East, southern and eastern Asia, and Latin America. Second, the *Industrial Revolution* associated with Europe affected virtually all parts of the earth. Many of our contemporary global problems have origins in the confrontation between modern European culture and the traditional systems employed elsewhere. This confrontation does not necessarily imply a deliberate attempt to erase traditional ways of life. Nevertheless, whether elements of the European system have been imposed or willingly adopted, the adjustments involved in adoption of the new ways have frequently meant difficult transitions.

Europe has a Mediterranean and Middle Eastern heritage. Technology such as the wheel and plow agriculture, social concepts, and religions are of Middle Eastern origin; they spread through the Minoan (based on Cyprus), Greek, and Roman civilizations and into northern Europe.

When Middle Eastern civilizations were ascending, Europe was a peripheral area inhabited by "barbarians" not yet receiving or accepting the innovations of the Neolithic period. Rome had a civilizing influence on Western Europe through the introduction of order, roads, and sedentary agriculture and the establishment of towns as market centers. Later barbarian invasions disrupted the Roman Empire, and several centuries of limited progress followed (the *Dark Ages,* the period from the fall of the Roman Empire in A.D. 476 to approximately the year 1000). Nevertheless, a long period of slow agricultural change stimulated the *Renaissance* and indirectly the Industrial Revolution.

By the thirteenth century, agricultural and manufactured trade was common along trade routes that crossed Europe. The *Crusades* (from the late eleventh through the fourteenth centuries) led to new contacts and stimulated thought and interest in learning. Spurred by internal competition and the new interest in science, exploration, and trade, the Europeans extended their influence and culture worldwide during the fifteenth to twentieth centuries. They explored, traded, conquered, and claimed new territories in the name of their homelands. Modern European states emerged with the capability of extending their power over other areas. European people resettled in newly "discovered" lands of the Americas, southern Africa, Australia, and New Zealand.

What took place has been aptly described as a "*European explosion.*" Professors J. E. Spencer and W. L. Thomas have argued that the European explosion has stimulated a process of *cultural convergence.*[3] Their thesis is that isolated cultures are largely relics of the past, with assimilation or extinction as their not-so-different options, and that methods of organizing politically, producing food, using inanimate sources of power, and consumption are becoming more similar. The blending of races and cultures is almost certain to continue. The technological aspects of European culture are the most readily accepted. Languages, religions, and attitudes adjust more slowly, however, and frequently become the sources of friction from which conflict arises.

This argument does not assert that contributions of traditional cultures will be discarded or that they will totally disappear. Rather, large and viable culture complexes will be modified by cultural blending. Japan provides a good example of how European traits can be accepted in modified form. Also, arguing for convergence does not assert that regional disparities in economic, social, and biological conditions will cease. Indeed, here comes the converse argument. Portions of the world provide evidence that the old axiom "the rich get richer and the poor get poorer" has basis in fact. Acceptance by a people of the European way of doing things (economic pursuits) means that particular areas will have unique and great value because of location, material resources, and human resources; other areas will not be so well endowed. Even in the United States, *regional disparity* has long been a fact of life and a continuing problem (the Appalachians). Some look at the world and see

---

[3] J. E. Spencer and W. L. Thomas, *Introducing Cultural Geography* (New York: Wiley, 1973), pp. 185–205.

rich and superindustrial powers continuing to widen the gap between haves and have-nots. Among the underdeveloped nations, they recognize a few countries progressing but most falling further behind as their populations continue to grow at unprecedented rates. Most certainly, regional disparity will remain a problem. There are conditions inherent in modern political and economic systems (and in human nature) that preclude simple solutions to these adjustment problems.

Special Elements of Culture

Language

Language is a set of meanings given to various sounds used in common by a number of people; it is the basic means of *cultural transfer* from one generation to the next. The relative isolation in which various societies evolved led to the use of a great number of languages, frequently with common origins but not mutually intelligible.

Language differences may function as a barrier to the exchange of ideas, acceptance of common goals, or the achievement of national unity and allegiance. Members of most societies are not bilingual and do not speak the language of neighboring societies if different from their own. Sometimes the linguistic differences are overcome through a *lingua franca,* a language used over a wide area for commercial or political purposes by people with different speech. The uses of Swahili in eastern Africa, English in India, and Urdu in Pakistan are examples. In these three areas many different mother tongues are represented.

The political leadership in many countries that acquired independence after World War II has found achievement of national unity and stability difficult. Internal problems and conflict frequently stem, in part, from cultural differences; *linguistic variation* is often one of these differences. An example is provided by East and West Pakistan. From 1947 to 1972 these two regions functioned as one political unit. The political leadership sought to promote a common culture and common goals. A single national language was deemed necessary, although several mother tongues were in use. Bengali, the language of Bangladesh (formerly East Pakistan) was derived centuries ago from Sanskrit. It provided the East a cultural element around which unity could be achieved. West Pakistan contained a number of languages: Baluchi, Pashto, Panjabi, Sindhi, and Urdu (a lingua franca). The problem resulting from this language diversity was just one of many that eventually led to the establishment of Bangladesh as a separate and independent state. Lack of a common language within a country is a factor in *pluralistic societies,* and pluralism frequently produces political instability or hinders the development process. Belgium, Canada, and Switzerland, however, have partly overcome the problem of language pluralism.

Religion

Religions have their origin in a concern for comprehension and security, a concern that seems as old as mankind. *Animism* (the worship of natural objects believed to have souls or spirits) is among the earliest religious forms and still exists in some isolated culture groups. It includes rituals and sacrifices to appease or pacify spirits but usually lacks complex organization. Most such religions were probably localized; certainly, those that have survived are held by isolated and small culture groups. The more important contemporary religions are codified, organized in hierarchical fashion, and institutionalized to assure transfer of basic principles and beliefs to following generations and other people.

Modern religions are classified in several ways. *Ethnic religions* are those that can, in origin, be tied to a particular area and group of people. Examples of ethnic religions are Shintoism (Japan), Judaism, and Hinduism (India). *Universalizing religions* are those considered by adherents as appropriate, indeed desirable, for all mankind. Buddhism, Christianity, and Islam all have ethnic origins but have become universal in nature over the centuries. Proselytizing is often considered a responsibility of adherents. Missionaries as well as traders, migrants, and the military have been the means by which religions spread from one people to another. Universalizing religions become such as they lose their association with a single ethnic group.

The impact of religious ideology on civilization is great. Sacred structures contribute to the morphology of rural and urban landscapes. The distribution of citrus growing in southern Europe is directly related to certain Jewish observances. Food taboos account for the absence of swine in the agricultural system of Jewish and Muslim people in the Middle East. The Hindu taboo on meat eating, a response to respect for life, has the opposite effect: an over-abundance of cattle requiring space and feed but returning little in the way of material benefits (manure, milk, draft power). In the United States the economic impact of religion is seen in taxation policies (organized churches are usually exempt from taxation), work taboos on specified days (blue laws), institutional ownership of land and resources, and attitudes toward materialism and work (the Protestant work ethic).

One implied function of religion is the promotion of *cultural norms*. The results may be positive in that societies benefit from the stability that cohesiveness and unity of purpose bring. Unfortunately, conflict also frequently arises from differences that have religious underpinnings; such differences may evoke intolerance, suppression of minorities, or incompatibility among different peoples. Examples include warfare between Catholics and Protestants in northern Ireland, the partition of the Indian subcontinent in 1947 as a response to Hindu-Muslim differences, and the Crusades of the Middle Ages in which Christians attempted to wrest control of the Holy Land from Islamic rule.

Political Ideology

Political ideologies also have major implications for societal unity, stability, and the use of land and resources. Political ideologies, however, need not be common to the majority of a population. People may be either apathetic or powerless to resist the imposition of a system of rule and decision making. Small groups (*oligarchies*) or in some instances individuals (*dictators*) have been guided by a particular philosophy and by imposition have made their thinking basic to the functioning of society.

Most modern governments assume some responsibility for the well-being of the people included in the state. Some *socialist governments* assume complete control over the allocation of resources, the investment of capital, and even the use of labor. Individual decision making consequently is limited. In other societies governments assume a responsibility for providing an environment in which individuals or people in corporate fashion may own and decide on the use of resources. National and economic development programs reflect differences in political systems. Socialistic governments take an approach different from that of capitalistic countries. These differ-

ences are treated in the chapters on Anglo-America, the Soviet Union, India, and China.

Social and Political Organization

Political organizations are control mechanisms that may spring from the common bonds of a few families or of millions of people. The promotion of specified behavior patterns and institutions is deemed necessary to stimulate common bonds. Responsibilities of political organizations include resolution of conflict between and within societies and control over the distribution, allocation, and use of resources. The ability to execute these functions in a satisfactory manner depends not only on those in authority and the political system used but also the unity, support, or passivity of the people.

The Band and Tribe

The *band* is the simplest and formerly most common political organization; it has always existed. A band usually consists of no more than a few dozen people occupying a loosely defined territory within which game, fish, insects, and vegetable matter are gathered. The community functions on a cooperative basis; no individual can for long direct group activity without voluntary cooperation. There are neither exclusive rights to resources and territory nor concentration of power in the hands of a specified number of persons. Groups of this sort are aptly called sharing societies.

Probably no more than 100,000 such people remain scattered over remote and isolated parts of the earth. Contact between these static societies and more advanced groups leads to extinction of the former. Of the few groups that have survived, the Bushmen of the Kalahari, the Motilón of Venezuela, and Jivaro of Ecuador are examples.

Increases in population density and advances in technology (agricultural production) require a more specific allocation of resources and space. A focus of authority and ranking of position evolve in which community structure is based upon kinship; elders or heads of extended families lead and direct tribal activity. Hundreds, or even thousands of people, may be included in a kinship-structured *tribe*. Boundaries may be either vaguely or specifically defined depending on external pressures, trespass by neighbors, or colonial imposition. The Ibo of Nigeria are an ethnically related group, that, prior to European penetration of Africa, was subdivided into hundreds of patrilineal clans or lineages, each with a specific territory. The Ibo are but one of the many tribal groups that Nigeria must unify in its effort to build a modern state.

States

As population, size of territory needed, and economic specialization increase, a more formal organization becomes necessary. The focus of power becomes more distinct, authority is channeled through a larger hierarchy, and society becomes more stratified. The importance of kinship as a medium for organization disappears. The *state* emerges as a specifically defined territory, occupied by a people with a distinct bond to territory (sense of territory). It is organized politically, if not economically, and controls the entire territory and people. Thus the state includes territory, people, and an organizational system. It might be said that the territory is the state and the people component elements, whereas, in lesser forms of political organization, the society was the state and the territory of secondary importance as far as identity is concerned.

City-states were among the earliest of states, but expansion based upon conquest led to the formation of empires. Preindustrial states emerged in the Mesopotamian, Indus, and Nile valleys, in ancient China, and likely in Subsaharan Africa. The Inca and Aztec civilizations of the Americas provide more recent examples. The difficulties in controlling diverse peoples with varied languages, religions, and economies were frequently beyond the capabilities of early governments. Traditional transport systems, of course, further handicapped control. Disruption and breakup of such states were frequent.

The evolution from tribal organization to modern states was a long and slow process in Europe. Finally, the emergence of England, France, Norway, and Russia, among others, during the later Middle Ages signified that differences in language, religion, and feudal organization were overcome. The nation-state was to become the model for much of the world.

The *modern state* is a product of the Industrial Revolution. The increasing complexity of the industrial society entails comparably more complex organizations for maintaining order, assuring communication, providing protection, and promoting the common culture. Increased interaction and interdependence of individuals also lead to allegiances to the larger organization rather than to kin or individuals. Nationalism is one result of industrialization and urbanization of society. Intense loyalty long directed to a state organization occupying a specified territory leads to the formation of a nation-state.

Nation-States

The *nation-state* and its territory are occupied by a people with a high degree of cohesiveness, common goals, and a culture that is transferred from generation to generation by a common language and who maintain a loyalty to the larger political organization. An emotional attachment is experienced that is not hindered by linguistic, ethnic, racial, or religious distinction. Such conditions are most difficult to achieve. Although the nation-state may be recognized as desirable from the standpoint of achieving stability, few states exist that do not exhibit some traits that detract from complete national unity or allegiance. Unity is more easily achieved when a degree of cultural homogeneity exists.

Problems of State Development

The existence of the modern state should not be taken to mean that states evolved along a single well-defined path. Indeed, strong tribal tendencies and even remnant band societies still exist within areas exhibiting the boundaries and organizational structure of the modern state (Figure 3–6). Political leaders strive to enhance the stability and visibility of the modern state in the world community of nations, but the internal confrontation between tribal loyalties and the modern state is one of the basic problems of the developing world. Tribalism in conflict with modern state structure is not a sign of inherent backwardness but rather the outgrowth of imposition of state organization by colonial powers on an existing tribal political structure.

Perhaps Africa provides the most appropriate examples. Upon acquiring independence the new states became rapidly aware, sometimes by violent experience, of the difficulty of forging stable political units from a dual heritage. On the one hand were *Europeanization,* modernization, and territorial definition from the colonial era, and on the other, the indigenous heritage of a great variety of tribal groups with poorly defined territorial boundaries, linguistic differences, and variations in cultural achievement and acculturation under the

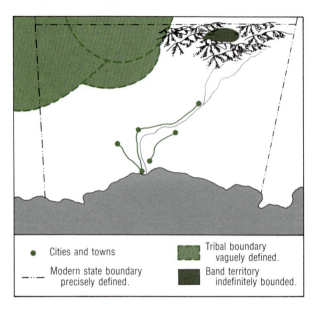

● Cities and towns	Tribal boundary vaguely defined.
─··─ Modern state boundary precisely defined.	Band territory indefinitely bounded.

FIGURE 3–6 Contemporary Political Systems. Tribal and band identity weakens and disappears as the modern state develops and gains strength. The process of modern state development may require a transition period of many generations before it is completed.

Europeans. The civil strife and secessionist attempts within Zaire during the 1960s and the more recent attempt of Ibo secessionists in Nigeria (Biafra) should not have surprised us. These newly independent states contain diverse ethnic and language groups and people at different levels of economic achievement. Such differences add greatly to the difficulty of achieving unity and stability. Although Europeans may have fallaciously thought of Nigeria as a nation, in fact it consisted of several "national" groups.

We must also guard against the notion that instability or divisiveness is characteristic only of recently independent and developing nations. Racial antagonism, for example, has been a problem in Malaysia, Uganda, Peru, and the United States. A strong culture group extending control over a weaker people rarely makes social, economic, or political equalization a priority. It may impose a lower order or position in a stratified society using physical traits as a method of identification. Sometimes it is formally aided by agencies and laws of the state (apartheid in South Africa). Reverse discrimination has been charged in instances where formerly subservient populations rule newly independent states. This reversal of roles has been a stimulus for migration of Asians out of east Africa in recent years.

An implicit suggestion that uniformity of religion ensures the unity necessary for political stability should be avoided. Religious unity led to the creation of Pakistan in 1947. But social and economic differences, combined with the physical fragmentation of the state's territory, ultimately led to the disintegration of Pakistan and the formation of Bangladesh in 1972.

Multinational Alliances

The formation of alliances among states for the promotion of common goals is not a recent phenomenon. The military alliance has been most common, usually with the objective of providing mutual aid and protection in the event of military attack from a specified source. A contemporary example is the *North Atlantic Treaty Organization* (NATO), founded in 1949.

The motivation for the *European Economic Community* (Common Market) was a concern for the economic growth of Western Europe. The ten members (six originally) have as their goal economic integration; the intent is that specific programs and policies assure economic development and prosperity for all member nations. The assumption is that one large economic unit can be more effective than numerous small countries. The Common Market represents a significant step toward unification. Similarly, the USSR and East European countries have formed the *Warsaw Pact* (military alliance) and the *Council for Mutual Economic Assistance,* or COMECON (economic union). Countries of Europe that formerly followed competitive and independent approaches now try interdependence and cooperation as a means for achieving progress. An important question is whether national interests of individual countries can best be met through the framework of large economic units or individually.

Economic Activity

Economic activities are those in which human beings engage to acquire food and other wants. They are the most basic of all the activities and are found wherever there are people.

Primary activity is the direct harvesting of earth resources. Fishing off the coast of Peru, pumping oil from wells in Libya, extracting iron ore from mines in Minnesota, and growing corn in Iowa are all examples of primary production. The commodities that result from these activities acquire a value from the effort required in production and from consumer demand. Processing these commodities is classified as *secondary activity*. A primary commodity such as cotton is processed into fabric or one of numerous other products. Fabric is cut and assembled as apparel. Textile and apparel manufacturing are two secondary activities in which an item is increased in value by the changing of its form to enhance its usefulness (*form utility*). Economic activities in which a service is performed are classed as *tertiary*. Wholesaling and retailing are tertiary activities by which primary and secondary products are made available to the consumer. Other tertiary activities include governmental, banking, educational, medical, and legal services, as well as journalism and the arts.

Transportation is a special kind of economic activity. Transportation links take numerous forms: pack animals, automobiles, rail, air, water, and pipeline. The cost of transportation is a measure of *economic distance;* if cost is too high, then movement of goods is not feasible. If resources and goods cannot be effectively exchanged, then self-sufficiency must be the basis upon which needs are met. An efficient transportation system is vital for a modern society. The nature of the transportation system in an area reveals much about the area's economic organization.

Each economic activity, including transportation, involves the creation of wealth. The value of primary commodities increases as they are transported to locations where they are needed (*place utility*). Value is also added by processing (*form utility*). Each move toward a market or additional manufacturing stage adds more value to the items involved. Value, however, does not accrue on an equitable basis with each step in the process. The primary levels of production are not great wealth producers—for some individuals, yes, but normally not for the society at large. Most of the final

Textile and apparel manufacturing occurs throughout the world. In some areas cloth and clothing are made in the home with few tools and often as an auxiliary activity. Elsewhere these products are made in modern factories where equipment is powered and labor is specialized. (United Nations)

value of a product comes from secondary, tertiary, and transportation activities.

Economic Organization

Numerous factors contribute to the specific manner in which economic activity is organized and accomplished. First, many facets of our physical world affect our decisions regarding economic activity. Second, the environment may be used in various ways, depending on technology available. Third, even when potential for land use is analyzed relative to technology, a spate of other economic, social, and political conditions influence our decisions.

Location relative to existing markets, or degree of accessibility, may encourage or preclude some forms of production. Long-established systems of land tenure may maintain an uneven allocation of the land resource. Governments aid in economic production by assuring proper *infrastructure* (credit systems, roads, education) or hinder by suppressing attempts to make necessary adjustments in the economic system. Traditions and attitudes have worked as both positive and negative forces in the functioning of economies.

Two extremes are recognized: the traditional and the modern commercial economic systems (Figure 3–7). The most traditional is the *subsistence economy* in which a family or small band engages in both the production and the limited processing required for local consumption. Group members function as producers, processors, and consumers of their own commodities. Self-sufficiency and sharing are the distinctive features. Present-day true subsistence societies are few, but there are the Tasaday of the Philippines, the Bushmen of Africa, and the Campa of Peru. Very few people are so isolated that they do not engage in some kind of exchange, even if only occasionally.

It is impossible to classify any country as a purely subsistence economy. Even in one that is obviously less developed, some people engage in subsistence activity, some produce and exchange on a reciprocal or *barter* basis, and others are clearly in an exchange or *money economy*. It is appropriate, therefore, to think of countries as existing along a continuum between the extremes of traditional subsistence and modern commercial. Although it is most difficult to locate a country precisely along such a continuum, there are features that are indicative of the level of modernization achieved.

Importance of Primary Production

The majority of the labor force in a traditionally oriented country is engaged in primary production, as much as 80% in some countries (many Subsaharan countries), the vast majority in agricultural labor. Any society that is restricted largely to the primary activities has limited wealth. Per capita gross national product for traditional economies is low. There are exceptions, of course; Kuwait, because of oil, is a good example. The modernized economy has a high proportion of its labor force in the secondary and tertiary activities. The primary sector may engage less than 10% of the labor force when extreme levels of industrialization and tertiary activities are reached. The concentration in processing and services means greater national income and has implications for the internal distribution and activity of the population.

Production Inputs

The production inputs in traditional society are largely labor applied to land—agriculture. The modernized society substitutes capital for labor; that is, industry is automated and agriculture is mechanized. The shift to *capital-intensive* production is based on the use of inanimate power sources. Therefore, per capita energy consumption is indicative of the extent of modernization.

	TRADITIONAL ⟷			MODERN
	Subsistence	Reciprocal	Peasant	Exchange
Features	Commodity sharing	Barter	Minor exchange for capital	Full commercial
	No urban foci			Major urban development
	Simple technology			Complex technology
	Animate power (muscle)			Inanimate power
	Localized economy			Regional specialization
Production Systems	Gathering			Commercial agriculture
	Nomadic herding			Commercial fishing
	Primitive agriculture			Commercial grazing
	Labor-intensive subsistence agriculture			Commercial forestry
				Manufacturing and commerce

FIGURE 3–7 Economic Organization.
Evolution from a traditional society into a modern society requires several fundamental changes. The affected institutions include the basis of exchange, settlement patterns, technology utilized, power development, trade and production systems.

Division of Labor and Regional Specialization

The division of labor reaches a high level in modern countries. It is evident at the local levels where members of a community perform a variety of tasks. Division of labor with increased modernization is also evident at the regional level. *Regional specialization* exists when several areas produce goods for which they have a particular advantage, or which are not produced elsewhere, and then contribute their specialized production to the larger economic system. Regional specialization occurs at both the national and international levels.

Regions engaged in specialized production illustrate the *principle of comparative advantage.* Comparative advantage of an area may stem from climate, soils, labor, power, capital and enterprise, transportation, institutional structure, or some combination of these. Whatever the advantage, the area gains by specializing in one product and trading for other commodities needed. Occasionally an area produces goods for which it has no distinct advantage. This type of production becomes economically feasible when other areas that could provide the same goods more efficiently choose not to do so because alternative possibilities yield an even higher return.

Urban Corelands

High levels of urbanization are also characteristic of modernized societies. Economies and savings are realized when the secondary and tertiary activities that dominate modern society are agglomerated; hence the basis for cities. The United States and Canada, the European countries, the Soviet Union, and Japan all have *urban-industrial corelands.* Numerous cities in proximity, high population density, a high proportion of national manufacturing, and high standards of living characterize such areas. Corelands also function as national educational, financial, political, and cultural centers. A distinctive feature is the high level of commercial and social interaction among the urban centers that are part of the coreland. Less modernized countries often have only one major city that functions as the coreland. The absence of a multiple-city coreland is indicative of the more traditional economic system.

Urban-industrial corelands have peripheral economic regions (extended *hinterlands*) that do not exhibit features of a coreland but that do have clear functional ties to the core. The economic hinterland may be located within the country in question or may include other regions of the world.

Trade Relationships

The trade relationship between corelands and their peripheral regions reveals their differing character. *Corelands* are suppliers of manufactured goods. Internal exchange and trade with other corelands of the world are at high levels. Some manufactured goods flow to less industrialized hinterlands, but the volume is far less. The *hinterlands* are suppliers of raw materials, energy fuels, and food products for corelands that often cannot supply all of their high per capita needs. This exchange is reminiscent of the *colonial relationship.* Figure 3–8 illustrates the trade relationship between rural areas specializing in primary production and urban regions specializing in secondary and tertiary activi-

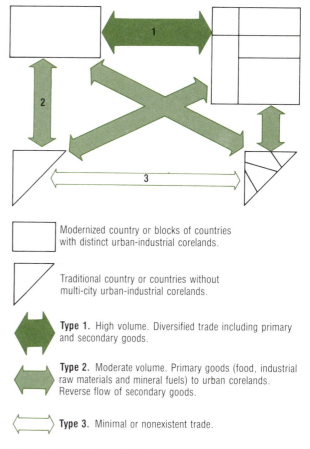

Modernized country or blocks of countries with distinct urban-industrial corelands.

Traditional country or countries without multi-city urban-industrial corelands.

Type 1. High volume. Diversified trade including primary and secondary goods.

Type 2. Moderate volume. Primary goods (food, industrial raw materials and mineral fuels) to urban corelands. Reverse flow of secondary goods.

Type 3. Minimal or nonexistent trade.

FIGURE 3–8 International Trade Flow Types. Petroleum accounts for an increasingly large proportion of international trade. Petroleum flow is largely from traditional countries to urban-industrial corelands.

47

ties. The relationship is the same, whether the exchange is domestic or international.

A problem, however, lies in the fact that such relationships often favor urban-industrial cores. The value of primary goods has generally not increased as rapidly as that of secondary goods. The likelihood of regional income disparity is great. If the exchange is between countries, it may cause balance-of-payment problems for less developed regions. The country that is a supplier of primary goods to industrialized regions outside its own territory also faces competition, price fluctuations, or political pressures over which it can exert little control. Furthermore, these countries often have few customers or depend heavily on two or three commodities for most of their exchange revenues. They face severe economic problems if trade is disrupted.

The more traditional countries often suffer the problem of dual economies. One part of the economy concentrates on commercial production of primary goods, much of which may flow to an external coreland. Other areas within the country with a large proportion of the population may remain traditional in economic structure. Economic dualism is commonly exhibited in distinct regional differences and contributes to internal social and political disunity.

SUMMARY

The intent of the foregoing discussion is not to suggest that all is well with modernized societies and point to the woes of those countries less progressive. Although the latter evidence poverty, health problems, dual economies, and regional disparities, the problems of modernized societies are also formidable. The high level of per capita consumption requires immense quantities of material and power resources, to such a degree that it is almost frightening to think of modern society as a model for the remainder of the world. Space, quality living, and suitable air and water are demanded at the very time when life-style places increasing pressure on our total environment.

Rich and Poor Nations: An Overview

For about 30% of the world's population, life's basic necessities are easily obtained. For this group the worry of obtaining sufficient food, clothing, and shelter is of secondary importance. The other 70% suffer from inadequate food, clothing, and shelter. Many must struggle constantly to survive. Life consists of a nearly empty stomach, a piece of cloth or a cheap set of clothing, and a small dirt-floored hut for a shelter. Others are a bit more secure, but food supply is barely sufficient, often of poor quality, and limited in variety. Clothing is adequate to protect the body but offers few frills. Housing provides shelter but little comfort.

THE WIDENING GAP

Over the past thirty years the gap between rich and poor nations has increased. The prospects are for a greater disparity. Statistical data are not available to show the *widening gap* fully, but there is enough information to illustrate some general dimensions and trends.

Trends in Per Capita GNP

Per capita gross national product (GNP) is considered one of the best indicators of economic well-being.[1] Since the end of World War II, nearly every nation has shown an increase in *per capita GNP,* although in many cases, because of inflation, the gain does not result in greater purchasing power. In fact, in some nations effective buying power has decreased. Figure 4–1 graphically demonstrates the increasing disparity between rich and poor nations. Note that the vertical scales (dollar values) of the two graphs are different. The rate of change in per capita GNP can be interpolated for individual nations. Ethiopia, representative of

---

[1]Gross national product is the total value of all goods and services produced and provided during a single year. GNP values are subject to some error, particularly in nations where a significant part of the population is engaged in subsistence activities. These activities are often undervalued or not reported.

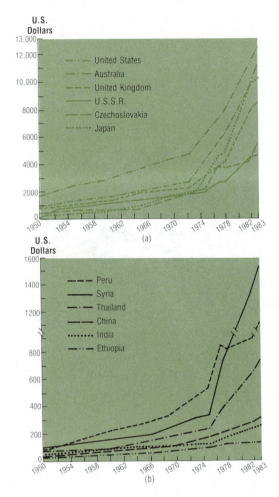

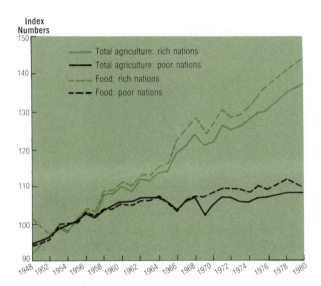

FIGURE 4-2 Per Capita Total Agricultural and Food Production Index Numbers (1952-56 = 100) for Rich and Poor Nations.
Both total agricultural production and food production show the widening gap between rich and poor nations. Remember that these are per capita values and that the poor nations have a much greater rate of population growth. An index number relates the production of a specific year to that of the base period. If the index number is 110, it means that production is 10% greater than the base period. In this graph of per capita data an index number of 110 means a per capita increase of 10%.
SOURCE: *Production Yearbook* (Rome: United Nations, 1982).

FIGURE 4-1 Per Capita Gross National Product for Selected Rich (a) and Poor (b) Nations, in U.S. Dollars. Most of the world's nations have increasing per capita GNPs. The increase for rich nations, however, is much greater than for poor nations.
SOURCES: Arthur S. Banks (comp.), *Cross-Polity Time Series Data* (Cambridge, Mass.: MIT Press, 1971); *Population Data Sheet* (Washington, D.C.: Population Reference Bureau, 1983).

many African nations, has increased its per capita GNP an average of 3% annually, equivalent to about $3.30 a year. India is only slightly better off. In the same period the United Kingdom, Australia, and the United States enjoyed average yearly increases of 8-10%. In absolute numbers the rich nations' gains are most startling.

Trends in Agricultural Production

Part of the reason for a widening gap between rich and poor nations lies in the different levels of *agricultural productivity*. Figure 4-2 charts the changes

in per capita agricultural and food production by means of index numbers.[2] For poor nations, per capita production has increased only slightly, and most of the increase was during the first part of the period covered. The lack of change in the poor nations does not mean that agricultural production has remained static; rather, it signifies that production has just kept pace with population growth. This rate of production increase is no mean feat when one remembers that the population growth rate in poor nations is two to three times that of most rich nations.

One reason why rich nations are experiencing greater farm production is that yields per unit of land for most crops have improved dramatically. Improved seed, more extensive use of fertilizers and pesticides, and better management tech-

---

[2]The index numbers reflect the level of total agricultural or food production for the years cited compared to the base period. An index number of 106 is 6% above the base-period level.

Improved high-yielding grains, often called "miracle grains," have been produced in research stations such as the one shown here. Rice plants have been genetically modified to increase yields to two or more times that of the traditional local varieties. (United Nations)

niques have led to increased yields. In the poor nations use of these *technological innovations* has not been so widespread. The trend in yields of the world's three principal cereals per unit of land area is demonstrated in Figure 4–3. These three crops represent an important element in the diet of most of the world's people and are cultivated widely. A three-year moving average is used to reduce the effects of weather and to show more clearly the broad trends in yields.[3] Figure 4–3(a) shows that wheat yields for most rich nations are higher and increasing more rapidly than those in poor nations. In the Soviet Union, Australia, and the United States production per areal unit is low because the crop is grown mainly in dry lands that are not well suited for high yields; however, mechanization permits the farmer to cultivate a large area so that production is feasible. In France and Hungary, wheat is raised under humid climatic conditions that favor high yields. In the poor nations production is not mechanized, and wheat is generally grown in more humid areas than those of the United States or the Soviet Union.

A more striking contrast in yields per land unit between rich and poor nations is observed with maize (corn) and paddy rice. Yields for these crops

have not changed much in the poor nations since 1951. Mexico and the Philippines are centers for research on maize and rice, respectively. Yet only in these two countries, and for rice in Brazil, is there an indication that cultivation of new varieties of these plants has had much impact. Even in these centers yield increases have been modest.

Trends in Industrial Production

Another indicator of the widening gap between rich and poor nations is the varied levels of *industrial production*. Although data on manufacturing are less prevalent than on agriculture, we can draw some general conclusions.

Most poor nations have a limited industrial sector; the contribution that industry makes to the GNP is normally small. In the rich world, however, manufacturing is very important and characteristically provides a large part of the GNP. Thus industry is a much more significant economic activity in rich nations than in poor nations. With that in mind, Figure 4–4 tells us two things. First, in relative terms poor and rich nations have for many years grown in industrial production at about the same rate. In recent years, however, the growth rate has been higher in the poor nations, but not enough time has elapsed to discern a true trend. Second, even if poor countries are growing in industrial production at a faster rate than rich countries, it is certain that the absolute increase in pro-

[3]A moving average is a smoothing method designed to show general trends, not comparisons, on a year-to-year basis. The yield value for each year is obtained by summing the yields for a consecutive three-year period, dividing the sum by 3, and assigning the result to the middle year.

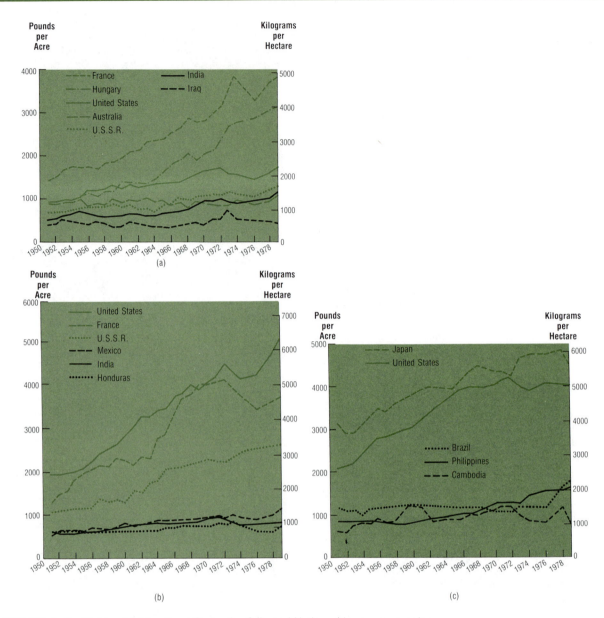

FIGURE 4–3 Yields of Major Food Grains for Selected Nations (three-year moving average). (a) Wheat, (b) Maize (corn), (c) Paddy Rice.

Grains form an important element in the diet. Wheat, rice, and maize (corn) are the most widely used grains. To illustrate the difference in yields per unit area a moving average is used. A moving average is a smoothing method to lessen the effect of fluctuations in individual years caused by weather or other erratic influences on production and is useful to show long-term trends. For a three-year moving average the values for three consecutive years are totaled, divided by three, and the result assigned to the middle year.

SOURCE: *Production Yearbook* (Rome: United Nations, 1982).

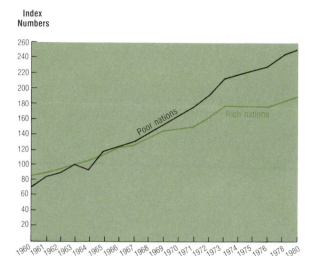

FIGURE 4–4 Industrial Production Index Numbers (1970 = 100) for Rich and Poor Nations.
Index values of industrial production do not show the widening gap as clearly as was noted for agriculture. Remember, however, that industrial production in poor nations is limited so that slight increases loom relatively large in relation to the base. One of the principal economic goals among poor nations is to become more industrialized.
SOURCE: *Statistical Yearbook* (New York: United Nations, 1982).

duction is much higher in the rich nations. This point is easily seen when we remember the industrial base in rich nations is much greater than in poor nations; thus an equal rate of increase of index numbers means a greater absolute value for the rich nations.

MEASUREMENT OF RICH AND POOR NATIONS

Most authorities on economic development differentiate rich and poor nations by one or more of three measures. They are per capita GNP, per capita consumption of inanimate energy, and percent of labor force in primary activities. Of these measures per capita GNP is most widely employed. Although this measure seems the best indicator of economic well-being, some caution is warranted. Per capita data are obtained by dividing total national values, in this case wealth generated, by the nation's population. Per capita data are thus average values. In many poor nations, average values do not apply to large segments of the population, because wealth is concentrated in the hands of a few. Per capita figures are therefore generally

higher than is actually the case for the majority of the population. In the rich nations, per capita values are more meaningful. Second, it should be recognized that GNP values, even under the best of accounting procedures, are only estimated. Caution is required in comparing nations; minor variations should be discounted. Finally, many poor nations—for example, Ethiopia—have per capita GNP values so low that it is inconceivable that the vast majority could physically survive, but they do. We must assume therefore that actual per capita wealth generated must be higher than reported or of a nature much different from that of the rich nations.[4]

Per Capita GNP

We have already compared, from a historical perspective, per capita GNP between selected rich and poor nations (Figure 4–1). It now remains to view the present *distributional GNP pattern* on a world scale (Figure 4–5). This map shows a striking regionalization. Areas of high per capita GNP ($2500 or more) include Anglo-America, most of Europe, the USSR, Japan, Australia, New Zealand, Israel, Gabon, and several of the major oil-exporting countries. Much of Latin America, parts of the Middle East, and Taiwan and Malaysia have per capita GNPs between $1250 and $2500. Countries with a per capita GNP of $625 to $1250 are few and widely dispersed throughout Latin America, Africa, and Asia. Poor areas (per capita GNP less than $625) include most of South, Southeast, and East Asia and of southern and central Africa, along with Haiti in Latin America.

Per Capita Inanimate Energy Consumption

One of the Industrial Revolution's characteristics is a shift from animate power (man and beast) to inanimate energy (mineral fuels and hydroelectricity). The degree to which a nation is able to supply inanimate energy from internal sources or to import fuel is an important indicator of the application of modern technology and consequently of productiv-

[4]Gross national product is often undervalued in poor regions because part of the economy is not commercialized. For example, in rich areas proceeds from the barber trade are counted as contributing to the GNP. In poor areas where barbering is done at home, no contribution to the GNP is acknowledged. In general, transactions in which money is not exchanged are not counted toward determining GNP, even though the work performed may be the same as one in which money was paid.

53

FIGURE 4–5 World Per Capita GNP.
Per capita gross national product is considered the best single measure of economic well-being. Note that high per capita GNPs are closely associated with areas of European culture and oil-exporting countries. Japan, Israel, and Gabon also have high values. Low per capita GNPs are associated with southern, southeastern, and eastern Asia and parts of Africa.
SOURCE: *Population Data Sheet* (Washington, D.C.: Population Reference Bureau, 1983).

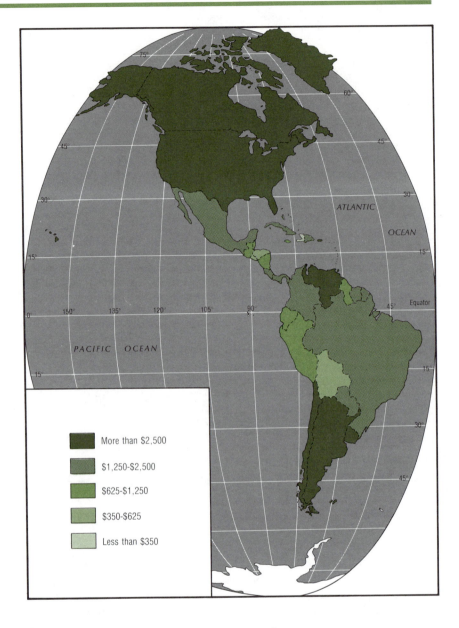

ity. Just as per capita GNP is a measure of productivity in terms of value, per capita inanimate energy use is a *measure of production* in terms of power expended. There is a close relationship between energy consumption and degree of economic activity. Low per capita energy consumption is associated with subsistence and other forms of nonmechanized agricultural economies; high per capita energy use is associated with industrialized societies. Intermediate levels of energy use characterize regions that have aspects of both the industrialized urban centers and the more traditional nonmechanized rural areas. The distributional pattern of per capita inanimate energy consumption is shown in Figure 4–6.[5] Its similarity to that of per capita GNP (Figure 4–5) should be noted.

Percent of Labor Force in Primary Activities

As we have seen so far, there is a relationship between economic activity on the one hand and wealth generation and energy use on the other. In general, nations with a large part of their labor force in primary activities are less able to produce

---

[5]One gallon of crude oil is roughly equivalent to one gallon of gasoline.

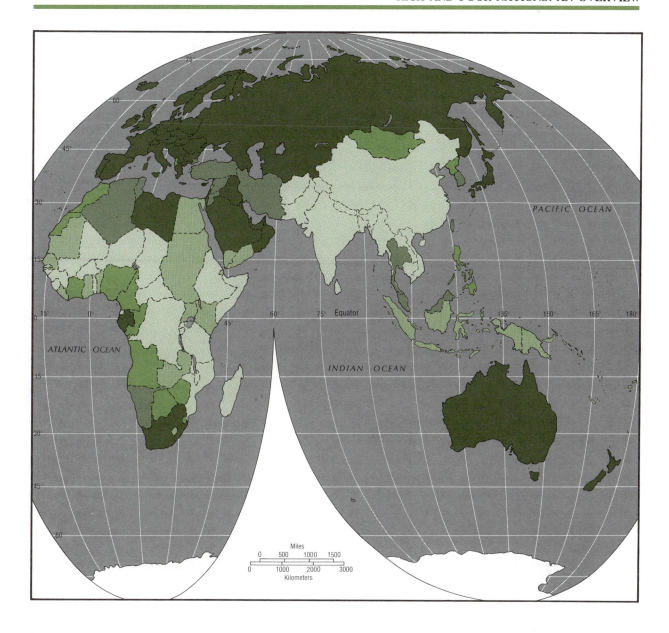

income and use relatively small amounts of power on a per capita basis. Conversely, a nation that has strong secondary and tertiary components in its labor force usually has a greater per capita GNP and consumes more energy. The use of percent of labor force in primary activities (mainly agriculture) as an indicator is based on these relationships. Furthermore, any nation with a labor force dominantly in primary production has a *limited opportunity for labor specialization.* In theory at least, labor specialization and production diversity are basic to economic growth. Certainly the prospects for high levels of individual production are diminished if the worker must not only grow his crops but also pro-

cess, transport, and market them and be a "jack-of-all-trades" in providing housing, tools, and clothing.

Of all the primary activities, agriculture is by far the most important. Roughly 98% of the primary labor force in the world is in *agriculture;* of the remainder, about 1% is engaged in hunting and fishing and 1% is engaged in mining. A map of the labor force in agriculture, for which data are most accurate, is a good representation of primary occupation dominance. The broad *pattern of the agricultural labor force* is similar to the distributions of per capita GNP and energy consumption (Figure 4–7, pp. 58–59).

FIGURE 4–6 World Per Capita Inanimate Energy Consumption.
Per capita energy consumption is a measure of development in that it gives us an idea of the use of technology. Most forms of modern technology use large amounts of inanimate energy. Countries that use small amounts of inanimate energy must rely principally on human or animal power.
SOURCE: *Statistical Yearbook* (New York: United Nations, 1982).

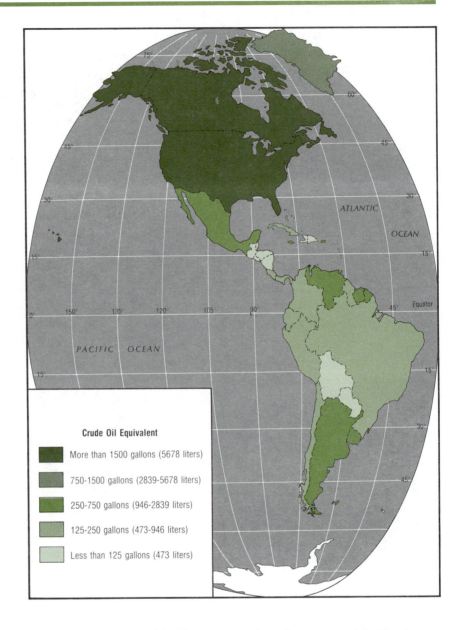

Crude Oil Equivalent

More than 1500 gallons (5678 liters)

750-1500 gallons (2839-5678 liters)

250-750 gallons (946-2839 liters)

125-250 gallons (473-946 liters)

Less than 125 gallons (473 liters)

Other Measures

Although GNP, energy, and agricultural labor force are most often used to determine rich and poor nations, other measures are occasionally used. Two additional measures worthy of recognition are life expectancy and food supply. The life expectancy measure would seem to be the ultimate indicator of development. After all, some cultures are less materialistic than others, so the standard measures mask some important cultural attributes of society. All cultures, however, place a value on preservation of life, although life is not so highly treasured in some as in others. Furthermore, *life expectancy* is a measure of the end result of economic activity: how well a system works to provide life support. A nation whose inhabitants can expect an average life span of but forty years has failed in its most important function. Figure 4–8 (pp. 60–61) shows the distribution of life expectancy at birth and correlates generally with the previous maps we have examined.

If we accept life expectancy as an ultimate measure of a rich or poor nation, then *food supply* is a measure of life quality. Two measures of food supply are fundamental. The number of calories available is an indicator of dietary quantity, and protein supply is a measure of dietary quality. Adequate quantity is considered at least 2400 available calories per person daily. Adequate protein supply is attained if at least 60 grams of protein are available

56

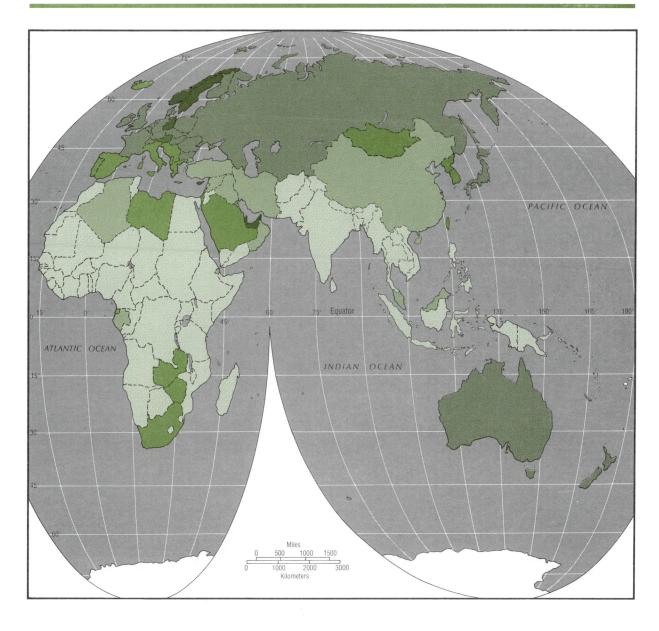

Major Rich and Poor Regions

per person daily. These two factors are combined in Figure 4–9 (pp. 62–63). Only one category provides both adequate calories and protein.

We can now combine the various measures of rich and poor nations into a single map and identify some broad *regional patterns*. This map is, of course, highly generalized and lumps some rich nations—Israel, for example—with poor nations. In a like manner, some poor nations, such as Albania, are included in the rich regions. Some nations are classified as rich by all measures, and many are equally classed as poor. A number of nations, however, fall among the rich in some categories and among the poor in others. All along the continuum from very poor to very rich are nations in various stages of development.

The map (Figure 4–10, pp. 64–65) shows a strong correlation between rich areas and populations of predominantly European origin. Anglo-America by its very name implies European roots. Latin America, a poor region, also has a European connotation, but this name is really a misnomer because many parts of the region are inhabited by peoples of Indian background. Mexico, northern Central America, and Andean South America have large and viable Indian culture components. Where people of European tradition do form the dominant population (Argentina, Chile, and Venezuela), the

FIGURE 4–7 World Percent of Labor Force in Agriculture.

Percentage of labor force in agriculture shows the degree of economic diversity in a nation. If a large percentage of the population is engaged in agricultural pursuits, manufacturing and services are limitedly developed. Conversely, if only a small percentage of the labor force is in agriculture, manufacturing and services are well represented.

SOURCE: *Production Yearbook* (Rome: United Nations, 1983).

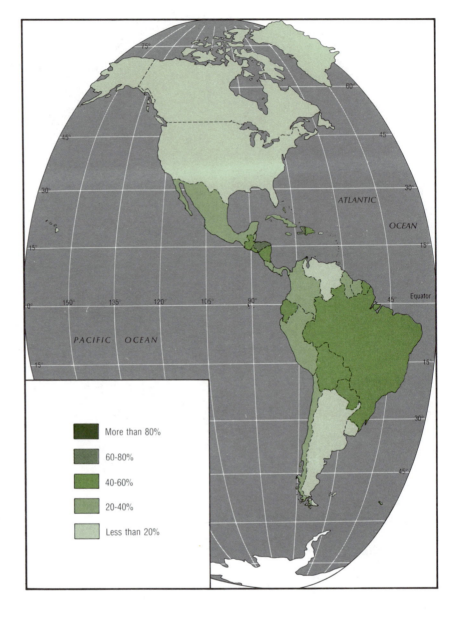

More than 80%

60-80%

40-60%

20-40%

Less than 20%

measurement indicators tend toward a rich index. Like Anglo-America, Australia's and New Zealand's populations are European. Of all the rich regions only Japan is an enigma, for it is truly non-European. The poor regions—Latin America, North Africa and the Middle East, Subsaharan Africa, and South, Southeast, and East Asia—have no common cultural background. Diversity of population and culture is, in fact, the rule.

SOME ADDITIONAL CHARACTERISTICS OF RICH NATIONS

We have already noted several characteristics of rich nations: high per capita GNP and energy use;

a small part of the labor force in primary activities and a consequent emphasis on secondary and tertiary occupations; and a longer life expectancy and a better and more abundant food supply. In addition, rich nations have a low rate of population growth. To a large degree, these nations have progressed through a demographic transformation.

Economic Characteristics

The most basic economic characteristic of the rich world is the widespread use of *technology*. The fruits of the agricultural and industrial revolutions are widely applied, and new techniques are quickly diffused and adopted. These new technologies mean new resources, a larger base with the

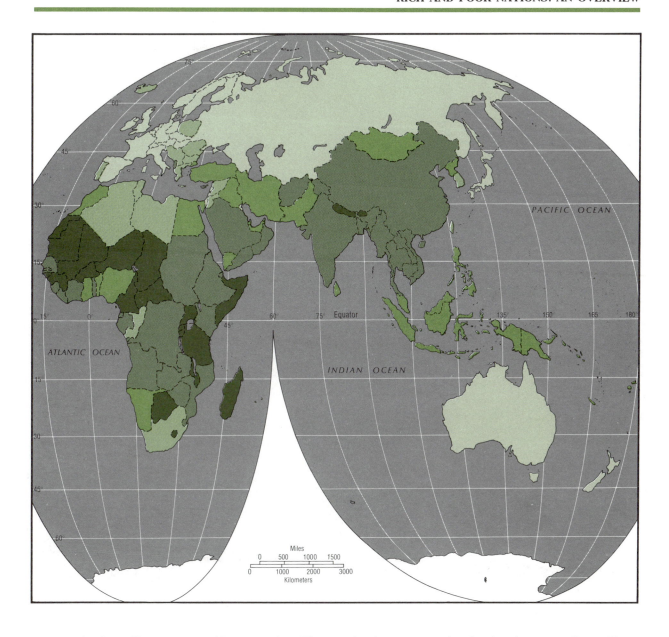

opportunity for still greater wealth generation. The use of advanced technology leads to increased labor productivity and the creation of an improved *infrastructure*. "Infrastructure" means support facilities such as roads, communications, energy and water supply, sewage disposal, credit institutions, and even schools, housing, and medical services. These support facilities are necessary for accelerated economic activity and are a prerequisite for specialization of production. Productivity is enhanced by labor and regional specialization.

The heavy dependence of rich nations on *minerals* further differentiates them from poor nations. Not only are we in an iron and steel age but also in a fossil fuel age, a cement age, a copper and aluminum age—the list is almost endless. To a greater and greater degree, rich nations are importing these minerals from poor nations; our heavy reliance on imported oil is a conspicuous example of the increasing *interdependence* of nations. This interdependence takes the form of trade in which the rich export manufactured goods, and in some cases food, and import raw materials. This trade arrangement has led to trade surpluses for most rich nations, since their exports have increased substantially in price, and to trade deficits for many poor nations, since the price of raw materials has increased more slowly—with the exception of petroleum. Moreover, the rich nations gain further revenues by banking and investment in poor nations. Removal of this wealth from

FIGURE 4-8 World Life Expectancy at Birth.
Life expectancy is a measure of how well a nation is able to care for its population. Note that long life expectancy is closely associated with the other indicators of developed nations.
SOURCE: *Population Data Sheet* (Washington, D.C.: Population Reference Bureau, 1983).

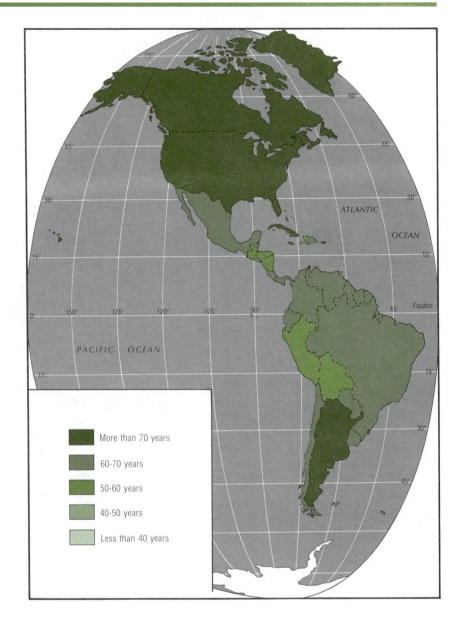

the poor nations has encouraged some poor nations to nationalize or exert greater control over foreign investments.

High productivity, a favorable trade balance, and new technologies result in higher personal and corporate incomes. Individuals only need to spend a part of their income for food and shelter. Other income is used for services, products, and savings. The more money spent, the more demand for economic growth, and money saved is usually invested in industries and businesses. For *entrepreneurs* (organizers who gather together labor, financing, and all those things necessary to construct an effective economic activity), the rich nations offer numerous advantages and are, in turn, benefited by the entrepreneurs' organizational skills.

Cultural Characteristics

None of the economic characteristics can stand alone. They are all intertwined. They are, in fact, but a part of cultural characteristics. It is probably no happenstance that all rich nations except Japan have a strong European heritage. People's attitudes and value systems are reflected in economic performance. The followers of the *Judeo-Christian ethic* consider the desire to work an important attribute and the accumulation of material things the representation of work performance. Success and

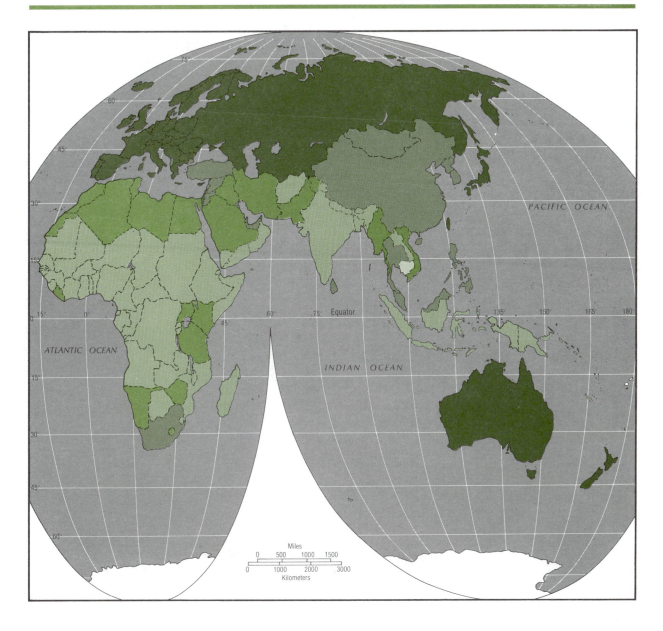

peer respect are measured largely by wealth. Japan does not have a strong Judeo-Christian tradition, but Japan's cultural ethic toward work and success is similar to the European example.

Another cultural attribute of rich nations is the importance given to *education*. Education is enhanced by communications infrastructure, and new ideas are quickly spread. Also, education is geared partly toward economic advancement; diverse disciplines such as engineering, geography, economics, agronomy, and chemistry have obvious direct applicability for resource development.

An educated population is an extremely important resource. In addition, a population that is educated and *urban,* as are most populations in the rich world, more readily accepts change. Acceptance of change means that technology is more easily adopted, new products and services are welcomed, and the population is more mobile.

SOME ADDITIONAL CHARACTERISTICS OF POOR NATIONS

Population Characteristics

We have already learned that poor nations are characterized by low per capita GNP and energy use, a high proportion of the labor force in primary

FIGURE 4–9 World Daily Per Capita Food Supply. Food supply is a measure of well-being in the most basic sense. Note that in large parts of the world the people have inadequate diets. Only one category listed in this map represents an adequate diet in both calories and protein. SOURCE: *Production Yearbook* (Rome: United Nations, 1974).

Adequate quantity and quality
(More than 2400 calories and 60 grams of protein)

Adequate quantity, inadequate quality
(More than 2400 calories, less than 60 grams of protein)

Adequate quality, inadequate quantity
(Less than 2400 calories, more than 60 grams of protein)

Inadequate quantity and quality
(Less than 2400 calories and 60 grams of protein)

pursuits, a relatively short life span, and a diet often deficient in either quantity or quality or both. Poor nations also have a high rate of population growth resulting from a continued high birth rate and a declining death rate. The *age structure* of a poor nation's population is consequently different from that of a rich nation. Figure 4–11 (p. 66) graphically illustrates this difference. For the poor nation, a large segment of the population is outside the most productive age group. Fully 40% of the population is less than fifteen years old. In a sense, per capita comparisons are unfair because all people are counted equally, yet in poor nations a smaller part (8% less) of the population is made up of mature laborers.

A *youthful population* also effectively reduces the ability of women to engage in economic activities because they must tend the young. Moreover, *women's* status in the diverse cultures of most poor nations excludes them from many occupations. In the rural sector, women may work with men in the fields during periods of peak labor requirement. Much of their time, however, is spent in the home and provisioning it with fuel wood and water. At home, women may engage in some craft industry such as weaving for household use and for sale in the local market. Many rural women in the poor world play a pivotal role in marketing the family's surplus on a daily or weekly basis. In the cities, where life is more cosmopolitan,

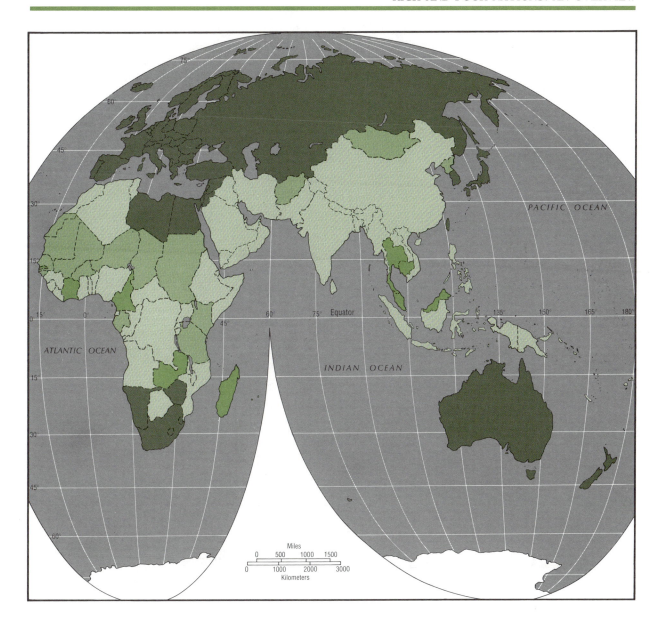

women find employment as domestics, secretaries, and more recently in industry. For many illiterate women, however, the two last-mentioned opportunities are not available.

Cultural Characteristics

Literacy in poor nations is generally low. Most rural inhabitants cannot read or write or do so at minimal levels. In the cities literacy is more prevalent, but, still, large numbers of urban poor lack effective command of the written word. The inability to read excludes a large mass of the population from learning new technologies and from engaging in more remunerative occupations. Despite some major campaigns to improve literacy, many nations have made little headway in educating their population.

Most cultures in the poor world are *conservative* and, therefore, resist change. Herein lies a fundamental paradox. These cultures by and large wish to preserve their customs and mores yet at the same time want to partake of the material benefits that Western society enjoys. Economic development leads to cultural change, the destruction of old traditions, and the acquisition of new behavior patterns. The family sometimes loses part of its cohesiveness; village orientation becomes subservient to the larger urban centers; labor and regional specializations lead to commercialization of the

FIGURE 4–10 World Rich and Poor Regions. The preceding maps that measure different aspects of development have a similar distributional pattern. This map generalizes these maps by dividing the world into rich and poor regions.

Rich nations   Poor nations

economy; and loyalties to the local community give way to national allegiance. These changes and others are almost inevitable if economic development takes place. Most poor nations are now somewhere along the continuum of economic development, and the disruption caused by the conflict between old and new ways of life is characteristic. This conflict accentuates the usually preexisting cultural pluralism of the poor nations of the world.

Cultural pluralism occurs where two or more ethnic groups exist within a single nation. Each group has its own set of institutions, language, religion, life-style, and goals. In some places these ethnic groups live apart, but in others they are intermingled. In both cases, however, if the

differences among the groups are great or if mutual antagonism exists, joint action for development is difficult. In fact, opposing goals may lead to inaction, and attempts to alter the status quo may cause conflict. Numerous examples of cultural pluralism are found within the poor world. The rural black of various parts of Africa clings to the traditional ways of life, but many blacks in the city have adopted the Western economic system and cultural goals. In the various parts of southeastern Asia, national unity is weak because of village orientation and the diversity of ethnic groups. In Latin America, cultural pluralism is evident in areas of large Indian populations. Cultural pluralism does not preclude economic development but does cre-

64

Miles
0 500 1000 1500
0 1000 2000 3000
Kilometers

ate an additional obstacle. Belgium is one example of a rich nation composed of two ethnic groups (Flemish and Walloon). In Eastern Europe, the diversity of ethnic groups has led to war and disruption. With the domination of the Soviet Union in the area, these disruptions have been suppressed.

Economic Characteristics

Economically the poor nations share many similar traits. Most have a *dual economy,* one part organized for domestic consumption and another for export trade. All nations, of course, produce for both internal and external markets, but in rich nations the farmer or factory worker produces the same items for both markets and rarely knows whether or not they are consumed locally. In the poor world, however, the producer for the local market provisions only the domestic market, and a different producer provides for the export trade. The domestic producer generally uses antiquated techniques and often is subsistence oriented. Export producers often have access to modern technological innovations and have the means to apply them. They are fully within the market economy and must remain competitive with rivals. The domestic producer does not have the same incentives or the same ability.

Poor nations rely mainly on *flow* (renewable) *resources,* principally those related to agriculture.

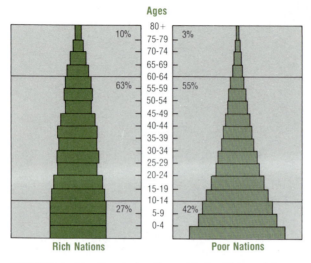

FIGURE 4–11 Population Pyramids for Rich and Poor Nations.
Population pyramids illustrate many demographic features. For example, rich nations have a greater proportion of their population in the most productive age group; poor nations have a much greater potential for rapid population growth since a large part of the population is young; economic growth in poor nations must be at high levels just to keep up with population growth.
SOURCE: *Trends in Developing Countries* (Washington, D.C.: World Bank, 1975).

This orientation does not necessarily mean that the people live in ecological balance with their surroundings. In many regions, increasing population pressures, coupled with static technology, have led to exploitation of agricultural resources to the point where soil depletion and erosion have become serious problems. Yet in parts of China where dense populations have lived for hundreds of years, agricultural resources have actually been improved. Fund resources—for example, minerals—are important export commodities in many poor countries, but they usually enter the domestic market in the form of finished goods imported from rich nations. Finished goods and food products are the main imports of poor nations, and raw materials and agricultural products are the principal exports.

Poverty is widespread. The vast majority of the population is within the lower class, and a small upper class controls the economic and political life. In most countries the middle class is embryonic, but in others it is growing as urbanization progresses. Limited income for the large bulk of the population means little savings. Accumulated wealth that does exist is in the hands of the upper class and often not available for investment in new activities. Limited income results from several factors: low levels of productivity, a limited infrastructure, a lack of applied technology, and a societal and economic structure not designed to facilitate economic developments. High levels of unemployment and widespread underemployment also contribute to low incomes because there are so few alternative occupations available to most of the labor force.

SOME DEVELOPMENT THEORIES

Control Theories

A number of theories have been advanced to explain why some nations are rich and others poor. None is totally accepted, and some have strong opposition. One of the earliest theories is *environmental determinism*. This theory has been discredited but continues to have advocates. In essence, the theory is based on the premise that the physical environment controls or channels what people do. Certain climatic regions, those in the mid-latitudes, are said to be more stimulating for economic activity than either the polar or tropical climate. A review of the maps measuring economic development does show more rich nations in mid-latitude locations and poor nations in the tropics.

Another "control" theory is *cultural determinism*. According to this theory, currently quite popular, a person's range of action is determined largely by culture. We can certainly see that parental guidance and peer pressure are strong influences and are reinforced by custom, mores, and laws. Should the culture place emphasis on a work ethic, most members of that society will work. Conversely, if the possession of worldly goods is not an attribute of a culture, economic performance, measured in such terms, will probably be low. Some previously cited examples of group emphasis on economic activities that are not particularly productive include the maize culture of Mayan Indians, Bedouin nomadic herding, and the adherence to subsistence-oriented village life and ethnic unity by numerous groups in southeastern Asia.

Colonialism and Trade

Colonialism and more recently the effects of foreign trade are related explanations that have numerous adherents. Colonialism in the traditional sense of the word is largely a thing of the past. In the years immediately following World War II, most colonial nations gave up, or were forced to give up, their possessions.

Mercantilism was the philosophy that governed the trade between most mother countries and their colonies. Under mercantilism, the colony supplied raw materials and foodstuffs needed by the mother country. The mother country in turn used its colonies as a market for finished products and other surpluses. To assure that colonies did not compete with the mother country, products the mother country had in abundance could not be produced in the colonies. Moreover, the colonies could trade only with the mother country. This arrangement obviously was to the advantage of the mother country and severely limited the economic options of the colonies.

Today trade among nations is controlled by *tariff structures*. Most rich nations have a tariff or duty on manufactured goods and low or no duties on raw materials and foodstuffs they need. The effect of this tariff structure is to perpetuate the trade patterns that existed in the colonial period (called dependency by some authorities). Most poor nations also have a similar tariff pattern. They use tariffs to protect their new manufacturing activities. Interestingly, the United States trade with Japan is similar to that of a poor nation with a rich nation, for the United States supplies foodstuffs and raw materials to Japan and in return receives manufactured goods.

Circular Causation

Another economic theory is *circular causation,* which results in either a downward or an upward spiral. In the downward case a farm family, for example, barely produces enough to feed themselves; they have little or no savings. Should a minor crop failure reduce the harvest, they have nothing to fall back on. They eat less, can work less, produce less, and so on. The upward spiral can also be illustrated by the farm family. Again the family produces only enough to feed themselves. By good fortune, they obtain some additional capital to buy fertilizer which increases crop yields; so they eat better, work harder to produce more, and sell the surplus, therefore increasing the family's capacity to buy more fertilizer and improved seed. Although our example is a single family, the theory is equally applicable to groups and nations. The circular causation idea reinforces the old adage that the rich get richer and the poor get poorer.

Rostow's Stages

Finally, Walt Rostow, from comparisons of historical economic data, has advanced the idea that economic growth occurs in five stages:

1 A *traditional society* exists; most workers are in agriculture, have limited savings, and use age-old productive methods. Indeed, all of the characteristics of a truly poor society are exhibited.

2 The *"preconditions for takeoff"* are established. This stage may be initiated internally by an awakening of the population to a desire for a higher standard of living or by external forces that intrude into the region. In either case, production increases, perhaps only slightly, but fundamental changes in attitude occur, and individual and national goals are altered.

3 *"Takeoff"* occurs when new technologies and capital are applied to increase production greatly. Manufacturing and tertiary activities become increasingly important and lead to migration from the rural environment to the bustling urban agglomerations. Infrastructural facilities are improved and expanded. During this stage, political power is transferred from the landed aristocracy to an urban-based power structure.

4 The *"drive to maturity"* is a continuation of the processes begun in stage 3. Urbanization progresses, and manufacturing and service activities become increasingly important. The rural sector loses much of its population, but those who remain produce large quantities with mechanized equipment and modern technology.

5 The final stage is *"high mass consumption."* Personal incomes are high and abundant goods and services are readily available. Individuals no longer worry about securing the basic necessities of life and, should they choose, can devote more of their energies to noneconomic pursuits.

SUMMARY

Whether or not any of the development theories are accepted, they provide a useful yardstick against which we can measure and examine the development process in various parts of the world.

67

Certainly a wide range of rich and poor nations span the globe. Some poor nations are undergoing economic development, but the gap between rich and poor is widening. For evidence we have shown the historical trends in some of the standard measures of economic well-being. Despite considerable diversity, all rich nations share certain characteristics, and the poor nations have their particular common features. Although most measures of rich and poor nations are economic, we can discern other cultural features that are charac-

teristic of each group. We must remember that the economy is but one aspect of culture, and to assign purely economic theory to explain the development process or lack of development may be misleading. Unfortunately, our knowledge has not progressed to the point where we can definitely rely on any theory to explain why some nations have developed extensively but others have made little progress. It is hoped that by analyzing individual countries more closely we will gain more insight.

FURTHER READINGS

The United Nations publishes several yearbooks of data on individual countries; among the most useful for our purposes are:

Department of Economic Affairs, *Demographic Yearbook* (New York: United Nations, 1948 to present). An annual compendium of useful population statistics including data on total population, population growth rates, urban-rural ratios, births, deaths, life tables, and population movements.

Department of Economic and Social Affairs, *Statistical Yearbook* (New York: United Nations, 1948 to present). An annual set of statistical data covering a wide range of topics. Particularly useful are data on economic activities including agriculture, forestry, fishing, mining, manufacturing, energy use, trade, transport and communications, consumption of selected items, and national accounts.

Food and Agriculture Organization, *Production Yearbook* (Rome: United Nations, 1946 to present). An annual compendium of agricultural statistics including area harvested, yields, and total production by country; also data on prices, pesticide and fertilizer consumption, and farm machinery.

AKINBODE, ADE, "Population Explosion in Africa and Its Implications for Economic Development," *Journal of Geography,* 76 (1977), 28–36.

BAKER, J. N. L., *The History of Geography* (New York: Barnes and Noble, 1963) and Fuson, Robert H., *A Geography of Geography* (Dubuque, Iowa: W. C. Brown, 1969) provide a brief but interesting review of the field of geography and its growth and development.

CHANG, JEN-HU, "The Agricultural Potential of the Humid Tropics," *Geographical Review,* 58 (1968), 333–61. An exploration of problems inherent in the use of a specific environmental realm.

CUFF, DAVID J., "The Economic Dimension of Demographic Transition," *Journal of Geography,* 72 (1973), 11–16. A brief discussion of the relationship

of demographic transformation stages to economic activity.

DALTON, GEORGE (ed.), *Economic Development and Social Change* (Garden City, N.Y.: Natural History Press, 1971). The impact of development from colonial times to the present, traced through case studies.

DETWYLER, THOMAS R. (ed.), *Man's Impact on Environment* (New York: McGraw-Hill, 1971). A volume about how people modify the environment.

"Economic Development," *Scientific American,* 243 (1980), 247 pp. The entire September issue of this journal is devoted to the question of development, including a projection of the world economy in 2000.

Economic Development and Cultural Change, a journal established to study the role of economic development as an agent of cultural change, contains numerous articles worthy of examination.

GROSSMAN, LARRY, "Man-Environment Relationships in Anthropology and Geography," *Annals, Association of American Geographers,* 67 (1977), 17–27. A good philosophical discussion, with lengthy and useful references, of the man-environment relationship as examined by geographers since 1900.

HUNTER, ROBERT E., and JOHN E. RIELLY (eds.), *Development Today: A New Look at U.S. Relations with the Poor Countries* (New York: Praeger, 1972). A collection of short articles covering a range of topics from the role of multinational corporations in economic development to the politics of foreign assistance legislation in the United States Congress.

MEADOWS, DONELLA H., et al., *The Limits of Growth: A Report for the Club of Rome's Project on the Predicament of Mankind* (New York: Universe Books, 1972). A sobering neo-Malthusian discussion of the race between population growth on the one hand and technology and pollution on the other.

PATTERSON, WILLIAM D., "The Four Traditions of Geography," 63 (1964), 211–216. Patterson shows that in the modern period four traditions have bound geography together.

ROSTOW, WALT W., *The Stages of Economic Growth: A Non-Communist Manifesto,* 2nd ed. (Cambridge: Cambridge University Press, 1971). An examination of economic development as a process and of the several stages through which a society must pass.

SPENCER, J. E., and W. L. THOMAS, *Introducing Cultural Geography* (New York: Wiley, 1973). An introduction to cultural geography written from a historical perspective.

STRAHLER, ALAN H., and H. STRAHLER, *Elements of Physical Geography* (New York: Wiley, 1976). An introductory textbook dealing with the entire spectrum of the physical environment.

WOOD, HAROLD A., "Toward a Geographical Concept of Development," *Geographical Review,* 67 (1977), 462–68. The major areas of concern in any development program—needs, territory, transport efficiency, environmental harmony, and quality of life—examined at local, regional, national, and multinational levels.

II

Anglo-America

James S. Fisher

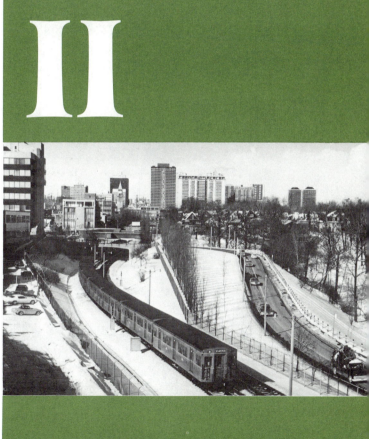

5

Anglo-America: The Basis of United States and Canadian Development

Anglo-America consists of two vast countries encompassing more than 14% of the land area of the world. This large size implies that a great variety of resources can be marshaled for the support of the 259 million people who live in Anglo-America, most of them in the eastern United States and adjacent areas of Canada.

Anglo-Americans have achieved a high level of development in an economic and technical sense. By virtue of its size, population, and economic, social, and political achievements, with their attendant problems, Anglo-America is an excellent example of the development process.

RESOURCES: STRENGTHS AND PROBLEMS

The initially small populations who developed the United States and Canada had the advantage of an immensely *rich environment.* They and their successors have often used this environment as though it contained an endless supply of resources. The two countries, particularly the United States, now consume vast quantities of resources and by all indications will continue to do so. Although it would be foolish to argue that Anglo-American achievement was based on a rich and bountiful environment alone, it would be equally unrealistic not to recognize that land, water, minerals, and numerous other environmental features have provided great advantages for the complex and interacting processes incorporated in the development experience.

Land Surface Regions

Anglo-America is composed of nine surface regions (Figure 5–1). The surface forms provide highly varied environments and have aided or inhibited agriculture, functioned both as a barrier and a routeway for movement, and in other ways contributed to the varied settling and use of Anglo-American space.

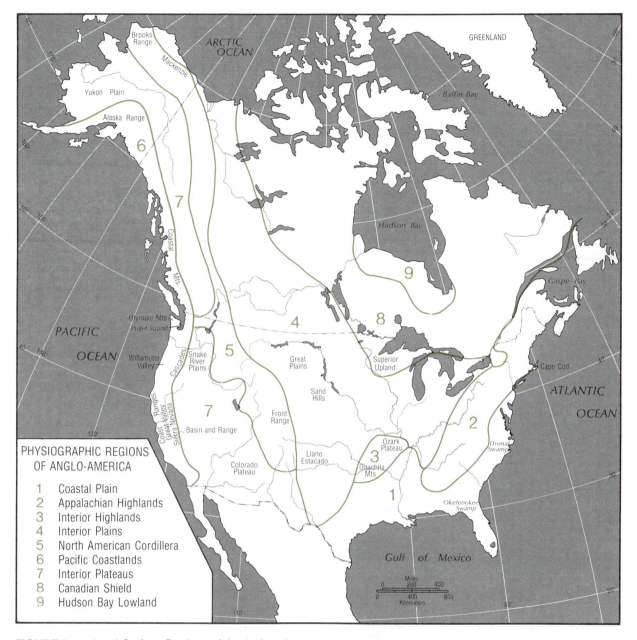

FIGURE 5–1 Land Surface Regions of Anglo-America.
The physiographic diversity of Anglo-America, together with climate, vegetation, and soils, provides highly varied environmental realms and the opportunity for a variety of natural resources.

The *Coastal Plain* borders Anglo-America from Cape Cod to Texas. The plain is of recent origin and composed of sedimentary materials. There are no major surface features that inhibited movement or settling, but some portions of the plain were given less attention as habitats for early settlement because of limited accessibility to the coast, infertile soils, or poor drainage. Typical areas of poor drainage are the Dismal Swamp, the Okefenokee Swamp, and the Everglades. Some parts of the Coastal Plain—for example, the black-soil prairies of Texas and Alabama and the alluvial Mississippi Valley—provide some of the best soil resources of the South.

The *Appalachian Highlands* cover a vast area extending from Alabama to Newfoundland. Although of a common geologic history, the highlands contain distinct topographic regions: the Piedmont, the Blue Ridge Mountains, the Ridge and Valley, the Appalachian Plateau, and the New England section. The *Piedmont* is a rolling upland plain that forms the eastern margin of the Appalachians from Pennsylvania southward. The *Blue Ridge Mountains* are distinguished by their relative height; some peaks attain elevations in excess of 6000 feet (1829 meters) in the wider southern portion in Georgia, North Carolina, and Tennessee. The *Ridge and Valley* is a landscape of parallel ridges and valleys that extends from Alabama to New York. The Great Valley is most distinctive and extends from Alabama to northern New York. The Coosa in Alabama, the Valley of East Tennessee, the Shenandoah Valley, the Cumberland Valley, and the Hudson Valley are some of the local names applied to parts of the Great Valley taken from the numerous rivers that drain the valley. Other valleys are neither as continuous nor as wide. The westernmost portion of the Appalachian Highlands, despite the name *Appalachian Plateau,* has been severely eroded into hill lands and mountains (West Virginia), while other areas, such as the Cumberland Mountains in Tennessee, retain a distinctive plateau character.

The *southern Appalachian Highlands* have had a peculiar role in the evolution of the social and economic geography of the United States. The highlands lie close to all of the early European settlements, yet many parts have remained isolated and even lagged culturally. The physiographic character of the area has variously hindered population movement, influenced the direction taken by settlers, contributed to isolation, and functioned as a roadway (the Piedmont) for settlers. At present, some of the nation's finest resources exist within the Appalachian area, yet it has long been one of the more severe economic problem regions in the United States.

The *Interior Highlands* have a geologic history, structure, and physiography similar to those of the Appalachian Plateau and the Ridge and Valley regions. The Arkansas River Valley separates the northern *Ozark Plateau* from the *Ouachita Mountains* to the south. A similarity of culture and problems further relates the southern Appalachians and Interior Highlands.

One of the largest continuous plains areas of the world, the *Interior Plains,* extends from central Tennessee and Kentucky in the southeast and Texas and Oklahoma in the southwest northward to the Mackenzie Valley of northern Canada. Because much of the area north of the Ohio and Mississippi rivers has experienced continental glaciation, glacial landscape features such as moraines, till plains, former glacial lake beds, and disarranged drainage are common.

Much of the Interior Plains is at low elevations. The western portion *(Great Plains),* however, is an exception. The eastern edge of the Great Plains is 2000 feet (610 meters) above sea level. Elevation gradually increases to the west and reaches 4000–5000 feet (1219–1524 meters) at the border of the Rocky Mountain system.

From a topographic standpoint, the vast Interior Plains provide one of the more favorable regions of the world for agriculture. Within the vast area, however, are also great variations in climate, which limit agricultural use in the drier western and colder northern segments.

In recent geologic time, great mountain-building processes disturbed part of western Anglo-America, forming the *North American Cordillera.* The results are high elevations and a mixture of landforms. Portions of the Rocky Mountains, such as the *Front Ranges* of Colorado, consist of intrusions of materials from deep within the earth that disturbed and uplifted the sedimentary layers of rock lying nearer the surface. The Canadian Rockies are formed of folded and faulted sedimentary materials. North and northwest of the Rockies are the *Mackenzie,* the *Richardson,* and the *Brooks* ranges. Altogether these several ranges form a mountain system that extends from the southwestern United States through Alaska. Within the larger system, however, are many valleys and basins where small settlements have been established.

The *Pacific Ranges* are a system of mountains and valleys that extend from southern California northward along the Canadian coast and westward to the Alaska Peninsula. In Canada, this region is formed by the *Coastal Mountains* of Canada and the *Alaska Range.* The western portion is formed of yet another chain of mountains with strong linear features (parallel mountains and valleys), the *Coast Ranges* of California. The linear character of these ranges becomes less distinct in Oregon and is totally absent in the *Olympic Mountains* of Washington. Between the Coast Ranges and the *Sierra Nevada* and *Cascades* lie several lowland areas. The *Great Valley of California,* a fertile alluvial trough, is one of the most productive agricultural regions of the United States. The Cascades and the Coast Ranges enclose the productive *Willa-*

mette Valley and the *Puget Sound Lowland.* The Coastal Mountains are discontinuous in the form of islands off the Canadian coast, and the lowlands are submerged. The lowlands reappear only in Alaska, as basins south of the Alaskan and *Aleutian ranges.*

Between the Pacific Ranges and the Rocky Mountains lie the *Interior Plateaus.* The essential character of this region is that of a level upland; elevations are normally in excess of 3000 feet (914 meters), but there are notable exceptions in the Yukon Plain of Alaska and in Death Valley of California, which is 282 feet (86 meters) below sea level. The Basin and Range segment of Nevada, California, and Utah is characterized by faulted mountain ranges half buried in alluvial debris.

The *Colorado Plateau* (Arizona and Colorado) displays the previous nature and geologic history of the area. Many layers of sedimentary rock now at 9000–11,000 feet (2743–3353 meters) above sea level have been exposed by the Colorado River, which has cut the Grand Canyon and provided spectacular evidence of the great depth of underlying sedimentary rock. In eastern Washington, Oregon, and Idaho (Columbia Plateau), formerly active volcanoes and fissure flows have deposited thick layers of lava, and the Snake River has cut a canyon through the lava that nearly rivals the Grand Canyon in splendor.

In northern and northeastern Canada lies the great *Canadian Shield,* which nearly encircles Hudson Bay and extends southward to the United States, forming the Superior Upland of Wisconsin, Michigan, and Minnesota. Continental glaciation has produced a relatively smooth surface. Unfortunately, thin soils, stony surfaces, and even the absence of soil are other results of glacial scour. Lakes, marshes, muskeg, and swamplands are characteristic in many parts of the shield. Although utility for agriculture is limited, the Canadian Shield has provided a wealth of furs, timber, and minerals.

The *Hudson Bay Lowland* is a sedimentary region along the southern margin of the bay. The lowland is a forested plain with little variation in elevation. Like the Canadian Shield, the Hudson Bay Lowland has experienced little development by mankind.

Climatic Regions

Anglo-America exhibits the climates expected on a large continent extending from subtropical to polar latitudes (see Figure 3–2). The eastern portion of the continent shows a strong latitudinal influence in the sequence of humid subtropical, humid continental (warm and cool summer varieties), subarctic, and polar climates. The most important differences among these climates are the shorter and cooler summers and longer and more severe winters as one proceeds northward from the Gulf of Mexico. The interior west, which is remote from major moisture sources and located leeward of major highlands, is dominated by dry climates (desert and steppe). Along the coast from California to Alaska is a subtropical-to-high-latitude west coast sequence—dry summer subtropical, marine west coast, subarctic, and polar climates. This sequence is similar to that of Western Europe (see Figure 3–2) except that the Anglo-American highlands are more effective in preventing the extension of these humid climates great distances into the interior.

Precipitation Patterns

Two humid regions are evident in Figure 5–2. One region extends from the Texas Gulf Coast to Hudson Bay and eastward to the Atlantic. The general pattern here is decreasing precipitation with increasing distance from coastal areas. Although the southern Appalachians receive somewhat more rainfall than adjacent areas, they are not effective as a barrier to the moisture-laden winds from either the Gulf of Mexico or the Atlantic Ocean. Maritime air masses from these water bodies are the moisture source for areas far to the interior.

A second humid region covers a smaller area along the west coast from California (the Sierra Nevada) to southern Alaska and the Aleutian Islands. The great variations in precipitation are a reflection of elevation and exposure to rain-bearing winds. Locations at lower elevations in sheltered positions receive much less rainfall.

A region of *low precipitation* (less than 20 inches [50 centimeters] per year) extends from the American southwest to the Northwest Territories of Canada. In the far north it spreads from Alaska to Greenland, but we normally do not think of this area as dry because of low *evapotranspiration.* The agricultural potential of the immense Interior Plains is considerably reduced by the deficiency of moisture over the Great Plains portion.

Water Resources

The distribution of precipitation is of major importance not only to agricultural activities but also to

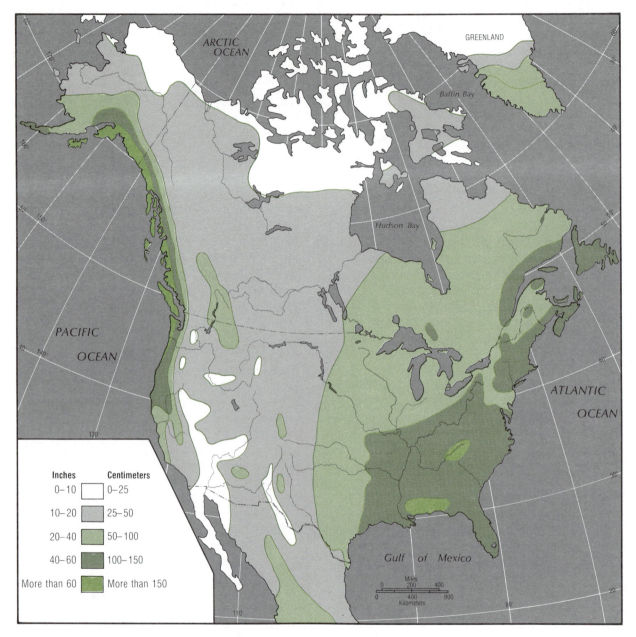

FIGURE 5–2 Annual Precipitation in Anglo-America.
A major humid region covers the eastern half of Anglo-America. A second, much smaller humid region is found in the far northwest. Much of the western interior between the two humid regions is moisture deficient.

many functions of an urban-industrial society. The fact that most of the Anglo-American population is found in areas where precipitation is most plentiful does not preclude *water problems,* for these are precisely the areas where the level of water consumption is high. One of the important problems facing the people in the northeastern United States is how to provide adequate and proper development of water resources for municipal, industrial, and recreational use, present and future.

The water problems of the American southwest are different. A high proportion of the much more limited water resources is used for *irrigation* and therefore is competitive with urban-industrial water demand. The importance of irrigation water for increasing agricultural productivity and income, as well as accelerated urban-industrial needs, is illustrated by a long history of legal disputes between California and Arizona and between the United States and Mexico for rights to Colorado River water.

As a society proceeds from an "underdeveloped" to a "highly developed" condition, the transformation of the economic system greatly reshapes the *demand for water*. Water needs increase with population growth both for direct consumption (domestic use) and for agriculture. Even more significant is the growth of water needs associated with industrial expansion. Industrial societies use immense quantities of water as a solvent, as a waste carrier, and as a coolant. In recent years in the United States less than 10% of the water removed from surface sources has been for domestic use. More than one-half has been for industrial use. The remainder, approximately 40%, has been used as irrigation water. Surface water is drawn from streams, rivers, or lakes as distinguished from groundwater, which is removed from wells.

Temperature

The amount of energy available for the conversion of nutrients and water into vegetable matter shows as much regional contrast as precipitation. Much of southern Anglo-America has a *growing season* in excess of 200 days, which allows the production of a great number of "subtropical" crops (cotton, peanuts, citrus, and even some sugarcane). The length of the growing season decreases with higher latitudes. Much of Canada and Alaska is not suitable for agricultural production. The highlands of western Canada and the thin soils and poorly drained glaciated areas of the east further detract from the utility of the Canadian environment for agriculture.

The *temperature regimes* of the United States and Canada are largely attributable to three climatic influences: latitude, altitude, and the marine influence. Plant growth seasons become shorter with increasing latitude. The effect of higher altitude is particularly noticeable in the western portion of the continent; the more severe minimum-temperature regions exist in the Rocky Mountains. The marine influence is reflected in the long growing season associated with coastal zones, particularly the west coast.

Vegetation Patterns

A narrow band of treeless *tundra* extends across far northern America from Alaska to Greenland (see Figure 3–3). In Alaska and northwest Canada, the tundra extends southward in highland areas. South of the tundra from Newfoundland to Alaska is a vast coniferous forest. This American *coniferous (boreal) forest*, along with the similar *taiga* of the Soviet Union, is one of the largest expanses of forest remaining in the world. The dominant species are spruce, fir, and pine. The southern margins of the forest and those areas adjacent to waterways have been intensively exploited.

The boreal forest gives way to *deciduous forest* through a broad transitional zone of white and yellow birch, poplars, and maples (broadleaf species) in the humid eastern half of the continent.

The original deciduous forest covered most of the northeastern United States. It consisted of oak, elms, hickory, beech, and maples. In less fertile areas, pines dominated. The deciduous forest is perhaps the most modified vegetation region in Anglo-America. Its very existence suggested to early European settlers advantages in climatic and soil resources. Utilization of such resources required removal of trees. This great deciduous forest extended much farther south than is generally realized. Today varieties of pine stand on the Piedmont where there were once immense hardwoods. The forests that have been reestablished by natural processes or planting are indeed part of the American cultural landscape.

The grasslands of interior Anglo-America change from tall grass *prairies* on the eastern margin of the Great Plains to short grasses on the drier western margin. Actually, the prairie extended eastward beyond the Great Plains into Iowa and Illinois at the time of European settling. Frequent "oak openings," forests interrupted by expanses of prairie, characterized the transition zone between the forest lands of the east and the grasslands of the west. This extension of the prairies eastward may have been caused by the Indians' repeated use of fire for improving their habitat.

Although grassland areas are indicative of moisture deficiency, there use for agriculture is not precluded. Excellent soils commonly are associated with prairies. Limited precipitation is a factor in the processes of fertile soil formation. Scant moisture supply also entails the risk of recurring drought. Even ranching is hazardous, for overgrazing has frequently damaged natural grasslands, which are then replaced by *woody shrubs* such as sagebrush or mesquite.

Soils

There is a spatial correspondence of soil with climatic and vegetative patterns. The associations are not unexpected, since climate and vegetation are intricately involved in the soil-forming processes. The gray-brown and red-yellow podzolic, chernozemic, and alluvial soils are most significant to Anglo-American agriculture because they are associated with areas in which other constraints such as temperature, precipitation, and topography are not severe. The fertility of the *gray-brown podzols,* merging with chernozemic soils to the west, the favorable topography, the moderate growing season, and reliable rainfall combine to make the area from central Ohio westward to Nebraska one of the world's most productive agricultural areas.

The *chernozemic soils* of the Great Plains of the United States and Canada are perhaps even more fertile as measured by nutrient content but are less productive because of limited precipitation and recurring drought. Utilization of these soils requires either farming systems that incorporate drought-resistant crops (wheat) or grazing systems based on pasturage.

The *red-yellow podzols (latosolic soils)* have experienced severe leaching and require substantial inputs of fertilizer to maintain productivity. Many problems of southern agriculture, however, are not inherent in the soils but are the result of farming methods. Row-crop production for many years on the hilly Piedmont accelerated severe erosion problems that eventually contributed to the abandonment of crop agriculture in many areas. There are exceptionally favorable soils in the South. The black-soil belts of Texas and Alabama, the limestone valleys of the Appalachians, and the alluvial Mississippi Valley contain some of the best soil resources of the United States.

Early use of the Anglo-American soil resources was often destructive for several reasons. Settlers occupying such vast areas were imbued with the notion of almost unlimited space and expansion possibilities. In older areas of settlement east of the Appalachians, much land had cycled in and out of agricultural use by the time of the Civil War. Elsewhere, large areas showed the destructive effects of *soil erosion* by water and wind by the 1930s. Destruction resulted from overgrazing, cropping of land subject to drought, water and

wind erosion, and improper row cropping. Hardly any major agricultural region of the United States or Canada was immune from the problem.

Since the 1930s, the Soil Conservation Service in the United States and comparable agencies in Canada have promoted both removal of *submarginal land* from agriculture and the use of *improved agricultural methods*. It is important to recognize, however, that during most of the period since the 1930s, Anglo-American agricultural problems have been related to surplus production and low prices. From the standpoint of land needed for food production, population pressure was low. Thus programs could be implemented that removed land from production or reduced the intensity of production and yet were not detrimental to the larger economic system. Few areas of the world have been in such a position. Since 1970, however, the United States has seen agricultural surpluses diminish and an increasing intensity of effort on agricultural land has become evident. Soil erosion problems have again become of major concern to those familiar with land-use problems.

RESOURCES FOR INDUSTRIAL GROWTH AND DEVELOPMENT

Energy and Power

Societies proceeding through industrial development become increasingly dependent upon inanimate power sources for both consumer and industrial uses. The United States experience illustrates the importance of an adequate power base for development. Power can be bought, but only at increasingly high prices and with potentially detrimental effects on balance of payments. Domestic availability of power resources is an index of a nation's potential vulnerability.

Coal. *Coal* was the power source for American industrial expansion. Even prior to World War II, however, coal's relative contribution to the energy supply decreased as petroleum and natural gas supplies were developed. Coal now provides less than one-fifth of the nation's energy supply (Table 5–1). Petroleum and natural gas are considered cleaner than coal and are preferred as power sources for heating and industrial use. The absolute production of coal has alternatively decreased and increased.

The major *coal-producing states* are Kentucky, West Virginia, Pennsylvania, Wyoming, and Illi-

Coal is one of the most abundant energy sources in the United States. Open-pit mining with mechanized equipment provides low-cost means of extraction yet creates problems of pollution and environmental degradation.

nois (Figure 5–3). Large quantities of coal are available in the western states, but production is handicapped by the small local need and the great distances from the major eastern markets. Tremendous quantities of bituminous coal remain, enough to supply United States energy needs for several hundred years—a comforting thought, perhaps, but use is not so simply implemented. The main use for coal currently is for generation of electrical power (about 80% of the market); the remainder serves mostly as industrial fuel. Use for home or transportation purposes would require costly conversion efforts and the application of new and expensive technology.

Canadian coal reserves are also large. Two Maritime Provinces (New Brunswick and Nova Scotia) and two Prairie Provinces (Alberta and Saskatchewan) contain most of the reserves. The great area of need for coal as a power resource is in the urban-industrial regions of Ontario and Quebec. Because of the great distances involved, Canadians find it more practical to import Appalachian coal. Another result of the unfavorable location of Canadian coal is that Canada has placed a correspondingly greater emphasis on both petroleum (cheaper to transport) and water as power sources.

Environmental concerns increasingly affect the feasibility of using various resources. Coal is a prime example. The supply of low-sulfur coal will depend on economical means of removing sulfurous pollutants. Any effort to expand coal use as a power source will be affected by the restrictions and regulations for controlling air quality placed on industry and individuals and by concern over the scarred landscapes created by open-pit coal mining.

Oil and gas. *Petroleum and natural gas* account for three-fourths of United States energy consumption. The largest single use of petroleum is for automotive fuel (52%). Approximately 13% is used for heating, and 26% as an industrial raw material (road oil, lubricants, etc.), for petrochemical in-

TABLE 5–1 Power Consumption by Source (1980)

Power Source	Percent Contributed	
	United States	Canada
Coal	19.2	8.9
Petroleum	47.1	44.0
Natural gas	26.2	18.6
Water	4.0	25.0
Nuclear	3.5	3.5
Total	100.0	100.0

SOURCES: U.S. Bureau of the Census, *Statistical Abstract of the United States* (Washington, D.C.: Government Printing Office, 1981); and *Canada Yearbook* (Ottawa: Dominion Bureau of Statistics, 1980–81).

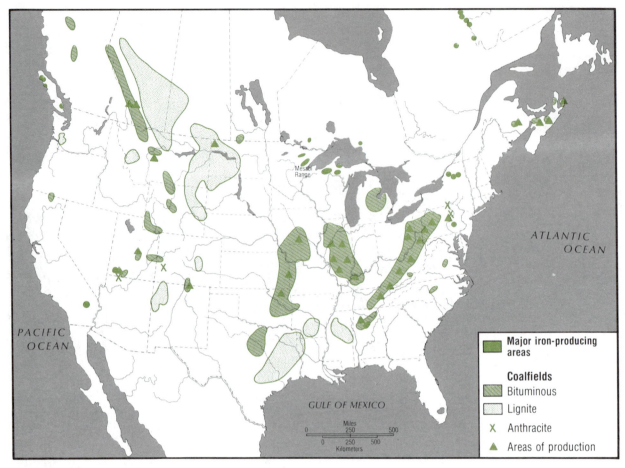

FIGURE 5–3 Coal and Iron Ore in Anglo-America.
Major coalfields are found in many parts of Anglo-America, but the far western
United States and eastern Canada show deficiencies. The Appalachian fields were
major contributors of power for United States industrial expansion during the late
nineteenth century.

dustries (ammonia, carbon black, synthetic rubber, plastics, and synthetic fibers), and for industrial fuel.

The *future of oil and gas* in the energy picture of the United States and Canada is difficult to assess. First, the United States is both a major producer and consumer. Although coal production and consumption have been nearly equal until recent years, the same cannot be said of petroleum. The United States has produced approximately 14% of the world's petroleum in recent years but has also accounted for nearly one-third of the world demand, more than 6 billion barrels annually. There is little doubt that demand will increase.

The term *"proven reserves"* refers to oil known to be available by actual drilling and removable at a given cost with existing technology. The ratio of

proven reserves to production has been approximately 9:1 in recent years. This ratio does not mean that the United States will run out of petroleum in nine years. Present estimates of proven and *unproven reserves* indicate that more than 500 billion barrels of oil remain; at present recovery rates, 250 billion barrels are therefore available. The recovery rate is likely to increase, and continued new information on the Alaskan reserves will require upward adjustment of these figures. Thus, depending upon the consumption level, enough oil is beneath the surface to last for some decades. Nevertheless, oil imports account for nearly 37% of the oil consumed in the United States. Saudi Arabia, Nigeria, Mexico, Libya, and Algeria are major suppliers. The degree of the United States dependence upon foreign oil will depend upon price, which increased greatly during the 1970s, and domestic and foreign availability, as other growth areas seek larger quantities of oil. The slight decreasing dependence on imported oil during the early 1980s may or may not continue in the future. It is likewise difficult to discern how much the decreasing energy use of the early 1980s is because of conservation effort and how much is attributable to repeated downturns in the national economy.

The present *major regions* of oil and natural gas production in the United States are in Texas, Alaska, Louisiana, California, Oklahoma, and Wyoming (Figure 5–4). Alaska's north slope has cata-

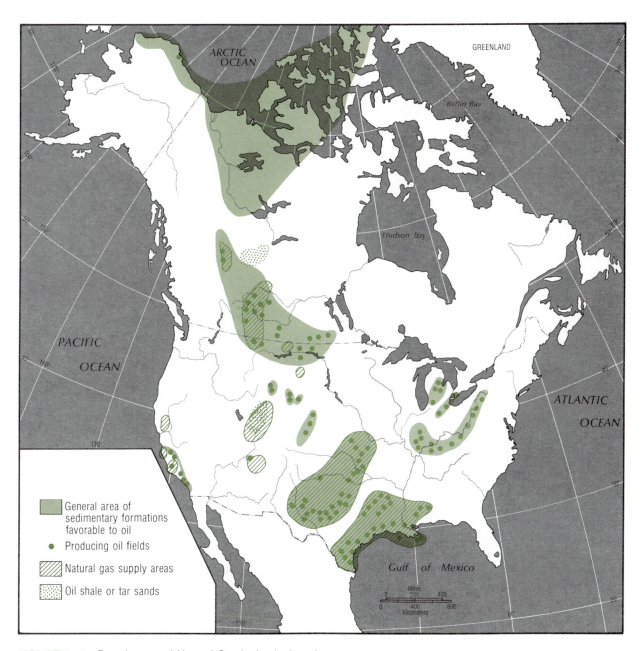

FIGURE 5–4 Petroleum and Natural Gas in Anglo-America.
The United States and Canada have had the advantage of large oil and gas supplies. Nevertheless, the level of development, the substitution of oil and gas for coal, and the high per capita consumption mean that the cost and availability of oil are a major problem, particularly for the United States.

pulted that state to a status of major importance as a source of petroleum for domestic use.

Canada, though not one of the world's oil giants, does produce significant quantities of oil and natural gas in the Prairie Provinces of Alberta and Saskatchewan. Only in recent years have oil exports exceeded imports.

Oil shale and tar sands. The solid organic materials associated with other minerals in *shale formations* of Utah, Colorado, and Wyoming (the Green River formation) represent one of the world's largest deposits of hydrocarbons; the energy potential is immense. The contribution that these resources will make to energy supply by the year 2000 may be limited, but advances in technology or changes in the price of petroleum could change this assessment dramatically. Technology for oil production from shale rock exists, but efficiency of production is far from competitive with other forms of power resources. Moreover, were it necessary and economically feasible to use oil shale, immense environmental problems would have to be overcome. Vast quantities of rock would have to be processed, and restoration policies minimizing environmental destruction would be significant factors in any decision regarding large-scale production.

Canadian tar sands along the Athabasca River in Alberta contain enormous quantities of oil. Realization of this potential, however, requires unprecedented capital investment in new technologies. Nevertheless, consortiums involving numerous firms are developing these resources, and the current modest production will likely increase significantly during the coming decade.

Waterpower. *Waterpower* provides 70% of electric power in Canada and 12.4% in the United States. Nevertheless, waterpower use continues to rank far below that of coal, oil, or gas. Despite the likelihood of continued expansion of waterpower development, the total share of power so provided, electrical and other, will probably continue to decline. The West Coast (Columbia River Basin), the Tennessee River Valley and southern Piedmont, and the St. Lawrence Valley are all areas where waterpower generation is proportionately high. Canada, with its limited quantity of coal, has placed great emphasis on the generation of electrical power with water. This power has contributed directly to the massive Canadian aluminum industry.

Nuclear power. The extent of uranium or thorium use in the future for energy purposes is difficult to predict. The continued advances likely in *nuclear technology,* the cost of alternative fuels, dependence upon and cost of foreign oil, and environmental concerns will all stimulate or inhibit the development of this energy resource. The growth of nuclear power use was rapid during the 1970s, but like petroleum and gas, showed a declining use by 1980. Whether or not nuclear power will become a major energy source in the future is uncertain. Uranium supply in the United States may not be adequate for large long-term demand. Nuclear technology, however, is in its infancy and continually advancing.

Metal Resources

The United States illustrates how *iron ore* (complemented by coal) underlies modern industrial structure. The United States is both a major producer (10% of the world total) and the major consumer (13%) of iron ore. Despite recent substitutions of aluminum and even plastic, iron remains the metal consumed in greatest quantity.

The great *Mesabi Range* (see Figure 5–3) began supplying ore in 1890. Despite increased United States dependence upon foreign ores, more than 60% of the iron consumed in this country still comes from the Lake Superior area. Other significant domestic sources include the Adirondack Mountains (New York) and the Birmingham, Alabama, area. Many scattered iron deposits are found in western states, notably California, Wyoming, and Utah. Distance from the major United States markets, however, dictates use in the smaller steel centers of the West (California and Utah).

It became apparent during the 1940s that high-grade ores (with an iron content of 60%) were becoming less readily available. Dependence upon foreign ores accelerated during the 1950s and 1960s. About one-half of the imported ore normally comes from Canada, which, considering its population size, must be considered one of the mineral-rich nations. *Canadian ore* is available in the Lake Superior district at Steep Rock Lake, Ontario, and major deposits have been developed in Labrador. High-grade ores are also available from Venezuela. The development and subsequent improvement of the St. Lawrence Seaway and the construction of large ore carriers have facilitated the use of higher-grade foreign resources. This trend corresponds to the *principle of the use of best resources first.* That is, other considerations aside, it is less costly to extract iron from ore with a high iron content than from low-grade ore, and water transport makes long-haul bulk shipment of raw material possible at low cost.

United States dependence upon foreign sources of ore was once near 50% but has decreased to 22%. Technological advancements now allow the use of low-grade ores such as *taconite,* a very hard rock with low iron content, which is concentrated into ore pellets (with more than 60% iron content). The taconite industry has expanded rapidly in Minnesota and Michigan, partially in response to favorable tax concessions. Steel manufacturers find some advantages in beneficiated ores, so taconite may one day be as economical to use as foreign ores. Recent declines in the steel industry undoubtedly also account for reduced dependence on foreign ores.

Aluminum. During the past three decades, *aluminum* has become an extremely useful and sought-after metal. It is used extensively in the transportation and construction industries. Although aluminum is a common earth element, its occurrence in the form (bauxite) that allows use for metal manufacture is limited. The United States produces only about 12% of its own ore needs, mostly in Arkansas, but consumes nearly 36% of the world's processed aluminum. Canada is the world's third-ranking aluminum producer and number-one exporter, yet the *Canadian industry* is based totally on the processing of imported ores and the use of substantial local hydroelectric power. The major consumers of aluminum are industrialized countries such as the United States, although most of the world's reserves exist in underdeveloped countries. The major Western world sources, which are also those for the United States and Canada, are Jamaica, Surinam, Guyana, and Australia. Greater resource independence will depend upon continued improvements in technology that allow use of lower-grade domestic ores.

RESOURCE SUMMARY

The United States and Canada are large countries with *great quantities of basic resources.* Anglo-America, however, also has a large population that consumes materials at high per capita rates.

The great resource base and ability to use it have contributed to both the high material level of living and the primary power position in world society of Anglo-America. Maintaining this position requires continued high resource consumption. To assume a major change in the basic nature of Anglo-America's social and economic system is unrealistic.

The varied climatic and soils regions of Anglo-America allow *diversified agricultural production* of food and industrial raw materials. Not only are domestic needs met, but large quantities of agricultural commodities (wheat, soybeans, cotton) are normally available for export. Such exports significantly affect the balance-of-trade position of both countries.

Anglo-America also is well endowed with *industrial resources;* many decades of high-level consumption and higher demand levels, however, put increasingly strenuous pressures on domestic resources. The result has been an increase in the dependence upon foreign materials and fuel. Even if programs aimed at energy saving are successful, it is probable that they will only slow the rate at which consumption increases.

United States *dependence upon foreign sources* for basic needs has been to a substantial degree economic. Technology or politics may change this situation but, if not, dependence on foreign areas may increase. Underdeveloped areas, however, are increasingly demonstrating a desire to control the price, production, and processing of their own resources (Jamaican bauxite) even when developed nations are the ultimate consumers. This trend will mean higher prices for resources. It could also provide incentive for domestic exploration and for development of the technology needed to use lower-grade domestic material.

The *Canadian situation* is somewhat different. Canada is deficient in high-grade coal except in the Maritime Provinces and British Columbia, but can with relative ease and low cost obtain coal for its industrial needs from the United States. In other respects Canada is mineral rich. The questions are:

What degree of dependence should the Canadian economy place upon primary production, of which petroleum and metals are a significant part?

To what extent should the much smaller Canadian population allow the neighboring industrial giant to draw upon its petroleum, gas, copper, iron ore, nickel, and other metals as well as forest resources?

83

EARLY SETTLEMENT

The Anglo-American environments have been used, and often misused, by numerous people with different cultural heritages and experiences. Furthermore, the New World experience, a relatively short one when measured against Old World cultures, did not have beginnings in a single location. At least *four early European settlements* served as source areas for cultural "imprints" that have lasted to the present (Figure 5–5).

Early exploration of Anglo-America was carried on by the Spanish, the Portuguese, and the French. The Spanish were the first to establish a permanent colony (St. Augustine, Florida, in 1565) but except for the United States Southwest never were successful in establishing a viable society from which settling and diffusion of culture could proceed inland.

The first permanent settlements from which distinctive American *culture traits* evolved were the English Jamestown Colony, established in 1607, and the French settlement at Quebec, established in 1608. Soon following were the Plymouth Colony (English, 1620) and settlement in New York (Dutch, 1625). Germans and Scandinavians also made their appearance, but the English cultural imprint was by far the most profound and lasting. The English came in great numbers and, over all, exercised the greatest control in the development process. English dominance should not preclude recognition of the value of native American, African, Asian, or other European peoples (the French in Quebec) to American culture.

More than a century was required to occupy the initial settlement core areas effectively. During the early settling phase, distinctive differences among the colonies began to emerge that ultimately contributed to significant regional differences within Anglo-America.

New England

Commercial activity was established early in *New England.* Agricultural efforts were necessary to sustain the populace, but there was no special crop that could provide great wealth or form the basis for trade as tobacco did in the southern colonies. Instead, wealth was accumulated by fishing, trade, and forestry. The white pine forest provided useful lumber for shipbuilding and trade. The codfish on the offshore banks were another resource that could be traded. These resources, plus the wealth generated by their exchange, became

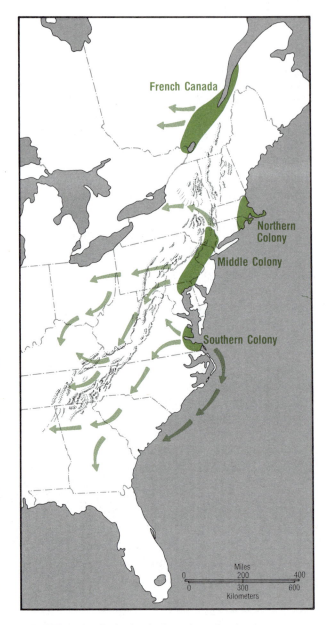

FIGURE 5–5 Early Anglo-American Settlement Focuses.
These four settlements were areas in which culture traits were evolved and later diffused to other parts of Anglo-America.

a source of capital and established *commercialism* early in the northeast. By the late eighteenth century, capital was available for industrial growth, and nonagricultural pursuits were already a tradition. Waterpower potential in mechanical form (waterwheel) was substantial, and the ocean-shipping capacity for movement of raw materials and manufactured goods existed. Shortly after independence, New England's *incipient industry* emerged as a competitor with Europe.

The Southern Colonies

The people of the *southern colonies* pursued a different path from the start. Tobacco became a commercial crop almost immediately; later, as settlements were extended southward at various coastal points, indigo, rice, and cotton were added. Many inland settlers, hampered by lack of access to water, were subsistence farmers. Where commercialism was feasible, the *plantation*—a distinct agrarian system in a spatial, social, and operational sense—began to evolve. Indentured and slave labor led to a distinction between labor and management, particularly when applied on a large scale. Thus emerged the commercial plantation, larger than "family size," with division between labor and management, social distinctions, and attendant forms of organization and layout of buildings.

The source of wealth for commercially oriented people was agriculture. The markets for agricultural products were in Europe, and the southern producer wanted no *tariff system* that might inhibit the movement of his source of wealth. Basic sectional differences appeared early, one of which was the attitude regarding tariff policies. Another arose over the socially *stratified society* that evolved along with the economic system based upon slave labor.

The plantation was established from tidewater Virginia to Maryland and southward at various coastal points to Georgia. Inland from tidewater and coastal agricultural colonies, and beyond easy water routes, were smaller free labor farms (yeoman farmers), particularly on the North Carolina and Virginia Piedmont. Later, when an improved cotton gin became available after 1800, both the yeoman-farmer and the plantation culture spread throughout the lower South. Generally the plantation system prevailed in the choice areas for agricultural settlement.

The Middle Colonies

New York and Pennsylvania, together with portions of New Jersey and Maryland, made a *third early settlement core* that contrasted with both New England and the southern colonies. Settling occurred by a great variety of people; English, Dutch, Germans, Scots-Irish, and Swedes were early participants. Neither the cash crops of the South nor the lumbering and fishing activities of New England were significant sources of income. Nevertheless, the area became an important American source region for both people and ideas. The settlers who moved southward into the Appalachians, and westward down the Ohio Valley and on into the Midwest, came from the middle colonies. They used the Indian grain maize (corn) to fatten hogs and cattle, a system that spread into the American Midwest. They made tools, guns, and wagons, and they worked the iron deposits discovered in eastern Pennsylvania. Although one might expect the middle colonies to have been transitional between the southern and northern colonies, they were not.

The Lower St. Lawrence Valley

Although English culture has dominated Canadian evolution, it has not done so exclusively. The United States has in fact functioned as a source area for Canada. Canada's most distinctive settlement, however, was French, located along the *lower St. Lawrence* between Montreal and Quebec. The French spread themselves over the vast area as fur trappers, traders, and missionaries, but it was only in the lower St. Lawrence Valley, where the French settled as farmers, that a lasting French cultural imprint was made. Their descendants, though now more urbanized, still give a distinctive French character to an entire Canadian province.

The Southern Appalachians

The *southern Appalachians* were not an original settlement area; nevertheless, the area from Virginia and West Virginia southward (including the Blue Ridge and the Ridge and Valley) functioned as a secondary settlement area from which distinctive cultural traits were eventually diffused. The region acquired settlers during the early eighteenth century who were descendants of the early Scots-Irish, German, and English settlers in the middle colonies. They moved southward along the Piedmont and the Appalachian valleys and westward to the plateaus of eastern Tennessee, Kentucky, and western Virginia. These settlers, who carved out small subsistence farms and were slaveless *yeoman farmers,* have contributed traits that distinguish the southern Appalachians from the rest of the lower South. The southern Appalachians have remained an area of small farms and an almost totally white population.

In the more isolated portions of the Appalachians, a distinct culture evolved, not from a spirit of progressiveness but as a *culture of archaism* resulting from isolation and an inability to change. As transport systems improved, the remote Appalachian coves and valleys remained unaffected, except when transport was necessary to remove a special resource such as lumber or coal. The distinctiveness of the mountain culture can be seen in Elizabethan speech and music, the use of distinctive suffixes attached to place names (cove, gap, and hollow), and mountaineer attitudes. Low education levels and poverty are widespread contemporary problems.

Labeling the southern Appalachians as a culture center implies that it functioned as a source area for other regions. It has done so for both the Interior Highlands of Arkansas and Missouri and the hill country of central Texas.[1] By the mid-nineteenth century, population levels in the Appalachians were such that out-migration was necessary. The surplus population found refuge in hill lands farther to the interior that provided some similarities in environment and the isolation for a culture that these people were not eager to change.

DEMOGRAPHIC EXPERIENCES

The seventeenth and eighteenth-century Anglo-American population grew slowly, even in New England and the Chesapeake Bay area where English interests and activity were most intensive. At the time of American independence the colonies contained only about 3 million people. Nearly a hundred years passed before Canada contained a similar population. The early size differential between the United States and Canadian populations was established and maintained largely because of differences in net migration.

United States Population Growth

The rapid *population growth* experienced by the United States after 1800 was a response to high birth rates, declining mortality, and immigration. The high United States birth rate declined to a low

[1]Terry G. Jordan, "The Imprint of the Upper and Lower South on Mid-Nineteenth Century Texas," *Annals, Association of American Geographers,* 57 (1967), 667–90.

TABLE 5–2 Demographic Features of Anglo-America. Population growth rates are determined both by birth rate and death rate differentials and net immigration. As a result, and despite belief to the contrary, United States population growth averaged more than 1% per year during the 1970s. The discrepancy between this figure and that suggested by 1980 data is explained by the compounding effect of population growth and illegal immigration.

	Canada (1981)	United States (1981)
Population (thousands)	24,100	229,800
Birth rate (%)	1.5	1.6
Death rate (%)	0.7	0.9
Natural increase (%)	0.8	0.7
Legal immigration rate (%)	0.7[a]	0.2[b]
Emigration rate (%)	0.3[a]	[c]

SOURCES: *Population Data Sheet* (Washington, D.C.: Population Reference Bureau, 1982); U.S. Bureau of the Census, *Statistical Abstract of the United States* (Washington, D.C.: Government Printing Office, 1981); and *Canada Yearbook* (Ottawa: Dominion Bureau of Statistics, 1980–81).
[a]Figures calculated from data for the years 1971–76.
[b]Value for 1979.
[c]An accurate emigration rate for the United States is not known; however, it is estimated to be negligible in the national population equation.

of 1.8% in the 1930s. The restraint on family size in the 1930s came to an end after World War II, possibly in response to wartime delays in family growth and the economic prosperity in the years following the war. Birth rates reached a new high of 2.7%, bringing a new era of relatively rapid population growth, or the *"baby boom."* The baby boom has been replaced more recently by what some refer to as a *"birth dearth"* (Table 5–2). During the mid-1970s the birth rate declined to an historic low of 1.5%. A recent modest upturn is probably a response to children of the "baby boom" attaining child-bearing age. The fertility rate, however, has shown a slight increase since its historic low in 1976.

The *declining growth rate* over the past century has not been in exact correspondence with changes in birth rates. Reduced mortality rates also are a factor. The decline of infant mortality and the extension of life expectancy beyond seventy years mean greater numbers of people alive at any given time. *Immigration* has been another important factor in the growth experience of the United States. Approximately 45 million people have immigrated to the United States since 1820. The immigrants enlarged the population as they came but also increased the population base from which future growth was derived. Recent legal immigration has been at the rate of about 400,000 persons annually and accounted for about 17% of U.S. growth during the 1970s. The amount of *illegal immigration* is not precisely known, but in 1978, United States governmental officials estimated 8 million to 12 million illegal aliens were living in the nation. It was estimated that nearly 1 million persons per year successfully enter the United States illegally.

The general trend has been declining growth rates during the last hundred years. The rate has declined from more than 3% during the early nineteenth century to less than 1% at present. The current low rates of growth do not mean small population increments, however, for now the population base is large. The population growth resulting from the natural rate of increase, though low, when added to legal immigration added approximately 2 million persons per year even during the early 1980s.

Canadian Population Growth

The *Canadian demographic experience* has been generally similar to that of the United States insofar as birth and mortality rates are concerned; however, a substantial difference in immigration has led to a Canadian population vastly smaller in size. A large proportion of Canada is not favorable for settlement and has attracted far fewer immigrants than the United States.

From 1867 (population about 3.5 million) to 1900, Canada grew mainly by *natural increase*. But even its natural increase was limited somewhat because of the low fertility rate, a response to the migration of youths to the United States. Net migration was negative during most of this early period. After 1900, however, there was a large influx of immigrants from Europe, raising fertility rates and slowing the decline of birth rates. Like the United States, Canada experienced a low birth rate during the depression and a sharp rise after World War II, despite urbanization and industrialization. The Canadian baby boom was also followed by a birth rate decline in the late 1960s and early 1970s.

Immigration to Canada has exceeded 9 million, but emigration is estimated to have been over 6 million. Immigrants to Canada have often returned to their country of origin or later continued on to the United States. Positive net migration has aided Canadian population growth in only two periods:

the first three decades of this century and during the post–World War II years. The decade of the 1950s was the decade of greatest Canadian population growth as a result of high birth rates applied to a larger population base and positive net migration. Net migration currently accounts for 34% of Canadian population growth.

Low Population Growth and the Future

Despite the fact that both the United States and Canadian populations are still increasing, the birth rates of these two countries have decreased to a historic low. The dramatic decrease in the birth rate is similar to what is being experienced in approximately thirty other highly developed nations. The developed countries constitute an atypical minority; most other nations are experiencing high birth rates and lowering death rates. Therefore, while some industrialized nations are rapidly approaching *zero population growth* (births balance deaths), other less developed countries are experiencing quite the opposite. Furthermore, some are not receptive to the view that their population growth is too rapid.

Zero population growth for the United States, if it is achieved, probably will not occur for some years. The United States, like any other country rapidly reducing its birth rates, will likely have a total population fluctuating around some base level or growing at a modest rate. The present youthful population, even with lowered fertility rates, has the potential to create another small "baby boom" simply because of the large number of people involved.

Figure 5–6 shows a contemporary population pyramid and one as it might appear in A.D. 2000. Obviously, the reduced birth rates have major implications for the age structure of the population. The present numerous youthful group will age and be replaced by a smaller youth group. The *changing age structure* has numerous economic and social implications, but people disagree on whether or not these changes will prove troublesome. Will the proportionately smaller youthful population mean possible labor shortages, especially for particular industries? Will some industries experience dramatic market declines? Will educational institutions experience an oversupply of facilities and personnel? Will the facilities required and the cost of caring for a disproportionately large aged population place a disturbingly high tax and social security

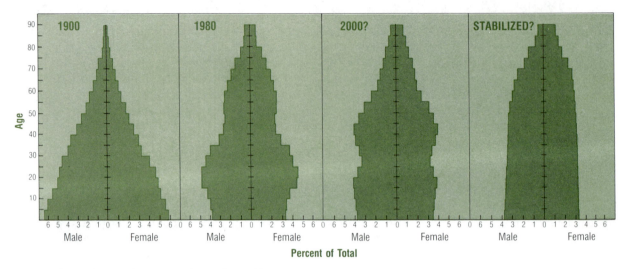

FIGURE 5–6 United States Population Pyramids.
Population pyramids for the United States since 1900 exemplify the changing age and sex structure of a country that has gone through the demographic transition. Predictions regarding future demographic conditions must be treated with caution, however.

SOURCE: Adapted from Charles F. Westoff, "The Population of the Developed Countries." *Scientific American*, 231 (1974), and U.S. Bureau of the Census, *Population Profile of the United States: 1980* (Washington, D.C.: Government Printing Office, 1981).

burden on the economically active population? Or will this new age structure allow an immense improvement and solid attack on some of the economic and social problems currently facing the United States?

Population Distribution

Most of the United States population is located east of the Mississippi River (Figure 5–7). The greatest density is especially concentrated in the *northeastern quadrant* bounded by the Mississippi and Ohio rivers, the Atlantic Ocean, and the Great Lakes. Population densities are somewhat lower in the South, except in the growth areas like the Piedmont. Most of the west coast population is concentrated in lowland areas, such as the Los Angeles Basin, the Great Valley of California, the valleys of the Coastal Ranges in the vicinity of San Francisco–Oakland, and the Willamette Valley and Puget Sound Lowland. The remainder of the western United States is sparsely populated, particularly west of the 100th meridian. Exceptions are the higher population densities found at the oasis-type locations exemplified by Phoenix, Arizona, and Salt Lake City, Utah (Table 5–3).

Most of the *Canadian population* is within 200 miles (322 kilometers) of the United States border.

The Prairie Provinces provide the only significant exception. If it were not for the sparse population on the barren Canadian Shield immediately north of Lake Superior, the distribution might be described as a long east-west ribbon, north of which lies most of the vast Canadian space, only sparsely inhabited. More than 60% of Canada's population is located on the Ontario Peninsula in the St. Lawrence Valley of southern Quebec. The Maritime Provinces contain approximately 10% of the population and the Prairie Provinces about 17%. The already populated urban-industrial areas of Quebec and Ontario have been experiencing positive net migration, in line with the continued concentration of population that is common in industrial societies (Table 5–4).

Population Redistribution

Population redistribution began even before the initial settling of the more habitable parts of the United States and Canada was complete. The shift from an agrarian to an industrial society began early in the nineteenth century. Industrial growth, particularly during the latter half of the nineteenth century, was the basis for major urban growth, which has continued during the twentieth century and has been spurred more by the expansion of tertiary activities than by industrial growth. Urban-

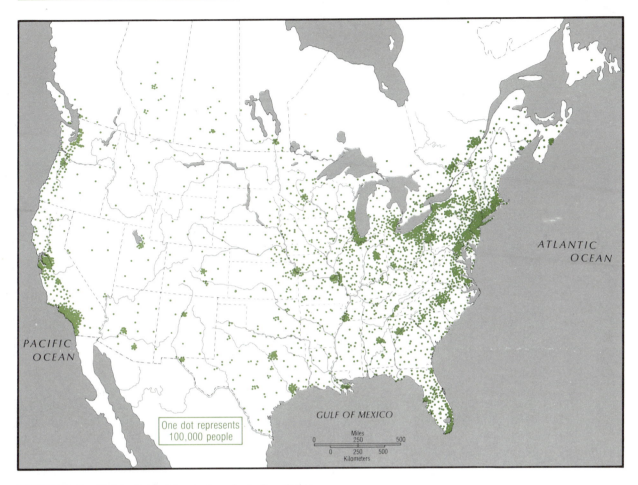

FIGURE 5–7 Distribution of Anglo-America's Population.
The unevenness of population in the United States and Canada reflects numerous
influences, including natural environment, early settlement areas, levels or urban
and industrial growth, and ongoing redistribution (mobility).

ization was the pervasive settlement process for more than a century.

The northeastern quarter of the United States led the country in *urban growth* until recent years. More than 50% of the northern population was urban in 1900. Now urban population exceeds 80% in some northern states. Even those people classified as rural are such only in residence, since most are urban workers (Table 5–5). The west coast and the Southwest have also become highly urbanized. The South and the Great Plains (the Canadian segment included) urbanized more slowly, as one would expect of areas with an agrarian orientation. In recent decades, however, urban growth in the South has been substantial as the region shifted from an agrarian to an industrial economy.

The United States and Canadian populations continued to concentrate in urban areas after World War II. The most rapid population growth occurred in the larger SMSAs (Standard Metropolitan Statistical Areas).

Another significant movement, that of population leaving the larger central cities and moving to the *suburbs,* adds even more to the land requirements of urban areas. Metropolitan area growth since World War II has been a response to population growth and population shifts within metropolitan areas.

During the 1970s two reversals of long-standing population growth patterns became evident. The basis for the reversals may have been established even earlier, but the most profound impacts are yet to occur.

First, evidence increasingly suggests that the post–World War II era of rapid urban growth may

89

TABLE 5-3   United States Population by Region and Race, 1980 (in thousands)

United States, Regions and Divisions	Total	White	Black	American Indian, Eskimo, and Aleut	Asian and Pacific Islander	Other
United States	226,504	188,340	26,488	1,418	3,500	6,756
Regions and divisions						
Northeast	49,136	42,328	4,848	78	559	1,321
New England	12,348	11,585	474	21	81	185
Middle Atlantic	36,788	30,742	4,374	56	478	1,136
North Central	58,853	52,183	5,336	248	389	695
East North Central	41,669	36,138	4,547	105	302	574
West North Central	17,184	16,044	788	142	86	121
South	75,349	58,944	14,041	372	469	1,521
South Atlantic	36,943	28,647	7,647	118	260	268
East South Central	14,662	11,699	2,868	22	41	31
West South Central	23,743	18,596	3,525	231	168	1,221
West	43,165	34,884	2,261	719	2,081	3,218
Mountain	11,368	9,958	268	363	98	679
Pacific	31,796	24,926	1,992	356	1,982	2,538

SOURCE: U.S. Bureau of the Census, *Race of the Population by States: 1980* (Washington, D.C.: Government Printing Office, 1981).

be coming to an end. Many central cities and even some metropolitan areas are losing population. Nonmetropolitan regions are now experiencing population growth rates which, while modest, nevertheless exceed those of metropolitan areas (Table 5–5). Also, the smaller metropolitan areas (less than 100,000) are growing at rates higher than the large centers. This reversal of long-standing urban growth patterns results from: (1) a dramatic decrease in natural rates of population growth; and (2) a reversal of net migration patterns which once favored the larger metropolitan areas. Particularly important is the *reversal of the net migration* pattern—the reasons for which are considerably debated but would seem to include:

1 Migration for purposes of retirement;
2 The unfavorable image of large urban centers;
3 A cultural predisposition for nonmetropolitan living by some people; and
4 A shift of new economic activity to small metropolitan and nonmetropolitan areas.

It may be that a significant *new settling pattern* is emerging and evident in the more rapid growth of the smaller metropolitan areas and nonmetropolitan areas. It almost would appear to be a discarding of the larger cities—but not of urbanization as a process. Rather, Americans may simply be building new urban centers at new places.

The second major reversal in population growth patterns involves *interregional migrations* (Table 5–6). The South, the Great Plains, and much of the interior West long experienced a net migration loss, particularly from rural areas. Figure 5–8 shows the major rural areas that had lost popula-

TABLE 5-4   Population and Percentage Change of Population by Canadian Province

Province or Territory, Canada	Population (1976)	Percentage Change (1971–76)
Newfoundland	557,725	6.8
Prince Edward Island	118,229	5.9
Nova Scotia	828,571	5.0
New Brunswick	677,250	6.7
Quebec	6,234,445	3.4
Ontario	8,264,465	7.3
Manitoba	1,021,506	3.4
Saskatchewan	921,323	−0.5
Alberta	1,838,037	12.9
British Columbia	2,466,608	12.9
Yukon	21,836	18.8
Northwest Territories	42,609	22.4
Canada	22,992,604	6.6

SOURCE: *Canada Yearbook* (Ottawa: Dominion Bureau of Statistics, 1980–81).

TABLE 5-5 United States Population by Residence and Race, 1980

	All Population			White			Black			All Other		
	Thousands	Percent	Change 1970–80 (%)	Thousands	Percent	Change 1970–80 (%)	Thousands	Percent	Change 1970–80 (%)	Thousands	Percent	Change 1970–80 (%)
SMSAs[a]	169.4	74.8	10.2	138.0	73.3	3.3	21.5	81.1	20.2	9.9	84.7	354.0
Central cities	67.9	30.0	0.1	47.0	25.0	−11.5	15.3	57.8	13.0	5.6	48.1	382.5
Outside central cities	101.5	44.8	18.2	91.0	48.3	13.1	6.2	23.3	42.7	4.3	36.6	321.4
Nonmetropolitan areas	57.1	25.2	15.1	50.3	26.7	13.9	5.0	18.9	6.5	1.8	15.3	153.7
Total	226.5	100.0	11.4	188.3	100.0	6.0	26.5	100.0	17.3	11.7	100.0	305.0

SOURCE: U.S. Bureau of the Census, *Statistical Abstract of the United States* (Washington, D.C.: Government Printing Office, 1981).
[a]Standard Metropolitan Statistical Areas.

TABLE 5–6 United States Interregional Migration (in thousands)

	Northeast	North Central	South	West
1965–70				
In-migration	1,273	2,024	3,124	2,309
Out-migration	1,988	2,661	2,486	1,613
Net migration	−715	−637	+638	+696
1975–80				
In-migration	1,106	1,993	4,204	2,838
Out-migration	2,592	3,166	2,440	1,945
Net migration	−1,486	−1,173	+1,764	+893

SOURCE: U.S. Bureau of the Census, *Geographical Mobility: March 1975 to March 1980.* (Washington, D.C.: Government Printing Office, 1981).

tion as of 1970. Areas that have experienced population decline include much of the Coastal Plain from Virginia to Texas, the middle and lower Mississippi Valley, the Great Plains from central Texas to the Canadian border, and portions of the mountain West. It is difficult to generalize about the Appalachian region. The significant areas of population loss during the 1960s include the plateau and hill portions of northern and eastern Tennessee, eastern Kentucky, West Virginia, Pennsylvania, and the hill lands of southeastern Ohio; to the east only a narrow string of Blue Ridge counties lost population. Obviously, much of the area gained in population, particularly the Piedmont and the Ridge and Valley areas of eastern Tennessee, northwest Georgia, and northeast Alabama. The Appalachian region is not uniform in its demographic experience.

While the west coast of the United States has experienced great population growth during the twentieth century, many areas in the West have been sparsely populated and experienced population loss and decline (mining centers). The traditional agrarian areas have been regions of population decline. Some attained their peaks in population by 1900. Meanwhile, urban areas gained in population. The areas peripheral to the Great Lakes in Canada and the United States (Snow Belt) constitute the older urban-industrial regions that have long experienced population growth from natural increase, immigration, and rural-to-urban migration. The Piedmont, the southern Ridge and Valley area, the Gulf Coast intermittently from southern Florida to Texas, and the Far West and Southwest are urban growth areas (Sun Belt). Population then is moving in several directions: to expanding urban regions outside the Northeast and North Central regions and to nearby rural areas (Figure 5–9). Areas remaining dispropor-

tionately rural such as parts of the Coastal Plain, the Great Plains, the Midwest, and Appalachia, which once were major areas losing population, now contain rural areas where population decline has reversed itself. Net migration now favors population growth in both the South and West. These continuing changes in American settling suggest that Anglo-America has not yet evolved a mature settlement landscape.

TRANSPORTATION AND DEVELOPMENT

The nineteenth-century development experience of Anglo-America is rooted in the Industrial Revolution. The commercialization of agriculture was stimulated by market expansion coinciding with that revolution. Raw materials needed for growing industry required improved and expanding transportation. With these needs came numerous *technological inventions* and improvements that were both contributors to and products of the Industrial Revolution. For example, an improved ginning technology was necessary to reduce the labor required for the removal of seeds from cotton. When improved ginning was achieved (Eli Whitney, 1793), much of the lower South became suitable for settling and cotton culture. Countless other technological changes aided in resource use and changed the value of particular places. Perhaps none, however, have been as important to resource use and life-style as those changes occurring in nineteenth- and twentieth-century transportation.

The development of a *transport system* with the capability of moving great quantities of materials at relatively low cost has been essential to the utilization of resources for industrialization in both the

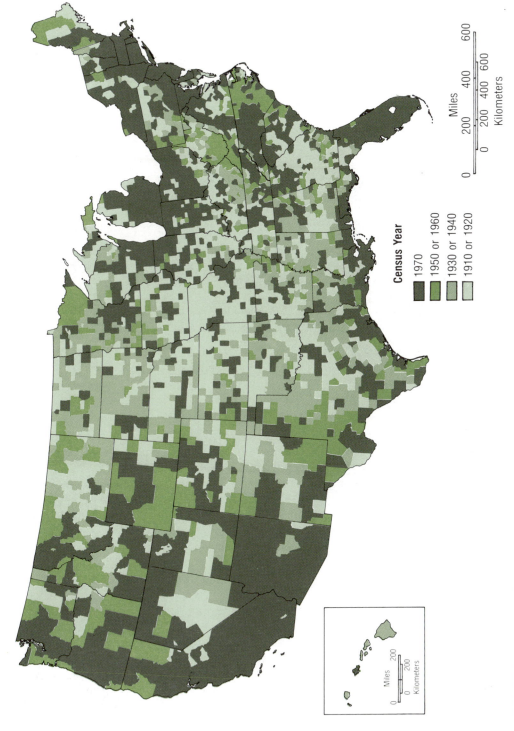

Census Year

- 1970
- 1950 or 1960
- 1930 or 1940
- 1910 or 1920

FIGURE 5–8 United States Population in 1970 as a Percentage of Maximum Population, by Counties.

Regions in which population is now 100% of the highest-ever population can be considered growth areas. Note in particular the Northeast, the Great Lakes area, portions of Appalachia, and the Gulf and West coasts. Other regions have smaller populations than they once had, for example, portions of the Coastal Plain and of Appalachia and much of the interior West.

SOURCE: U.S. Bureau of the Census, 1970.

93

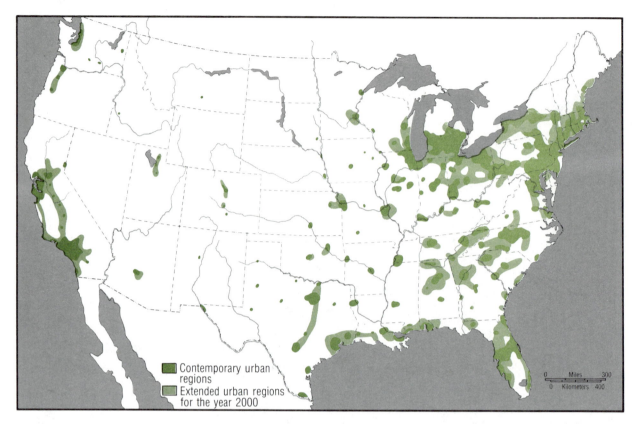

FIGURE 5–9 Urban Regions of the United States, 1980 and 2000.
The continued growth of metropolitan regions suggests further evolution of megalo-
politan areas, not only in the Northeast but also along the margins of the Great
Lakes and the West Coast.

United States and Canada. Frequent changes in transport technology have lowered the cost and increased the feasibility of moving goods and also frequently altered the significance of various places.

Overland wagon transport and water transportation by raft, barge, and sailing vessel were the major forms of transport available until the 1830s. The great cost advantage of *water transport* had a major effect on early settlement, both for those people inclined toward agriculture and for those with a commercial bent.

By the early nineteenth century water transport had increased even more in importance, particularly inland. With the development of steam power, the utility of the Ohio-Mississippi and other river systems appreciably changed. A *canal-building era* was ushered in by the Erie Canal (1825). The cost and time required for shipping from Buffalo to New York via the Erie Canal and Hudson Valley were reduced from $100 per ton and twenty days to $5 per ton and five days. The

Erie Canal allowed the marketing in the East of efficiently produced agricultural commodities from beyond the Appalachians. Furthermore, later connecting canals on the Great Lakes complemented the Erie Canal and facilitated the movement of grains, forest products, and minerals. The general effect of improved transportation was to extend commercial activities far to the interior, drawing upon new and rich resources. The value of New England as an agricultural resource area, which was not rich to begin with, declined. Production became concentrated on some specialized commodities. The vast new hinterland centered on New York, giving that city a distinct lead over other eastern cities, particularly Boston.

The *railroad era* began even before the short canal boom ceased. By the mid-nineteenth century, new focal points were identified. The railroad, though occasionally competitive with canals, generally complemented water transportation and greatly increased the significance of the Great Lakes as an interior waterway. Rail networks

Improved forms of transportation have aided urban growth and industrialization. Toronto, on the edge of Lake Ontario, has access to both land and water transport systems. (United Nations/Toronto Transit System)

were focused on selected coastal ports; New York grew much faster than Philadelphia or Boston. The networks also converged on "interior ports," such as Chicago and St. Louis.

The rails connected farm, forest, and mineral resource areas with ports and cities. Continued improvements in *rail technology* made long-haul transport feasible (and concentration of raw materials at a few selected points). The rails aided in the opening of agricultural land in the West and the tapping of copper, lead, and zinc in the Far West. Grains could be moved to and from ports. Coal was hauled to Great Lakes ports, and iron ore from Great Lakes ports to inland cities (Pittsburgh). By the late 1860s a major turning point had been reached in urban-industrial growth that would not have been possible without improved transportation. Transport technology provided special focus on selected cities, most often located by water (Figure 5–10). Massive urban and industrial growth proceeded for the next half century, and major or national routeways were established among these selected cities. A common result of such transport technology was selective growth. Some port facilities declined in importance, while others (fewer in number) became centers of attention.

The railroad has been partly replaced by the *truck and plane*. Railroads have retained importance as long-haul freight carriers but have lost a large part of their passenger and short-haul freight carrier functions. Both automotive and air transport

have captured passenger traffic, and the auto-highway system is extremely important for freight movement. The national routeways and flows established during the railroad era, however, were essentially maintained during the automotive and air era, and so the growth of established large centers continued.

The settled portions of the United States and Canada contain intensive networks of paved roads. In the United States these range from paved county roads to the limited-access Interstate Highway System. The present highway system will have a "selecting" effect in localized areas but will not reroute national flows. Rather, the system will aid in the integration of existing economic regions.

In recent decades *international waterways* have received renewed attention. The completion of improvements on the St. Lawrence Seaway in 1959 extended the Great Lakes system from the west end of Lake Superior (Duluth-Superior) and southern end of Lake Michigan (Chicago) to the Atlantic via the St. Lawrence and the New York State Barge Canal (formerly the Erie Canal). Thus, not only are the Great Lakes important as an internal waterway, but they now also function as an international water route connecting cities such as Cleveland, Detroit, and Chicago with Europe. Canals also connect Chicago, the nation's most important inland urban-industrial concentration, with the Mississippi, including its numerous tributaries (Figure 5–10). The waterways are used for mas-

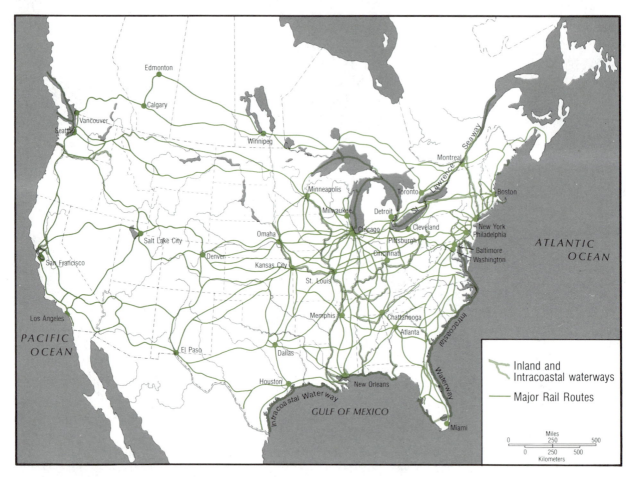

FIGURE 5–10 Major Rail Routes and Inland Waterways of Anglo-America. Of great advantage to the United States has been a system of natural waterways, coastal and interior, that when improved and linked by rails contributed much to the economic integration of a large resource-rich area.

sive movements of grain, coal, iron ore, and petrochemicals, that is, for bulk commodities. Waterways in effect surround the highly urbanized and industrialized northeastern United States and penetrate more than 2000 miles (3208 kilometers) along the southern border of Canada, providing major transport advantages to the continental portions of each country.

6

Anglo-America: Economic Growth and Transformation

Commercial economic activity began in Anglo-America shortly after European settlement. Wherever transportation was suitable, usually meaning accessibility by water, *commercialism* rapidly became the norm. Agricultural commodities, lumber, furs, and fish were produced and gathered for exchange. Primary production and tertiary activities (trade) were important long before settling was complete or manufacturing became significant. Of all activities, agriculture became the mainstay and remained so for more than two centuries.

Shortly after independence the growth of population, domestic markets, and transportation stimulated manufacturing in the United States. Much of the industrial expansion of the middle and late nineteenth century was based on domestic needs and potential, whereas earlier commerce had been more externally oriented.

A complete change in *economic emphasis* occurred during the century after 1850 (Table 6–1). Few Anglo-Americans today are farmers—only 3.4% in the United States are employed in agriculture. About 22% are employed in manufacturing, and the remainder, instead of being producers of commodities, are engaged in the distribution of goods and provision of services. Since the turn of the century and particularly after World War II, the tertiary sector has rapidly become the dominant source of employment and the basis of continued urbanization. The employment changes illustrate the transformation from agrarian to highly developed urban-industrial societies.

ANGLO-AMERICAN AGRICULTURE

Availability of Agricultural Land

Anglo-American agriculture evolved in a *large and rich environment*. Of the total land area of the United States, approximately one-fifth is classified as cropland, not all of which is cultivated in any given year. Land in crops actually has decreased by 45 million acres (18 million hectares) in recent dec-

TABLE 6–1 Employment by Economic Sectors in Anglo-America (percentage)

Economic Sector	United States (1980)		Canada (1978)	
Primary		4.5		7.3
Agriculture	3.4		4.7	
Other primary	1.1		2.6	
Secondary		28.4		25.9
Manufacturing	22.2		19.6	
Construction	6.2		6.3	
Tertiary		67.1		66.8
Total		100.0		100.0

SOURCES: U.S. Bureau of the Census, *Statistical Abstract of the United States* (Washington, D.C.: Government Printing Office, 1981); and *Canada Yearbook* (Ottawa: Dominion Bureau of Statistics, 1980–81).

ades. Pasture and range lands, also part of the food-producing resource base, make up about 29% of the land area. Nearly 5 acres (2 hectares) per person exist for agricultural production and industrial raw materials. Obviously the land resources useful for agriculture are substantial relative to population.

Canada has only 4.4% of its land in cropland and pasture. The Canadian population, however, is substantially smaller than that of the United States, and so the ratio of people to agricultural land is much the same (4.58 acres, or 1.85 hectares, per person). The export of primary commodities, which include farm products (wheat in particular), is important to the Canadian economy despite the small proportion of total land area used for agriculture.

The figures cited above are based upon land actually used for production in recent years and do not include land that might be added by clearing forests or draining wetlands. Both countries have in fact experienced a decline of agricultural land use in marginal areas that are not essential for food supply, areas that were settled during earlier expansion periods when agriculture was the dominant economic activity. Much additional land not now used for agriculture could be converted into cropland. For example, in the lower Mississippi Valley and the Southeast, 50 million acres (20.2 million hectares) are considered convertible from woodland and pasture to cropland.

Agricultural Regions

The largest expanse of highly productive land is that part of the Interior Plains known as the *Corn Belt* (Figure 6–1). This region, extending from western Ohio to Nebraska, is a large expanse of moderately rolling to flat plains, highly suited to mechanized agriculture. The outstanding climatic conditions are the reliable rainfall and a moderately long growing season. The gray-brown podzolic and chernozemic soils are among the best of Anglo-America. This favorable combination of features provides a resource base that can be used for the production of a great variety of grains, forages, and vegetables. Individual farmers, however, tend to specialize. The single most important crop throughout the region is corn, sometimes produced as a *feed grain* and marketed commercially. In other instances it is fed to cattle and hogs to fatten them on the farms where the grain is produced. Farms specializing in fattening of livestock are referred to as *feedlot farms.* Farms producing corn for commercial markets commonly include other crops such as soybeans or wheat. Other production systems include dairying and vegetables.

The naturally favorable position of the Corn Belt region is enhanced by the location of transportation routes and urban-industrial districts. Various parts of the region lie between or adjacent to the inland waterways of the United States and Canada and have been laced with a dense network of railroads. The *transport system* facilitates easy marketing of efficiently produced commodities in nearby eastern markets and worldwide.

Surrounding this superlative agricultural region are several others that are not quite as productive. The *Northeast,* where Anglo-American agriculture had its beginnings, does not provide such large expanses of favorable terrain and fertile soils. Rather, productive land—the *Connecticut Valley,* the *Hudson Valley,* and limestone soils of *southeast Pennsylvania* (Lancaster County)—is the exception. Topography is commonly hilly to mountainous, and the soils are thin and frequently infertile. As better interior lands became accessible, northeastern farmers found it necessary to

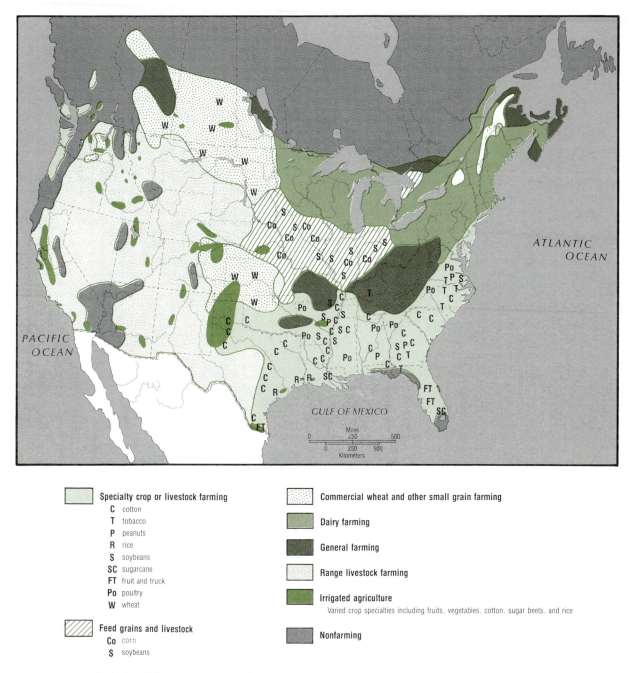

FIGURE 6–1 Agricultural Regions of Anglo-America.
A diversity of natural environments has contributed to great variety in types of agricultural specialties possible in the United States and Canada.

specialize in the production of items needed in nearby cities to make good use of their one great advantage—proximity to growing northeastern markets. Thus evolved the vegetable and poultry production of the *Delmarva Peninsula* and New Jersey and the *dairying* of New England, New York, and Pennsylvania. These commodities are of high value but require great capital inputs, have high transport costs, and sometimes are perishable. Closeness to *urban centers* can also mean disadvantages such as high taxes, pressure to relocate, or complaints from urban dwellers concerning odors and noise. Although agriculture may be profitable in such a setting, it is also commonly unstable; farms and farmland frequently are lost to rapid urban growth. The common presumption, how-

99

ever, that urban growth is the only reason for loss of farmland over the past several decades is incorrect. Far more land has been idled or shifted to less intensive use (grazing or forestry) because of lack of profitability.

The *southern United States,* from southern Maryland and Virginia to Texas, is a large area with *variable resources* for agriculture. Large parts of hill land in the South, the Appalachians, and the Ozarks have slopes too steep for sustained use as cropland. Where farming does occur, it is often quasi-subsistence. Other areas, such as the *Piedmont,* have been used for agriculture during favorable economic periods, such as when cotton prices were high, but the land has deteriorated with continued row-crop production. Many *coastal areas* of swamp and sandy "pine barrens" also are undesirable for intensive agriculture. Within the southern region, however, are subregions that, because of topography and soils and the subtropical environments, provide some exceptionally good agricultural resource areas. Included are the alluvial *Mississippi Valley,* the *Nashville* and *Bluegrass basins,* the *black-soil belts* of Alabama and Texas, and the limestone valleys of Appalachia.

The *South* is a diverse farming region based upon numerous specialties in different areas. Cotton, which was so important across much of the lower South, is now produced mainly in the western sector and the far western states of California and Arizona. In the east, the Mississippi Valley (Tennessee, Arkansas, Mississippi, and Louisiana), the coastal plain of Texas, and the lower Rio Grande Valley remain areas of specialized cotton production. In many other parts of the South, cotton has disappeared or has become of secondary importance. The diversity of southern agriculture is illustrated in Figure 6–1.

The United States's *Great Plains* and the *Prairie Provinces* (Alberta, Saskatchewan, and Manitoba) of Canada have an environment in which farmers face recurrent drought, and even "normal" rainfall may mean water deficiencies. Characteristic of wetter parts of the plains states and provinces is wheat farming, sometimes diversified to include livestock raising (usually cattle). The drier margins are given over almost exclusively to cattle raising. Areas such as the *Sand Hills* of Nebraska, which are stabilized dunes with poor moisture retention, and the rolling *Flint Hills* of Kansas are devoted to specialized ranching. Only where irrigation water is available is the pattern of dry farming and ranching broken. Irrigation allows the *high plains* of Texas to

function as one of the major cotton-producing regions of the United States. Eastern Colorado produces irrigated sugar beets and corn for nearby livestock-finishing farms similar to those of Illinois and Iowa. Among the plains states Kansas and Nebraska in particular have shown major recent increases in irrigation. The plains are *food surplus areas,* and much of the product is transported elsewhere to supply domestic and international demands. The products of the Canadian Great Plains provide the most competition for United States farm output in the international marketplace.

A *dairying and general-farming region* extends from Minnesota across the Great Lakes states and Canada's Ontario peninsula to New England and Nova Scotia. Much of this dairy region has a decided advantage because of its location adjacent to one of the larger urban regions of the world. The humid and cool summer of this region is favorable for forage crops and feed grains, but to the north the environment becomes harsh, the growing season short, and the soils thin and infertile. The Canadian section of the region is narrow in latitudinal extent and represents a transition to the nonagricultural northern lands.

While the productivity of the plains states is hampered by dry conditions and distance from eastern markets, areas farther west are even more handicapped. Most of *the West* is nonagricultural because of rugged terrain, aridity, or inaccessibility and is used extensively as grazing land. The exceptions to this generalization are extremely important. Aside from the Pacific Northwest (northern California, western Oregon, and Washington), agriculture, when carried on at all, is intensive at *oasis locations.* The valleys east of the Cascades, the Snake River plains, the Salt Lake oasis, and California's Imperial Valley, Los Angeles Basin, and Great Valley are some of the areas with highly productive agriculture. Farmers in these areas make use of fertile soils and level terrain and take advantage of national transport systems that allow marketing in the East, but they are utterly dependent upon irrigation. These conditions have made California the most diversified and most productive agricultural area in either the United States or Canada. The *Pacific Northwest* has a humid climate and is therefore less dependent upon irrigation. The *Puget Sound Lowland* and *Willamette Valley* contain a productive agricultural industry that derives farm income from such varied sources as dairy products, other livestock products, grains, orchard crops, and berries.

Commercialized Agriculture

The *commercial emphasis* in Anglo-American agriculture began early. Tobacco was the first successful venture, but rice, indigo, and cotton followed as colonies were established southward along the Atlantic coast. The frontier farmer was generally subsistence oriented, although he might also have produced items such as livestock that could be driven to markets. Later the urbanization and industrialization of Europe stimulated expansion of commercial agriculture in various parts of the world, including Anglo-America, during the nineteenth century. Railroad expansion and improvement allowed the Anglo-American settler to participate in an expanding commercial system.

As a result of transportation improvement and expanding settlement, land with greatly varying capabilities became accessible for many kinds of production. The *principle of comparative advantage,* therefore, is allowed to operate, leading to widespread regional specialization. The principle of comparative advantage means that some locations have a definite advantage over other areas in the production of one or several items. If adequate transportation exists to allow commercial exchange, *regional specialization* will result. Farmers having an advantage in several forms of production will choose the form providing the greatest return. They may concentrate on one or two crops or some combination of crops and livestock. Areas with no advantage then may concentrate on the production of items ignored by the others. Areas, therefore, specialize in limited types of production, either because of advantage or by default, and through exchange contribute their products to the larger economic system. The high degree of regional specialization is possible because an efficient transport system allows the use of resources best suited to particular kinds of production.

Continued Adjustment of Agriculture

As one of the world's major agricultural regions, Anglo-America is young; one would expect *change* in an area that undergoes rapid settlement for agricultural purposes. For example, the "corn and livestock" region of the Midwest uses a system of grain production and cattle fattening that originated in the East where lean cattle driven from frontier areas to market areas were fattened before slaughter. Today more and more cattle are fattened in the Great Plains, and over half the cattle fattened in the Midwest are raised in the Midwest. The system spread westward along the Ohio Valley and into the Midwest, which is an excellent area for the production of feed grains. The original cotton-producing areas of the southeast experienced a decline in cotton acreage because of government production controls and acreage reductions, soil erosion, and the inability of farmers to adopt better production methods. Cotton production is now concentrated on the better lands of the Mississippi Valley and westward.

Initial settlement patterns often were not permanent. Farmers settled Appalachian hilly land where the steep slopes were easily eroded. Oth-

Irrigated cotton farming on the high plains of Texas is highly mechanized and productive. Such a farming system minimizes the use of labor and maximizes capital inputs in the form of fossil fuels, fertilizers, and machinery. (USDA photo by Fred White)

ers occupied steppe margins (the western edge of the Great Plains) where drought meant disaster. The early nineteenth-century retreat of agriculture in the New England uplands is an event that has been repeated in many areas. Stated otherwise, we have withdrawn some of our agricultural efforts to concentrate on other, better lands. Land clearing, for example, for cropland expansion continues in the Mississippi delta.

During the nineteenth century a substantial immigrant and native-born population was available as a labor force. Later, however, urban-industrial growth combined with the availability of land for individuals made *farm labor scarce*. Labor-saving devices, such as mechanical reapers (nineteenth century), were significant technological advances even before the power revolution in agriculture. The growing industrial capacity of Anglo-America created the supply of implements.

Mechanization of agriculture means more than the substitution of capital for labor, which in itself often requires difficult adjustments. Mechanization frequently requires operational reorganization, larger capital expenditures, and greater scale of production. Mechanization or other technological improvements usually help only those who have the wherewithal (land or money) to make the required adjustments. From time to time many farmers have found it impossible to make the required changes and have not survived as farmers (Figure 6–2). For those who can find alternative employment, cessation of farming is not necessarily bad. A shift of inefficient and low-income farmers into other activities can be beneficial for the larger economy. The problem, however, is that not all farmers have been able to find satisfactory alternatives. Large numbers of poor tenant farmers and farm laborers have migrated to cities, ill-prepared for urban life, jobs, stresses, and the urban discrimination that may exist for both blacks and poor whites.

Agricultural Productivity

Anglo-American agriculture is extremely *productive*. The high degree of regional specialization involves the production of many crops using the best resources for a particular crop. By world standards, yields per land unit for many commodities rank high (see Figure 4–3). Furthermore, productivity has increased greatly during the past three decades and includes major increases per unit of land and per person (Table 6–2).

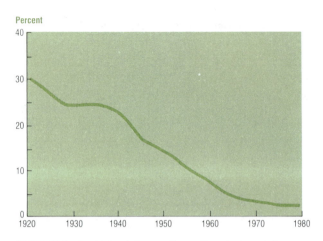

FIGURE 6–2 United States Farm Population as Percent of Total Population.
The percentage of the United States population living on farms has continually decreased. This decrease results both from increased mechanization and other advances in farming methods and from the growth of urban centers with their diversity of economic opportunities.
SOURCE: U.S. Bureau of the Census, *Statistical Abstract of the United States,* 1981.

The *increased production* has several causes, one of which is mechanization. Although mechanization may aid in increasing yields per unit of land, this is not always true. A more important result is the increase in output per person who, with the aid of machinery, can do the work of many. More significant causes of productivity increase are improved seed and plant varieties, more effective fertilizers, new pesticides, and chemicals for disease and weed control. Indeed, even now rapid progress is being made in the development of "hybrid wheats" that will probably mean a great increase in yield and production similar to that which occurred with corn over a number of decades.

Paradox: Productive Agriculture and Poor Farmers

As is frequently true in a time of rapid change and innovation, many *adjustments are required* of farmers who wish to participate in improved production methods. Many farm laborers no longer needed because of mechanization or farm reorganization have migrated to cities, particularly from southern areas where the landless farm population was large. Other farmers remain, however, who are marginal producers with low incomes. Such farmers contribute little to the national agricultural

TABLE 6–2 Agricultural Productivity in the United States

Product	1935–39 Average	1950–59 Average	1975–79 Average
Corn for grain			
Bushels per acre	26.1	39.4	95.2
Hours required per 100 bushels	108.0	34.0	4.0
Cotton			
Pounds per acre	226.0	296.0	485.0
Hours required per 500-pound bale	209.0	107.0	8.0
Wheat			
Bushels per acre	13.2	17.3	33.4
Man-hours per 100 bushels	67.0	27.0	9.0

SOURCE: U.S. Bureau of the Census, *Statistical Abstract of the United States* (Washington, D.C.: Government Printing Office, 1974, 1981).

economy, and in fact most agricultural output occurs on a small number of farms that are generally large and capital intensive. It cannot be presumed that the marginal farmers will necessarily contribute more if food demand increases. Their shortage of capital and land may prevent their increased participation even if needed.

ANGLO-AMERICAN MANUFACTURING

Manufacturing activities are basic to the modern Anglo-American economies. Nearly 20 million people (22% of the labor force) are now employed in manufacturing in the United States. The more than

The Salinas Valley, long famous for its lettuce production, exemplifies the fertility and productivity of California agriculture. The flat valley bottoms with alluvial soils are rich agricultural lands within the Coastal Ranges. (USDA photo)

2 million so employed in Canada represent about 20% of the Canadian labor force. Those employed in manufacturing and the great numbers required in the distribution of goods and needed services are the basis for massive Anglo-American urbanization.

The proportion of the total labor force employed in manufacturing has decreased somewhat as relatively faster growth has occurred in tertiary activities, a characteristic of developed societies. Manufacturing, however, remains basic to the stability of the American economy. Approximately 23% of the income of the United States is directly derived from manufacturing, but even more important is the great number of employment opportunities provided in the tertiary realm as goods are distributed to consumers and services provided for manufacturers and others.

Evolution of American Manufacturing

Industry developed first in *southeastern New England* and the *Middle Atlantic* area from New York to Philadelphia and Baltimore, two of the early American settlement cores. Prior to the railroad era, access to water meant great transportation advantages. Industrial power was available as mechanical waterpower. New England had become an important agricultural and settlement area out of necessity, even though soils and slopes were basically poor for agriculture. Great wealth had been accumulated from other pursuits such as lumbering, fishing, and ocean shipping. People seeking financial backing for new industrial activity could find entrepreneurs with capital and a venturesome spirit. The surplus rural populations of New England, suffering from the economic competition of newly opened land in the interior, also facilitated industrial growth during the nineteenth century. These people readily stopped farming for work in the textile mills and leather and shoe factories that dominated New England industry by 1900.

Shipbuilding, food processing, papermaking and printing, and ironworking were other *early industries* of New England and the middle colonies. Iron-working became an important industry in eastern Pennsylvania, though the iron industry was scattered and localized in numerous communities. Prior to the steel age, iron ore and charcoal (from hardwood trees) were the basic components for ironworking. With high transport costs on bulky and heavy goods, many small iron deposits were used in conjunction with local hardwood forests. New technology was to change that during the mid-nineteenth century.

Rostow's early or "traditional" *stage of economic development* (see Chapter 4) never really existed in the Americas for many people of European descent. Settling and development were begun by a people with experience from areas in Europe where "preconditions for takeoff" existed. The United States and Canada were created by British and other European elements already in transition. The traditional society (feudal) was decaying, and the worth of the individual was emerging; democratic ideals were taking root. Capitalism was developing as those with excess wealth diverted funds into new economic enterprises: trade, transport, power development, and manufacturing. Thus the people who built Anglo-America brought the United States to the "takeoff" stage by the 1840s and Canada by the 1890s. In both cases independence was necessary to allow the pursuit of industrial growth.

The *takeoff* began an era of great industrial expansion, aided by railroads that made possible the movement of materials over great distances and gave areas and places an entirely new locational significance. The railroads themselves became a major market for steel (available from the 1850s).

By 1865 the United States was experiencing its *drive to maturity* achieved, according to Professor Rostow, by 1900. During this time the improvements in railroads (steel rails), continued immigration, and an extremely favorable population-resource balance aided growth. From 1865 to World War I, the immense growth of industry was directly related to urbanization; it was the primary basis of city growth. During this formative period (1840–1900), the northern states experienced the bulk of the industrial growth, but the southern lag is not accurately explained as simply "southern indifference" to industrialization. As stated previously, the differences between northern and southern colonies dated from the days of early settlement. Southern colonies early developed a commercial agricultural system, financially and socially rewarding for a select group.

In contrast, *New England,* partially because of a less rewarding agricultural resource base, concentrated on commercial shipping and fishing and along with the *Middle Atlantic* area developed traditions in nonagricultural as well as agricultural pursuits. When railroads became the backbone of the transportation system and the dominant market for steel, specific northern locations took on a *new locational meaning* and value. For example, New York became the primary focus of movement from the interior eastward by way of the Mohawk Valley and the Hudson River. The new and growing steel industries required great quantities of coal and iron ore. As Appalachian bituminous coalfields and Mesabi iron ore increased in importance, the locations with utility were those between the coal and iron ore: Pittsburgh, at the junction of the Monongahela and Allegheny rivers, and Cleveland, Erie, and Chicago among others, on the Great Lakes. Rails complemented the lakes by moving coal westward to meet iron ore moving eastward by water. The areas between the Appalachians and the lakes, and along the periphery of the lakes, gained importance for the assembly of materials for production. Furthermore, as the process of growth proceeded, the agricultural goods of the rich interior moved eastward to market and stimulated domestic industry in the market area.

The *South,* remote from the new national routeways for most materials, continued its agricultural production for external markets. It was to some extent like a colonial appendage. Manufacturing did exist in the antebellum and postbellum South but never achieved the rate of growth or dominance of manufacturing in the North. Furthermore, after the Civil War, when the weakness of southern industrial strength was evident, other conditions made a reversal of the traditional economy even more difficult. The South had embraced an agrarian philosophy that many of its leaders continued to advocate after the Civil War.

Thus, by the 1930s when the economic development process was well into its maturity, great *regional variations* existed. Part of the United States had evolved as a major urban-industrial region, characterized by numerous specialized urban-industrial districts interspersed with agricultural regions. The industrial coreland was an area of relatively high urbanization, high industrial output, high income, immense internal exchange, and interaction. The South was a region of low urbanization, limited industrial growth, and poverty for great numbers of whites and blacks in overpopulated rural areas.

Manufacturing Within the United States Coreland

The *Anglo-American manufacturing region* is a large area that consists of numerous urban-industrial districts within which certain types of industrial specialities can be associated (Figure 6–3). Measured by employment, southern *New England* is the most industrialized area in the United States. The district's prominence is based upon the nineteenth-century growth of the textile and leather-working industries. During this century, however, the region has suffered some severe economic problems. First, an area that is dependent upon one or a few products runs the risk of severe economic consequences should competition in the form of more efficient producers arise. Second, New England and the meaning of its location changed with industrial maturity. New England's highly unionized labor markets have become a high-tax area that relies upon imported resources for power. In addition, population increases farther west have adversely affected New England's location relative to national markets. Eastern New England is in some respects now peripheral to the core of national markets. The areas's once dominant industries (textiles and shoes) failed to grow nationally and experienced a regional shift to the South. New England has tried to emphasize high-value products such as electronic equipment, electrical machinery, firearms, machinery, and tools that can withstand high costs for transport, power, and labor. New England, however, illustrates that the industrial structure of a region may change but not always by conscious choice.

Metropolitan New York contains the largest manufacturing complex in the United States. Its location at the mouth of the Hudson, its function as the major port for the rich interior, and its own huge population have combined to generate and support nearly 11% of United States manufacturing. The tendency is toward diversified manufacturing that includes printing, publishing, machinery, food processing, metal fabricating, and petroleum refining. A heavy concentration of garment manufacturing and the lack of primary metals processing are other features characterizing Greater New York's industry. The functions of New York and Montreal are notably parallel.

There are three manufacturing districts in which *steel industries* are characteristic. Inertia, immense capital investments, and linkages with other industries assure considerable locational stability for such industries. The first district is the area extending from Baltimore to Philadelphia, Bethlehem, and Harrisburg, Pennsylvania. Massive steel-producing capacity exists near all of these cities and supports shipbuilding (along the

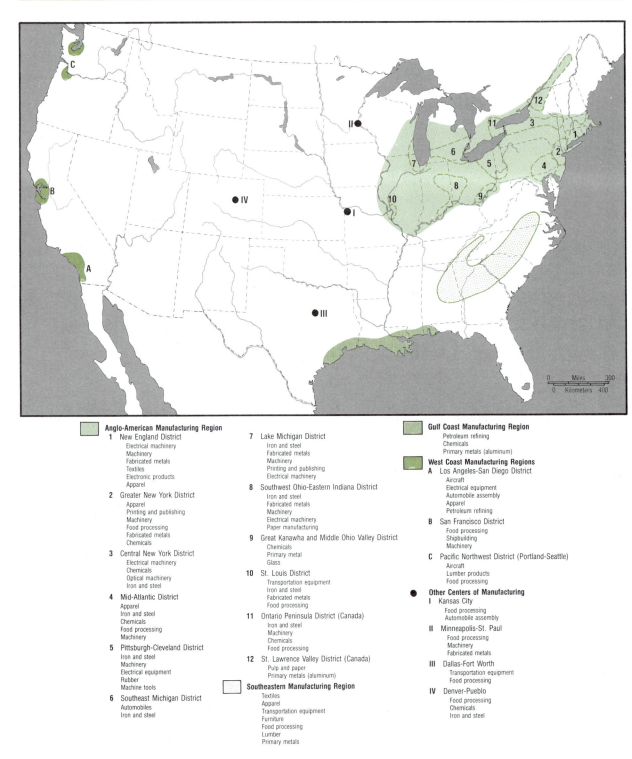

Anglo-American Manufacturing Region

1 New England District
   Electrical machinery
   Machinery
   Fabricated metals
   Textiles
   Electronic products
   Apparel

2 Greater New York District
   Apparel
   Printing and publishing
   Machinery
   Food processing
   Fabricated metals
   Chemicals

3 Central New York District
   Electrical machinery
   Chemicals
   Optical machinery
   Iron and steel

4 Mid-Atlantic District
   Apparel
   Iron and steel
   Chemicals
   Food processing
   Machinery

5 Pittsburgh-Cleveland District
   Iron and steel
   Machinery
   Electrical equipment
   Rubber
   Machine tools

6 Southeast Michigan District
   Automobiles
   Iron and steel

7 Lake Michigan District
   Iron and steel
   Fabricated metals
   Machinery
   Printing and publishing
   Electrical machinery

8 Southwest Ohio-Eastern Indiana District
   Iron and steel
   Fabricated metals
   Machinery
   Electrical machinery
   Paper manufacturing

9 Great Kanawha and Middle Ohio Valley District
   Chemicals
   Primary metal
   Glass

10 St. Louis District
   Transportation equipment
   Iron and steel
   Fabricated metals
   Food processing

11 Ontario Peninsula District (Canada)
   Iron and steel
   Machinery
   Chemicals
   Food processing

12 St. Lawrence Valley District (Canada)
   Pulp and paper
   Primary metals (aluminum)

Southeastern Manufacturing Region
   Textiles
   Apparel
   Transportation equipment
   Furniture
   Food processing
   Lumber
   Primary metals

Gulf Coast Manufacturing Region
   Petroleum refining
   Chemicals
   Primary metals (aluminum)

West Coast Manufacturing Regions
A Los Angeles-San Diego District
   Aircraft
   Electrical equipment
   Automobile assembly
   Apparel
   Petroleum refining

B San Francisco District
   Food processing
   Shipbuilding
   Machinery

C Pacific Northwest District (Portland-Seattle)
   Aircraft
   Lumber products
   Food processing

● **Other Centers of Manufacturing**
I Kansas City
   Food processing
   Automobile assembly

II Minneapolis-St. Paul
   Food processing
   Machinery
   Fabricated metals

III Dallas-Fort Worth
   Transportation equipment
   Food processing

IV Denver-Pueblo
   Food processing
   Chemicals
   Iron and steel

FIGURE 6-3 Manufacturing Regions and Districts of Anglo-America.
Industrial regions and districts show as much variety in specialties as does agriculture. The industry specialties of each region or district reflect the influences of markets, materials, labor, power, and historic forces.

Delaware River and Chesapeake Bay) and many other machinery industries. The steel industry has expanded here because of proximity to large eastern markets (other manufacturers) and accessibility to external waterways. Waterway accessibility is a factor of growing importance as dependence on foreign sources of iron has increased. In addition, those cities on waterways have become major petroleum refining centers and petrochemical manufacturers.

The second major steel district is a large triangle with points at Pittsburgh and Erie, Pennsylvania, and Toledo, Ohio. This district is the oldest steel-producing center of the United States. Initial advantages were derived from location between Appalachian coalfields and the Great Lakes, by way of which iron ore came from the Superior Ranges, especially in northern Minnesota. The locational significance of Pittsburgh and its steel-producing suburbs has changed. Now South American iron ore is moved to eastern coastal works, and Canadian ores come by way of the St. Lawrence and move even farther inland along the St. Lawrence Seaway (Great Lakes). The eastern district (Baltimore-Philadelphia) is nearer eastern markets and foreign ores; Detroit and Chicago are more easily reached by Canadian ore and closer to Midwestern markets.

The southwestern Lake Michigan area (Gary, Chicago, and Milwaukee) has a vast array of machinery manufacturing that is supplied by the massive steel industry at the southern end of Lake Michigan. The steel industries built in Chicago, Illinois, and Gary, Indiana, have benefited from a superb location. Ore moving on the Great Lakes meets coal from Illinois, Kentucky, and West Virginia. Chicago had become a major transportation center by the late nineteenth century, where rails met and complemented water transport, making the southern Lake Michigan area an excellent location to assemble materials and distribute manufactured products. The St. Lawrence Seaway has simply given renewed importance to the location, for now Chicago and other inland cities can function as midcontinent ports from which ships can sail an almost *great circle route* to Europe. A great circle route is the shortest distance between two points on the surface of the earth. Southern Michigan and adjacent areas in Indiana and Ohio are distinguished by the emphasis on automotive production, both parts and assembly. These industries are linked not only to the Detroit steel industry but also to steel manufacturers in the Chicago area and

along the shores of Lake Erie (Toledo, Lorain, and Cleveland). The automotive industry serves as a huge market for major steel-producing districts on either side.

Figure 6–3 shows several *other districts* within the Anglo-American manufacturing region. This large region of urban-industrial districts contains nearly 60% of the United States and Canadian manufacturing capacity and the majority of the Anglo-American market. The complementary transportation system of water, roads, and rails, recently made even stronger for manufacturing districts adjacent to the Seaway, provides great advantages for both assembling materials and distributing finished products.

Major problems now jeopardize the vitality of the Anglo-American manufacturing region. First, some American industries (steel and manufacturing) now face intense *competition* from foreign producers who find advantage in lower labor costs and more modern equipment. The problems of these industries, when integrated into an entire complex of industries, are extended as economic and social stress to the entire industrial community of which they are a part. Second, recurring national recessions or *economic fluctuations* have provided economic environments in which it has been difficult to achieve the needed effort to remain competitive. Third, *other problems* that may jeopardize the Anglo-American manufacturing region are less directly those affecting the functioning of industries and more directly concerns of residential quality, social conflict, air and water pollution, urban water supply, and the governing and integration, or lack of it, of numerous but contiguous political units. The result of the above is both a restructuring of American industry in which some industries grow and others decline and a spatial adjustment which is manifest in differential growth of industrial employment by region (Table 6–3).

The Southern Economic Revolution

It is difficult to identify the beginning date of the *southern economic revolution*. The revolution probably started in the 1880s in the attitudes of "New South" advocates who believed industrialization was necessary. Thus, while the recent increase in the southern (and western) share in American manufacturing is the result of both absolute decline of manufacturing in the industrial coreland and recent real growth experienced in the South,

TABLE 6-3 Manufacturing Employment Share in the United States by Geographic Division, Selected Years (percentage)

Division	1899	1929	1954	1967	1977
New England	18.1	12.4	9.0	8.1	7.1
Middle Atlantic	34.1	29.0	26.6	22.6	18.7
East North Central	22.8	28.8	28.6	26.7	25.3
West North Central	5.6	5.4	6.0	6.2	6.6
South Atlantic	9.7	10.3	11.0	12.9	14.3
East South Atlantic	3.8	4.3	4.5	5.7	6.7
West South Central	2.4	3.4	4.5	5.6	7.3
Mountain	.9	1.2	1.1	1.6	2.4
Pacific	2.6	5.3	8.8	10.6	11.6

SOURCES: U.S. Bureau of the Census, *Census of Population* and *Census of Manufactures*.

these events are but a continuation of a process of industrial dispersion which has been ongoing for many decades.

The first major manufacturing activity to become distinctly identified with the South was the *textile industry*. The textile industry had evolved as the dominant force in New England's nineteenth-century industrial growth, but by the early twentieth century the industrial maturity of the New England area was reflected in high wage rates, unionization, costly fringe benefit programs, high power costs (imported coal), and obsolete equipment and buildings. The response was a regional shift in the textile industry. New England plants closed, and new plants were established in the South. Since the textile industry is not a growth industry on a nationwide basis, this regional shift has benefited one region at the expense of another. Firms locating within the South found a major advantage in the quantity of labor available at relatively low cost; the agrarian South had a surplus of landless rural inhabitants willing to accept alternative employment. Other advantages were better location with respect to materials used (cotton), lower cost of power, and lower taxes. By 1930 more than one-half of the United States textile industry was located in the South. At present, more than 90% of cotton, 75% of synthetic fiber, and 40% of woolen textiles are of southern manufacture.

The *rate of southern industrial growth* has increased, particularly since World War II, but with distinct differences. In addition to the labor-oriented textile and apparel industries, material-oriented pulp and paper, food processing, and forest industries have grown rapidly. In Texas and Louisiana petroleum refining and petrochemical industries have contributed much to Gulf Coast in-

dustrial expansion. The economic transformation has now reached a stage where the South itself provides a significant regional market that generates further industrial growth ("*multiplier effect*"). This regional market orientation is exemplified by the automotive industry; auto assembly operations exist in Louisville, Atlanta, and Dallas to serve regional markets. The transformation from an agrarian to an urban life-style means higher incomes and new consumption patterns that greatly increase the market importance of a formerly rural population.

Southern Manufacturing Regions

A distinctive manufacturing region coincides with much of the *southern Piedmont* and adjacent areas of Alabama. This region, from Danville, Virginia, to Birmingham, Alabama, is characterized by *light industry* (textiles, apparel, food processing, and furniture). The chief attraction has undoubtedly been the availability of suitable labor at costs below industry wage scales elsewhere. As an industrial region, the Piedmont has traits quite unlike the core districts of the Anglo-American manufacturing region. It is comprised of industries that for the most part are located in small cities, in towns, and not infrequently in rural areas. Atlanta and Birmingham are two major exceptions.

Southern material resources have also been the basis for substantial industrial growth. The *Gulf Coast* as a manufacturing region is distinguishable from the Anglo-American manufacturing region in that density of manufacturing centers is low; that is, a series of distinctly separate industrial nodes extend from Corpus Christi, Texas, to Mobile, Alabama. The material base for much of the region's industry and recent growth includes petroleum, natural gas, salt, sulfur, and agricultural products. Coastal location facilitates the exchange of goods for both domestic and international trade. Industries include petroleum refining, petrochemicals, and other chemicals; alumina smelting and aluminum refining based upon Jamaican and Guyana ores; processing of sugar and rice; and steel manufacture based upon local and imported ores. The Gulf, with industries based upon local petroleum and natural gas resources, has benefited and can benefit more from high-growth industries (petrochemicals) capable of generating many linked industries.

The *Birmingham-Gadsden,* Alabama, steel industry began with the unique circumstance of coal, iron ore, and limestone all available in proximity, and this area has become the major steel center of the South. The *Coastal Plain* is also not without industry. Pulp and paper industries and pine plywood industries have expanded greatly since World War II, mainly because of advantages in rapid forest growth. *Atlanta* and *Dallas–Fort Worth* are noted as centers of aircraft manufacture but are atypical of southern industrial centers. They are cities whose growth is more the result of their functions as regional centers than of manufacturing.

Manufacturing Growth on the West Coast

Approximately 10% of the manufacturing of the United States is located on the *Pacific Coast.* The largest single concentration is in the greater Los Angeles area, where aircraft, defense industries, food processing, petrochemicals, and apparel are dominant.

The productivity of California's agriculture and commercial fishing stimulated the *food processing* that became the state's first major industry and was dominant until the 1940s. World War II generated the defense industries, aircraft, and shipbuilding, and defense has continued to be a major employer in the Los Angeles area in postwar decades. In addition, automobiles, electronic parts, apparel, and petrochemicals have achieved importance in *California's industrial structure.* Much of California's industrial growth is not based on local material resources or access to eastern markets but on a rapidly growing local market. California, particularly Southern California, has received a large number of migrants from other parts of the United States. Industries, especially those in which material and power costs are not high, have followed the migrants.

Manufacturing in the *Pacific Northwest* includes food processing (dairy, fruit, vegetable, and fish products), forest products industries, primary metals processing (aluminum), and aircraft factories. The emphasis, however, is on the processing of *local primary resources* and use of *hydroelectric power* from the Columbia River system. The region's great distance from eastern markets and the smaller size of local markets has inhibited growth.

Canadian Industrial Growth

Secondary activities (manufacturing) are relatively as important in Canada as in the United States. The *Canadian manufacturing* economy is in fact closely integrated with the United States economy, as indicated by trade flows and the high level of United States investment in Canadian industry.

The *takeoff period* in Canadian economic development began in the 1890s and coincided with an immense boom in economic growth. The relatively later start (United States takeoff began in the 1840s) is probably attributable to several factors:

1 The harsh physical environment attracted fewer immigrants.
2 Since political independence was later, there was less concerted effort to industrialize until late in the nineteenth century.
3 Economic ties with the United States were limited until "maturity" was approached in the larger economy to the south, which then stimulated Canadian development.

Canadian industry and exports have been tied closely to the production and processing of staples: first, fishing and furs; later, wheat, forest products, and metals. A *maturing of the Canadian economy* since Wold War II has reduced this primary commodity dominance, particularly in Ontario, but it nevertheless remains an identifying feature of the Canadian economy. Wheat, primary metals (raw or partially processed), forest products, and tourism are still the major dollar earners.

The *early twentieth-century* growth was a response to the United States market and capital. Canadians feared that their location relative to the United States would make them simply a supplier of materials for the United States, an economic colony. The *tariff* became a protective device to encourage manufacturing in Canada, and then exportation, thereby assuring primary production and secondary processing in Canada. Some Canadians have argued that the tariff policy has meant higher prices for commodities they consume and, therefore, a lowered level of living. Tariffs, however, have forced the use of resources, human and material, by encouraging manufacturing at home, as is evident in the immense investment of capital by United States and other foreign companies in the Canadian economy. There is a high degree of United States–Canadian economic integration.

The Distribution of Canadian Manufacturing

The St. Lawrence Valley and the Ontario Peninsula form the *industrial heart of Canada* and are contiguous with the United States industrial core (79% of output). The area may be thought of as the northern edge of the Anglo-American industrial core, specializing in the production and processing of materials from Canadian mines, forests, and farms.

Montreal contains approximately 13% of Canadian industry. In a sense it parallels New York. Both produce a variety of consumer items intended for local and national markets (food processing, apparel, publishing). Both function as significant ports for international trade. Outside Montreal, the industrial structure is more specialized.

The immense hydroelectric potential of *Quebec* is a major power source for industries along the St. Lawrence and has been the basis for Canada's importance in aluminum production. The Saguenay and St. Maurice rivers (both tributaries of the St. Lawrence) have provided power for aluminum refining and smelting: Alma, Arvida, Shawinigan Falls, and Beauharnois. Bauxite is brought by water from Jamaica and Guyana. The aluminum produced is more than Canada consumes, and the country is the world's leading exporter. The aluminum industry illustrates Canada's role as a processor and supplier for other nations, using national resources. In this case the national resource is the power from the St. Lawrence and tributary streams. Other metals-processing industries located near production centers are copper and lead at both Flin Flon and Noranda, nickel at Sudbury, and magnesium at Haleys.

The valleys of the *St. Lawrence* and its tributaries are also Canada's major area of wood pulp and paper manufacturing. Canada is the world's leading supplier of newsprint, most of it sent to the United States and Europe.

Outside Montreal and vicinity, the most intense concentration of industry is found in the *golden horseshoe,* which extends from Toronto and Hamilton around the western end of Lake Ontario to St. Catharines. This region produces most of Canada's steel (Hamilton) and a great variety of other industrial goods such as auto parts, assembled autos, electrical machinery, and agricultural implements. It is also one of the most rapidly growing industrial districts in Canada. Two factors have contributed to the past growth and present advantages of the region:

1 Over 60% of Canada's market is found along the southern edge of Ontario and Quebec, and market orientation of industry appears to be strengthening.
2 Canadian industry has evolved behind a protective tariff.

The *tariff* was initially important when foreign capital, especially United States and British capital, was invested in industries processing Canadian resources for foreign use. It has become even more important as Canadian industry has matured, as industries have sought to serve expanding Canadian markets. Foreign companies have found it necessary to locate in Canada to avoid tariffs, but in doing so they have tended to locate in larger Canadian industrial centers (Windsor, Hamilton, Toronto, and Montreal) and close to the city containing the parent United States firm. For example, many Detroit firms with subsidiary operations in Canada have located immediately across the river in Windsor.[1] Had the tariffs not been used, Canadian industry probably would be even more oriented to primary materials processing. The economic integration is further suggested by the fact that each is the other's most important trading partner, a situation not unlike that which is found among Western Europe's Common Market members.

INDUSTRIALIZATION AND URBANIZATION

The industrial growth of the United States and Canada during the late nineteenth and early twentieth centuries stimulated employment in *tertiary activi-*

---

[1]D. Michael Ray, "The Location of United States Manufacturing Subsidiaries in Canada," *Economic Geography,* 47 (July 1971), 389–400.

ties. New industrial jobs have a *multiplier effect* by generating employment opportunity in the service activities needed to support the new workers. Employment expands in wholesaling, retailing, education, government, the professions, and a host of other urban-oriented activities. Thus, secondary and tertiary activities expand, and the cumulative growth may generate new industry and other economic activities because a larger market exists. Economies accrue to secondary and tertiary activities when they are located in towns and cities. The growth of these activities, therefore, has been the basis for tremendous *urban expansion.* Manufacturing activities were probably the principal stimulus for urban growth until the 1920s, but during the past fifty years tertiary growth has become the major stimulus.

The rural population of the United States in 1900, when industrialization was well under way, was 46 million, 60% of the total population. By 1970 the rural population had grown to 54 million but was only 25% of the total United States population. *Rural-to-urban migration* and the natural increase of population in cities expanded the urban population to nearly 160 million people in the United States alone and another 15 million in Canada. Furthermore, the vast majority of rural residents are nonfarm people who live in rural areas and small towns but work in urban areas. They send their children to school and shop in urban areas; they are urban in most ways except residence.

Present and Future Regions

From a population standpoint, both the United States and Canada consist of a number of *urban regions,* many contiguous, which continue to intensify and increase. Figure 5–9 shows the present urban regions of the United States and the urban regions projected for 2000. Urban development is so great in some areas as to justify the use of the term *megalopolis.* Professor Jean Gottmann used this term to refer to the massive urban region extending from northern Virginia to southern New Hampshire.[2] The core of this region is formed by the cities of Washington, Baltimore, Philadelphia, New York, and Boston. Together with a host of nearby cities, they comprise a massive urban region with more than 40 million people. Megalopol-

itan areas include numerous cities, adjacent or in such proximity that the boundaries between them are almost indistinguishable. The interaction between these high-density centers becomes so great that single-city identification is almost impossible, yet administratively this is how they continue to function. Within the *northeast megalopolitan region* are contained massive industrial complexes with much of the nation's manufacturing capacity and the nation's most intense market concentration.

Gottmann's designation of "mainstreet and crossroads of the nation" implies far more functional significance for the region than just manufacturing.[3] Indeed, it is a center of political and corporate management and decision making. The ports—particularly New York but also Boston, Philadelphia, Baltimore, and others—are the focuses of international and national connectivity, as well as the major termini of inland rail and auto transport routes. Furthermore, the nation's most prestigious financial and educational centers are located within the region. This massive urban region indeed functions as "downtown U.S.A."

Even agriculture within and adjacent to a megalopolis has a distinctive character. Farmers occupy land that is high in value because of urban land-use potential, not its inherent food-producing capabilities. Dairy products, vegetables, poultry, and other specialty items are produced in quantity for the adjacent markets. Although production costs are high, transport cost to market is low. Farmland, however, like recreation land, faces tremendous pressures from urban encroachment. Proximity to the urban region means advantages to the farmer and an increased value for land, but also problems. High taxes and inflated offers from developers lead to reduced numbers of farms, often of highly productive ones.

The *eastern megalopolis* may extend from southern Virginia to southern Maine by the year 2000. The projection for the year 2000 also shows a *Great Lakes megalopolitan* region connecting with the eastern megalopolis. The Great Lakes megalopolitan region is already well advanced and by 2000 may be a major part of a massive urban-industrial coreland extending from Illinois-Wisconsin to the Atlantic shore. A series of less intense, but nevertheless distinct, urban regions will extend along the Piedmont and into north Georgia and northern Alabama. Florida, the Gulf Coast, and the rapidly growing California megalopolis ex-

---

[2]Jean Gottmann, *Megalopolis: The Urbanized Northeastern Seaboard of the United States* (Cambridge, Mass.: MIT Press, 1961).

[3]Ibid., pp. 7–9.

tending from San Diego to San Francisco are other extended urban regions.

Some Urban Problems

There exist common problems with which all of these urban regions must deal. One major *problem is administrative.* The actual city, in a functional sense, remains subdivided into many independent political units. Governments exist and try to provide services and maintain jurisdiction over the inner city, while numerous suburbs attempt the same for themselves. Frequently the administrative functions overlap those provided by county governments. Consolidation of services and government remains a major issue in urban regions. Planning and problem solving in urban areas are complicated by differences between cities and suburbs. Economic, social, and racial segregation means that communities have unequal abilities to generate funds for the provision of services. The problem becomes particularly acute in central cities that experience in-migration of poor whites and blacks and out-migration of relatively more prosperous people. Consolidated urban political regions, or political boundaries coincident with "real cities," would mean a more equitable tax base. Not surprisingly, however, this idea is often opposed.

Population numbers are one aspect of urban growth; another is *land area absorbed.* Cities and suburbs continue to absorb substantial amounts of land. Farmers who, with some frustration, view themselves as having been here first find a new life-style, system of land use, and tax scale encroaching upon their domain. Although some "suffer" the consequences of developers' prices with tongue in cheek, others genuinely resent being squeezed from some of the nation's best agricultural land. Urban growth frequently has occurred at such a rapid pace that planning has failed to prevent unsightly development and traffic congestion.

Our seeming cultural predisposition to emphasize the new and spread ourselves outward from the old cities suggests an intensification of some of the preceding problems. The *economic base* of many central cities (and even of some metropolitan areas) continues to weaken. Many cities lose population. Yet areas adjacent to the cities and even more remote rural areas and smaller cities experience higher rates of population growth. Hence we experience at once the problems of *growth and decline.* Our settling experience then continues—but with a reversal of the long-standing trend to emphasize the big city. Indeed, a major American dilemma may be evident in the issue of whether an explicit policy of restoring the central cities should be established or whether to allow by benign neglect the evolution of new urban forms and the dispersion of population and economic activity.

7

Anglo-America: Problems in a Developed Realm

Rich and developed countries such as Canada and the United States are not without significant and sometimes pressing problems. The complexities of supply and use of enormous quantities of resources have already been discussed. Unbalanced economic growth, the ineffective integration of various regions into a national economy, and the social, economic, and political situation of minority groups are other fundamental problems with which Canada and the United States must deal.

INCOME DISPARITY AND REGIONAL PROBLEMS

It is no surprise that there are large numbers of people who have not been able to acquire the material benefits of the average Anglo-American (Table 7–1). How many *poor* exist in America depends, of course, on one's definition of "poverty," but possibly 12% of the population should be included. The United States Bureau of the Census estimates that some 30% of the nearly 27 million black Americans are in poverty. This group represents approximately 31% of all poor in the United States. Although a higher proportion of all blacks live in poverty than whites, the absolute number of poor whites is double that of the blacks.

The poor in the United States were once almost equally divided between metropolitan and nonmetropolitan areas. Currently, however, 62% of the poor are found in metropolitan areas. The urban poor tend to be concentrated in central cities, though not exclusively so. The poor in metropolitan areas are often a smaller proportion of the total metropolitan population than the poor in nonmetropolitan areas, but they are more intensely concentrated in *ghetto communities*. The nonmetropolitan poor are dispersed areally, which adds to the problem of employment and provision of services. The greatest concentrations of poor are found in the states of California, Texas, New York, Florida, Illinois, and Pennsylvania. Poor peo-

TABLE 7-1 Poverty in the United States by Residence and Race

Residence	Percent of Families Below Poverty Level 1980[a]		
	White	Black	Spanish Origin
Region			
Northeast	6.8	24.4	30.4
North Central	5.7	27.1	15.4
South	8.0	30.0	20.7
West	6.5	22.1	14.8
In metropolitan areas	5.8	25.4	19.3
In central cities	7.9	28.5	23.8
Outside central cities	4.6	17.3	13.2
Outside metropolitan areas	8.9	35.4	21.6
All areas	6.8	27.6	19.7
All families	9.1		

SOURCE: U.S. Bureau of the Census, *Statistical Abstract of the United States* (Washington, D.C.: Government Printing Office, 1981).

[a]"Poverty level" is indexed to the consumer price index and varies with size of family, number of children, and the age of householder. The income cutoff at 125 percent of poverty level in 1979 was $9,265 for a family of four. The number of "persons" in poverty in 1980 was 12.5% of the United States population.

ple living in small towns and rural areas are less concentrated and are therefore less visible in most respects. Numerous poor are found throughout the southern Coastal Plain from Virginia to Texas and in the upland South, particularly the Appalachian Plateau and Ozarks. Among the rural population, however, poverty is not limited to the South, though it is most widespread there. Smaller areas in peripheral New England, the upper Great Lakes (Michigan, Wisconsin, and Minnesota), the northern Great Plains, and the Southwest also have been identified as *poverty regions.*

It is unwise to assign a simple, single cause or solution to a problem that exhibits such variation in social, economic, and physical setting. Certainly racial biases and cultural attitudes have created immense barriers for blacks, Mexican-Americans, American Indians, and Appalachian whites in both rural and metropolitan areas. These biases not only affect employment opportunity directly but have also contributed to unequal education and training. Other factors have also contributed to poverty. In portions of the Appalachians agriculture has been the basic activity since initial settling. As agriculture evolved into a commercial and profitable enterprise elsewhere, farmers in isolated areas with small units and poor land (slope and soil) have been unable to adjust to modern ways. The populace slips into poverty as it *fails to modernize*—a major problem in a dynamic and changing society.

Another contributing factor is an imbalance in the *supply and demand for labor* with particular qualities and skills, a problem that has a spatial implication. Large parts of the South in the nineteenth and twentieth centuries had a relatively high population density in rural areas where agriculture remained labor intensive; agricultural labor rarely realized more than a low-to-modest income. Other portions of the country experienced rapid urban-industrial growth and the development of labor skills that provided higher incomes. The result was distinct *regional income variations* and greater proportions of the total southern populations in the poverty categories. Such regional income disparity contributes to migration, but unfortunately the migrants are often inadequately prepared for participation in those sectors of urban life that are materially rewarding. Individual progress is difficult when people are faced with limited resources, social bias, and the frequent concentration of the poor.

The changing *significance of location,* reduced need for specific kinds of resources, or the decline in particular kinds of economic activities also contribute to unemployment and poverty. New England experienced a decline in the dominant textile industry. Miners from Pennsylvania to Kentucky saw jobs disappear as the mines were automated and the demand for coal failed to increase. Farmers in the South witnessed a declining and chang-

Poverty in metropolitan areas is concentrated in ghettos. These ghettos are often occupied by distinct ethnic groups, as exemplified by this Puerto Rican scene in New York City. (United Nations/John Rabaton)

ing agricultural scene that had major implications for laborers, tenants, and farm owners. Indeed, the industrial restructuring currently in progress leaves many skilled urban factory workers facing unemployment because of a reduction in demand for their particular skills. The results are often pockets of economic problems if not widespread poverty. Migration may concentrate poor people in cities, but often many who remain as residual farm people, miners, or unwanted factory laborers are unable to adjust to a changing society and also exist in poverty. The solution to poverty is usually far more complicated than merely changing locations.

The numerous causes contributing to poverty may operate independently or in concert; more often it is the latter. Moreover, these causes are of varying importance from one region to another. Although we cannot study all the poverty regions with the thoroughness needed for ample understanding, it is beneficial to examine one area and the approach taken to solve its problems.

Appalachia

The problems of *Appalachia* are attributed to the long period of *isolation* and an inability to participate in the modernization and commercialization of agriculture. The prosperity of the coal-mining era was short-lived and created more wealth elsewhere. After 1940 *mining automated* rapidly and concentrated on fewer mines, leaving behind unemployed people and scarred landscapes. Lumbering operations that removed the wealth and beauty of an area in a single generation can be viewed in much the same way. Both the physical and human ecology of the region have been severely affected. *Soil erosion,* the scars of strip mining, floods worsened by poor agricultural techniques, and the removal of forest vegetation from watersheds were major problems before the early twentieth century. Human poverty, low expectation, violence, and human stagnation have been a way of life for generations.

Appalachia consists of uplands that extend from

the Maritimes of Canada to central Georgia and Alabama and includes several distinct physiographic regions. The area with severe economic and social problems is usually considered to be restricted to the Plateau, Ridge and Valley, and mountain areas south and west of the Mohawk and Hudson valleys. But even here, the hard-core problem area is not spatially coincident with a physiographic region. Figure 7–1 identifies the Appalachian corridor or problem area as defined by the *Appalachian Regional Commission.* The Applachian Regional Commission administers the funds allocated under the Appalachian Regional Development Act of 1965, which provided for a concerted effort to eliminate

poverty. The greatest emphasis thus far has been on *highway development.* It is believed that part of the reason for Applachia's problems has been isolation, the inaccessibility of the area's interior. Early subsistence farmers in the more remote areas failed to make the adjustment to commercial systems, possibly because of isolation from markets as well as poor resources. The lumber and coal resources were harvested here, but the lucrative markets were elsewhere. Coal and lumber stimulated transportation development only to the degree necessary to haul the product outside the region.

If advantages can initiate a circular process of *causation and growth,* it may well be that disadvantages can do the same in reverse fashion (Chapter 4). Isolation, poverty, low education levels, and limited incentive can "feed on" each other and contribute to further problems. Those who believe that transportation improvements will stimulate economic growth argue that industry will locate near highways, and the potential for recreation and tourism can be exploited.

Others argue differently. Economic growth potential may be greater for cities that have good *potential for interaction* with other cities or population regions. Thus those centers with the greatest potential for growth and economic improvement are not the interior cities of Appalachia but rather the cities peripheral to or fringing on the region where accessibilty to other national centers is better. Furthermore, small and isolated Appalachian communities contain a small labor supply, part of it in sparsely populated rural areas. It is easy for such areas to become overindustrialized relative to labor supply, even where low-wage, labor-intensive industry is involved.

The greater number of *growth centers* are peripheral to the Appalachian corridor. The northern edge, in Pennsylvania and New York, may be thought of as relatively prosperous. The same applies to the peripheral Piedmont extending from Virginia to Alabama. It is possible that highways may simply aid some of the corridor people in leaving for the more advantageous periphery without really changing conditions within the corridor. Despite effort expended toward regional development to date, Table 7–2 suggests that the Appalachian Region continues to lag behind the national economy. Not unexpectedly, the effectiveness and the future of this *regional planning* effort may be in question.

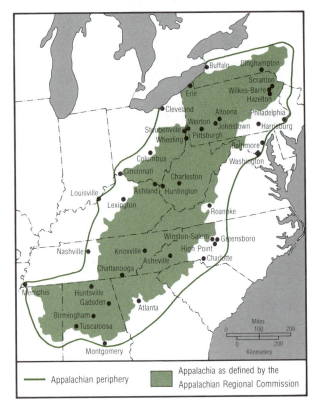

Appalachian periphery

Appalachia as defined by the Appalachian Regional Commission

FIGURE 7–1 Appalachian Region of the United States. The Appalachian region has long been an area of economic and social problems. It should be recognized, however, that any such large region is not uniform. Some portions are areas of growth and economic progress and other portions continue to lose population. One of the most persistent and severe problem areas has been the Kentucky and West Virginia portions where the declining need for coal miners contributed major problems after World War II. Perhaps renewed demand for coal in response to high oil prices will bring a new era of prosperity to portions of Appalachia.

TABLE 7–2 Population and Personal Income of the Appalachian Region

	Population		Per Capita Personal Income		
	1980 (thousands)	Percent Change 1970–80	1979	Percent of U.S. 1970	1979
United States	226,505	11.4	$8,757	100	100
Appalachian states	80,854	6.3	8,248	97	94
Appalachian region Appalachian portion of:	20,234	11.1	7,333	81	84
Alabama	2,427	13.6	7,260	77	83
Georgia	1,104	35.6	6,696	74	76
Kentucky	1,077	22.9	5,931	57	68
Maryland	220	5.2	7,197	84	82
Mississippi	483	15.3	5,931	62	68
New York	1,083	2.5	7,155	91	82
North Carolina	1,218	17.2	7,248	80	83
Ohio	1,263	11.7	7,103	79	81
Pennsylvania	5,995	1.1	8,094	91	92
South Carolina	792	20.6	7,506	83	86
Tennessee	2,074	19.6	6,857	75	78
Virginia	550	16.9	6,755	67	77
West Virginia	1,950	11.8	7,402	78	85
Non-Appalachian portions	60,620	4.7	8,548	102	98

SOURCE: U.S. Bureau of the Census, *Statistical Abstract of the United States* (Washington, D.C.: Government Printing Office, 1981).

BLACK AMERICA

It is not unusual for political units to contain a number of *subgroups* distinguishable by race, ethnic and linguistic difference, or economic achievement. Such divisions can provide major obstacles to the achievement of unified political organization and social and economic satisfaction. The difficulty of integrating such groups into a larger society stems not only from outward cultural differences but also from human nature. The United States illustrates this problem explicitly in the *black-white relationship* that has been so difficult to resolve.

In the United States, the initial patterns of black residence and the black-white relationship were an outgrowth of the *diffusion of the plantation system* across the lower South. The southern plantation was a land-based social and economic system, engaged in commercial production on relatively large holdings and using slave labor. The original center for this system was tidewater Virginia and adjacent portions of Maryland and North Carolina. As settlement proceeded southward, slavery accompanied commercial crops such as rice and indigo to the coastal cities of Wilmington, North Carolina;

Charleston, South Carolina; and Savannah, Georgia. In the hill lands, and northward into the Middle Atlantic and New England colonies, slavery remained of minor importance and was eventually declared illegal. Economic emphasis there was often on less labor-intensive efforts.

The plantation, once established, became the basic system used in those areas of the lower South considered best for agricultural production. The diffusion of the plantation system went hand in hand with the *diffusion of slave labor* and established the initial distributional pattern of black residence across the South. Selected areas such as the outer Piedmont, the inner Coastal Plain, a narrow coastal zone of islands and river banks in Georgia and South Carolina, the black-soil belt of Alabama, the Tennessee Valley of northern Alabama, the Mississippi Valley (Arkansas, Tennessee, Mississippi, Louisiana, and Missouri), and, by 1860, portions of Texas had dominantly black populations. The Nashville Basin and Bluegrass Basin were areas of some plantation occupancy and, therefore, black residence. That distribution left much of the South at an opposite extreme. The southern Appalachians, including the inner edge of the Piedmont, the Interior Highlands (Ozarks), portions of the Gulf Coast, and lower Texas were either without blacks or contained few in proportion to the white yeoman farmers. A small percentage, possibly one-seventh, who were freedmen, lived outside the South or in southern urban areas where slightly less rigid social pressures allowed their existence as free black artisans.

The *distribution of blacks* did not change immediately after the Civil War. Although no longer slaves, the freedmen had not changed in other respects. They were, after all, agricultural laborers with little training other than in farming, with no land and no capital. In the aftermath of the Civil War, they did not migrate in large numbers but instead entered into a system of tenancy with the white landowners in the same locations previously important for black residence. With a few exceptions, the rural black population of today's South identifies the former plantation regions.

By 1900 a *large rural black population* lived as a landless tenant labor force. By World War I the great century of white immigration from Europe (1814–1914) was over. The black could find employment outside the South as a substitute for the no longer available European immigrants. Although the *black migration* slowed somewhat during the 1930s when economic conditions in urban areas were also bad, the migration continued into the 1970s (Table 7–3). The black population is now *northern urban* as well as *southern rural and urban* (Figure 7–2). An interesting feature of Table 7–3 is the apparent stabilization of the southern regional share of black population. This condition results from another reversal of long-standing migration tendencies; that is, more blacks are now moving into the South than are leaving the South (Table 7–4).

The migration has *implications* that go far beyond population redistribution. Many blacks have clearly improved their economic and social position in urban areas, but it was a group progress not won easily and not without failure for many individuals. The migrants have often been the better educated and more motivated persons, meaning an economic and social loss for the area of origin, but paradoxically they have also lacked the skills needed in urban areas. Furthermore, racial and social bias added to the difficulty of obtaining suitable housing, a proper education, and access to economic opportunity. The result for many has been existence in a *ghetto* with its distinctive structure and seeming hopelessness for the future. The result is an unemployment rate for blacks which is consistently higher than for other segments of society.

The *spatial pattern of residence* for urban blacks stems from their economic weakness and social position. Hence we derive the American phenomenon of highly concentrated black neighborhoods in the older residential portions of cities, often vacated as economically progressing whites flee to city peripheries and suburban communities. The process has progressed so far that numerous cities have become more black than white. The strength of black numbers is already evident in the increasing numbers of black mayors—but noticeably central city mayors, not suburban.

The 1954 Supreme Court ruling that official segregation of schools is unconstitutional is spotlighted as a landmark in black-white American history. The process that led to a new social and economic position for American blacks, however, began much earlier. The late-nineteenth-century South contained advocates of a "New South," envisioned as urban, industrial, and economically

TABLE 7–3 Black Population of the Conterminous United States

| Year | Regional Distribution (%) | | | | Total (millions) |
	Northeast	North Central	South	West	
1980	18.3	20.1	53.0	8.5	26,505
1970	19.2	20.2	53.0	7.5	22,580
1960	16.1	18.3	60.0	5.7	18,860
1950	13.4	14.8	68.0	3.8	15,042
1940	10.6	11.0	77.0	1.3	12,886
1930	9.6	10.6	78.7	1.0	11,891
1920	6.5	7.6	85.2	0.8	10,463
1910	4.9	5.5	89.0	0.5	9,828
1900	4.4	5.6	89.7	0.3	8,834

SOURCE: U.S. Bureau of the Census, *Census of Population* (Washington, D.C.: Government Printing Office, 1910–80).

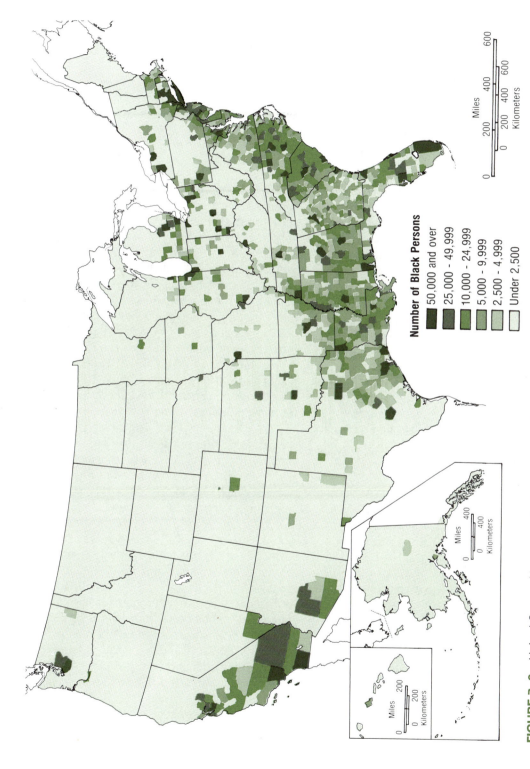

FIGURE 7–2 United States: Number of Blacks, by Counties, 1970.
The black population in northern communities is almost totally in urban areas.
Throughout the South, blacks are found in rural areas once associated with planta-
tion agriculture and in urban centers.
SOURCE: U.S. Bureau of the Census, 1970.

Number of Black Persons

- 50,000 and over
- 25,000 - 49,999
- 10,000 - 24,999
- 5,000 - 9,999
- 2,500 - 4,999
- Under 2,500

TABLE 7–4 Black Migration to and from the United States South[a] (in thousands)

	Northeast	North Central	West	Total
1965–70				
To the South[a]	69	57	36	162
From the South	120	164	94	378
Net migration	−51	−107	−58	−216
1975–80				
To the South	192	121	102	415
From the South	50	94	76	220
Net migration	+142	+27	+26	+195

SOURCE: U.S. Bureau of the Census. Current population reports, Series P.20, No. 368, *Geographical Mobility: March 1975 to March 1980* (Washington, D.C.: Government Printing Office, 1971).
[a]All values for age 5 and over.

strong. Opponents of the new doctrine believed that a social and cultural disaster would be born of an urban-industrial society. It would mean a racially mixed labor force, a new black consciousness, and education systems that would undermine the status of an agrarian southern society. The opponents of the New South were correct to an extent, for out of urbanization, new economic gains and opportunity for some, and better education (even if slowly attained) has arisen a new black consciousness. The new status, hopes, and demands for a participating role by blacks could come only with a break from the old agrarian system. Migration, whether to northern or southern cities, was symptomatic of that break and was stimulated by numerous other economic, social, and political forces.

HISPANIC AMERICA

Persons of Spanish origin, or *Hispanics,* are numerically the second-ranking American minority (14,605,883 people in 1980). The proportion of the United States population considered Hispanic increased from 4.5% in 1970 to 6.5% in 1980, and the absolute number of Hispanic persons increased by 60% in the same decade. The majority (more than 60%) of the Hispanic population resides in the southwestern United States. The state of Texas is 21% Hispanic, New Mexico 36.6%, Arizona 16.2%, and California 19.2%. People with Spanish surnames almost completely dominate many smaller communities and even some sizable cities (Corpus Christi is 49% Hispanic, San Antonio 45%, and Brownsville 77%).

The initial Hispanic *cultural infusion* into the southwestern United States resulted from expansion of the Spanish empire in the late sixteenth century and seventeenth century. The general area into which this cultural infusion occurred has become known as the *borderland.* Thinly scattered settlement eventually extended from Texas to California, but the Spanish were unable to prevent a flood of American settlement during the nineteenth century. The initial growth of Anglo-American settlement diminished the share of Hispanic population but did not erase long-established cultural imprints. The twentieth century, however, has witnessed a considerable legal immigration from adjacent *Mexico* and an especially large illegal immigration since 1950. This migration in combination with a fertility rate higher than the national average is contributing to a renewed increase in Hispanic influence in the borderland.

A significant Hispanic population is also found in several large *metropolitan* areas outside the southwest borderlands. In New York and Chicago (8% and 16% Hispanic), Hispanics are most often persons of *Puerto Rican* descent who participated in a major migration to the mainland during the post–World War II years or who are children of the migrants. In the case of *southern Florida* (Miami is 36% Hispanic), the large Hispanic minority is Cuban and results from the significant middle-class migration which followed the Cuban revolution.

Though significant numbers of Hispanics work as agricultural laborers, most are urban residents. Whether rural or urban, the Hispanics experience problems similar to those of black Americans: low income and educational levels, limited economic opportunity, and often substandard and crowded housing conditions.

The Hispanic American culture region identified on Figure 7–3 is also an area occupied by the largest *American Indian* population—nearly 50 percent of the United States's 1,361,869 indigenous population. Despite the southwestern Indians' greater success in tribal survival, the socioeconomic dis-

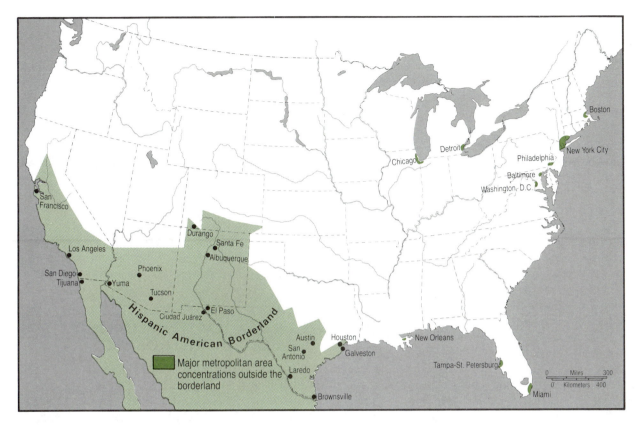

FIGURE 7–3 United States: Hispanic Population.
The distribution of Hispanics exhibits two distinctive patterns. The concentration in the southwest borderlands results from a long tradition of Spanish and Mexican influence. The concentrations in urban centers of the eastern United States result from post–World War II migrations of Puerto Ricans and Cubans.

parity endured by these groups is greater than for any other minority.

CANADIAN IDENTITY AND UNITY

French Canada

A major concern of Canadian political and community leaders has been *national unity and identity.* Canada's political organization as a federation (more than a hundred years ago) was at the insistence of French descendants that any system for union preserve French identity and influence. Thus was promoted a confederation of colonies with already distinctive cultural differences—French and English Canada.

The French settled first in the *lower St. Lawrence Valley* and later in Acadia, the area around the Bay of Fundy. Great numbers of French settlers did not follow, however, and furs and fishing remained major economic activities. By 1763 the

British had overcome French control in Anglo-America. Despite vast holdings of France used for extracting furs, it was only the lower St. Lawrence that was to remain French in culture. The British allowed the French agriculturists to remain. The approaching United States Revolution, however, caused "loyalists" from the British colonies farther south (New England and the Middle Atlantic) to move to the loyal colonies of Nova Scotia and Newfoundland, in numbers great enough to justify a new colony (New Brunswick). Others settled farther inland in Ontario. The rapid extension of people of British descent further weakened the position of the French except along the lower St. Lawrence, where they remained firmly established.

The French have maintained their identity and are intent on continuing to do so. Their distinctiveness is not only linguistic, though that in itself is enough to promote a *separate cultural identity,* but also religious. French Canadians are Roman Cath-

121

olic, in contrast to Protestant English Canada. In other respects, the French of Canada have been stereotyped to a misleading degree even by other Canadians. It is a misconception to characterize them as quaint, rural, agrarian, unchanging people with high birth rates. Quebec, which is about 80% French, is 75% urban, part of industrial Canada, second only to Ontario in income, and clearly integrated into the Canadian core in an economic sense. It once was more agrarian and experienced high population growth rates, which engendered some of the fears of French culture overtaking English culture. With modernization and an urban lifestyle, however, birth rates have decreased sharply below the national average. Continued immigration will further reduce the French proportion of Canada.

The French-Canadians, though 70% urban, do not yet experience an occupational structure similar to that of English Canadians. Although they have a social hierarchy and political power, occupationally they remain overrepresented in the primary and unskilled areas. The French insist on identity—their language remains—and so Canada must forge a national unit in a *bilingual framework,* not an easy task. However, violence and the political separation suggested by extreme French nationalists are not supported by the majority of French Canadians.

The French and English may be thought of as *charter ethnic groups.* Immigration patterns of the present century, however, have seen significant numbers of Poles, Dutch, Germans, and Italians, sometimes to the consternation of the French, who often oppose immigration because of a diminution of their numerical strength.

Canada and the United States

Canadian identity is further affected by Canada's proximity to the United States. Canada is one of the largest countries in the world, but all its vast area contains only 24 million people. It is rich in forest, water, and mineral resources, but its value for agricultural production is modest because of harsh environments. Considering available resources and proximity to the industrialized United States, it is not surprising that *major economic linkages* between the two countries have evolved. Canada and the United States have become each other's most important trading partners. This economic relationship is further evident in the high proportion (nearly 40%) of Canadian industries that

are controlled by United States parent firms. The trade relationship has, to a large degree, been based on the removal and processing of Canada's special material resources. Canada's concern over a "colonial" relationship generated a long-standing *tariff policy* that has had the effect of encouraging United States firms hoping to market in Canada to establish industry within Canada. Paradoxically, the success of the policy in turn generates the concern over foreign influence and control. Economic prosperity in Canada is significantly tied to United States prosperity, but this fact can only contribute to the problem of Canadian identity. Political efforts to redirect trade relationships toward other areas may be in direct conflict with the normal economic relationships expected between two large and well-endowed countries in proximity.

CANADA AND THE UNITED STATES IN RETROSPECT

Canada and the United States are two large countries in area and resources. Each has achieved a *high level of living* for the majority of its populace and a powerful position in the world. The motivations of these two nations are such that they will certainly attempt to maintain their positions. Out of their development experience has come a *technology* that can be of great benefit to the remainder of the world. In fact, however, resources of underdeveloped area have frequently been used to promote the welfare and economic expansion of the United States and Canada. As other countries develop and consume more material resources, the question of how much resource material the United States and Canada can—or should—gather from elsewhere becomes difficult to answer and one of the important future issues.

The processes of *immigration and settlement* have affected the two nations differently. The larger population of the United States has aided in the accumulation of greater wealth through the processing and use of the nation's resources. The United States, however, is now consuming on such a scale that the cost and availability of basic goods may well encourage or require *limitation of growth* or even stabilization. On the other hand, the economic sluggishness of the Anglo-American economies during the early 1980s and the problems of no-growth communities suggest that limiting growth is fraught with pitfalls. For less populous Canada, greater economic independence and *internal industrial growth* may require domestic population growth and expansion. The value of population growth for developing and using resources is even now reflected in Canadian immigration policies. Although we may view stabilization as necessary for some large and developed countries, for others substantial economic development may occur in the context of population growth and increasing demand.

Internal problems for both countries include the need for integration of minority groups into the larger society. Indeed, as the United States progressed, entire regions as well as minority groups lagged in the acquisition of wealth and position within the system. Effective means are needed for integrating the poverty pockets into the larger economy.

American society also will have other adjustments to make. North Americans are becoming an even *more urban society,* though in form perhaps different from the past. The large urban formations, "megalopolitan regions," may require new approaches to government and planning. These are such highly integrated urban systems that many needs such as water, transportation, revenue, and recreation facilities must be planned in a regional framework that extends far beyond the traditional political city.

FURTHER READINGS

General texts useful for an overview of Anglo-America are C. Langdon White, Edwin J. Foscue, and Tom L. McKnight, *Regional Geography of Anglo-America,* 5th ed. (Englewood Cliffs, N.J.: Prentice-Hall, 1979), 608 pp.; Stephen S. Birdsall and John W. Florin, *Regional Landscapes of the United States and Canada,* 2nd ed. (New York: Wiley, 1981), 497 pp.; J. H. Paterson, *North America: A Geography of Canada and the United States,* 6th ed. (New York: Oxford University Press, 1979); and Richard S. Thoman, *The United States and Canada: Present and Future* (Columbus, Ohio: Charles E. Merrill Publishing Company, 1978), 471 pp. Data of great variety are found in the U.S. Department of Commerce, Bureau of the Census, *Statistical Abstract of the United States* (Washington, D.C.: Government Printing Office, annual); and *Canada Yearbook* (Ottawa: Dominion Bureau of Statistics), an annual publication of statistics with narratives on the resources, history, institutions, and social and economic condition of Canada.

Information on the resource position of Anglo-America is included in U.S. Department of the Interior, Bureau of Mines, *Mineral Facts and Problems* (Washington, D.C.: Government Printing Office, annual). See also Trevor M. Thomas, "World Energy Resources: Survey and Review," *Geographical Review,* 63 (1973), 246–58.

Wilbur Zelinsky, *The Cultural Geography of the United States* (Englewood Cliffs, N.J.: Prentice-Hall, 1973), 164 pp., is a short but excellent treatment of the cultural geography of the United States. See also, by the same author, "North America's Vernacular Regions," *Annals, Association of American Geographers,* 70 (1980), 1–17. The Commission on Population Growth and the American Future, *Population and the American Future* (Washington, D.C.: Government Printing Office, 1972), 186 pp., provides a useful discussion of the United States demographic experience and

the implications for the future. See also Phillip D. Phillips and Stanley D. Brunn, "Slow Growth: A New Epoch of American Metropolitan Evolution," *The Geographical Review,* 68 (1978), 274–92; and James S. Fisher and Ronald L. Mitchelson, "Forces of Change in the American Settlement Pattern," *The Geographical Review,* 71 (1981), 298–310.

A special issue of *Annals, Association of American Geographers,* 62 (1972), 155–374, is devoted entirely to the regional geography of the United States. The several articles individually or as a whole provide an excellent supplementary source of information on various United States regions. John F. Hart, *The Look of the Land* (Englewood Cliffs, N.J.: Prentice-Hall, 1975), 210 pp., provides a good synthesis of the origins and character of the American landscape. Other useful papers that deal with American agriculture include the following: Howard F. Gregor, "The Large Industrialized American Crop Farm: A Mid-Latitude Plantation Variant," *Geographical Review,* 60 (1970), 151–75; John F. Hart, "Loss and Abandonment of Cleared Farmland in the Eastern United States," *Annals, Association of American Geographers,* 58 (1968), 417–40; and Everett G. Smith, Jr., "Americas Richest Farms and Ranches," *Annals, Association of American Geographers,* 70 (1980), 528–41.

For readings that focus on the process of American metropolitan and industrial growth, see John R. Borchert, "American Metropolitan Evolution," *Geographical Review,* 57 (1967), 301–32; and Allan Pred, "Industrialization, Initial Advantages, and American Metropolitan Growth," *Geographical Review,* 55 (1965), 158–85. See also William J. Lloyd, "Understanding Late Nineteenth-Century American Cities," *The Geographical Review,* 71 (1981), 460–71. A somewhat older but classic description of America's first metropolitan region is Jean Gottmann's *Megalopolis, the Urbanized Northeastern Seaboard of the United States* (Cambridge,

Mass.: MIT Press, 1961), 810 pp. See also John R. Borchert, "America's Changing Metropolitan Regions," *Annals, Association of American Geographers,* 62 (1972), 352–73.

A general work dealing with poverty is Richard L. Morrill and Ernest H. Wohlenberg, *The Geography of Poverty in the United States* (New York: McGraw-Hill, 1971), 148 pp.

A special issue of *Economic Geography,* 48 (1972), 1–134, is entitled "Contributions to an Understanding of Black America" and includes seven articles dealing with geographical aspects of the black experience in America. Richard L. Nostrand, "The Hispanic-American Borderland: Delimitation of an American Culture Region," *Annals, Association of American Geographers,* 60 (1970), 638–61.

III

Western Europe

Louis De Vorsey, Jr.

8

Western Europe: A Varied Home for Humanity

Western Europe (Figure 8–1) is the home of more than 350 million of the earth's most productive and prosperous people. The twenty-four political units they inhabit cover an area that is a good deal smaller than the United States or Canada; it makes up just 3% of the earth's total land surface. Moreover, many of Western Europe's most populous and advanced countries are at the same latitude as the thinly populated north of Canada. How is it that so many people have managed to create such productive and comfortable life-styles in this relatively small and northerly region? There is no simple answer to this question, but even a partial answer will give us a better understanding and appreciation of this important part of the developed world.

Western Europeans often appear closer to being the masters of their physical environment than the people of any other great culture region. As a result of their accomplishments, modern Europeans feel the effects of many elements of their physical environment in more indirect and subtle ways than the people in much of the rest of the world.

WESTERN EUROPE: THE CENTER OF THE LAND HEMISPHERE

The countries making up Western Europe occupy a *strategic location.* By moving far into space, astronauts and cosmonauts have been able to gain a unique view of our planet. In one glance they have seen and photographed a hemisphere at a time. If a photograph were taken from a space station directly above the small *x* in western Germany on the map shown in Figure 8–2, it would reveal the most important hemisphere of all, the *Land Hemisphere.* This half of the earth's surface contains about 90% of the inhabited land area and about 94% of the total population and economic production of the world.

As a result of their *central location,* Western Europeans enjoy relatively easy contact with almost

FIGURE 8–1 Nations and Principal Cities of Western Europe.
Western Europe is composed of many nations. Most are small in area, but some
have large populations. Home of Western culture and the Industrial Revolution, this
region plays a major role in world affairs.

the entire habitable world and its resources. This advantageous position has existed since Europeans mastered the arts and technology of distant oceanic navigation in the fifteenth and sixteenth centuries and the *sea-lanes* to the New World, India, Africa, and East Asia began to function as circulation routes for people, materials, and ideas. The colonial empires of many Western European

127

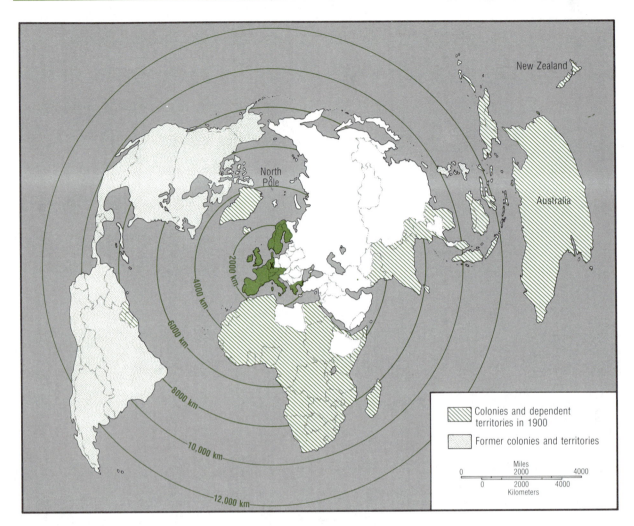

FIGURE 8–2 Western Europe, the Center of the Land Hemisphere and a World-wide Colonizer.
The impact of Western European culture has been so great that the phrase "European explosion" is used to characterize the region's control over many parts of the world. Western European countries used their colonies as sources of raw materials and as markets for mother-country products.

nations were partly a reflection of Europe's central location. Later, the advent of *railroad technology* opened the vast continental interiors and enlarged the *world circulation network* that focused on Western Europe. Now long-range aviation and high-speed oceanic travel further emphasize the tremendous significance of Western Europe's centrality. The Western Europeans have consistently been in the forefront as pioneers of commercial aviation. Count Ferdinand von Zeppelin of Germany organized the world's first commercial airline in 1910, employing his famous airships, or zeppelins, as the lighter-than-air craft were known. By 1919, the first regular international airmail service

was established between London and Paris. British Overseas Airways began the first regularly scheduled jet airliner service in 1952. Nor is it surprising that the British and French combined to develop the world's first supersonic commercial airliner, the controversial Concorde.

A recent decision by the Soviet Union has opened that nation's Arctic Ocean route to foreign ships. As ships begin to ply this new route, the distance between the ports of northwestern Europe and eastern Asia will be considerably shortened. It seems safe to predict that Western Europe's position at the *center of the Land Hemisphere* will become even more important as

128

transportation technology improves in the years ahead.

MARITIME ORIENTATION

Western Europe, viewed at the continental rather than the global scale, forms an irregular Atlantic fringe to the vast Eurasian landmass. Western Europe's *sheltered coasts and many harbors* provided an almost perfect setting for the development of *maritime-oriented* economies as the Europeans extended their world trade and political linkages through the past five centuries. Western Europe's irregular and fragmented outline provides a higher ratio of coastline to total land area than any other major culture region of the world.

Only a small handful of the countries of Western Europe lack direct access to the 70% of our planet covered by the oceans, and no part of Western Europe is far from the sea and its challenges. A glance at a map or globe reveals that numerous seas, gulfs, and bays—such as the North, Bothnian, Baltic, Irish, Mediterranean, Adriatic, Biscay, Ligurian, and Aegean—provide Western Europeans with matchless opportunities for *ocean-borne contacts and trade.* It is no wonder that flags of the Western European nations from Norway to Greece are commonplace in ports the world over.

THE "CONTINENTAL ARCHITECTURE" OF WESTERN EUROPE

The term "continental architecture" is a bit unusual. It includes the many landform areas such as the plains, uplands, and mountains that form the physical framework or "skeleton" on which the Western Europeans have built their landscapes. At first glance the physical map of Europe seems complicated and confusing (Figure 8–3). There are, however, certain broad landform similarities over extensive areas, recognition of which can help us to understand the diverse human and economic patterns that make up modern Western Europe.

Western Europe occupies a portion of each of the *four great physiographic subdivisions* of Europe:

1 The Northwestern Highlands.
2 The Great European Plain.

3 The Central Uplands.
4 The Alpine Mountain system of southern Europe.

A glance at Figure 8–3 reveals that these divisions follow a general *west-to-east trend* across Europe. This orientation is in sharp contrast to Anglo-America, where the major physiographic divisions are aligned from north to south. The west-east orientation in Europe has had a profound influence in the development of the region through time. For one thing, it has allowed for the relatively easy *penetration of marine climatic influences* into much of Western Europe. A lofty range of north-south trending mountains, such as the Rockies, across the breadth of Europe would have severely limited the penetration of these influences. It is safe to state that Western Europe would have developed in a far different way had such been the case. One might even hazard the suggestion that had the Alps and Northwestern Highlands formed a continuous barrier from the Mediterranean to the Arctic, Western Europe would be in the underdeveloped rather than the developed world. Such an alignment in the northerly latitudes of Western Europe would most certainly create a far more severe climate than is now found. As a result, the range of agricultural opportunities would be greatly limited and transportation extremely difficult. Space does not allow a discussion of how Western Europe might have developed under such hypothetical conditions. Such speculations are worthwhile, however, because they assist in demonstrating the significance of basic geographic patterns to human economic development.

The Northwestern Highlands

The *Northwestern Highlands* include much of the northern countries of Sweden, Norway, and Iceland as well as a large portion of Finland, the British Isles, and the Brittany Peninsula in northwestern France. Generally speaking, the area is underlain by very hard and geologically ancient rocks. In a few areas like Iceland, volcanism is still actively building masses of igneous rock. Rugged, hilly uplands and windswept plateau surfaces dominate the landscapes. *Isolation and ruggedness,* plus a northerly latitudinal position and exposure to gales and winds, create many serious problems for the people living there.

Several thousand years ago, the climate of this part of the earth differed significantly from the

129

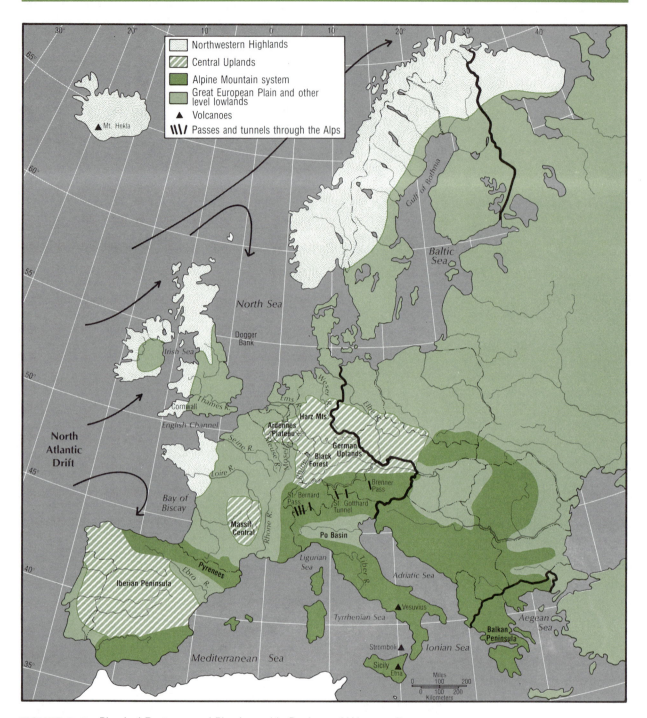

FIGURE 8–3 Physical Features and Physiographic Regions of Western Europe. Most of Western Europe's population lives on the fertile Great European Plain and other lowland areas. Many of the region's minerals, however, are located in the Central Uplands and the Northwestern Highlands. The Alpine Mountains are a hindrance to north-south transport and serve as a climatic barrier.

Glaciation accentuates the rugged character of the Northwestern Highlands in Norway. Many glaciers have scoured broad and deep valleys (fiords) that, along the coast, have been inundated by the sea. In many fiords small fishing harbors are common. (Norsk Telegrambyrå)

present. Annual snowfalls accumulated on highland surfaces and failed to melt in the cool summers. Gradually the great thicknesses of the accumulating snowfields led to the formation of massive *glaciers* of continental scale. These ponderous masses flowed from higher areas toward the sea. In their slow but relentless progress, because of the incredible weight of the ice accumulated over many centuries, they ground and gouged the land over which they flowed. As a result, many areas in the Northwestern Highlands are practically *devoid of soils* and exist as vast stretches of barren rock waste that continue to defy man's attempts to put them to productive uses (Figure 8–4). These landscapes go a long way toward explaining why Norwegians, Icelanders, and other northern Europeans have traditionally *turned to the sea* and its resource potential in their efforts to find a rich and rewarding life-style. Certainly the location of some of the world's most *productive fishing grounds* just off these coasts has encouraged their residents to choose fishing rather than farming as a way of life.

Other visible relics of the glacial heritage of the Northwestern Highlands are the vast areas of marshland and countless *lakes* that dot the landscapes of much of Finland, Sweden, and Ireland. The many lakes serve as reservoirs for *hydroelectricity,* the region's chief source of power. Except for United Kingdom coal, the fossil fuels (petro-

leum and coal) are almost entirely absent from the geologically ancient Northwestern Highlands. Consequently, hydroelectric power installations have played a very significant role in the industrial development there (Figure 8–4). Norway and Sweden are among the world's leading producers of hydroelectricity.

Still other reminders of the glacial heritage are the deep coastal *fiords* that characterize northwestern Europe's coastal zone. These glacially excavated, now-flooded valleys allow the sea to penetrate deeply into the land. In Norway and Scotland, the stark beauty of the fiords has become the basis for a lucrative tourist trade.

The *agricultural potential* of the Northwestern Highlands region is severely limited by ruggedness of topography, thin and infertile soils, remoteness, and the excessively cloudy, wet, cool climate. Agricultural effort is largely *restricted to grazing* of sheep and cattle. In more sheltered inland locations, the grass cover is replaced by hardy coniferous forests, the westernmost extent of the great taiga forest belt that girdles subarctic Eurasia from the Sea of Okhotsk to the North Sea (see Figure 3–3). The trees are cropped to provide the raw materials for a host of forest-related industries such as paper-pulp manufacturing, lumbering, and construction. Large segments of the national economies of Sweden, Finland, and Norway are based on forest-related industries. In fact, Sweden and

131

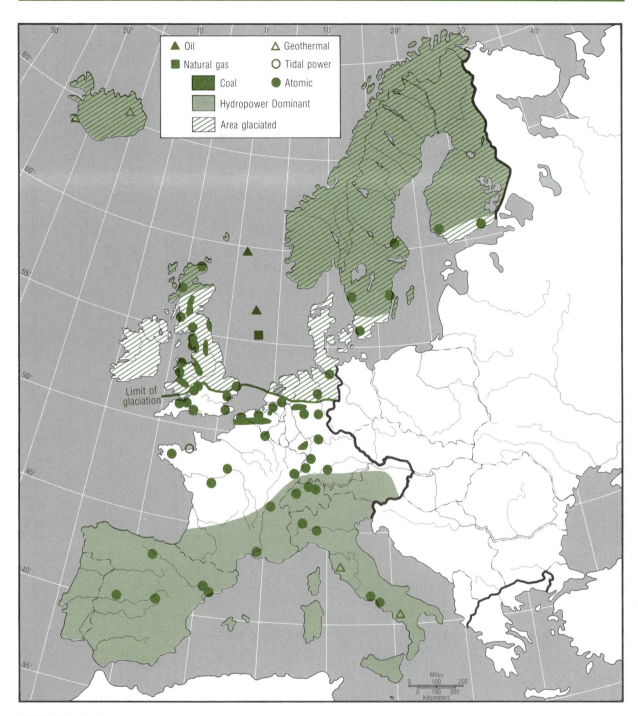

FIGURE 8–4 Energy Sources in Western Europe.
The glacial heritage of the Northwestern Highlands presents limited economic opportunities, but glacial scouring has created a landscape with hydroelectric power potential. In the United Kingdom and adjacent northwestern Europe thick, easily worked coal seams have provided the energy source for modern manufacturing and urban life. Much of southern Europe is power deficient, relying largely on hydroelectric plants in the Alpine Mountain system. Throughout Western Europe atomic power is increasingly important.

The Rhine River is a major transport artery of Western Europe heavily used by Switzerland, France, West Germany, and the Low Countries. Along the Rhine are roads, railways, and manufacturing plants. (United Nations)

Finland consistently rank as Europe's first and second most important exporters of both lumber and wood pulp.

Metallic minerals are found in some of the ancient rock masses of the highlands and frequently support important mining operations. By far the best known of these mining centers is north of the arctic circle near the Swedish cities of Kiruna and Gällivare. Here, ultramodern cities have been built to support the mining of one of the world's great deposits of very high grade iron ore. Such cities in the interior of the Northwestern Highlands are the exception rather than the rule. This region is one of Europe's most thinly settled areas.

There is little in the habitat of the Northwestern Highlands to encourage people to settle here in large numbers. It is largely a land of glaciers, rugged uplands, quiet crystal lakes, rapid-flowing rivers, and dark forests that yield their wealth grudgingly to a small but hardy population.

The Great European Plain

The *Great European Plain* stretches from the Pyrenees Mountains on the west to the Ural Mountains in Soviet Russia on the east as a broad, undulating lowland south of the Northwestern Highlands. In Western Europe it is considerably narrower and is interrupted in places by relatively shallow bodies of water such as the Baltic Sea, North Sea, and English Channel. Generally, the Great European Plain is underlain by geologically younger rocks including many sedimentary layers. Many of these sedimentary rocks are relatively soft and are weathered more easily than the harder crystalline rocks of the adjacent Highlands. The land has a "softer" and more *gently rolling surface* even where local folding and warping have thrust up ranges of hills to break the plain. Although usually of only moderate elevation, these ranges are of considerable local importance because of differences in soil and moisture. On the whole, the region has presented mankind with a stimulating degree of challenge rather than with impenetrable barriers. The plain also is richly endowed with a *wide range of resources*. The Great European Plain in Western Europe is one of the earth's most populous and highly developed regions.

During the glacial periods, part of this region was covered over by the continental ice sheets that slowly flowed down from the Northwestern Highlands. Where the glaciers reached their outer limits, they deposited great quantities of earth materials. As a result, the topography, drainage, and soils in these areas are more varied and complex than those beyond the reach of the glaciers. These differences contribute to the *agricultural variety* that marks much of the Great European Plain today.

As Figure 8–3 shows, the Great European Plain is bounded by high land to the south and seas to the north. Many *large rivers* (Seine, Rhine,

Ems, Weser) flow across the region in a northerly direction. These rivers have provided the people of Western Europe with natural routeways for travel and trade. The Great European Plain itself has served as Europe's major east-west routeway. Today the highways, railways, rivers, and canals of the Great European Plain are integrated into one of the world's most dense and efficient transportation networks.

Many of the world's greatest *cities and industrial centers* have grown here (for example, Paris and Berlin). Perhaps even more fundamental to the growth of these great industrial centers and cities have been the *coal* deposits below ground (Figure 8–4). As a fuel resource, coal remained of purely local significance for a very long time. It was not until the middle decades of the eighteenth century that coal became a factor of major importance to large numbers of Western Europeans. During this period the steam engine was perfected, and coke was also utilized in blast furnaces to process iron ore in large quantities.

The *agricultural potential* of the Great European Plain in Western Europe is immense. In Denmark and the Netherlands, for example, almost 75% of the total land area is used for some form of agricultural activity. The range of crops is also broad, including most of the temperate-climate foods and fibers known today. Contributing to the agricultural potential is the mild, moist *marine climate* that prevails. Rainfall, though not heavy, is evenly distributed throughout all seasons of the year, and temperatures are moderate in summer and winter.

From almost every point of view, the Great European Plain area of Western Europe is richly endowed to provide the basis for the development of modern technologically advanced societies. In such a habitat, alert, inventive, and energetic populations have been able to create some of the earth's richest and most varied life-styles.

The Central Uplands

The *Central Uplands* stretch as a belt of hilly and rugged plateau surfaces across the central part of Europe generally south of the Great European Plain. They are composed of geologically ancient rocks and resemble portions of the Appalachian Mountains of the United States. Like the Appalachians, they are often rounded and of moderate elevation and frequently heavily forested.

Many of Europe's great rivers cross the Central Uplands in *deep valleys* that make movement across the region difficult. Most *routeways* and important *settlements* are located in these valleys and form the chief focus for life in the Central Uplands. Compared to the Great European Plain, the Central Uplands are not as productive or densely populated.

The Central Uplands of Europe are important for their extensive and varied *mineral deposits*. Mining and metalworking had an early start in this region. *Coal* was found near the surface where the river valleys passed from the Central Uplands to the Great European Plain. Typical of this coal-producing zone is the *Ruhr Valley* of Western Germany. The Ruhr ranks among the world's greatest mining and industrial regions today.

The Central Uplands are somewhat less well suited for agricultural enterprises than the Great European Plain. Much of the *terrain is rugged* and steeply sloping. The higher elevations of the Central Uplands have cooler temperatures and more abundant precipitation than the lower European Plain. Cool wet conditions favor the growing of grass and other fodder crops that support herds of grazing animals. Densely populated industrial regions lying in and around the margins of the Central Uplands are ready markets for agricultural products.

The Alpine Region

The *Alpine Region,* which stretches across the southern flank of Europe, is a region of high mountains, rugged plateaus, and steeply sloping land. Parts of *three major peninsulas*—the Iberian, Italian, and Balkan—are included in the Alpine region, as are a number of Mediterranean island groups.

Geologically, the lofty folded Alpine mountains are relatively young. In many respects they are similar to the Rocky Mountains of the United States. Like the Rockies, the Alps have many mountains more than 10,000 feet (3048 meters) above sea level, and some are over 15,000 feet (4572 meters). Breaks in the Alpine system are few but extremely important. These breaks, or "passes," in the mountains have served as traditional *focal points for routes* connecting northern Europe with the Mediterranean Basin. Certain of these Alpine passes, such as the Brenner, St. Bernard, and St. Gotthard, have been among the most important routes of human movement in all history. Today, modern motor and railway tunnels speed travelers through the passes in comfort at all seasons. Petroleum pipelines, conveying oil from the ports on the Mediterranean Sea to the industrial regions of central Europe, are built through the same historic Alpine passes that saw the march of Caesar's armies.

The modern inhabitant of the Alpine region is frequently reminded of the geological youthfulness of the area. *Earthquakes,* a symptom of geologic activity, are common. *Volcanic activity,* too, is a sign of geologic unrest and youthfulness. In the province of Tuscany in modern Italy, volcanic hot springs and steam geysers are tapped to provide *geothermal energy* for industrial uses.

Industrial activity in southern Europe is concentrated in centers such as Milan, Turin, Barcelona, Marseilles, Genoa, and Athens. These centers are located on coastal plains or enclosed lowland basins within the Alpine region. Coal and petroleum are largely absent from this part of Europe. As an alternative source of energy, the rapidly flowing rivers of the Alps have been extensively used to produce *hydroelectricity* (Figure 8–4).

High elevations, thin soils, and steep slopes severely limit agriculture over large areas of the Alpine region. In other areas, however, agriculture flourishes. Vineyards, fruit orchards, and olive groves lend a distinctive character to the *agricultural landscapes* found along the southern flank of Europe. The proportion of people engaged in agriculture is far higher in the countries of this southern region of Western Europe than in the three regions to the north. For example, only about 2% of the labor force in the United Kingdom is employed in agriculture. In Greece, on the other hand, almost 40% of the labor force is engaged in farming.

The Alpine region is quite different from the three other physiographic divisions making up Western Europe. It is a region of tremendous challenge, frequent disappointment, and sometimes even disaster to its inhabitants.

THE CLIMATES OF WESTERN EUROPE

The Alps can be viewed as a *"climatic divide."* To the north, most of the heavily inhabited area enjoys an extremely temperate and moist marine climate. South of the Alps is the dramatically different dry summer subtropical climate. To the north, the marine climate produces a *lush green landscape.* Ireland's nickname, "Emerald Isle," emphasizes this characteristic regional greenness. To the south of the gleaming glaciers and snowfields that crown the lofty Alps, the landscape changes rapidly to the *browns and yellows* of parched Mediterranean fields and purple-gray of mountains.

When the ancient Romans left their Mediterranean homeland to conquer most of Western Europe, they found a strange and hostile world. It is easy to sympathize with Tacitus, who, in about A.D. 100, wrote: "The climate in Britain is disgusting from the frequency of rain and fog." He did find, however, that "the cold is never severe." Countless millions of vigorous and productive humans have flourished in both climatic regions in the past and continue to flourish there at the present time. It would be unwise to say, as some have, that one of the climates is superior to the other. More realistic is the recognition that the *two climatic regimes* are very different and so provide differing challenges and opportunities for the people living in them.

Marine Climate

The *marine climate* owes its essential characteristics of moderate temperatures and abundant supplies of moisture to the Atlantic Ocean, which lies to the west of Europe. Water heats and cools more slowly than land and has a *moderating influence* on climate. In addition, Western Europe lies in the earth's prevailing-westerly-wind belt and is affected by conditions originating over the Atlantic. Finally, the surface waters of the Atlantic to the west of Europe are warmer than one might expect, for they receive a great drift of warm tropical water pushed in a clockwise direction by the prevailing winds of the northern hemisphere. Along the coast of the United States from Florida to Maine, this warm water forms the Gulf Stream. In the Atlantic off Nova Scotia and Newfoundland, the Gulf Stream becomes diffused into a broad area of relatively warm water drifting to the east. This relatively warm surface water is called the *North Atlantic Drift.* Thanks to the prevailing westerly winds and North Atlantic Drift, northwestern Europe has mild moist winters and cool moist summers. These conditions result in a lush green landscape. The climate of Dublin typifies the cool moist conditions that prevail over much of Western Europe north of the Alps (Figure 8–5).

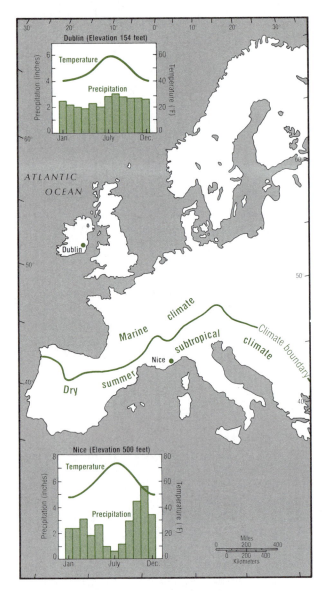

FIGURE 8–5 Average Monthly Temperature and Precipitation at Selected Western European Cities. Climate and weather conditions in Western Europe vary markedly between the north and the south. The moderating influence of the ocean is felt especially in the north. In the south rainfall is less and both rainfall and temperatures vary much more from season to season. Note the differences in average monthly weather conditions between Dublin and Nice.

SOURCE: U.S. Department of Agriculture, *Agricultural Geography of Europe and the Middle East,* 1948.

Dry Summer Subtropical Climate

The southern margins of Europe share a climate that is common to the Mediterranean Basin and often called the Mediterranean climate. It is characterized by clear, dry, hot summers and moder-ately moist, mild winters. *Dry summer subtropical climate* is also found in California, central Chile, southernmost Africa, and southern Australia.

The Mediterranean lies close to the thirtieth parallel of north latitude, which marks the approximate center of the belt of relatively high pressure known as the *Subtropical High.* In summer, this belt of desert-making high pressure shifts a few degrees to the north to cover the Mediterranean Basin. When it comes, it brings the clear sunny skies and dry air that are common to the Sahara, the earth's largest desert. In winter, the Subtropical High pressure belt shifts a few degrees to the south, and the Mediterranean Basin is influenced by the marine air masses and cyclonic disturbances of the prevailing westerly-wind belt. These *marine air masses* and storms bring cloudy skies, cooler temperatures, and moisture to provide the winter rains common to the region. The city of Nice is representative of the pattern of hot dry conditions that characterize Mediterranean summers (Figure 8–5). Notice particularly how dry the three summer months—June, July, and August—are in Nice. This *seasonal drought* is far more important to plant life than is the fact that Nice receives a total amount of precipitation slightly greater than that of Dublin in an average year.

The result of this *wet-dry climatic rhythm* is seen in almost every aspect of the landscape in the southern part of the Alpine region of Europe. The summer drought is extremely hard on many plants that are common elsewhere in Europe. The natural vegetation of the Mediterranean Basin is made up of plant species that resist excessive evaporation and loss of moisture. Some have thickened stems or bark, thorns, waxy coatings, small leaves, or hairy fibers. Succulent (water-storing) plants such as cactus also do very well here.

Human life, too, is geared to the alternation of wet and dry seasons. Farmers plant and tend crops during the winter and spring and harvest them in early summer. *Fruit* and other deep-rooted tree crops such as the *olive* are well adapted and form an important part of the Mediterranean agricultural scene. The olive is so representative of the region that its limits of cultivation are taken by some as the limits of the dry summer subtropical climatic region. *Wheat* probably originated in the area of the eastern Mediterranean and remains the chief grain crop of the dry summer subtropical region. The moisture of the Mediterranean winter is ideal for its germination and growth, the aridity of the summer for its maturation and harvest.

136

The reliably sunny summer weather of Mediterranean Europe has become one of the region's greatest modern economic assets. It is the basis of a flourishing *tourist industry* that provides a large and profitable income. Hundreds of thousands of prosperous Europeans from the northern industrial centers of Western Europe enjoy their yearly vacation holidays in the sun and blue waters of the Mediterranean coast.

CONCLUSIONS

In this chapter several of the more significant aspects of Western Europe's location and physical geography have been discussed to show how they are related to the region's rich and varied lifestyles. On the whole, the view has been continental rather than regional or local in scale. This view has been necessary but not entirely satisfactory, since the meaning of these patterns and elements to the individual living in Western Europe has not been dealt with directly. To understand Western Europe and its position in our unfolding modern world, some insight into local situations is imperative. The following passage, written by one of Western Europe's great humanists, Salvador de Madariaga, is particularly well suited. In his Introduction to the splendid book *Europe from the Air,* Madariaga observes:

. . . Europe is a continent of modest dimensions. There are no boundless open spaces, no mountain ranges towering into the sky, no rivers resembling inlets of the sea, no icy cold and no torrid heat. Everything is moderate—not too hot nor too cold. . . . The physical shape of our continent is exceedingly complicated, mountain ranges, inland seas, and the configuration of its Atlantic coastline divide and subdivide it into numerous areas—rather like a large building with several wings. It is an important point that these different "rooms" in Europe are separated from one another by obstacles which are just big enough to make the division clear, but not big enough for complete isolation. This circumstance may well be at the root of the main features of the European character. For in the "rooms" all that is best in the European tradition has accumulated over the centuries. Here lies the origin of the strongly pronounced local characteristics, whereas the comparatively easy traffic between the "rooms" has at the same time made possible a certain intermingling, a duologue of the blood. And it is probably to this duologue, to this tension between characteristics, that we owe the wealth of intellect and willpower which marks the European. . . . Such is Europe. A landscape of quality, not of quantity, rich in nuances and tensions, where humanity has achieved clear definition not only in the individual but also in the nations. . . .[1]

[1]Emil Egli and Hans Richard Müller (translated from the German by E. Osers), *Europe from the Air* (London: George D. Harrap, 1959), pp. 10–11.

9

Western Europe: Landscapes of Development

Western Europe is a world *culture hearth,* a place from which fundamental changes in human life have flowed. By comparison with most other world culture hearths, Europe flowered much later but spawned many profound and far-reaching changes. Also, it is acknowledged that many of the world's most serious contemporary problems result from the tensions generated as an *exploding European* culture, particularly its technology, confronts traditional non-European societies around the globe. Professor Terry Jordan has suggested:

> The world is in the process of being Europeanized in numerous, fundamental ways. . . . European culture may one day be world culture, as regional differences fade in an increasing acceptance of the European way of life.[1]

EUROPEAN CULTURE

Certainly, "Europeanization" is increasingly visible in the material and tangible aspects of life in even the farthest corners of the inhabited world. The acculturated bulldozer-driving or radar-operating Eskimo of the polar north now has more in common with his American, Canadian, Danish, or Russian co-workers than with his tradition-bound hunting uncles and father. A recent observer at Thule, Greenland, noted:

> Within a generation or two the Polar Eskimo hunting culture may die slowly through attrition. Or the outside world of technology may finally engulf it. Recently, a major oil discovery was made on Ellesmere Island, just west of Thule. Two mining companies are planning explorations on Peary Land, north of Thule. . . . no longer will Thule be the farthest of lands.[2]

---

[1]Terry G. Jordan, *The European Culture Area: A Systematic Geography* (New York: Harper & Row, 1973), p. 15.

[2]Fred Bruemmer, "The Northernmost People," *Natural History,* 83 (February 1974), 33.

Similar accounts could be written concerning the Bedouins of the oil-rich Old World deserts or the Stone Age Indian tribes of Latin America's rain forests—in fact, about primitive and traditional societies around the world. What is it about the Europeans and their culture that makes them such a potent force for world change? Many geographers and others have concerned themselves with providing answers to this question. Although none has been entirely successful, their attempts can be helpful in an effort to gain an understanding of Western Europe, one of the most influential of the world's developed regions.

Professor Jordan has defined the European culture area as all Old World areas in which the people:

1 Have a religious tradition of Christianity.
2 Speak one of the related Indo-European languages.
3 Are of Caucasian race.

To these three basic traits he added ten more that he found necessary to form a detailed areal definition of present-day Europe:

4 A well-educated population.
5 A healthy population.
6 A well-fed population.
7 Birth and death rates far below world averages.
8 An annual average national income per capita far above the world average.
9 A predominantly urban population.
10 An industrially oriented economy.
11 A market-oriented agriculture.
12 An excellent transport system.
13 Nations that are old.

It has already become apparent that most of these additional traits of "Europeanness" are also key *traits of the developed or rich nations* of the world. By evaluating the countries of Europe in terms of an index of these culture traits, Jordan produced a map of their Europeanness similar to Figure 9–1. It is noteworthy that almost all of the countries possessing the highest degree of Europeanness are located in Western Europe. If one accepts Jordan's criteria, it follows that a country's degree of Europeanness is positively associated with *modern technological development* and *widespread wealth* for its citizens. By this reasoning

Spain appears to be less European than Italy. Is this a valid conclusion?

The total area of Western Europe is only about one-third that of Canada. Yet it is divided into eighteen fully autonomous and independent countries plus six semi-independent "micro states": Andorra, Liechtenstein, Malta, Monaco, San Marino, and Vatican City. It is no wonder that Western Europe appears to be *politically fragmented.* Has this pattern of political fragmentation played any role in Western Europe's emergence as a developed region? This question, like many others in these discussions, can only be partially answered, but even partial answers are of value in that they suggest modes of thought and lines of inquiry for further exploration.

In many ways the individual countries have served as *cultural cradles* for the various national groups living in Western Europe. Until a little over a century ago, when steam power was put on wheels to create the first rail transportation systems, the great mass of people spent their lives almost within walking distance of their places of birth. Even modest ranges of hills helped to define natural regions or compartments where *distinctive patterns* of living, working, and speaking developed in relative isolation. In the period before the Industrial Revolution, however, slow improvements in transportation and communications gradually brought many of these neighborhood-like regions together to form larger groupings.

The larger groupings adopted patterns of living that were typical in particularly influential regions called *core areas.* The characteristically English way of life originated in the basin of the Thames River in southeastern England and gradually spread over much of the British Isles. Similarly, the French pattern of life and culture first developed in the fertile region drained by the Seine River and its tributaries. As Professor Norman J. G. Pounds wrote:

> A core-area must have considerable advantages in order to . . . perform [its] role. Simply put, it must have within itself the elements of viability. It must be able to defend itself against encroachment and conquest from neighboring core-areas, and it must have been capable at an early date of generating a surplus income above the subsistence level, necessary to equip armies and to play the role in contemporary power politics that territorial expansion necessarily predicates.[3]

[3]Norman J. G. Pounds and Sue S. Ball, "Core-Areas and the Development of the European States System," *Annals, Association of American Geographers,* 54 (1964), 24.

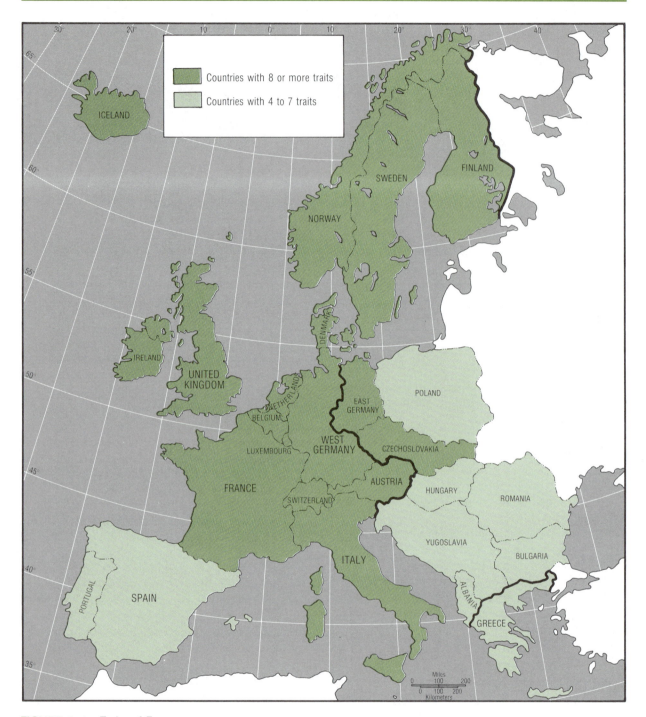

FIGURE 9–1 Traits of Europeanness.
European cultural characteristics are most highly developed in Western Europe. Many European cultural traits have been spread over the earth, and some authorities believe that other cultures will eventually be replaced by a modified European culture.

The *formation of nation-states* as we know them today is a relatively recent development in human organization. In the period following the withdrawal of the Roman Empire from Western Europe, the major language groups evolved along several paths, and these eventually led to the present political and language patterns. Certain states, like Portugal and Spain, unified early. They led the

140

Western Europeans to a position of world influence through ocean exploration. Others, such as Germany, remained unconsolidated and divided until only a century ago. Some, like France, grew through the efforts of strong kings, whereas Switzerland grew by the voluntary association of small regions, the Swiss cantons or counties, into a confederation.

Whatever their origin, the many countries of Western Europe represent the will and ambitions of the people living in them. The *political fragmentation* that has characterized Western Europe in the past several centuries has had a profound influence on the lives and activities of the millions of people living in the culture area. These differences have resulted in an exceptional degree of variety in the ways that the local resources have been developed.

Since World War II, however, the leaders and people of Western Europe have increasingly recognized the *advantages of greater unity.* This recognition has been particularly true in such areas as economic development and military defense. If and when this union comes about, the pattern of political fragmentation that has characterized Western Europe for so long may change drastically. Many serious observers feel that a true "United States of Europe" is already evolving. Many others are equally convinced that the roots of European nationalism reach too deep, and that such a development is more chimerical than real.

THE TREND TOWARD UNITY

In a memorable speech delivered at Zurich University in 1946, Britain's flamboyant wartime leader, Winston Churchill, called for the creation of a "European Family . . . with a structure under which it can dwell in peace, in safety and in freedom." He went on to say:

> We must build a kind of United States of Europe. In this way only will hundreds of millions of toilers be able to regain the simple joys and hopes which make life worth living.[4]

His was only one of countless voices that urged the *unification of Europe* in the dark days that followed the destructive havoc of World War II. A sense of common misfortune resulting from World

War II developed into a resolve that Europe must never again experience the indiscriminate horror of modern warfare. Along with this resolve grew a conviction on the part of many European leaders that the traditional national frameworks of the European states were too narrow. Europe, many felt, must revive in an economic framework of a sufficiently large scale to allow it to compete with the world's "superpowers," the United States and the Soviet Union.

Perhaps the most important force of all encouraging this development was the highly successful *European Coal and Steel Community* (ECSC), which included France, West Germany, and the three small Benelux states—Belgium, the Netherlands, and Luxembourg. Among other goals, the ECSC was organized to ensure that the full potential of the Ruhr coal supplies in West Germany and the Lorraine iron ore deposits in France be realized as Western Europe's vital steel industry was rebuilt. Industrial and economic interdependence replaced traditional rivalry. The economies of modernized, large, efficient mills and mines brought benefits to the workers in both of these important industrial regions as production and profits soared.

It was no coincidence that the members of ECSC became the founders of the current move for Western European unity, the *European Economic Community (EEC)* or "Common Market." The long-standing tradition of a customs union between the Benelux countries (Belgium and Luxembourg since 1921, the Netherlands since 1947) was coupled to the obvious success of the ECSC in the 1950s. In 1955 the governments of Belgium, Luxembourg, and the Netherlands urged the member states of the ECSC to take a new step on the road toward European integration. They stated that they

> consider that the establishment of a united Europe must be sought through the development of common institutions, the progressive fusion of national economies, the creation of a large common market and the progressive harmonization of social policies.[5]

Meetings that followed led to general agreement on the idea of the Common Market. It was ratified in the *Treaty of Rome* signed in March 1957, and the European Economic Community or Common Market became a major fact of European life. In 1958 the European Common Market began actively to remove gradually the differences that ex-

[4]S. Patijn, *Landmarks in European Unity* (Leiden, The Netherlands: A. W. Sijthoff, 1970), p. 29.

[5]Ibid., p. 93.

isted among the economic policies of its six members and bring about prosperity and harmonious development.

In June 1979 the citizens of these nine wealthy countries took part in the world's first truly international election when they voted to choose the members of the *European Parliament.* The European Parliament's function is to exercise democratic control over the executive and administrative institutions of the Common Market.

The European Parliament does not sit in continuous full session. This would be impossible because many of its members are also elected representatives in the respective national parliaments of their home countries. Instead, sessions are held for, on average, one week in each month. Although important, these full meetings form only a small part of the Parliament's work. As in the United States Congress, much of the work is carried on in specialized standing committees and political groups. Full sessions of the European Parliament alternate between the European Parliament headquarters in Luxembourg and the Palace of Europe in Strasbourg, France. Much work is also carried on in Brussels, Belgium, where parliamentary committees conduct their meetings.

With the acceptance of Greece as a full member of the European Community on January 1, 1980, the Common Market grew to a total of ten member states. The four largest members, France, West Germany, Italy, and Britain, have 81 seats each in the Parliament. The Netherlands has 25 seats with 24 each for Belgium and Greece. The three smallest states, Denmark, Ireland, and Luxembourg, have 16, 15, and 6 seats respectively.

Although it does not enjoy all the powers that are usual to the national parliaments of its member states, the *Parliament of Europe* is steadily increasing its powers in managing the affairs of the European Community. In the normally crucial area of budget, for example, it has the last word on all items which are classed "nonobligatory" and total about one-quarter of all funds appropriated. The Parliament also has the power to reject the Community's budget as a whole and thus force a new one to be drawn up and proposed. It will be interesting to observe how the Parliament of Europe evolves in the years ahead.

The creation of the parliament was not easy to accomplish in the event-filled decade of the 1960s. Bitter partisan political battles and even riots took place as the Common Market countries moved from traditional and *strongly nationalistic policies*

toward the ideal of *economic and political unity.* Nevertheless, as the 1970s opened, the economic success of the Common Market was clear to all.

Norway, Denmark, Ireland, and the United Kingdom began to seek membership actively in 1972. In a referendum held in September of that year, however, the Norwegians voted against the move. As a result, only the United Kingdom, Ireland, and Denmark completed membership negotiations and became fully fledged partners in the Common Market with Belgium, France, the Federal Republic of Germany, Italy, Luxembourg, and the Netherlands on January 1, 1973.

POPULATION PATTERNS

Although less familiar than the political pattern, Western Europe's pattern of *population distribution* is even more important to our understanding of this highly developed region. The complex interaction of a wide range of human and physical factors is responsible for where people choose to live and center their activities. It might even be argued that if one could know where and how the people of any area live, one would be a long way toward achieving a full understanding of the area's physical geography and cultural history, as well as of the people's way of life.

In Chapter 8 mention was made of the significance of environmental conditions to the pattern of population. For example, the rather harsh environments of the Northwestern Highlands have not attracted the vast numbers of people who have chosen to live on the richly endowed Great European Plain. It should be kept in mind, however, that modern Western Europeans are not mere passive pawns moving in response to their environments. On the contrary, the Western Europeans are probably closer to being the *masters of their physical environment* than are the people of any other great culture region of the world. Still, it is true that most human decisions of a spatial nature reflect to some extent the physical environment.

Population Distribution

Although by world standards Western Europe is *densely populated,* its 350 million inhabitants are unevenly distributed within their homelands. As the map of population distribution (Figure 9–2) shows, northern Europe and the Alps have sparse populations. On the other hand, the Great Euro-

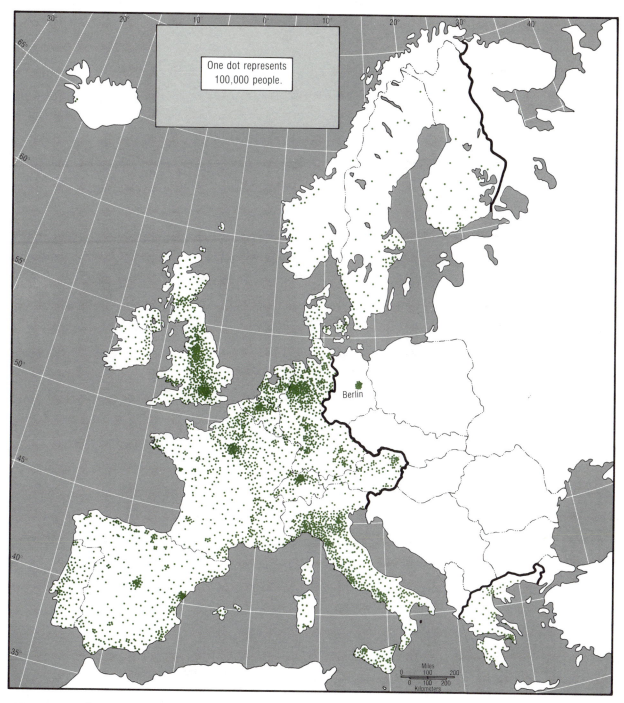

FIGURE 9-2 Distribution of Western Europe's Population.
Western Europe is one of the most densely populated areas in the world. In contrast
to other principal population centers (China and the Indian subcontinent), Western
Europe's population is primarily urban and has a high level of living.

pean Plain, particularly near coal sites, has a very
high density. Belgium and the Netherlands, two of
Western Europe's most densely settled countries,
lie here. The Po Valley and adjacent northern and

central Italy are also densely settled. So, too, are
the areas flanking the Rhine River corridor from
Switzerland northward. Other densely populated
areas include the central lowlands of Scotland,

portions of central England, the London Basin, the Rhône-Saône corridor in southern France, and coastal Portugal and Spain. These areas of dense productive population help to form the *economic and political core* of Western Europe.

Population Change Through Time

It is significant that *population distribution has changed through time*. For example, at the beginning of the Christian era about 2000 years ago, conditions were very different. Then the areas of densest population in Western Europe were located along the Mediterranean Sea. It was in southern Spain, France, Italy, and Greece that the great classical empires of Greece and Rome developed flourishing agrarian economies and an urban focus to life. Most of Europe north of the Alps was thinly peopled. Table 9–1 gives a summary of population change by region from A.D. 1 to the present.

Notice the *decrease in total population* during the first ten centuries of the Christian era. After the *decline of the Roman Empire* and its well-integrated economic system, Europe suffered a long period of economic collapse, famines, epidemics, and invasions of barbarians from the north and east. The toll in human life was enormous. In the 700s, the great Muslim empire of the Middle East and North Africa conquered the Iberian Peninsula and threatened to overrun Western Europe. It is no wonder that the term *"Dark Ages"* is used to identify this period of Western European history.

Gradually population began to grow again as relative stability returned to Europe. By 1340, it had reached 57 million, double what it was at the beginning of the era. In 1348, disaster struck in the form of the *Black Death,* as the bubonic plague was called. Between 1348 and 1350, it is estimated that one-quarter of all Europeans died from the plague. From 1348 to 1379, England's total population dropped from an estimated 5.7 million to 2 million, more than 50% in the course of one lifetime. *The Hundred Years' War* added to the toll of lives in the period following the first outbreaks of plague. Wars should not be overlooked as a factor in contributing to Western Europe's high death rates. The barbarian invasions and Hundred Years' War contributed to a net decline in total numbers. In the period 1618–48, the *Thirty Years' War* once again made war ravages a factor in overall population growth.

The Demographic Transformation

With the *Peace of Westphalia* in 1648, Europe once again achieved relative peace and stability. Families were large, with an average of from six to eight children sharing in the labor associated with the almost universal agricultural way of life. The Black Death also disappeared as a scourge during this period. From 1650 to 1750, Western Europe's population grew by 20 million to place the total at approximately 100 million on the eve of the Industrial Revolution. The year 1750 is a momentous date for another reason, since it marks the point at which Western Europe began the *demographic transformation* (see Chapter 2). Figure 9–3 presents the actual population statistics for England and Wales in the form of the demographic transformation model (compare Figure 2–3).

Stage I is characterized by rather static population growth brought about by the very high death rates that counterbalanced the high birth rates of the still *traditional agrarian way of life* in early eighteenth-century England and Wales. At one point in the late 1730s, the birth rate was about 3.9% and the death rate also was 3.9%. As a result, there was no natural increase in the population in that year. By 1750, a clear trend of continuing high birth rates accompanied by sharply dropping death rates shows that England and Wales had entered *stage II* of their demographic transformation with the consequent accelerating growth of total numbers. A huge surplus of births over deaths could spell disaster to traditional agricultural society, but in Western Europe it coincided with the *Industrial Revolution* and the opening of *overseas empires* and other emigration opportunities on a world scale. Rather than disaster, it provided the substance of Europe's most important export of all times—people. Western European explorers and colonizers followed by immigrants carried their culture to the far corners of the world and set in motion processes of change which are still going on.

By 1880, the population of England and Wales had grown to approximately 26 million, and a noticeable decline in the birth rate began. The *urban-industrial way of life* had come of age, and attitudes toward family size were changing *(stage III)*. Children were no longer viewed as economic assets, as child labor laws were enacted and formal education became widespread. More women joined the work force and sought freedom from the burdens of numerous pregnancies and large

TABLE 9–1 Estimated Western European Population, A.D. 1–1983 (Selected Years)

Year	Population (millions)	Iberia, Italy, Greece		France, Benelux		British Isles		Scandinavia, Finland, Iceland		Germany, Austria, Switzerland	
		Millions	Percent	Millions	Percent	Millions	Percent	Millions	Percent	Millions	Percent
1	27.0	16.5	61.2	6.6	24.4	0.3	1.1	0.3	1.1	3.3	12.2
350	18.9	10.0	52.9	5.2	27.5	0.3	1.6	0.2	1.1	3.2	16.8
600	13.2	7.2	54.5	3.1	23.5	0.7	5.3	0.2	1.5	2.0	15.2
800	22.0	11.6	52.7	4.9	22.3	1.2	5.4	0.3	1.4	4.0	18.2
1000	25.9	14.1	54.4	6.1	23.6	1.5	5.8	0.4	1.5	3.8	14.7
1200	37.7	17.6	46.7	9.8	26.0	2.9	7.7	0.5	1.3	6.9	18.3
1340	57.3	21.0	36.6	18.9	33.0	5.6	9.8	0.6	1.0	11.2	19.5
1400	32.2										
1500	43.1	15.1	35.0	16.2	37.6	3.9	9.0	0.6	1.4	7.3	16.9
1650	76.0	26.0	34.2	30.0	39.5	7.0	9.2	2.0	2.6	11.0	14.5
1700	80.3	26.0	32.4	27.2	33.9	7.9	9.8	4.5	5.6	14.7	18.3
1750	96.6	30.8	31.9	32.2	33.3	9.8	10.2	5.6	5.8	18.2	18.8
1820	132.3	42.0	31.7	35.7	27.0	21.0	15.9	6.3	4.8	27.3	20.6
1900	235.1	70.5	30.0	54.9	23.3	39.2	16.7	11.7	5.0	58.8	25.0
1930	280.0	80.0	28.6	60.0	21.4	50.0	17.8	15.0	5.4	75.0	26.8
1950	324.5	99.0	30.5	71.5	22.0	60.5	18.6	16.5	5.1	77.0	23.7
1983	351.4	114.5	32.6	79.3	22.6	59.5	16.9	22.5	6.4	75.6	21.5

SOURCES: Terry G. Jordan, *The European Culture Area: A Systematic Geography* (New York: Harper & Row, 1973); *Population Data Sheet* (Washington, D.C.: Population Reference Bureau, 1983).

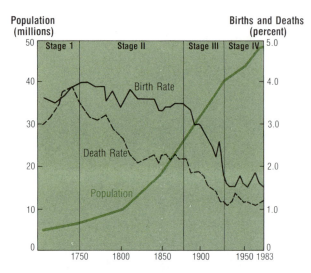

Population (millions) — Births and Deaths (percent)

Stage 1 | Stage II | Stage III | Stage IV

Birth Rate

Death Rate

Population

1750 1800 1850 1900 1950 1983

FIGURE 9–3 Demographic Transformation of England and Wales.

The model of demographic transformation appears as Figure 2–3. Here we see not a theoretical diagram but what has actually happened. In fact, the idea for a demographic transformation model was based on the analysis of Western European birth and death rates.

families. The result was a shift toward smaller families with a consequent sharp drop in birth rates. At the same time, growing affluence and improved medical knowledge and public health care brought about an equally sharp decline in death rates.

These trends, which began in England and Wales, diffused across Europe from west to east until, by the 1930s, most of the nations of Europe were clearly in *stage IV* of the demographic transformation.

PATTERNS OF INDUSTRIALIZATION

The American industrial geographer E. Willard Miller has observed:

> The Industrial Revolution was largely a revolution in energy consumption, and coal predominated as the source of energy for mechanical power until after World War I.[6]

A pattern of *industrial coalescence* in the vicinity of easily worked coal deposits began in Britain and spread to the rest of Western Europe in the nineteenth and early twentieth centuries.

[6]E. Willard Miller, *A Geography of Manufacturing* (Englewood Cliffs, N.J.: Prentice-Hall, 1962), p. 130.

146

Locational Shifts in Industry

Figure 9–4 shows the *spatial shift* that has taken place in the iron and steel industry of south Wales. An impressively large charcoal-iron industry was flourishing in the mid-eighteenth century with no relationship to the coal seams that underlay the region. The charcoal-iron industry of 1750 was so large, in fact, that the wood supply was growing short, and the industry was beginning to drift to the rugged and still forested valleys of western

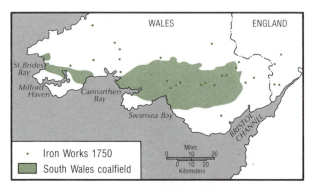

WALES — ENGLAND

St.Brides Bay — Milford Haven — Carmarthen Bay — Swansea Bay — BRISTOL CHANNEL

• Iron Works 1750
South Wales coalfield

Miles 0 10 20 / 0 10 20 Kilometers

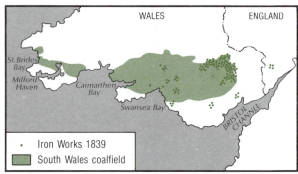

WALES — ENGLAND

St.Brides Bay — Milford Haven — Carmarthen Bay — Swansea Bay — BRISTOL CHANNEL

• Iron Works 1839
South Wales coalfield

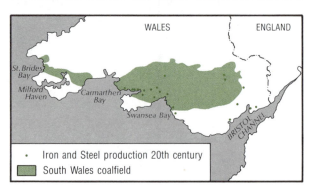

WALES — ENGLAND

St.Brides Bay — Milford Haven — Carmarthen Bay — Swansea Bay — BRISTOL CHANNEL

• Iron and Steel production 20th century
South Wales coalfield

FIGURE 9–4 Industrial Coalescence in Southern Wales Since 1750.

The Industrial Revolution brought many changes to Western Europe. One change was the gradual coalescence of manufacturing around coalfields. The change in the distributional pattern of iron and steel plants in southern Wales illustrates this coalescence.

Wales and the interior. The map for 1750 shows this dispersed pattern quite clearly.

Abraham Darby's successful experiments with *coke* as a blast-furnace fuel were studied by other ironmasters and led to the widespread adoption of the new technique. By the end of the eighteenth century, scarcely any charcoal was being used in the blast furnaces of south Wales. The occurrence of *coal, iron ore, and limestone* in proximity to one another became the crucial factor in the *location of the iron industry.* The 1839 map shows a distinct clustering of ironworks on the northeastern outcrop of the coalfield near easily mined deposits of iron ore and limestone.

In the modern period the *Welsh iron and steel* industry is increasingly dependent on imported iron ore and steel scrap for its operations. This shift in raw material supply has resulted in a *peripheral pattern,* with the steelworks being located near harbors. These kinds of locational shifts did not take place without having severe impacts on both people and the environment. Derelict buildings and mine-head works, spoil heaps, and serious subsidence often blight the landscape in the older industrial regions of Western Europe.

Recent Industrial Development

In more *recently industrialized countries,* coal has played a much less significant locational role. Italy and Sweden are among the best examples of industrialization in the absence of local coal supplies. In both countries *hydroelectricity,* a more recently perfected energy source, has played a strong role

Turkish migrant coal miners at a German mine in Essen. Coal served as the major power source for the Industrial Revolution. The post-World War II years have witnessed major intra-European migration for purposes of work. (United Nations/Y. Nagata)

in *industrial location.* In Italy about three-fourths of all manufacturing takes place in the northeast in the Po Valley. Within this region no single factor accounts for the industrial pattern. Most industrial materials must be brought into the area from outside, thus giving a distinct advantage to cities that are well served by transportation facilities. Often these are very old places, like Turin and Milan, with rich cultural traditions of an earlier age. Also, hydroelectric power distributed cheaply over wide areas lends a high degree of flexibility to industrial location choice. Similar conditions prevail in central Sweden around Stockholm on the Baltic and Göteborg on the Atlantic. In both this lake-studded northern area and Italy's northeastern plain, a diversity of *light industrial activities* blends in the urban and rural landscape with a greater degree of visual harmony than is usually found in the older and more concentrated coalfield-oriented industrial belts of Britain, France, and Germany.

Tariffs and Boundaries as Locational Factors

Another set of *locational factors* has been traditionally of great importance to Western European industrialization but now seems to be declining in significance. They are *tariff policies* and the location of international boundaries, aspects of the political fragmentation discussed earlier. *High tariff walls,* which were a widespread feature in Western Europe prior to World War II, encouraged the development of a broad range of industries within each European country. Often these industries had relatively *inefficient* plants that could not have competed with larger foreign producers. They were literally kept alive through the protection of high tariffs that forced the price of imported goods to levels higher than would otherwise have been necessary. Likewise, a factory located near an international political boundary often suffered a distinct disadvantage because the boundary acted as a *barrier* to the distribution of products when import duties and tariffs were imposed. The effect was a great addition to the cost of the factory's products at the boundary and consequent lowering of their competitive position in the market of the neighboring country. Also, fears of invasion or war damage in time of tension made border-zone locales less attractive in the past.

These problems are greatly minimized in present-day Western Europe. The member nations of the Common Market now form one economic unit, thanks to the elimination of barriers to movement of raw materials, labor, capital, and finished goods.

THE PATTERN OF AGRICULTURE

In Western Europe, cultural factors and physical environmental conditions have interacted to produce an extremely *varied pattern of agricultural activity*. Despite the region's high state of industrial and associated development, agriculture remains the dominant form of land use.

Three fairly distinctive types of agricultural systems can be identified:

1 Mediterranean market gardening and orchardry.
2 Dairy farming.
3 Mixed livestock and crop farming.

In each case a large area of the region is dominated and characterized by the particular system (Figure 9–5).

Mediterranean Market Gardening and Orchardry

The region dominated by the *Mediterranean system* of *market gardening* and *orchardry* is shown as forming a fairly continuous ribbon along the Mediterranean coast. Agriculture has an ancient heritage in the Mediterranean Basin where, since very early times, *wheat* has grown through the moist, mild winter season to provide a harvest in the spring. Several drought-resistant, deeply rooted *vines and trees,* particularly the grape and olive, also characterized classical Mediterranean landscapes. In many respects agricultural activities here closely resemble those in California, where a similar climatic condition prevails. Most of the activity is extremely *labor intensive*. The resulting high proportions of the total labor forces of countries like Spain, Greece, Portugal, and Italy that are engaged in agricultural pursuits is shown on Table 9–2.

Recent developments in the Mediterranean market gardening and orchardry region result from accelerating *specialization and commercialization*. These developments are in response to the huge market found in the industrialized and urbanized countries to the north. Improved transportation facilities make it possible for Mediterranean growers to speed highly *perishable fruits and early vegeta-*

bles to the affluent markets of Germany, France, Britain, Benelux, and Scandinavia. A notable trend has been for whole districts to specialize in one or a few crops. Within the French Mediterranean belt, for example, vast areas have been turned into a huge *mer de vignes,* or "sea of vines," so dominant are the vineyards. Along the Mediterranean belt in coastal Spain, irrigated areas called *huertas* present a similar scene, with citrus trees in place of vines. Olives, too, dominate extensive areas, lending a distinctive gray-green appearance to the landscape. Early tomatoes are cultivated extensively for the northern European market. Perishable melons, apricots, and table grapes flow to the same market. Tobacco, another labor-intensive crop, is also an important source of income, particularly in Italy and Greece.

Goats and sheep have traditionally provided the milk for the regions's famous cheeses. In some mountainous areas, large flocks are still maintained by small numbers of herders who follow a seasonal herding rhythm called *transhumance*. Transhumance involves the grazing of animals in high mountain pastures during the dry Mediterranean summer and in the lower areas during the moist winter when grass flourishes.

The future of Mediterranean agriculture appears bright, since the farmers enjoy a considerable advantage over other subtropical producers in the world. They are near to the rich and growing markets of urbanized northern Europe. Also, local industrialization promises to increase steadily and provide growing markets within the region. Yet a history of deforestation and exploitive farming has resulted in serious erosional problems. Conservation measures such as terracing and reforestation receive strong support from governmental and private agencies.

Dairy Farming

As the map of Western European agriculture indicates, much of the region to the north of the Mediterranean is devoted to *dairy farming*. Dairying is usually associated with a rich, urban population of Western culture. The dairy belt stretches from the British Isles and Brittany Peninsula along the shores of the North and Baltic seas to Finland. A second large area of dairying has developed in the Alps of Switzerland and Austria. This region is one of the cloudiest and coolest portions of Western Europe. Many field crops, such as wheat, often fail to mature or suffer fungus attack under these

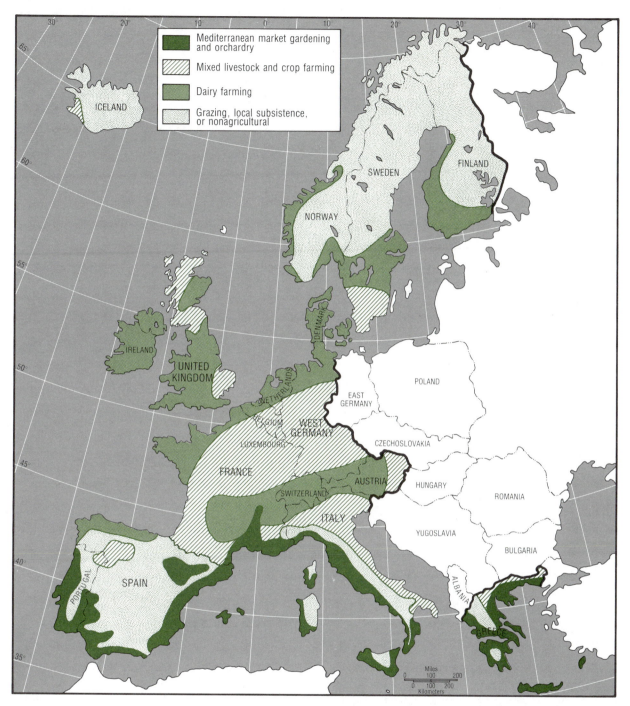

FIGURE 9–5 Agricultural Regions of Western Europe.
Although Western Europe is a center of industrialization, agriculture has always been an important activity. Today most of Western Europe's agricultural activities are highly productive and intensive.

conditions, whereas hay and a number of other fodder crops flourish. In the rugged Alpine terrain, the steep, grass-clad slopes can be grazed or cropped for hay with little or no danger of soil erosion.

The long tradition of dairy farming in northern Europe is clearly evidenced in the fact that almost every major *dairy cattle breed* originated here. The breeds carry names that indicate their origin. The

149

TABLE 9–2 Distribution of Western European Labor Forces in the Main Employment Sectors

Country	Agriculture (%)	Industry (%)	Services (%)
Belgium	3.0	34.8	62.2
Denmark	7.0	33.0	60.0
France	8.8	35.9	55.3
Germany, West	6.0	44.8	49.2
Greece	30.3	30.2	39.5
Ireland	19.2	32.4	48.4
Italy	14.2	37.8	48.0
Luxembourg	6.3	38.4	55.3
Netherlands	4.6	32.0	63.3
Norway	8.5	29.7	61.8
Portugal	28.6	36.1	35.3
Spain	18.8	36.1	45.1
Switzerland	7.4	39.3	53.3
United Kingdom	2.6	38.0	59.4

SOURCE: *Basic Statistics of the Community* (Luxembourg: Statistical Office of the European Communities, 1981).

large Brown Swiss was originally bred in Schwyz Canton in central Switzerland. The popular Jersey and Guernsey breeds originated on the small British Channel Islands near the coast of France. The Holstein originated in the area known as Schleswig-Holstein in northern Germany. These breeds and others less well known were carried with the colonists who traveled overseas in more recent centuries. Herds of these distinctly northern European animals are a familiar sight in New Zealand and Australia as well as in the United States and Canada.

The nature and intensity of dairying vary somewhat from region to region in northern Europe.

Thatched farm building in Denmark. Most Danish farms are small and produce products of high value. Butter, ham, and bacon are among the country's principal exports. (Danish Ministry of Foreign Affairs)

Denmark, for example, has developed a strong *specialization in butter* production. Over 16% of all butter that is involved in international trade originates in Denmark. Skim milk, which is a by-product of butter production, is used to fatten pigs. The pigs in turn are utilized in the production of world-famous Danish ham and bacon. In the Netherlands much grazing takes place on permanent man-made *polder pastures* where high water tables preclude the cultivation of many crops. Here specialization in the production of *cheese* and *condensed and powdered milk* allows the Dutch to export much of their output to markets throughout the world. In the United Kingdom much *fluid milk* is marketed locally for human consumption.

The proximity of Western Europe's most densely settled urban and industrial areas seems an ideal insurance that dairy farming will continue to prevail as a vital agricultural activity. Changes will continue, however, as small-farm operators diminish and mechanization and specialization of the industry continue.

Mixed Livestock and Crop Farming

The intensive crop specialization of the Mediterranean market gardening and orchardry region is separated from the animal-dominated dairy farming belts to the north by a broad zone where a *mixed form of farming* characterizes the landscape. Here environmental conditions facilitate a wide range of agricultural activities, and an extremely mixed and varied agricultural landscape is found. Diversity in this area is also promoted by political fragmentation that has discouraged the evolution of regional belts of agricultural specialization such as those found in the United States, Canada, Australia, and the Soviet Union.

Mixed livestock and crop farming is based on the *subsistence farming systems* that evolved here during the medieval period. Crops came to dominate the scene on fertile, easily tilled soils like the loess lands stretching across northern France to central Germany. Livestock assumed significance in valley areas of heavy clay soils or on cool, moist uplands. *Swine* were particularly important among Germanic groups, who grazed them in herds in heavily forested areas.

To maintain soil fertility in the absence of chemical fertilizers, *crop-rotation* schemes played an important role. So, too, did animal manures. In many respects the medieval farming scene was one that would have delighted the modern ecology-conscious conservationist.

As population pressure increased in postmedieval Europe, this traditional way of agriculture began to change. New high-yielding crops such as the *white potato* and *maize (corn)* were introduced from the New World. Each found an important niche in Western Europe's agricultural system. The potato was ideally suited to the cool, moist conditions of the north, and maize to the sunnier conditions of the midsouth.

Table 9–3 illustrates a striking *change* that is occurring in Western European mixed agriculture—the rapidly declining importance of small farms. Notice the dramatic drops in the percentages of the traditional small, family-operated type of farm in the fifteen-year period from 1960 to 1975 in highly developed countries like France, Belgium, and Germany where mixed livestock and crop farming are widespread. It is clear that many problems of adjustment lie ahead for Western Europeans, as agriculture ceases to be a way of life for many and becomes a technologically complex form of business enterprise.

THE PATTERN OF URBANIZATION

Urbanized societies dominate the Western European scene today. Such societies, in which the majority of people live in towns and cities, result from a new step in humanity's *social evolution*.

TABLE 9–3 Percentage of Western European Farms Under 25 Acres (10 Hectares) in Size

Country	1960	1975
Belgium	75	53
Denmark	47	32
France	56	36
Germany, West	72	55
Greece	96	N.A.
Ireland	49	38
Italy	89	86
Luxembourg	40	32
Netherlands	54	47
Spain	79	N.A.
Switzerland	77	N.A.
United Kingdom	N.A.	27

SOURCE: *Basic Statistics of the Community* (Luxembourg: Statistical Office of the European Communities, 1975).
N.A. = not available.

Modern cities form large and dense agglomerations involving their populations in a degree of human contact and social complexity never before witnessed. Few people seem to comprehend fully either the newness of such great *urbanization* or the speed with which this process has been taking place. In the words of an eminent sociologist, Kingsley Davis:

Before 1850 no society could be described as predominantly urbanized, and by 1900 only one—Great Britain—could be so regarded. Today, only 65 years later, all industrial nations are highly urbanized, and in the world as a whole the process of urbanization is accelerating rapidly.[7]

One should not, however, jump to the conclusion that cities as such are new to Western Europe. They are not. What is new is the overwhelming influence of the city form and function that have become the norm of modern Western European societies.

Early Cities

The *city of the classical age* was first introduced into Europe by the ancient Greeks and Phoenicians as they spread their commercially oriented civilizations around the Mediterranean Basin. The Romans carried the urban form of life deep into Western Europe as they founded new cities in their conquered provinces to the west, east, and north.

[7]Kingsley Davis, "The Urbanization of the Human Population," *Scientific American*, 213 (September 1965), 41.

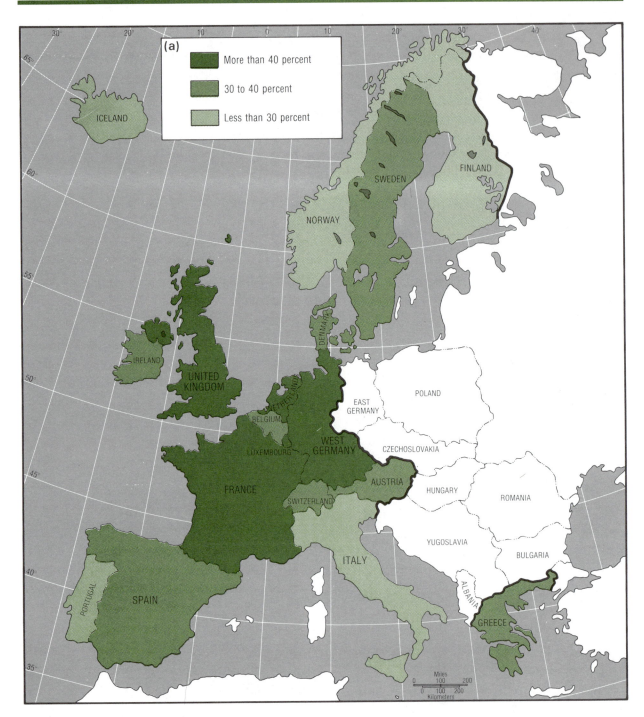

FIGURE 9–6 Urban Population of Western Europe.
Western Europe is one of the world's most highly urbanized regions. Whether one uses only urban areas of more than 100,000 inhabitants, shown in (a), or a smaller threshold (over 5,000 or over 20,000) as used by the various national governments (b), it is clear that urban life-styles and economic activities dominate the region.

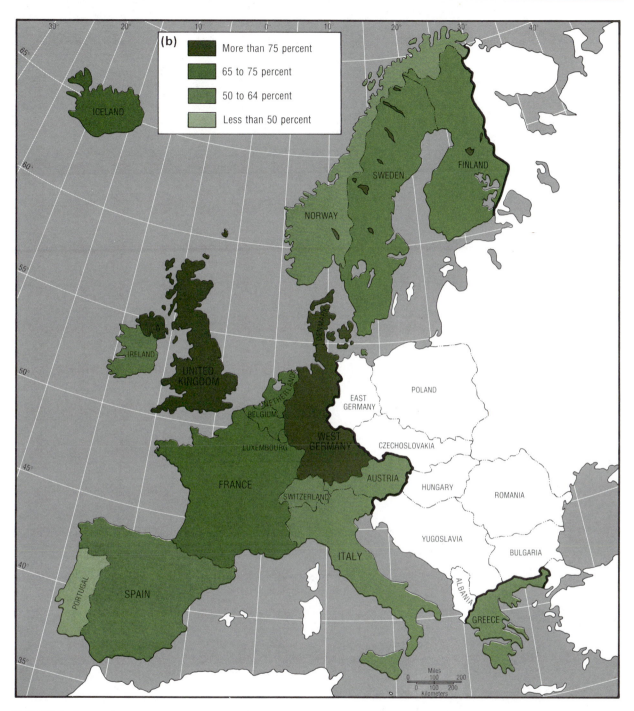

FIGURE 9–6, *continued*

With the breakup and decline of the *Roman Empire* in Europe, a period of urban decline was experienced. When the Pax Romana disappeared from Western Europe, commerce declined sharply, and the *subsistence societies* that arose in the wake of Roman withdrawal had little need for cities. Some commercial towns reverted completely to the status of agricultural villages, while others were abandoned and forgotten. A few with considerable religious or political significance survived the *Dark Ages,* but by and large the urban way of life ceased to be a dominant feature of the Western European scene until the Middle Ages when commercialism once again became important.

During the *Middle Ages,* many of the moribund Roman towns were revived and once more became the centers of *urban commercial activity.* Also many new towns were established both within and without the limits of the old empire. Often the new towns grew up around a fortified preurban core where a feudal lord or ecclesiastical authority assured a degree of security and protection. As they grew, the towns sought and gained *political autonomy* that was spelled out in formal charters from the ruling prince of the area. These medieval cores continue into the present in many of Western Europe's towns, and the *narrow streets* cause tremendous problems for automobile users. A study of 141 West German cities indicated that well over three-quarters of the total urban street mileage is too narrow for safe two-way traffic in any significant amount. The pressure to create parking lots and wider streets is growing in cities across the region. In some areas, such as

Western Europe's love affair with the automobile has created serious problems of traffic congestion, especially in the old urban centers. (United Nations)

Florence and Rome in Italy and Athens in Greece, *air pollution,* primarily resulting from automobile exhaust, is threatening priceless works of art and sculpture as well as ancient marble temples. Coping with the automobile is one of the major problems facing most Western European city authorities at the present time. Pollution and *congestion* are the unwanted by-products of Western Europe's current love affair with the automobile.

Industrialization and Contemporary Urbanization

Clearly, it was *industrialization* that brought the present pattern of *urbanization* to the people of Western Europe. In about the year 1800, nearly 10% of the people in England and Wales were living in cities of 100,000 or more. This proportion doubled to 20% in the next forty years and doubled once more in another sixty years. By the year 1900, as we have seen, the British were an urbanized society, the first in the modern world. Western Europe as a whole has a population that is slowly increasing at the present time. Its urbanized population, on the other hand, is growing more rapidly. One commonly used *index of urbanization* is the percentage of a country's total population that lives in large urban areas of more than 100,000 population. The pattern that emerges when these data are mapped (Figure 9–6a, page 152) shows that only a handful of the most peripheral Western European countries have fewer than 30% of their people living in large metropolitan centers. If, however, different criteria were used, such as how many people live in towns and cities of more than 5,000 or 20,000 in population (often used by various national census bureaus), a much higher percentage of urban living emerges. Notice on Figure 9–6b (page 153), which portrays urban percentages as they are defined by the individual national government, that only Portugal claims less than 50% of the people as urban dwellers.

Suburbanization

It is important to note that much of modern European urbanization is really what is better described as *suburbanization.* Modern trends of urban living in Europe, as in other areas of the developed world, bring about decreasing population densities in central cities. As a result, ever-increasing amounts of rural and agricultural land are being converted to urban uses. In England and Wales,

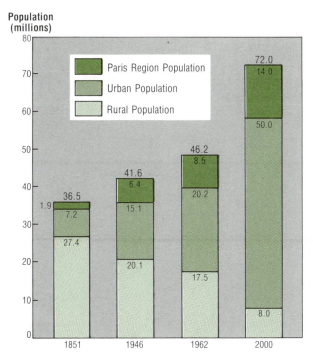

FIGURE 9–7 Populations of Urban and Rural France and of Paris, 1851–2000.

Western Europe is becoming increasingly urban. In this chart the changes in the rural and urban populations of France illustrate the urbanization process since 1851. Paris, the primate city of France, has shown spectacular growth. It is predicted that an ever-increasing proportion of France's population will be urban.

SOURCE: Service de Press et d'Information, *France, Town and Country Environment Planning* (New York: Ambassade de France, 1965).

lem in environment planning" nationally.[8] During the period 1954–62, France's total population rose 8.1%, but its urban population rose by 13.8%. Figure 9–7 shows the breakdown of French population in the years 1851, 1946, and 1962, and as it is projected for the year 2000. In their national planning to cope with the problems of increasing urbanization of the Paris region, French officials have adopted a scheme of encouraging growth in selected peripheral cities that are designated as *"Regional Metropolises"* (Figure 9–8). Great effort is being put into equipping these cities with high-level facilities in the areas of research, higher education, medical care, government, culture, and communications in an attempt to counterbalance the lure of Paris, the *primate city.* As now scheduled, these cities should soon function as full-scale regional metropolises. It is felt that planning on such a comprehensive scale is necessary to

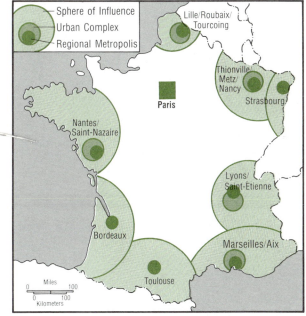

FIGURE 9–8 Regional Metropolises of France and Their Spheres of Influence.

In an effort to diminish the population growth of Paris and to spread urban economic activities to other cities, French officials have devised a plan to encourage growth in selected regional centers. These regional centers receive special aid to stimulate their growth.

SOURCE: Service de Press et d'Information, *France, Town and Country Environment Planning* (New York: Ambassade de France, 1965).

for example, only about 5% of the total land area was classified as urbanized in 1900. Fully 11% of England and Wales was classified as urban in 1972. This slow but inexorable process is continuing to change Britain's famed "green and pleasant land" to an *increasingly urbanized,* built-over, and paved landscape. To most Britons urbanization is a disconcerting prospect, to say the least. In West Germany and the Netherlands, the prospect is even worse. By the year 2000, less than a generation away, one-quarter of the Netherlands will be urbanized, while the totals for Britain and West Germany will be about 15% and 18%, respectively.

Sprawling suburbs so common in Anglo-America are now increasingly obvious features in Western Europe as more and more *agricultural land disappears* before the bulldozer's onslaught. In France, urbanization is officially considered "the main prob-

[8]Service de Presse et d'Information, *France, Town and Country Environment Planning* (New York: Ambassade de France, 1965), p. 14.

preserve an acceptable life-style as France looks forward to the prospect of its population increasing to more than 57 million by the year 2000.

In this review of the pattern of urbanization, we have been limited to only a few facets of both the process and the problems that it poses for the countries of Western Europe. Needless to say, a whole spectrum of changes and potential problems will confront the people of Western Europe as they continue to alter their life-styles to fit in with the tempo and demands of urbanized life in the last years of the twentieth century.

Western Europe: National Groupings to Meet Modern Challenges

Thus far, the many countries of Western Europe have not been discussed individually. A review of the current realities of modern economic and political life suggests that most countries of Western Europe fall into two groupings, as shown on Figure 10–1:

1 The ten full members of the European Community, known as the Common Market: Belgium, Denmark, France, Federal Republic of Germany, Greece, Ireland, Italy, Luxembourg, Netherlands, and United Kingdom, plus Spain which is scheduled to join; and
2 The seven states of the European Free Trade Area (EFTA): Austria, Finland, Iceland, Norway, Portugal, Sweden, and Switzerland.

This chapter is devoted to these groups and the individual countries that they include.

THE COMMON MARKET TEN

As the map (Figure 10–1) shows, the ten member states forming the group officially known as the *European Community (EC) or Common Market* stretch from Greece and Italy on the south to Ireland and the United Kingdom on the north. Together these ten countries form the *world's richest market* with a total population of more than 270 million. The Community's population is affluent and larger than that of either the United States or the Soviet Union. In terms of world trade, the Common Market is easily the *world's largest importer and exporter,* accounting for more than 20% of all world trade. After Canada, the Common Market is America's largest trading partner. In recent years about one-quarter of all United States exports have been purchased by the European Community while only about 6% of the Community's exports went to the United States.

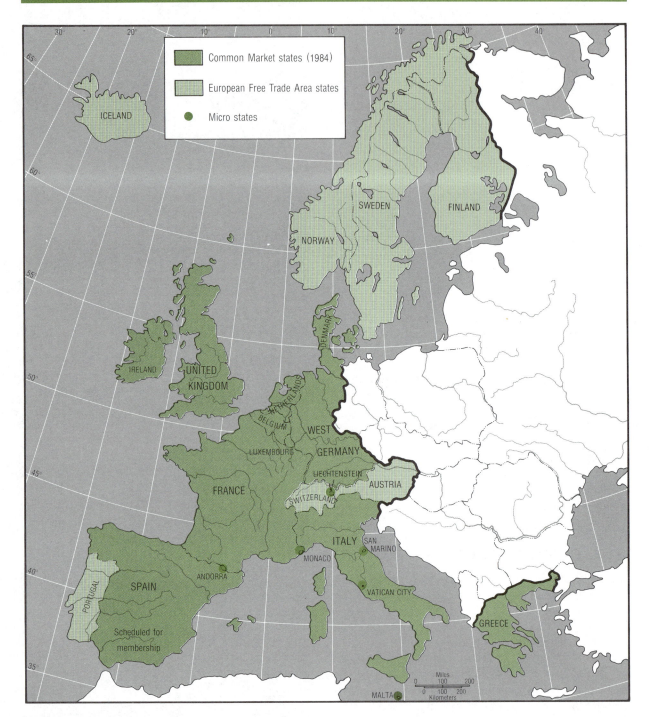

FIGURE 10–1 Groupings of the States of Western Europe.
Ten European nations have joined together in an economic union called the European
Economic Community. Spain is scheduled to join the Community in the near future.
Another economic union is the European Free Trade States, which includes northern
Scandinavian countries and Portugal and Austria. A number of micro states still exist
and have special functions and relationships with neighboring countries.

To help move their prodigious volume of foreign trade, the Common Market countries support a *merchant fleet* which is more than five times larger than that of the United States. In terms of *steel,* the most basic industrial product, the Common Market countries outproduce both the United States and the USSR by considerable amounts. The Common Market Ten also lead the world in the production of several *important agricultural commodities* such as hogs, wines, milk, butter, and cheese. Moreover, the Common Market produces about 2 million more passenger cars than the United States in an average year. All of these statistical facts provide some basis for an appreciation of the tremendous significance of the Common Market in today's world of sharpening economic competition.

Since World War II *individual mobility and mass communications* have enabled young Europeans to experience much more in common than ever before. Eurovision, an international television broadcasting organization, makes possible the transmission of important events to audiences throughout Western Europe simultaneously. Most of the elements of modern "pop" culture are universalities crossing boundaries with ease in modern Europe. People living within the Common Market need *no passports* to visit other member countries. In addition, border-crossing formalities and requirements have been greatly lessened throughout Western Europe, making it simple for hordes of tourists to travel about, supporting an increasingly important tourist industry. Most Common Market governments have joined in plans and programs for *large-scale youth exchanges* deliberately designed to assist in breaking down the old psychological barriers created by intense *nationalism* of the past. Historians, geographers, and other Common Market academics are producing school textbooks with European rather than nationalistic stress and interpretation. European schools in Brussels and Luxembourg have worked to develop truly European curricula and outlooks, and their rapid growth and commendable results have led to the basic idea being adopted in many other cities. To encourage pupils to approach their studies on a *European rather than a national basis,* a large part of their time is devoted to reading and speaking other European languages.

We now undertake a brief review of each Common Market country. As has already been mentioned more than once, the roots of nationalism and individuality run deep in Western Europe. It would be foolhardy ever to forget this fact while attempting to understand Western Europe's role in either today's or tomorrow's world.

The Federal Republic of Germany

The Federal Republic of Germany, better known as West Germany, is a *highly developed,* prosperous, *industrialized nation,* ranking among the world's leading economic powers (Table 10–1). After defeat in World War II by the allied powers—chiefly the Soviet Union, France, Britain, and the United States—Germany was divided into *occupation zones* administered by the military authorities of the four chief allies. Originally it had been agreed that for economic purposes Germany would be treated as a single unit. This plan for an economically unified Germany failed as the ever-increasing postwar differences between the Soviet Union and the Western allies made themselves felt. In 1948 the Soviets withdrew from the four-power governing organizations of Germany and Berlin, which had also been divided. Berlin became cut off from the area under allied control, since it was located deep within the Soviet-controlled zone.

As the differences between the Soviet Union and the United States and its allies crystallized into what became known as the *"Cold War"* of the late 1940s, 1950s, and into the 1960s, *two separate and distinctive political units were formed:* West and East Germany (more properly the Federal Republic of Germany and the Democratic Republic of Germany). Winston Churchill struck a prophetic note in a telegram he sent to President Truman in the summer of 1945 by declaring:

> An iron curtain is drawn down upon their [the Russian] front. We do not know what is going on behind. There seems little doubt that the whole of the regions east of the line Lübeck-Trieste-Corfu will soon be completely in their hands.[1]

The term *"Iron Curtain"* was soon to become a household word along with "Cold War" in Anglo-America as well as Europe. Not until the 1970s did the tensions that so seriously divided Europe and the world for three decades begin to disappear.

On the economic side, West Germany made a *recovery* after World War II that can only be termed miraculous. There have been periods of

---

[1]Quoted in Norman J. G. Pounds, *Divided Germany and Berlin* (New York: Van Nostrand Reinhold, 1962), p. 5.

TABLE 10–1 Common Market Ten: Selected Economic Characteristics

Country	Total Area		Percent of Area in Farms	Percent of Farmland Cultivated and in Temporary Pasture	Total Labor Force (millions)	Percent of Labor Force		
	Sq. Mi. (thousands)	Sq. Km. (thousands)				Primary Activities	Secondary Activities	Tertiary Activities
West Germany	96	249	54	61	26	6.0	44.8	49.2
France	210	544	59	60	23	8.8	35.9	55.3
Italy	116	301	61	65	22	14.2	37.8	48.0
Netherlands	16	41	68	42	5	4.6	32.0	63.3
Belgium	12	31	49	54	4	3.0	34.8	62.2
Luxembourg	1	3	51	46	0.2	6.3	38.4	55.3
United Kingdom	94	244	77	38	25	2.6	38.0	59.4
Ireland	27	70	69	25	1	19.2	32.4	48.4
Denmark	17	43	69	91	2	9.6	32.3	58.1
Greece	51	132	73.4	44	3	30.3	30.2	39.5

SOURCE: *Basic Statistics of the Community* (Luxembourg: Statistical Office of the European Communities, 1980).

business recession, but high levels of employment and a stable currency have been maintained more successfully than in most developed countries. The average rate of growth of West Germany's gross national product during the first ten years of the Common Market was an impressive 6.6%. Contributing to West Germany's *economic good health* has been the steady productivity of the coal industry. Nevertheless, West Germany, like all members of the Common Market, is a *net importer of energy* supplies, and its dependence on foreign energy sources, largely petroleum, is increasing dramatically (Table 10–2). Natural gas, which is also being used in growing quantities, is increasingly being imported from the nearby Netherlands. Agreements with the Soviet Union have been reached to purchase large amounts of natural gas from that source. West Germany's large energy requirement is met as shown in Table 10–3.

West Germany's *major industries* are steel and iron, chemicals, machinery, electrical equipment, and automobiles. The industrial section of the German economy requires approximately one-third of the total foreign oil imports, transportation about one-quarter, and residential-commercial use approximately one-third; the remainder is consumed as a nonenergy raw material. It can be seen that West Germany's *reliance on imported oil* represents the most serious threat to its economy's continued development.

West Berlin

Countless Germans living in both West and East Germany still regard Berlin as the rightful national capital and believe that it will ultimately be restored to its former position. At present, however, *East Berlin functions as the capital* of the Democratic Republic of Germany, while *West Berlin is an enclave* of West Germany deep inside Communist-controlled East Germany. West Berlin has a population of 2.1 million and covers an area of 186 square miles (482 square kilometers). It shares a 28-mile (45-kilometer) boundary with East Berlin along which the famous *Berlin Wall* was constructed in 1961 by the Communist regime to keep East Germans from moving to the economically more attractive West.

France

Almost four-fifths the size of Texas, France is the largest Western European state and fulcrum of the Common Market. The country combines *great natural wealth* with a *central location* and is endowed with a wide variety of terrain, two-thirds of which is nearly level or gently rolling. Without doubt, France has exercised the strongest influence on the Common Market to date. Somewhat unique among the major industrial countries of Western Europe, France boasts *an important agricultural sector* with substantial domestic resources of primary raw materials, a *diversified and modern industrial plant,* and a capable labor force.

France's gross national product ranks fifth in the world, and per capita GNP is equally impressive. French exports total well over $100 billion with agricultural products contributing more than one-fourth of the amount. The *principal products of France's agricultural sector* are grains, sugar beets, wine grapes, dairy products, livestock and meat, and fruits and vegetables. About 9% of the total labor force is engaged in agricultural pursuits. As these figures demonstrate, agriculture contributes much to France's economic well-being.

France is amply provided with such *raw materials* as iron ore, soft coal, bauxite, and uranium, and hydroelectric power sources are well developed. As a result, France stands as one of Western Europe's major suppliers of metals and minerals and one of the world's leading producers of iron ore, coal, and bauxite. In addition, it possesses large deposits of antimony, magnesium, pyrites, tungsten, and some radioactive minerals.

Recent years have seen a *declining production* of the low-iron-content French ores in favor of high-grade imports from Africa. Similarly, French coal production lagged because of the attractiveness of competitive fuel oil and cheaper coal imports. Serious *problems of readjustment* have been created in several traditional French mining regions. As Table 10–2 shows, France is increasingly dependent on foreign energy sources. More than three-fourths of its energy requirements must be met by imports.

Historically *one of the world's leading manufacturing countries,* France is active in all major branches of industrial activity. The aluminum and chemical industries rank among the largest in the world, as do the mechanical and electrical sectors. The French automobile industry produces more than 3 million vehicles yearly. The electronic, telecommunication, and aerospace industries have all contributed to the present *high level of technological development* that characterizes much of French industry. Military use of atomic energy, as well as

TABLE 10–2 Common Market Dependence on Foreign Energy Sources (Percentage of Total Supply)

Year	West Germany	France	Italy	Netherlands	Belgium	Luxembourg	United Kingdom	Ireland	Denmark	Greece
1960	9.6	41.0	58.1	50.9	32.6	99.8				
1961	12.8	43.7	62.8	49.6	35.8	99.4				
1962	17.7	44.3	68.1	58.6	40.9	99.5				
1963	22.4	52.6	70.2	63.2	49.4	99.7	26.3	71.9	98.0	
1964	27.2	54.1	71.5	65.8	54.7	99.7				
1965	32.6	55.0	73.5	62.8	59.3	99.3				
1969	42.8	63.8	78.2	47.7	73.5	99.3	43.8	73.0	99.6	
1970	47.9	71.0	81.8	42.3	81.8	99.2	45.1	76.1	100.0	83.5
1971	50.6	73.1	81.8	26.2	84.6	99.5	49.5	80.5	100.0	82.3
1972	53.2	75.0	81.1	15.6	82.7	99.5	49.7	77.8	96.6	91.5
1973	54.8	78.0	83.0	6.3	86.4	99.6	48.2	80.7	99.6	91.4
1974	51.1	82.3	83.1	−9.1	91.2	99.4	50.2	82.2	99.5	96.6
1975	55.0	73.8	79.1	−24.8	84.7	99.5	43.2	84.6	99.1	68.3
1980	58.1	81.3	90.4	4.8	89.2	99.6	5.7	81.7	98.7	85.3

SOURCE: *Basic Statistics of the Community* (Luxembourg: Statistical Office of the European Communities, 1981).

TABLE 10–3 European Economic Community: Energy Sources as Percentage of Total Energy Production

Country	Petroleum	Solid Fuel	Natural Gas	Nuclear	Tidal, Hydro, Geothermal
West Germany	48	31	16	4	1
France	59	17	12	9	3
Italy	71	8	17	1	3
Netherlands	45	6	47	2	0
Belgium	50	24	19	6	1
Luxembourg	30	51	12	0	7
United Kingdom	39	35	20	5	1
Ireland	70	20	9	0	1
Denmark	69	31	0	0	0
Greece	77	21	0	0	2

SOURCE: *Energy Statistics Yearbook, 1982* (Luxembourg: Statistical Office of the European Community).

certain civilian applications, has achieved an impressive level of sophistication. France has joined with the United Kingdom in producing the first supersonic commercial jet air transport, the Concorde. France launched its first earth satellite in 1965 and has pursued an active space research program.

Italy

Compared with many other Western European countries, Italy is *poorly endowed by nature.* Much of the country is *unsuited for farming* because of mountainous terrain or unfavorably dry climate. There are *no significant deposits of coal or iron ore.* Most other important industrial mineral deposits are widely dispersed and generally of poor or indifferent quality. Natural gas deposits, found in the Po Basin in the post–World War II period, constitute the country's most important mineral resource, but the gas supply is being rapidly depleted and does not constitute a long-term energy supply. As a result of these natural deficiencies, most raw materials for manufacture must be imported. Other obstacles to Italian development have been *low levels of productivity in agricultural* as well as in certain industrial sectors and the need to upgrade the skill of much of the labor force.

The south of Italy, a region known as the *Mezzogiorno* (Land of the Midday Sun), comprises one of the Common Market's major regions of *underdevelopment.* It covers an area the size of Pennsylvania and Delaware and includes the Italian Peninsula south of Rome and the islands of Sicily and Sardinia. Here a *rugged landscape, poor soils,* and *Mediterranean climate* have had a most profound impact on the nature of the Italian economy. Only

15% of this vast area is considered level enough for cultivation, 50% is categorized as hilly, and the remainder is mountainous. Unfortunately these mountains are not high enough to ensure year-round snow accumulations. As a result the rivers

The *Mezzogiorno* section of Italy, a region of poverty, suffers from thin, stony soils and a lack of water. Major efforts have been exerted to improve the economy of the region. (United Nations)

are dry for long periods of the year, flowing only during the moist Mediterranean winter season. Ironically winter is often a period of flooding, waterlogged fields, and serious soil erosion. Irrigation schemes are consequently very difficult and exceedingly expensive undertakings in the *Mezzogiorno.*

Although still lagging, the *Mezzogiorno* is now making *some economic progress.* The effects of massive investments made since World War II in regional agricultural improvement schemes and industrial development are beginning to be felt. It will, however, be a long time before the Italian south begins to equal the north in terms of productivity and prosperity for its workers. People still leave the area at a disturbing rate in search of a better life.

Conditions are considerably different in *northern Italy.* Here Italy's essentially private enterprise economy has been flourishing. Despite a severe economic recession in 1971 and the relative stagnation of the early 1970s, Italy ranks as the world's seventh most important industrial power.

As Table 10–2 indicates, Italy is even more dependent on foreign energy sources than France. What is especially disturbing for the Italians is the overwhelmingly important role of oil in meeting their energy requirements (Table 10–3).

The Netherlands

With approximately 890 people per square mile (344 per square kilometer), the Netherlands ranks as Western Europe's most *densely populated country.* The Netherlands is even more densely populated than Japan. Less surprising is the fact that the Dutch people have been energetically adding to their low-lying territory through *land reclamation and drainage schemes* for centuries. Dike building appears to have begun about A.D. 1000 in the extensive tidal marshes of the southern Netherlands. The dikes protected land that lay above the low-tide level. Sluice gates, which allowed surplus water to drain out of ditches at low tide, were closed to keep out the incoming tidal flow. The protected lands that resulted, known as *"polders,"* were still subject to periodic flooding during high-tidal storm conditions. It was not until the perfection of the windmill for pumping and later steam-driven pumps that the Dutch were able to begin creating agriculturally productive polders in areas that lay below sea level.

In 1953 the Dutch suffered a severe disaster when storm winds of up to 100 miles (161 kilo-

meters) per hour drove the waters of the Rhine-Maas-Scheldt across farmlands and towns. An estimated 1800 persons died, and another 100,000 had to be evacuated. The flood damage was conservatively estimated at more than $500 million. To prevent a recurrence of such a flood, the Dutch have undertaken the immense *"Delta Project,"* that consists of four huge sea dikes to close most of the mouths of the Rhine-Maas-Scheldt distributary system.

It is not without good cause that many observers of the Netherlands speak of the *"Dutch Miracle."* They are properly impressed by the fact that the Dutch have literally created almost 40% of their country by diking and draining polders. Here more than anywhere else in Europe it is appropriate to speak of a *"human landscape."* Productive *agricultural enterprises* flourish on much of the reclaimed land, and in many areas new towns and industrial areas have been developed.

The Dutch economy is characterized by *private enterprise,* but the central government plays a strong role. A *limited base of natural resources,* primarily gas, oil, and coal, plus the *strategic location* have been developed into an economy unusually dependent on foreign trade. *Industry is modernized* and characteristically competitive, backed by a diligent and highly *skilled labor force* and a vigorous and adaptable business community.

Although almost completely modernized and characterized by very high crop yields, *agriculture plays a small role* in the Dutch economy. As an industry, agriculture relies heavily on imports, particularly livestock feed. The *livestock industry* accounts for about two-thirds of the total farm output. The Netherlands ranks as the Common Market's most efficient dairy and livestock producer. Expansion in this sector has taken place as a result of the Market's *Common Agricultural Policy (CAP),* which favors efficient producers. Only about 4% of the Dutch GNP is contributed by agricultural enterprises. The service sector of the economy, primarily transport and financial, accounts for more than half of the national income. Industrial activities, primarily the metalworking, chemical, and food processing industries, account for another quarter.

Belgium

The kingdom of Belgium has long been geographically and culturally at the *crossroads of Western Europe.* Today the Belgians are sharply divided along ethnolinguistic lines between *French-*

speaking *Walloons* in the southern half of the country and the *Flemish, speaking a Dutch dialect,* in the north. In effect, Belgium is a state formed of *two distinct national groups.* Many observers feel that its continued existence will depend on how successfully the central government can manage to allow each group to achieve the almost complete autonomy it demands.

Belgium emerged from World War II much less damaged than its neighbors. The immediate postwar era saw rapid reconstruction, trade liberalization, and high economic growth rates. The pace began to slacken, however, and it was the establishment of the Common Market in 1958 that brought about a new surge in the Belgian economy. Confidence in the opportunities provided to Belgian producers in the huge new market led to bold new investments in manufacturing plants and equipment.

Agriculture plays a relatively minor role in the Belgian economy. In recent years agriculture has been responsible for slightly less than 3% of the GNP, employing roughly the same percentage of the labor force. Livestock and poultry raising are the dominant agricultural activities, although the traditional crops—sugar beets, potatoes, wheat, and barley—still form an important part of the rural scene.

At the hub of a major West European crossroads with a dense concentration of industry and population, Belgium has played an impressive *role in international trade.* The country's *industries* contribute $50 billion in exports and about 33% of the total GNP. Belgium does not, however, possess any significant stores of natural resources and *must import most raw materials,* as well as fuel, machinery, transport equipment, and about one-fourth of its food. The country's development and prosperity are, as a result, largely the product of a highly *skilled labor force* and *managerial expertise.* The cornerstone of Belgian industry remains the iron and steel and metal-fabricating sectors, which supply approximately 42% of the exports by value.

Luxembourg

Closely linked with Belgium in an economic union for over a half century, tiny Luxembourg bears many similarities to its larger neighbor. Its neutrality also violated by Germany in the two world wars, Luxembourg is a staunch charter member of NATO. The government is a constitutional monarchy with executive power in the hands of the hereditary Grand Duke and a cabinet. Internally, Luxembourg is divided into 126 communes, each administered by an elected council in a system closely patterned on that of Belgium.

A *high level of industrialization* provides the inhabitants of this Rhode Island–sized country with one of the highest per capita GNPs in the Common Market. Despite large-scale attempts to diversify, Luxembourg's *industrial scene is dominated by iron and steel.* These basic enterprises account for about one-half the country's total industrial production. Steel output is about 5.4 million tons (about 1% of world production), a real feat for a country of only 365,000 people.

Agriculture absorbs about 8% of the labor force, mostly in small-scale livestock raising and mixed farming. The vineyards of the Moselle Valley provide the raw material for a lucrative wine industry. Luxembourg's excellent dry white wines are exported widely.

More than 40% of the workers in the country are engaged in Luxembourg's flourishing *financial activities.* There are more than sixty large banks in this small country, making it one of the world's important financial centers. Trading in *multinational corporate securities* is also an important feature of Luxembourg's active commercial and financial sector. Favorable tax treatment by the government has encouraged many large holding companies to establish their headquarters here.

The United Kingdom

There can be little doubt that Britain[2] has the *oldest industrialized economy in the world.* In the eighteenth century, the island-based British gained mastery of the seas and established an empire scattered on all the continents. The nineteenth century saw this empire enlarged and reinforced to make Britain supreme as a world power. Changing world conditions punctuated by two costly world wars severely weakened Britain's position in the twentieth century. The old empire gave way to the *Commonwealth* (a loose political and economic association between the United Kingdom and many of the former colonies) as political independence was achieved by the colonies in the interwar and post–World War II periods.

When the Common Market was formed, there was considerable concern on the part of several

---

[2]The terms *"United Kingdom"* and *"Britain"* are used synonymously to refer to The United Kingdom of Great Britain and Northern Ireland, which consists of England, Wales, Scotland, and Northern Ireland.

Western European countries that did not join. It was feared that a division would develop between the countries that were in the Common Market and those that remained outside. Britain, in particular, was faced with several problems because of its long-standing connections with the Commonwealth.

Seven countries—Austria, Denmark, Norway, Portugal, Sweden, Switzerland, and the United Kingdom—formed the *European Free Trade Area,* or ''outer Seven,'' during the 1960s. Britain, with more than one-half of EFTA's total population and nearly 60% of its total production, clearly dominated the organization.

By the mid-1960s, however, it was becoming clear that Britain's brightest *economic prospects lay in the direction of the Common Market* rather than the Commonwealth and EFTA. The information presented in Table 10–4 helps explain how this opinion gained wide support and backing. As can be seen, *Commonwealth trade was declining* while trade with Western Europe was increasing. The importance of Western Europe in Britain's future economic outlook became obvious. Once the Common Market was crystallized, a formidable barrier to British imports would result. For instance, German cars sold in France and French cars sold in Germany would not be subject to any import duty. British cars sold to Common Market countries, on the other hand, would be subject to the *common external tariff* of the community. Also, it was obvious that Commonwealth countries were intent on rapidly developing their own industrial sectors and could no longer be considered as ready markets for British goods. In 1973, Britain, together with Denmark and Ireland, *joined the Common Market.*

Leaders in both Britain and the Common Market enthusiastically hailed its entry as a new renaissance. They saw it as a step that could immensely increase the security and stability of the Market and the prosperity and quality of life for the peo-

The petroleum and natural gas in the North Sea are being extensively explored despite the cost of drilling at sea and the roughness of the North Sea waters.

ples of Western Europe. Only time will reveal how accurate these hopes and aspirations will be. The present ten Common Market nations are certainly a more impressive economic unit than were the original six, thanks largely to Britain's large population and productive capacity.

The most notable feature of Britain's economy is the *importance of industry, services, and trade.* It ranks fifth in world trade behind the United States, West Germany, France and Japan. Britain takes 6.29% of the world's exports of primary products and contributes 6% of the world's exports of manufactured goods. In its share of *invisible world trade* (such as financial services, civil aviation,

TABLE 10–4 Destinations of United Kingdom Exports (percentage)

| Year | Commonwealth | Western Europe | | | | United States |
		EEC	EFTA	Other	Total	
1958	38	14	11	2	27	9
1960	37	15	12	2	29	10
1965	28	19	14	4	37	10
1966	26	19	16	4	38	13

SOURCE: *Basic Statistics of the Community* (Luxembourg: Statistical Office of the European Communities, 1977).

travel, and overseas investment), Britain ranks second behind the United States.

Coal, the paramount energy source for the early industrialization of Britain, has been declining in importance during the present century, and an alarming reliance on imported petroleum had developed. During the mid-1960s, however, huge deposits of *petroleum and natural gas* were discovered under the bed of the North Sea (Figure 10–2). In 1972, natural gas from the North Sea fields accounted for 88% of all gas available in Britain. More than one-half of all gas sold in Britain is for industrial and commercial uses with the remainder going to domestic consumers.

Reliable sources of energy are imperative if Britain and other Western European countries are to maintain and improve their standards of living in the future. The stakes are high; astronomical sums must be invested now in exploration and production as a guarantee that future economic development will be possible in Western Europe. The "energy crunch" brought about by the Arab petroleum-producing nations' boycott in 1973–74 and the 1979 Iran revolution have made this fact painfully obvious. In the case of Britain, the future outlook for energy production seems bright. In 1975, the first offshore petroleum from the Argyll field came on stream and moved Britain into the ranks of the producing nations. By 1980, Britain had become *self-sufficient.* Although it is hazardous to speculate too far into the future, it does seem that *an energy renaissance* is in store for the world's senior developed country. It will be interesting to see how this renaissance is translated to the day-to-day lives of the people of Britain and Western Europe and the world at large.

Ireland

The Irish Republic occupies about five-sixths of the island of Ireland, once entirely controlled by England. The northern portion of the island is occupied by the six counties making up Northern Ireland, or Ulster, which is still an integral part of the United Kingdom. Unlike predominantly Protestant Northern Ireland, the Irish Republic is about *94% Roman Catholic.* Recent years have seen Northern Ireland embroiled in a *bitter civil dispute* that often erupts in the form of bombings and street warfare.

English is the common language, but Irish Gaelic or Erse, currently encouraged by the government, is spoken in some places; both are an official language. The curious combination is very obvious to the tourist or visitor from abroad these days, since highway directional signs are printed in English and Gaelic.

Ireland is notably *lacking in most industrial natural resources.* What industrial development Ireland has had until very recently was oriented primarily toward the domestic market. At the present time, efforts are being exerted toward accelerating industrial development. To this end, Ireland's entry into the Common Market is seen as a great boon.

Not surprisingly, *agriculture remains a major factor in national development* schemes. Agricultural output has failed to keep pace with industrial growth in recent years. Exports, however, are still largely agricultural in origin.

Although economically very closely *linked with*

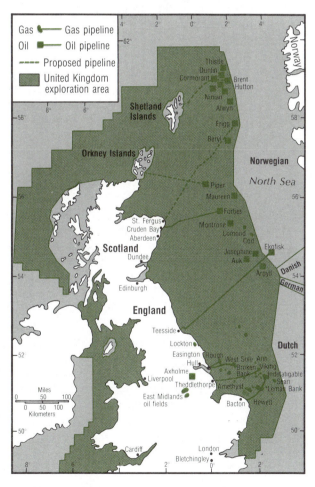

FIGURE 10–2 North Sea Oil and Gas Areas. The North Sea petroleum deposits are now being developed and provide substantial energy supplies to Western Europe. Dependable energy supplies are especially critical to areas of industrial activity and to densely populated areas with high levels of living.

167

the United Kingdom, Ireland scrupulously adheres to a policy of *political independence* from its large and more affluent neighbor. In recent years Irish troops have served with distinction in a number of United Nations peacekeeping operations around the globe. Ireland remains outside NATO, however, and in 1969 the prime minister stated that Ireland's international policy was neutrality like that of Sweden, Austria, and Switzerland. This view was reaffirmed when Ireland joined the Common Market in 1973.

Denmark

Long an *agriculturally dominated economy,* Denmark shifted its *emphasis to industry* about twenty-five years ago. As a result, industrial exports outstripped agricultural sales for the first time in 1964. In recent years industrial products have accounted for about 65% of its total export trade. More than 37% of Denmark's workers are engaged in industrial pursuits. Manufacturing contributes about 39% of the total GNP.

Denmark is *poorly endowed with fuel resources.* Some low-grade iron ore is found in the south, but on the whole Denmark depends on imports for its principal industrial raw materials.

One of the reasons for the successful development of the Danish economy in the postwar period has been its *adaptability,* an adaptability that has resulted in a wide range of *specialization.* The Danes boast of being able to "one-up" their trading partners by selling such things as fine smoking pipes in England, whiskey in Scotland, and chewing gum in the United States. Danish beer is sold widely throughout the world; Europe's two largest breweries are located in Denmark. Clearly, Denmark's one really *important domestic "raw material"* is its skilled labor.

The most extensive use of Denmark's 16,619 square miles (43,044 square kilometers) of territory remains agricultural. By far the largest portion of the agricultural land is cultivated for animal foodstuffs. Ninety percent of farm gross income derives from animal production, chiefly butter, cheese, bacon, beef, veal, poultry, and eggs.

Greece

Greece, the ancient cradle of democracy and European culture, became the tenth member of the Common Market on January 1, 1981. Application had been made along with Spain and Portugal in the 1970s when these three Mediterranean countries emerged from long periods of dictatorship in their government. In responding to their requests the Commission of the European Communities stated:

> When Greece, Portugal and Spain, newly emerging as democratic States after a long period of dictatorship, asked to be admitted to the Community, they were making a commitment which is primarily a political one. Their choice is doubly significant, both reflecting the concern of these three new democracies for their own consolidation and protection against the return of dictatorship and constituting an act of faith in a United Europe, which demonstrates that the ideas inspiring the creation of the Community have lost none of their vigour or relevance.[3]

Thus it can be seen that *political concerns* rather than economic factors were crucial to Greece's membership.

Concern over politics was probably a very good thing because Greece's level of economic development was relatively low compared to other Common Market countries. For example, only 8.4% of the Common Market's labor force is engaged in agriculture, whereas well over *one-third of Greek workers are involved in agrarian activities* for their livelihood. Greece's *location and long maritime tradition,* however, give the nation an impressive strength in commerce. The Greek merchant ship fleet is almost 10% of the world's total, and with the Greek fleet, the Common Market now controls approximately 15,000 of the world's 70,000 large merchant ships.

Greece's fabled natural beauty and pleasant Mediterranean climate have been combined with reasonable prices and good facilities to provide the country's chief source of foreign exchange, *tourism.* More than 7 million tourists visited Greece in 1980 and brought in the equivalent of $1.8 billion in foreign currency. The increased stability and ease of travel which membership in the Common Market guarantees should make tourism an even more important element in the Greek economy in the years ahead.

Spain

Spain, about equal in size to the states of Arizona and Utah combined, is a large country by Euro-

[3]Commission of the European Communities, *The Second Enlargement of the European Community* (Luxembourg: Office for Official Publications of the European Community, 1979), p. 5.

pean standards. Negotiations for Spain's accession to the European Community (Common Market) began during the 1960s, culminating in full membership in the mid-1980s. With Italy, France, Greece, and Spain to be eventually joined by Portugal, the European Community will become the *major force in the Mediterranean Basin*. Such a presence will surely have profound impacts on the countries of Northern Africa and the Middle East which share the shores of this great inland sea.

Spain's entry into the European Community became possible when the country emerged from the long *period of dictatorship* that ended with the death of General Francisco Franco in 1975. With Franco's death Prince Juan Carlos de Borbón y Borbón assumed the position of king and chief of state. Political freedom was restored, and in 1977 Spain's *first elections since 1936* were held. Since that time the Spanish *Cortes,* or Parliament, has proven itself to be an effective branch of government.

Along with the adoption of *democratic government* Spain has experienced significant changes in its economy. The protective centralized economic policies which were followed during Franco's dictatorship were unworkable in the *open society* of the late 1970s, and competitive economic forces have taken their place. Although high energy costs and world economic uncertainties pose enormous problems, the outlook for Spain's economic growth is promising, and entry into the European Community will assure that the country's long-term development aspirations will be more easily achieved.

In recent years about 46% of Spain's *exports* have gone to the countries of the European Community, while 35% of Spain's *imports* have come from the EEC. Traditionally about two-thirds of Spain's agricultural exports have gone to the Common Market area of Europe. Crops and products typical of the Mediterranean such as citrus fruits, wine, fresh vegetables, soybean oil, olives, nuts, and olive oil form a major part of these exports.

THE EUROPEAN FREE TRADE ASSOCIATION

The *European Free Trade Association (EFTA)* is a group of governments—Austria, Iceland, Norway, Portugal, Sweden, and Switzerland are members and Finland is an associate—which ad-

ministers *free trade in industrial products* among the seven countries involved. As such, EFTA forms an important part of the process of economic integration which has been taking place in Western Europe since the aftermath of World War II. Free trade agreements between the European Community and EFTA countries have promoted a *growth of trade* for the benefit of all concerned. The most important outcome of these agreements is the creation of *a free trade zone* for industrial products throughout Western Europe. In the words of the Stockholm Convention which established EFTA in 1959, EFTA's objectives are:

a) to promote . . . a sustained expansion of economic activity, full employment, increased productivity and the rational use of resources, financial stability and continuous improvement in living standards;

b) to secure that trade between member states takes place in conditions of fair competition;

c) to avoid significant disparity between member states in conditions of supply of raw materials produced within the area of the Association; and

d) to contribute to the harmonious development and expansion of world trade and to the progressive removal of barriers to it.

We now turn to a brief discussion of each of the seven member countries making up the European Free Trade Association. As can be seen on the map (Figure 10–1), they form a rough ring around the heart of the Common Market.

Austria

The *people* of Austria are about 99% German and about 90% Roman Catholic. Two small but significant minority groups are found: about 20,000 Slovenes live in the region known as Carinthia in south central Austria, and about 25,000 Croatians are found in Burgenland on the border with Hungary. They serve as reminders of the days, prior to World War I, when Austria was the center of the great *Austro-Hungarian Empire,* which controlled a vast multinational territory in Europe.

The *terrain* of Austria varies from hilly to mountainous over practically all of the country. The Alps dominate the southern and western provinces and provide an important basis for the country's lucrative tourist industry. As a result of historical factors and uneven terrain, Austrian farms are *small and fragmented*. Generally speaking this smallness and fragmentation causes farm products to be relatively expensive. Consequently, agriculture's share of the national economic output has been steadily declining and now is less than 5% of the gross domestic product.

Austria's most important *trading partners* are members of the European Community, which take more than 40% of its exports, while EFTA countries take about 27%. On the import side, the European Community accounts for 56% and EFTA for approximately 20%. Trade with Eastern Europe is significant with about 13% of Austrian exports going to the Communist bloc and about 10% of imports coming from that area.

As can be seen, Austria is both geographically and economically in a *strategic position* between the East and West. During the allied occupation after World War II, Austria was required to deliver goods worth more than $150 million plus 10 million tons of petroleum to the USSR. In 1955, when all occupation forces were withdrawn and the Austrians once more became a sovereign state, the parliament passed a constitutional law making the country perpetually *neutral in international affairs.* As a result Austria actively pursues policies which are best described as ''bridge-building to the East.'' These policies are designed to foster increasing contact at all levels with Eastern Europe and the Soviet Union. In this way the Austrians feel they can make a contribution toward easing East-West tensions far beyond that which might otherwise be possible for a country of their small population.

Iceland

Iceland is an island country slightly smaller in area than the state of Kentucky. Located on an active volcanic summit to the east of Greenland and immediately south of the arctic circle, Iceland was, until the advent of modern air transport, one of the most remote countries of Western Europe. Thanks to the relatively warm waters of the North Atlantic Drift (Figure 8–3), Iceland's lowland *climate is surprisingly mild.* Summers are damp and cool and winters very windy but with relatively mild temperatures. Reykjavík, the capital, for example, has an average temperature of 52°F (11°C) in July and 30°F (−1°C) in January.

Iceland's population of 227,000 is made up of a remarkably *homogeneous* group descended from Norwegian and Celtic peoples who first colonized the island during the late ninth and early tenth centuries. The modern Icelandic language is reputed to be relatively unchanged since the twelfth century

and thus very close to the language spoken by the Vikings of old.

Chief among the resources on which the Icelanders have built their prosperous economy are *fish, hydroelectricity, and geothermal energy.* The main agricultural products are livestock, hay, fodder, and cheese. The work force numbers close to 100,000 and is almost entirely *literate*—Iceland boasts one of the highest literacy rates in the world. As might be expected, a fairly high percentage of the work force is engaged in the primary economic activities, with 26%, mainly herdsmen, engaged in agriculture and 14% employed in the all-important fisheries. About 30% of Iceland's *exports* go to the European Community and another 15% are purchased by EFTA partners. The United States is also a large customer, taking some 28% of Iceland's exports. More than three-quarters of the exports are products of the highly developed Icelandic fishing industry.

During World War II, Iceland was occupied by British and American military forces. After the war Iceland became a charter member of the North Atlantic Treaty Organization (NATO). At the request of NATO during the early 1950s, the United States and Iceland agreed that the United States should assume responsibility for Iceland's defense. Since that time Iceland has been the only NATO country without its own military forces.

Norway

Like Iceland, Norway is a northern land. Almost one-third of its 1100-mile (1770-kilometer) length is north of the arctic circle. Even its southernmost point at Lindesnes reaches only 58°N, approximately the latitude of Juneau, Alaska. Roughly the size of New Mexico and large by European standards, Norway suffers from a *lack of land suitable for agriculture.* In fact only 4% of Norway's total area of 125,000 square miles (343,750 square kilometers) is suitable for farming. About one-quarter of the country is forested while an amazing *72% is classified as mountainous* with bogs, glaciers, and barren rock waste dominating the scene. Adding to the difficulty of developing this land is the character of its coasts which are deep *fiords.* The *fiords* are steep-sided valleys resulting from past glacial action that are now flooded by the sea. Where Norway is narrowest the fiords virtually cut the country into segments so there is no continuous land route from southern Norway to the provinces of the north. The far-flung towns and settlements of the north can only be reached by planes, ships or, with difficulty, highway-ferry links.

Norway's *chief natural resources* are fish, timber, hydroelectric power, and in recent years, oil and natural gas from the North Sea fields. Offshore oil was discovered in the 1960s and development began in the 1970s. Thanks to the rapid rise in oil prices Norway's economy has enjoyed a considerable boost from this new and technologically demanding industry. In 1980 about one-quarter of the value of all Norwegian exports came from oil and gas sold abroad. A policy of moderate petroleum production has been adopted to make certain that petroleum will remain an important part of the Norwegian economy for many decades into the future.

Portugal

In size, Portugal resembles the state of Indiana, but in terms of population it is almost twice as large, Indiana having approximately 5.5 million citizens and Portugal close to 10 million. The northern portion of the country is rugged and mountainous, and the area south of the Tagus River is largely rolling plains. Thanks to its position fronting the Atlantic Ocean, most of Portugal has a more *moist and temperate climate* than is found in the Mediterranean Basin to the east. The combination of terrain and climate allows almost one-half of Portugal's total land area to be utilized for some agricultural purpose or other. Of that large area almost 90% is classified as *arable land* on which crops, orchards, and vineyards can be developed.

Although the percentage of the work force engaged in *agriculture* has declined in recent decades from 42% in 1960 to 25% in 1980, Portugal still lags behind most other EFTA countries in this respect. As the agricultural component has declined with respect to the total work force, so the *industrial and service sectors* have shown the increases that usually accompany economic development in the modern world. In 1960, for example, the industrial work force accounted for only 21% of the total, whereas today approximately 40% of the workers are employed in the industrial sector. The shift into the service sector has been equally dramatic.

Portugal is one of Europe's oldest countries, tracing its modern roots to A.D. 1140 when Alfonso I became king. The present-day boundaries of Portugal were achieved in 1249 after a long period devoted to expelling the Moors. During the *Age of Discovery,* Portuguese explorers pushed to almost every part of the world and laid the foundations for a *great territorial empire* in Africa, Asia, and South America where they colonized Brazil. In the 1970s Portugal fought expensive wars in a futile effort to retain its African colonies in places like Angola and Mozambique. The drain on the Portuguese economy was disastrous and ultimately led to the downfall of the dictatorial regime which had been in power since the 1930s.

In 1974, Portuguese voters chose an assembly to draft a democratic constitution and launched the country on its present path of representative *parliamentary government.* This fact was crucial to the Portuguese when, in 1977, their government sought entry into the Common Market. In the opinion submitted by the European Commission recommending the eventual acceptance of Portugal it noted that democracy was an established fact in Portugal and that, "the community cannot leave Portugal out of the process of European integration." The Commission continued: "The resulting disappointment would be politically very grave and the source of serious difficulties. The accession of Portugal, which set its face firmly towards Europe almost as soon as its democracy was restored, can only strengthen the European ideal."[4] It is clear that Portugal's long-term future and economic development will be closely tied to that of the expanded European Community.

Sweden

Sweden shares the Scandinavian peninsula with neighboring Norway, and like its neighbor, Sweden is a long and narrow country. If Sweden were superimposed on a map of the United States, it would spread from the southern tip of Florida to Washington, D.C., or a distance of slightly under 1000 miles (1610 kilometers). Although Sweden lies in approximately the same latitude as Alaska, the warming influences of the North Atlantic Drift help provide a climate like that of New England over much of its southern half. The winter days are noticeably short with sunset occurring near 3 P.M. during Stockholm's midwinter weeks. Similarly, however, days are long during midsummer.

Sweden's per capita gross national product is consistently among the three or four highest in the world, and higher than that of the United States. As might be expected, Sweden is basically *an in-*

---

[4]Commission of the European Communities, *Portugal and the European Community* (Brussels, 1980).

dustrial country with agriculture and forestry contributing about 5% of its annual gross national product. In recent years only about 5% of Sweden's work force has been employed in agriculture. Extensive *forests,* rich deposits of *iron ore,* and *hydroelectric power* along with a highly *literate and well-trained population* are Sweden's basic resources on which a productive economy has been built. As a leading trading nation Sweden *exports* more than 50% of its industrial output. The chief markets for Sweden's exports of wood products, iron ore, machinery, metals, bearings, ships, instruments, and automobiles are Norway, the United Kingdom, Denmark, and Finland.

An *important problem* facing Sweden's society is one which is beginning to concern many rich nations: the adjustments required as increasing numbers of citizens pass into retirement and join the over-65 age group. Life expectancy, a good measure of development, is over 75 years in Sweden. Just what sorts of social institutions and programs will best suit the increasing ranks of senior citizens promises to be a question to which all developed countries will need the answer in the future. Without doubt Sweden's solutions to the problem will be studied closely by leaders and planners in every rich nation.

Switzerland

Switzerland, in the *heart of Western Europe* and bounded by West Germany, Austria, Liechtenstein, Italy, and France, has traditionally controlled the major routeways between southern and northern Europe. The modern country of Switzerland traces its roots back to 1291 when three county-sized districts called cantons signed the "Eternal Alliance" that bound them together in resistance to foreign rule. Through time, other cantons joined the confederation, and in 1848 a federal constitution partially modeled on the United States Constitution was adopted. Swiss neutrality was guaranteed by the great powers of Europe early in the nineteenth century, and at the present time Switzerland is a staunchly neutral country composed of several distinctive ethnic groups.

Reflecting this *ethnic diversity,* Switzerland has four national languages. Romansh, based on Latin, is spoken by a small minority of Swiss living in the Alpine valleys of the southeast. Italian is the language of Ticino Canton in the south. The western cantons facing France speak French. German is the language of approximately two-thirds of all

Swiss, and the highly developed urban-industrial core of the country is in the German-speaking region.

After World War II, Switzerland was one of a few Western European countries whose economies were virtually unscathed. Since most industrial raw materials must be imported, the Swiss concentrate on those products containing *high labor and skill inputs* such as watches and precision instruments. Other *specialized quality products* like chemicals, pharmaceuticals, cheese, and chocolate, as well as items which do not lend themselves to mass production such as power generators and turbines, are produced by the Swiss.

The small domestic market can consume only a limited portion of the industrial output, so Switzerland *depends on the world economy* for its prosperity. Exports, along with Switzerland's highly developed *tourist industry, international banking,* and the insurance and transport industries, form the foundation for one of the world's highest standards of living.

Only the belief that membership in the European Community would be incompatible with traditional strict neutrality keeps Switzerland outside the Common Market. The countries of the Common Market provide well over two-thirds of Switzerland's imports and buy one-half of her exports. In common with the other EFTA members, Switzerland cannot escape a future closely tied with the fortunes of the European Community.

Finland

The Finnish name for Finland is *Suomi,* which means land of marsh. It is well chosen to describe the *heavily glaciated lake-strewn landscape* which comprises most of Finland. Like neighboring Sweden and Norway, Finland is a northern land with about one-third of its 724–mile (1165-kilometer) length to the north of the arctic circle. It shares lengthy land boundaries with Sweden and Norway in the north and the Soviet Union in the east.

Until 1809 when it was conquered by the Russians, Finland was closely associated with Sweden. Swedish is still an official language, and about 6% of the population of Finland speak it as their mother tongue. From 1809 until 1917, Finland was connected with the Russian Empire as a grand duchy. Taking advantage of the Bolshevik Revolution in Russia, the Finns declared their independence in 1917. During the tense period leading to World War II, the Soviet Union requested

Finland to surrender territory north of Leningrad and permit the Soviet navy to establish a base on the coast of the Gulf of Finland in exchange for territory in eastern Karelia. The Finns refused, and in late 1939 Soviet troops invaded Finnish territory and several cities were bombed by the Soviet air force. After several battles the Finns were forced to accede to the USSR's demands. At the conclusion of World War II, Finland's territorial losses amounted to 17,800 square miles (46,102 square kilometers), roughly the area of Massachusetts and New Hampshire combined. The entire population of the lost territories chose to move to Finland rather than become Soviet citizens. Relocating and integrating these 400,000 displaced Finns was an enormous challenge. To experience the magnitude of the problem facing Finland's people in the late 1940s, the United States would have to lose 12%

of its total area while accommodating a sudden influx of 16 million immigrants. On top of this the USSR exacted a huge indemnity that the Finns were forced to pay in ships, machinery, and other manufactured goods.

In the face of these monumental obstacles, the economy of Finland made an almost *miraculous recovery* in the postwar decades. This recovery is verified by the fact that in recent years the *industrialized economy* of Finland, which totals only 0.1% of the world's population, has been producing about 0.3% of the world's output and contributing just under 0.9% of the world's total trade. An associate member of EFTA since 1961, Finland's future development and prosperity will depend on its continued ability to secure foreign markets for its exports in an increasingly competitive world.

FURTHER READINGS

BEAUJEU-GARNIER, J., *France* (New York: Longman, 1976). An examination by a noted geographer of the diversity of the French landscape with an analysis of the future of French regional planning.

BURTENSHAW, D., M. BATEMAN, and G. J. ASHWORTH, *The City in West Europe* (New York: Wiley, 1981). An excellent review of the region's urban geography.

CATUDAL, HONORE M., "The Plight of the Lilliputians: An Analysis of Five European Microstates," *Geoforum*, 6 (1975), 187–204. All that most people ever wanted to know about Western Europe's microstates is presented in this readable article.

CHANDLER, T. J., and S. GREGORY (eds.), *The Climate of the British Isles* (London: Longman, 1976). An overview of climatological factors that incorporates the findings of recent research to describe the climatic regimes of the British Isles.

CHAPMAN, KEITH, *North Sea Oil and Gas: A Geographical Perspective* (North Pomfret, Vt.: David & Charles, 1976). An informative assessment of the first decade of exploitation of the North Sea hydrocarbons.

CLOUT, H., *The Regional Problem in Western Europe* (London: Cambridge University Press, 1976). A brief volume that identifies some of the regional problems existing in Western Europe.

JONES, PETER, "The Geography of Dutch Elm Disease in Britain," *Transactions, Institute of British Geographers*, 6 (1981), 324–34. Traces the spread of a new strain of Dutch elm disease that killed the vast majority of elms in southern Britain. An important and traditional component of many rural English

landscapes has disappeared in the past twenty years as the result of this disease.

KORMOSS, I. B. F., *The European Community in Maps* (Brussels: Commission of the European Communities, 1974). An excellent set of maps illustrating many facets of the European community.

LEE, R., and P. E. OGDEN, *Economy and Society in the E.E.C.: Spatial Perspectives* (Farborough: Saxon House, 1976). A variety of essays dealing with the integration of the economies of the nations within the Common Market.

LE ROY LADURIE, EMMANUEL, *Times of Feast, Times of Famine: A History of Climate Since the Year 1000* (translated by Barbara Bray) (Garden City, N.Y.: Doubleday, 1971). A fascinating reconstruction of climatic fluctuations in Europe and their effect on the human community.

MANNERS, G., et al., *Regional Development in Britain* (London: Halsted Press, 1973). Examinations of the results of the national government's attempt to redress unbalanced spatial development.

MELLOR, ROY E., *The Two Germanies: A Modern Geography* (London: Harper & Row, 1978). The centrality of the two Germanies on the continent makes an understanding of their problems essential. In a lucid and highly readable study, the author contrasts the separate geographical development of these once-unified states.

RILEY, R. C., and G. J. ASHWORTH, *Benelux: An Economic Geography of Belgium, the Netherlands, and Luxembourg* (New York: Holmes & Meier, 1975). The economic transformation of the Benelux nations with specific reference to the significance of governmental decisions in the process of regional development.

SALT, JOHN, and H. CLOUT (eds.), *Migration in Post-War Europe: Geographical Essays* (London: Oxford University Press, 1976). A discussion of the effects of the emerging patterns of migration in Europe, which remains the migratory continent par excellence.

SUNDQUIST, JAMES L., *Dispersing Population: What America Can Learn from Europe* (Washington, D.C.: Brookings Institution, 1975). An analysis of the national population policies of five of the European nations. The author is hopeful that American planners will gain useful information from the collective experiences of these European nations.

IV

Eastern Europe and the Soviet Union

Roger L. Thiede

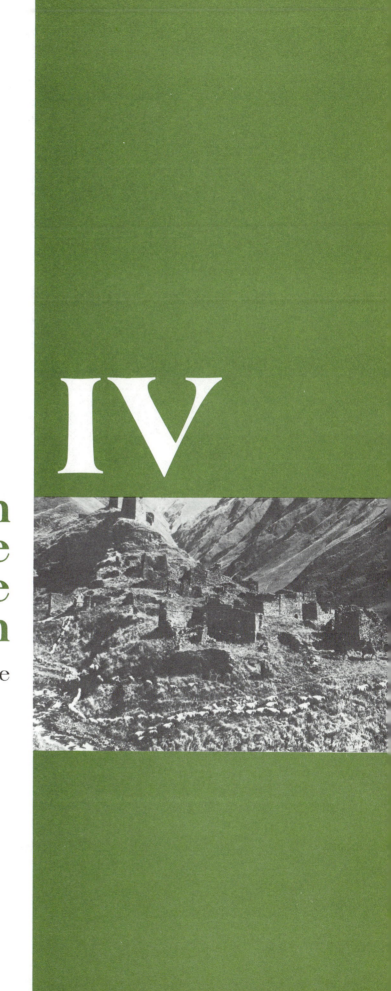

11

Eastern Europe: The Land Between

The Eastern Europe that emerged from the destruction of World War II is a region whose course of political, economic, and social development has been abruptly altered. Soviet-trained Communists stepped into positions of leadership. Leftist groups, often with support from the Soviet army, established satellite "people's republics" or "people's democracies." Within three years after the end of the war, *Soviet-style institutions* were imposed in all eight Eastern European countries, and, to use Winston Churchill's expression, an *"Iron Curtain"* was drawn between east and west, dividing Europe into two parts (Figure 11–1).

The Communist party became the sole political party. Land, industries, banks, and all but the smallest commercial establishments were nationalized. The traditional churches came under close governmental supervision and control as the *Marxist-Leninist-Stalinist philosophy* became the basic code of life.

A primary objective of these new governments was *economic reconstruction and industrial development.* The ideological goal was to create a Communist society in which there was no private property and the people owned all the means of production. To work toward this ideal goal of *communism* where all were to "give according to their ability and take according to their need," Soviet Marxist dogma specified a period of *"socialism"* where the remnants of the former economic, political, and social systems would be eliminated, and the state would own all the means of production. The communist parties of Eastern Europe were to direct these new *command* (i.e., centrally controlled and planned) *economies* under Moscow's guidance.

It was not long, however, before *dissent* appeared in this Soviet-dominated region. The Communist-nationalist leader of Yugoslavia, Marshal Tito, broke away from the Stalinist Soviet Union in 1948 and embarked on his own variant of socialism. After Stalin's death, the process of *de-Stalinization* under Khrushchev and revision of many of

FIGURE 11–1 Nations and Major Cities of Eastern Europe.
Eastern Europe has often been called a shatter zone since many ethnic groups have sought sovereignty over the areas they occupy. As a consequence the region has a history of political instability and conflict.

Stalin's practices in the Soviet Union led to Albania's withdrawal from the Soviet bloc.

During the post-Stalin years, the six countries that remained as satellites were able to follow more distinctive and independent ways. The Soviet Union recognized that there were *"different paths to the building of communism."* Eventually countries like Hungary, Czechoslovakia, and Bulgaria were able to introduce economic reforms that the Soviet Union considered too "Western" for its own tastes. Poland has ceased the collectivization of agriculture. The Catholic Church speaks with authority to many millions of Poles, especially with the voice of Pope John Paul II, the first pope of Polish origin. Romania has balked at plans that would lead to greater integration of the economies of the *Council for Mutual Economic Assistance (CMEA or COMECON).* CMEA is a supranational organization which consists of several Soviet-dominated countries that have joined forces to solve certain economic, cultural, or political problems they have in common. The Council for Mutual Economic Assistance is designed to coordinate and promote increased economic cooperation among the Soviet, Polish, East German, Bulgarian, Czechoslovakian, Hungarian, and Romanian economies. Vietnam, Cuba, and Mongolia are also members of CMEA, and Yugoslavia is an associate member.

Moscow, however, still considers Eastern Europe vital to its *national interest.* Unlike the European Economic Community, which consists of equal partners, CMEA is dominated by the USSR. Foreign trade of the six Eastern European members of CMEA is largely with the Soviet Union and other CMEA states. Soviet troops are stationed in East Germany, Poland, Czechoslovakia, and Hungary. Although the Soviet Union is willing to accept some expression of independence from the "people's democracies," it will not tolerate a challenge to its political control of Eastern Europe. Proof of this policy is seen in the Soviet invasion of Hungary in 1956 and Czechoslovakia in 1968 and its pressures on the Polish government to clamp down on the independent labor union, *Solidarity.*

PHYSICAL GEOGRAPHY

The landform map of Eastern Europe (Figure 11–2) shows the major contrast between the north and the central and southern sections. The low-lying *Great European Plain* widens as it crosses the north. Most of East Germany and Poland lies in this plain. Central Eastern Europe is a region of *hills and low mountains* separated by fertile plains. Most notable of these relief features are the low, nearly circular mountain ranges of the Bohemian Massif, whose crests form the boundary of western Czechoslovakia. Enclosed within the Massif is the Bohemian Plain, wherein lies Prague, the capital of Czechoslovakia. Farther south is the higher *mountain system* of the Alps. The Carpathian

177

FIGURE 11–2 Land Surface Regions of Eastern Europe.
In Eastern Europe most of the population lives on the plains. It is there that most of the large cities are located and where much of the manufacturing occurs. Hills and mountains serve as sources of minerals, water, and wood and, especially in the south, as areas of livestock raising. The mineral resources of the Central German Uplands, Bohemian Massif, and northern Carpathian Mountains are particularly important.

Mountains are an eastward extension of this system. Mountain landscapes dominate most of the Balkan Peninsula with the major exceptions of the Hungarian and Wallachian lowlands.

It is essential to recall that studying the topography of an area is not an end in itself but a means to understand more significant features of human geography. For example, the complexity of the mountain ranges of the Balkan Peninsula not only contributes to the diversity of ethnic groups found there but also complicates the construction of a well-integrated transportation network over much of the region. In the north, on the other hand, the easily traversed Great European Plain more readily exposes the Poles to conquerors from both east and west.

Climatically, Eastern Europe may be viewed as a bridge connecting the *marine climates* of northwestern Europe and the severe *continental climates* to the east. The continental influence is felt most strongly in eastern Poland, which has the coldest winters as well as the longest periods of snow cover. Western East Germany lies within the marine west coast climate zone. Here summers are cool, winters are moderate, and precipitation is adequate.

Central Eastern Europe has a *warm summer humid continental climate.* On the western and southeastern margins of the Balkan Peninsula are mild winter climates. In the interior of western Yugoslavia, except in the mountains, the climate is humid subtropical, with winter temperatures above freezing, and rainfall is plentiful. Along the western coast of Yugoslavia, over most of Albania, and in southern Bulgaria, a *dry summer subtropical climate* exists with moderate moist winters and hot dry summers.

CULTURAL DIVERSITY

The expression "shatter belt" frequently is applied to Eastern Europe. This term describes the *political instability* of the region: its inability to resist the greater military and political power of its past and present neighbors, particularly Germany, Austria, Russia, and the Ottoman Empire. Consequently, throughout much of modern European history, the boundaries of Eastern European countries have frequently changed. Eastern Europe's vulnerability to its stronger neighbors was and is promoted by its *diverse cultural character.* The absence of cohesiveness among the peoples of Eastern Europe has resulted in division and subjugation.

Varieties of Language

One measure of the cultural diversity is the *multiplicity of languages* spoken in the region (Figure 11–3). All of the peoples of Eastern Europe except the Hungarians and small Turkish minorities speak a language belonging to the diverse family of *Indo-European languages.*

The majority of the Eastern Europeans speak *Slavic* languages, which are divided into three

Indo-European Family of Languages

SLAVIC GROUP
Western Slavic languages
- P   Polish
- C   Czech
- S   Slovak
- L   Lusatian

South Slavic languages
- Slovene
- Serbo-Croatian
- Macedonian
- B   Bulgarian

Eastern Slavic languages
- Ru   Russian
- Be   Belorussian
- U   Ukrainian

GERMANIC GROUP
- G   German
- Danish
- Swedish

ROMANCE GROUP
- R   Romanian
- I   Italian
- V   Vlach

ILLYRIAN GROUP
- A   Albanian

HELLENIC GROUP
- Gr   Greek

BALTIC GROUP
- Latvian
- Lithuanian

GYPSY -Gy

Uralic Family of Languages

UGRIC GROUP
- H   Hungarian

Altaic Family of Languages

TURKISH GROUP
- T   Turkish

(Symbol indicates where a significant minority of each language is found.)

FIGURE 11–3 Language Patterns in Eastern Europe. The large number of different languages in Eastern Europe illustrates the region's ethnic diversity. Cultural differences such as language and religion inhibit political stability and, consequently, economic development.

subgroups: the eastern, western, and southern. Eastern Slavic languages (Russian, Ukrainian, and Belorussian) are spoken in the USSR, whereas the western and southern Slavic languages are found in Eastern Europe. Polish, Czech, and Slovak, which prevail in northern Eastern Europe, are the major Slavic languages of the western subgroup. Southern Slavic languages, Serbo-Croatian, Slovene, Macedonian, Montenegrin, and Bulgarian, are spoken in the Balkans.

Non-Slavic Indo-European languages of Eastern Europe are German, a Germanic language; Romanian, a Romance language related to Italian and French but with strong Slavic influences; Albanian, probably the oldest of the Indo-European tongues spoken in Europe; and the language of the Gypsies, Romany. Romany surprisingly does not belong to the European branch of the Indo-European languages. Its origin is found in India, from where the original gypsies migrated into Eastern Europe in the fourteenth century.

Although small minorities of various language groups are found in all countries of Eastern Europe, the overwhelming majority of the people speaking German, Polish, Hungarian, Romanian, Bulgarian, and Albanian are found in their respective states. Czechoslovakia consists largely of the Czech speakers in the western part of the country and those using Slovak in the east. Yugoslavia is a *multilingual state* with Serbo-Croatian, Montenegrin, Slovene, and Macedonian all official languages.

Pattern of Religious Heritage

Greater complexity is added to the cultural diversity map when traditional religious affiliations are considered. Although the Marxist governments have imposed restrictions on religions throughout Eastern Europe, they continue to exist. Significantly, the *religious tradition* has been an important element in the formation of a national character and the antagonisms that developed among cultural groups.

Not only are the three branches of Christianity (Roman Catholicism, Protestantism, and Eastern Orthodoxy) found in East Europe, but there are also adherents to Islam and Judaism. Numerically, Eastern Europeans who profess religious beliefs are predominantly *Roman Catholic*. The majority of the Poles, Czechs, Slovaks, Hungarians, and Croats are Roman Catholic. Protestant minorities are found in Czechoslovakia, Hungary, Romania,

179

and Yugoslavia. The East German population is overwhelmingly Protestant (Lutheran).

The second largest religious group in the region is the *Eastern Orthodox* Christians. Orthodoxy is predominant among the Serbs, Macedonians, Bulgars, and Romanians. Whereas the Roman Catholic religion was brought to Eastern Europe by missionaries from Rome, Orthodoxy was disseminated from Constantinople (Istanbul), the seat of eastern Christianity. Missionaries brought the alphabet and consequently a written language for the people they converted. Today, the language of the Serbs, Macedonians, and Bulgars is written in the *Cyrillic alphabet,* as are the eastern Slavic languages. The Romanians, although traditionally Orthodox in faith, use a Roman or *Latin alphabet* derived from their pre-Christian Latin heritage.

On the other hand, the Roman Catholic cultural areas of Eastern Europe and the Protestant areas (once Roman Catholic) use the Latin alphabet. Consequently, although the Serbs and the Croats speak the same language (Serbo-Croatian), they use different alphabets. The Serbs, principally of Orthodox heritage, use the Cyrillic alphabet, whereas the Catholic Croats use the Latin alphabet. The animosity between the Serbs and the Croats has at times been among the most intense of Eastern Europe. The possibility exists that a loosening of the tightly controlled government of Yugoslavia, as well as of those in several other countries of Eastern Europe, could reawaken old cultural enmities.

POPULATION

The total population of the eight Eastern European countries is about 137 million, or about one-half the population of the Soviet Union. The most populous of these states is Poland. With 36.6 million people the Polish Republic is the sixth largest country in Europe (after West Germany, Italy, the United Kingdom, France, and Spain), whereas Albania with 2.9 million people is less populated than all the European countries except Iceland and Luxembourg (Table 11–1).

Distribution

The *distribution of the population* throughout the region is highly variable (Figure 11–4). The largest area of *high population density* is found in southern East Germany, northwestern Czechoslovakia, and

TABLE 11–1 Population of Eastern European Nations

Nation	Population (millions)
Poland	36.6
Yugoslavia	22.8
Romania	22.6
East Germany	16.7
Czechoslovakia	15.4
Hungary	10.7
Bulgaria	8.9
Albania	2.9

the most southern part of Poland. A smaller high-density zone surrounds Budapest, Hungary. *Moderate population densities* are found in much of the remaining Great European Plain, the Wallachian Lowland, the Hungarian Lowland, and adjoining areas in southern Czechoslovakia, northeastern Yugoslavia, and central Romania. *Lower population densities* prevail in the mountainous areas that cover much of Yugoslavia, Albania, and Bulgaria, as well as in the Carpathians.

Population Growth

Much of World War II was fought on the territory of Eastern Europe. As a result of this conflict (the military and civilian losses and the migration of peoples), Eastern Europe's total postwar population was 10% less than before the war. During the early postwar years population increased at generally high rates, except in East Germany where emigration to the West proceeded briskly before it was sharply curtailed by the Berlin Wall.

With the exception of Albania, the countries of Eastern Europe today fit into the last stage of the *demographic transformation:* low birth rates, low death rates, and low rates of natural increase, 1% or less per year. Hungary has attained *zero population growth* (ZPG), where the population is replacing itself—births equal deaths. Albania's high birth rate (2.7%) and low death rate (0.7%) result in an annual rate of population growth of 2.0%. This figure has been slowly declining during the last decade but is still comparable to many countries in the developing world such as Mexico, Angola, and India with growth rates of 2.6%, 2.5%, and 2.1%.

Ideologically, the countries of Eastern Europe subscribe to the *Marxist theory of population growth.* This theory contends that population growth can only enhance economic benefits to the

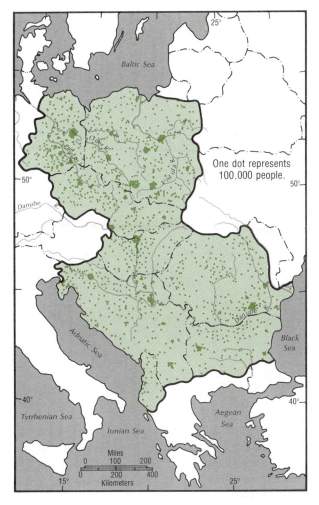

FIGURE 11–4 Eastern Europe's Population. Population distribution in Eastern Europe is quite uneven. Population density is higher in the north than in the south and higher on the plains than in the uplands.

population as long as the economic system has created an even distribution of resources. Policies such as *subsidies* to families and the availability of *day-care centers* affirmed the Marxist's pronatal philosophy. Other governmental policies, however, operated to slow the rate of population growth. For example, the *role of the woman* was changed in the new order. Women were needed in the growing labor force and gained rights to increased education and training. Liberalized *birth controls* were enacted that greatly increased the availability of abortions. These factors, along with the family's need for the woman's income and the shortage of adequate housing, helped reduce the birth rate. By the mid-1960s the governments of East Germany, Czechoslovakia, Hungary, Bulgaria, and Romania feared that dropping growth

rates would lead to an inadequate future labor supply. While the Western industrial economies see zero population growth as an ideal to be worked for, the Eastern European countries tend to rely more heavily on human labor resources and are not as highly automated and mechanized as their Western neighbors. Therefore, population policies were modified in Eastern Europe to *encourage an increase in the birth rate.* These measures included restricted availability of abortions, increased subsidies for families, extended maternity leave, and improved social welfare programs. The countries of Eastern Europe, however, continue to experience general declining growth rates as they become more urban and industrial and as their people demand better living conditions and more consumer goods.

Urban Population

Compared to Western Europe, Eastern Europe has a smaller proportion of its population living in cities. Out of nineteen countries in Western Europe, only Portugal and Norway have less than one-half their people classified as urban dwellers, whereas in Eastern Europe three out of the eight countries have less than 50% of their population living in cities. These relatively low proportions reflect the greater predominance of agriculture as a way of life in the region and the smaller role of industrial and service activities in the economy.

There is, however, considerable variation in the *levels of urbanization.* East Germany is one of the more urbanized states in Europe with 77% of its people living in cities. At the other end of the scale is Albania with 37% of its population living in cities. With the exception of Portugal, Albania is the least urbanized country in Europe. The level of urbanization for other Eastern European countries ranges from 42% for Yugoslavia to 67% for Czechoslovakia. Seven Eastern European cities have a population greater than 1 million (Budapest, Bucharest, Warsaw, East Berlin, Prague, Belgrade, and Sofia). These cities, excluding Belgrade, but including Tirane, Albania, are good examples of *primate cities,* or cities that are disproportionately large compared to other cities in their country (generally, more than twice as large). A primate city dominates the economic, cultural, and political life of its country and tends to be a more frequent occurrence in less developed countries where scarcer capital and resources tend to be more geographically concentrated.

181

An East German iron and steel plant. Construction of the plant and the accompanying new town of Eisenhüttenstadt began in 1950. Iron ore is imported from the USSR, and coking coal is brought from Poland. A barge canal, a highway, and a railroad serve the plant and the growing town. (Eastfoto)

The Communist regimes of Eastern Europe place a high priority on *industrial growth.* As a consequence urban development proceeded briskly after World War II. In addition to the reconstruction and expansion of existing cities, several new *industrial towns* were built. Some of these new towns are Eisenhüttenstadt (iron and steel) in East Germany, Nowa Huta (iron and steel) in Poland, Victoria (chemicals) in Romania, and Havirov (mining) in Czechoslovakia.

LEVELS OF ECONOMIC DEVELOPMENT

The emphasis on industrialization in Eastern Europe has meant a concentration of *heavy industry,* the manufacturing of *producer goods,* and the development of the mineral-resource and energy base of the state. Before World War II only Germany and Czechoslovakia had well-developed industrial sectors. Today, despite considerable progress in industrial production, and in agriculture as well, Eastern Europe still has the least industrialized and poorest economic landscape of Europe.

The levels of economic development and of living are varied. This diversity is most evident when the five criteria of development discussed in Chapter 4 are applied to the region, particularly the three standard indicators (per capita GNP, per capita consumption of energy, and percent of the labor force in agriculture). The two supplementary measures (daily food supply and life expectancy) are more uniform.

The three standard measures suggest more convincingly that the term *"developed" needs qualification* for some of the countries of the region. East Germany, Czechoslovakia, Hungary, Bulgaria, and Poland may readily be included among the richer states. *Per capita GNP* ranges from $7,286 for East Germany to $4,187 for Poland. The per capita GNP of the United States is $12,530 and the USSR $4,701. However, Yugoslavia, Romania, and most notably Albania show greater contrasts to the richer states of Europe. Per capita GNP is lower ($2,789 for Yugoslavia, $2,546 for Romania, and $840 for Albania). *Per capita energy consumption* is also relatively low, and employment in agriculture is relatively high compared to their more economically developed neighbors in the west. Perhaps it is best to think of Yugoslavia and Romania as transitional, or developing—neither rich nor poor. In the case of Albania, however, convincing arguments may be made for its inclusion among the ranks of the poor nations.

AGRICULTURE

The historical legacies, physical environments, and local reactions to Soviet policies have all contributed to the creation of a *diverse agricultural landscape.* At the end of World War II much of Eastern Europe suffered from chronic rural overpopulation. The existence of large estates and the shortage of land for the peasants in the best agricultural areas in the central and northern sections were a major problem. In the Balkans, however, agriculture was at a subsistence level on small, highly fragmented farms.

Collectivization

Numerous attempts at land reform were made throughout Eastern Europe before World War II. None could compare, however, with the drastic changes attempted after the postwar socialist regimes were established. Large estates were nationalized, and the land was redistributed among the peasant. The next step was the *collectivization* of agriculture. Land was consolidated into either large-scale, Soviet-style *state farms* or *collective (communal) farms.* The state farms are funded and directed by the central governments. The operation of the collective farms is directly the responsibility of the farmers within the framework of state directives.

The *objectives* sought through collectivization were the same used to collectivize the land in the Soviet Union A pure Marxist-Leninist form of agricultural organization would make it easier to control this important sector of the economy and thereby facilitate the enactment of programs of industrial development. It was anticipated that the availability of foodstuffs for the growing number of industrial workers would be increased, for it was expected that large-scale agricultural units would be more productive through greater application of mechanization and specialization of labor.

The *results* of the collectivization are varied. There was large-scale resistance throughout the region, particularly in Poland and Yugoslavia. In the other countries, collectivization has proved much

more durable and successful. All of Albania's farms are in the socialist sector (state and collective farms). In Hungary, Romania, East Germany, Czechoslovakia, and Bulgaria the socialist sector controls 84–92% of the agricultural land.

General State of Agriculture

On the whole, *agriculture* plays a larger role in the economy of Eastern Europe than in Western Europe. One indicator is the percent of the labor force of each country employed in agricultural pursuits.[1] As expected, the lowest proportion of farm workers is encountered in the most industrialized and wealthiest states: East Germany and Czechoslovakia. Agriculture is more mechanized in these two countries than in the rest of the region. *Levels of productivity* in both East Germany and Czechoslovakia are below the standards of Western Europe, and neither is self-sufficient in foodstuffs. Everywhere, however, agricultural equipment is scarce. Other limiting factors have been a shortage of farm labor and the chronic problem of lack of work incentives.

Compared to twenty-five years ago, the proportion of agricultural workers in Hungary has been cut in half. Today, one-quarter of Hungary's labor force is so employed. Hungary has long been important as an agricultural producer and today remains a net exporter of farm goods. Of all the countries with a predominantly *socialized agricultural organization,* Hungary has gone the furthest in allowing collectives to make their own production decisions. In Hungary, however, the growth of agricultural production has suffered from poor financial incentives for the farmer.

Only Poland is dominated by *private farm* holdings. These private farmers, however, are closely regulated by the central government. The production of foodstuffs in Poland is insufficient to meet domestic needs, and Poland must import food.

The four *Balkan states*—Yugoslavia, Albania, Romania, and Bulgaria—are more agricultural than the northern Eastern European states, with the possible exception of Poland. In Albania more than six out of every ten workers are engaged in agriculture. For Yugoslavia, Bulgaria, Romania, and

---

[1]The use of percent of labor force employed in agriculture overemphasizes the contribution of agriculture to the economy. In terms of dollar value added to the GNP, Eastern European farmers are only one-third to one-half as productive as their fellow industrial workers.

Poland between 30% and 40% of the work force is in agriculture. Romania and Bulgaria are net exporters of foodstuffs. Yugoslavia is generally self-sufficient in its food needs; however, during years of poor harvest, some food must be imported. Compared to the levels of pre–World War II agricultural production, the Balkan states have made substantial gains. Despite these improvements, their per capita and per area productivity remain among the lowest in Europe.

Types of Agricultural Production

The cooler *northern section* (East Germany, Poland, and Czechoslovakia) is a major *grain-growing* region with rye and wheat the principal grains. There has been an increased emphasis on *industrial crops* (sugar beets, oils, and fiber plants) in the north, and the important staple, potatoes, is widespread throughout the area. Vegetable and fruit growing is prominent near urban areas.

The shortage of meat has been a chronic problem in the northern agricultural zone. The area planted to fodder crops has increased generally throughout the region but is still insufficient to meet the needs of the meat and dairy industries. Cattle and hogs are the principal livestock in this part of Eastern Europe.

In the *southern part* of Eastern Europe, the principal *grain-growing regions* are the Hungarian and Wallachian lowlands. Wheat and maize are the most important grains here and in other parts of Bulgaria, Yugoslavia, and Albania. Emphasis has been placed on the cultivation of industrial crops, most notably sugar beets. Bulgaria is Eastern Europe's major producer of cotton, tobacco, and vegetables. Bulgaria, Romania, and Hungary all export vegetables, both fresh and processed, principally to the Soviet Union and northern Eastern Europe. The dry summer subtropical climate zone of Yugoslavia and Albania produces a variety of *specialized crops* such as citrus, olives, and vegetables. Vineyards, primarily for wine grapes, are a significant feature of the dry summer zone of Yugoslavia, the Hungarian Lowland, and parts of Romania and Bulgaria. Livestock production has encountered the same difficulties in this part of Eastern Europe as in the north, largely because of an inadequate feed base. The raising of sheep for both meat and wool, partially reflective of the area's Islamic heritage, is an important phase of the livestock industry in the four Balkan states.

INDUSTRY

Resource Base

Eastern Europe is not well endowed with industrial resources (Figure 11–5). It is *deficient in energy resources,* iron ore, and other minerals. The few resources that do exist are unequally distributed and many are of inferior quality and expensive to utilize.

The only *high-quality coal deposit* of substantial size is the Silesian-Moravian field. Most of this deposit lies in Poland, but a small part extends into Czechoslovakia. The production of coal in Poland exceeds that of all the other countries in Eastern Europe. A coalfield of secondary importance is found in western Czechoslovakia, and small scattered deposits are found elsewhere in Eastern Europe. *Low-quality coals* are widely scattered throughout Eastern Europe. With the exception of the East German lignite fields, these fields are small. Low-grade coal is used for the production of electricity and for domestic heating.

There is also a shortage of *oil and natural gas* in Eastern Europe. Romania's Ploesti field has the

Romania's Ploesti oil field contains continental Europe's most extensive petroleum deposits. Although the field is old, exploration and drilling continue, because present reserves are calculated to last only a few more years. (Eastfoto)

Industrial districts
• Important industrial cities

■ Major hard coal deposits
□ Minor hard coal deposits
● Lignite and brown coal
▲ Petroleum
★ Natural gas
U Uranium

Fe Iron ore
Cu Copper
Pb Lead
Zn Zinc
Al Bauxite
Cr Chromium
Ni Nickel
Sn Tin
K Potash

FIGURE 11–5 Mineral Resources and Industrial Districts of Eastern Europe.
Eastern Europe has three principal industrial districts: the Saxony District of East Germany, the Silesia-Moravia District of Poland and Czechoslovakia, and the Bohemian Basin in Czechoslovakia. All these industrial centers are located on or near coal deposits and have access to other nearby minerals.

oldest commercial oil well in the world. The total reserves of Romania—the Ploesti field, the newer Bacau field in the eastern part of the country, and other scattered small deposits—are expected to last only a few more years at the present rates of production. Yugoslavia's reserves of petroleum and natural gas are adequate for its needs at present, and production is increasing. However, because consumption of these fuels is rapidly growing, there are serious doubts that Yugoslavia will be able to meet its domestic requirements for oil and gas in the near future. Albania is now a small exporter of petroleum products. The remaining Eastern European countries have little or no oil and gas reserves. They are dependent on imported petroleum and gas, principally from the Soviet Union. Oil and gas consumption is increasing throughout Eastern Europe, and dependence on foreign sources, particularly the Soviet Union, will inevitably become greater.

Hydroelectric potential is limited in Eastern Europe, except for the Balkan states. At present there is little developed hydroelectric power, and most plants are quite small. The potential for development of hydroelectricity is greatest in Yugoslavia. Extensive construction is under way to more than double 1970's output by 1985.

Because of the poor availability of good quality energy resources, the countries of Eastern Europe have expressed widespread interest in *nuclear power*. Generation of electricity by nuclear power is still limited, but increased output is expected in the future. Uranium, the fuel for nuclear power, is mined throughout much of the region, but details of production are closely guarded secrets.

Ferrous and nonferrous metals are also in short supply in Eastern Europe. Most critical is the limited amount of iron ore. Even Poland, which has the largest iron and steel industry in Eastern Europe, must rely on the USSR for iron ore. Yugoslavia is in the best overall position. Although most of its metallic deposits are relatively small and production is limited, it possesses a wide variety of raw materials, including iron ore, copper, lead, zinc, chromium, and bauxite. Among the most important metals in other Eastern European states are bauxite in Hungary, chromium in Albania, and copper and tin in East Germany. Most notable of the region's nonmetallic minerals are Poland's sulfur and East Germany's potash deposits. In both countries these minerals provide the raw materials for chemical industries.

Industrial Growth

The greatest strides made in the economic development in Eastern Europe during the postwar period have been in *manufacturing.* The greatest gains have been posted by the least industrialized countries, particularly Bulgaria. Although the gap between the richer and poorer industrial states has been narrowed, East Germany and Czechoslovakia remain the most industrialized countries.

In the early postwar years, emphasis was on increasing production. Tight *centralized management* of the economy often resulted in inefficiency, waste of raw materials, and low levels of labor productivity. Since the mid-1960s, there has been increasing concern over *production efficiency.* Governments have become cost conscious and have attempted, with the notable exception of Albania, to reduce central direction and increase the profits of industry. As a consequence, growth rates have generally been lowered in comparison to the early years of development, but efficiency has been improved. Hungary has been very successful in introducing *economic reforms* that give plant managers and labor unions real decision-making power. Even orthodox Communist Bulgaria has begun reforms that should reduce central bureaucratic control of the economy.

Yugoslavia has gone the furthest in reducing centralized control over industry. It has attempted to establish a *Western market-type economy* in place of the Soviet-style economy in which prices are set arbitrarily. Plant managers and workers have a greater decision-making role. These reforms have reduced waste and made Yugoslavian industry more competitive with the industries of Western Europe. The Soviet Union has condemned Yugoslavia for being nonsocialistic.

The Location of Industrial Activity

Soviet and Eastern European Marxists contend that because capitalism involves the exploitation of humankind by humankind, one can also expect the *exploitation of one region by another region,* or the uneven distribution of economic activity. Theoretically, according to Marxists, *equal and balanced regional economic development* throughout a country is possible only in a planned socialist society that has eliminated capitalism. Eastern European planners have tried to stimulate the development of backward areas such as eastern Poland and the Slovakia area in Czechoslovakia. While there has been some success in the geographic *decentrali-*

zation of industrial activity, the older, established centers have also continued to grow and still dominate the industrial geography of the region.

A large *industrialized region* extends from Western Europe across the northern part of Eastern Europe. The western parts of Czechoslovakia and Poland and all of East Germany are part of this area of concentrated industrial activity. The principal industrial areas of Eastern Europe found within this part of the region are the Silesian-Moravian district of Poland and Czechoslovakia, the Bohemian Basin of Czechoslovakia, and the Saxony district of East Germany (Figure 11–5).

The *Silesian-Moravian district* was formerly Germany's second heavy industrial region. After World War II, Poland acquired the largest part of Silesia; a small section was given to Czechoslovakia. Silesia is Poland's major heavy industrial district based on local coal resources. In addition to coal mining and the production of iron and steel, manufacturing includes agricultural machinery, machine tools, and various chemicals. The Czech section of Silesia is centered in the city of Ostrava, the largest iron and steel center of the country.

The Bohemian Basin in western Czechoslovakia is also an iron- and steel-producing district. Small coal mines are worked in this area. Czechoslovakia's capital city, Prague, is located within the basin and is Czechoslovakia's largest city and center of industrial production. The capital city's industry is diversified and includes important chemical, food, and machinery industries.

East Germany's highly urbanized *Saxony district* contains the country's greatest concentration of industrial production. There is some iron and steel production, but the region is highly diversified and famous for its manufacture of chemicals, textiles, and engineering industries.

Manufacturing outside the industrial region is chiefly in the large cities. For example, cities like Berlin, Budapest, Warsaw, and Belgrade are major focal points of diverse industries. Other centers of production are generally smaller and often associated with the location of a raw material (e.g., chemical industries at Ploesti).

THE UNCERTAIN FUTURE

Eastern Europe has traveled a long road since those early post–World War II years when it was enclosed behind the Iron Curtain. Yugoslavia has gone the furthest in establishing relationships with the West. Westerners may travel in Yugoslavia with relative ease. Yugoslavians are permitted to migrate to Western Europe to hold temporary jobs. More so than elsewhere, the Yugoslav Communist Party has relaxed its grip on the economy and society; yet the party is still the sole legal political force. Yugoslavia's future course depends on the ability of its political leaders to hold the diverse elements of the country together after the death of the popular Marshal Tito. Latent internal ethnic hatreds and boundary disputes with Albania and Bulgaria could threaten the security and economic progress of Yugoslavia.

Albania, on the other hand, remains tightly within the grasp of a conservative Stalinist-type Communist Party. Although the government has recently opened the door to a trickle of outsiders, Albania remains the most closed society of Europe.

The six CMEA countries are closely tied to the Soviet Union, although they increasingly are at variance with the policies of the USSR. Attitudes toward the Soviet Union range from that of Bulgaria, which has proposed becoming part of the USSR, to those of the defiant Polish workers whose independent trade union movement has directly challenged Communist Party rule and Soviet domination.

Economic realities guarantee a continuing close relationship with the Soviet Union. Dependence is still on the Soviet Union for raw materials, capital goods, and credit essential for economic growth. At the same time, the decrease in the rate of economic development and the rising expectations of the citizenry of Eastern Europe have made these countries increasingly interested in greater contacts with the West. The CMEA countries want technology and capital from the industrialized West, and increased markets for their goods in the West could stimulate production at home. Progress in this direction depends on détente between the United States and the Soviet Union. Warmer relations between the Soviet Union and the United States would most likely result in greater Western contacts with the European CMEA countries. Particularly meaningful in this respect are the recent improved relations between East Germany (one of Moscow's most loyal allies) and Western states, notably West Germany. The internal stability of the Eastern European governments reflects the economic welfare of their peoples. This welfare relates to the ability and desire of the governments to bring about changes necessary for greater economic efficiency and progress. The changes themselves still must have the tacit approval of the Soviet Union. The nature of the modifications of the economic and political life of Eastern Europe in the end will reflect changes in Soviet policy toward its satellites.

12

The USSR: Land and People

The sheer physical magnitude of the Union of Soviet Socialist Republics, *one-sixth of the earth's land surface,* is in marked contrast to the size of other European states. One passes through eleven time zones traveling the 6000 miles (9700 kilometers) from the Baltic Sea to the Bering Straits. At its maximum north-south extent the Soviet Union is almost as wide as the United States is long, approximately 2900 miles (4667 kilometers).

To many the size of the USSR evokes the image of unlimited raw materials awaiting exploitation. To others it conjures a picture of vast tracts of virgin lands awaiting the settler. Unquestionably, the size of the country enhances the potential occurrence of vital natural resources. Indeed, the USSR is extremely *well endowed with raw materials* requisite for its world-power status. The presence of these natural resources does not, of course, mean that their development is assured. Technology and sufficient capital for their exploitation must be available, and the resources must be economically accessible to their potential consuming regions. Many of the Soviet Union's raw materials are found in regions remote from population concentrations, and their exploitation requires a large capital investment that the government may be unable or unwilling to pay. Furthermore, much of the Soviet land presently is considered inhospitable for settlement and unsuitable for agriculture because it is too cold, too wet, or too dry.

NATURAL REGIONS

The magnitude of the Soviet landmass and its high latitudinal location are important elements in the *severe continental climates* that dominate the country. The southernmost part of the country lies approximately at 35° N, about the same as Memphis, Tennessee. Moscow is farther north than Edmonton, Canada. More than 75% of the USSR lies poleward of the 49th parallel, the northern boundary of the United States. A useful device for studying the physical environment is the *"natural re-*

gion'' (Figure 12–1). These regions are essentially vegetation zones with related and generalized climatic and soil characteristics. ''Natural region,'' however, is a misleading term. Large parts of the Soviet Union have been altered by mankind and are not in a natural or primordial state.

Tundra

Northernmost of the Soviet natural regions is the *tundra.* This zone, together with tundra conditions that exist elsewhere in the mountains, covers about 13% of the Soviet Union. The mean temperature of the tundra's warmest month is between 50°F (10°C) and 32°F (0°C). This region's very short growing season and poor soils result in sparse veg-

etation characterized by such hardy plants as reindeer moss, lichens, and shrubs. The tundra is treeless because of the limited heat received in the high latitudes, the high winds, and the presence of *permafrost* (permanently frozen ground that restricts root growth). Needless to say, the tundra, with its bleak and cold climate, has been little affected by humans. Widely scattered indigenous tribes, hunters, trappers, and miners are among the few who have penetrated this remote and inhospitable region.

Taiga

Forests cover almost 50% of the Soviet territory. These forests are divided into three natural re-

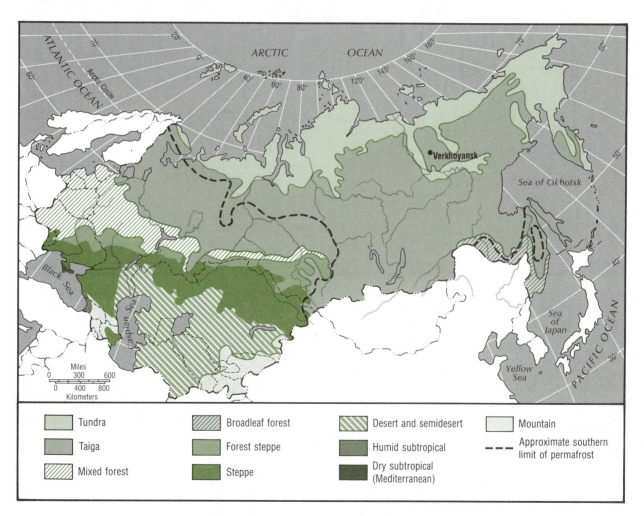

FIGURE 12–1 Vegetation Regions of the USSR.
The Soviet Union is so large that it contains many different environments. Environmental diversity provides an opportunity to grow many different types of crops, and the large area enhances the possibility of a varied mineral resource base. Much of the USSR, however, is located at high latitudes and little used.

The taiga in the USSR contains one of the world's greatest forests, mainly conifers. In western Siberia the land is a flat to rolling plain, but it is only sparsely inhabited because of poor soils, permafrost, and a short growing season. Trees grow very slowly, often taking more than 100 years to reach a size suitable for lumbering. (TASS from Sovfoto)

gions: the vast *taiga* (the coniferous forest of the north), the mixed forest of European Russia, and the broadleaf forest of the far east. The taiga lies south of the tundra, and its climate is primarily *subarctic.* Lower temperatures have been recorded in the taiga than in the tundra. A record low of −96°F (−71°C) makes Verkhoyansk the coldest place in the world outside Antarctica. Summers are short and cool. In general, the growing season is less than a hundred days, and consequently *agriculture is very limited.*

The dominant soil of the taiga is *podzolic.* Grayish in color and leached of many nutrients, these podzolic soils are infertile and highly acidic. Areas of permafrost cover virtually all of eastern Siberia except the extreme southeast portions. In western Siberia, only the northern portion of the taiga is affected by these large areas of frozen soils. Besides retarding root growth, permafrost can contribute to excessive waterlogging of the topsoil in the summer. Waterlogging may lead to lateral soil movements that can cause havoc to railroad lines, roads, and even buildings if they are not firmly anchored in the frozen subsoil.

Coniferous forests are the most common of the taiga vegetation. However, extensive swamps and meadows are frequently interspersed among the forest lands.

Under Soviet programs nomadic native groups of the region are turning to more sedentary forms of *herding* and some *agricultural pursuits.* The traditional wealth of the taiga—*timber, fur-bearing animals,* and *precious metals*—is still significant.

However, newly developed industrial resources, such as the *oil and gas fields* in northwestern Siberia, are growing in importance.

Mixed Forest

Between the taiga and the grasslands of southwestern Russia lies the triangular-shaped region of the *mixed forest.* The name "mixed forest" indicates the occurrence of both coniferous and deciduous broadleaf trees. From north to south, deciduous trees more and more displace the conifers and eventually take over completely. Temperatures and the length of the growing season increase toward the south, but precipitation decreases, and the native forest cover thins as the open meadows increase. This region more than any other in Russia has been *culturally modified,* for within the mixed forest evolved the modern Russian state. Centuries of settlement have resulted in land clearing for agriculture and the rise of cities. The longer growing season and the *gray-brown forest soils* make the mixed forest more suitable for agriculture than the taiga.

Broadleaf Forest

The third forest region is the *broadleaf forest* of the southern part of far eastern USSR. This region has cold dry winters and hot humid summers. The vegetation is mostly broad-leafed deciduous trees of Asiatic origin with some conifers and open grassy meadows.

Steppes

The *steppe* zone of the USSR consists of two subregions: the *forest steppe,* a region of transition from the forested land to the north, and the treeless grassy *true steppe.* The forest steppe is characterized by woods separated by extensive grasslands. In the true steppe, trees grow only in the river valleys, and grasslands stretch as far as the eye can see. The soils of the forest steppe are more leached than in the true steppe, where higher temperatures and higher evaporation rates reduce the effectiveness of precipitation. Consequently, in the true steppe there is a greater buildup of humus and minerals in the topsoil, which produces rich *chernozemic* (from the Russian meaning "black earth") soils. The entire steppe zone is *important for agriculture.* Conditions for farming are best in the forest steppe, where the soil is fertile and precipitation generally sufficient and reliable. In the true steppe, arid conditions and variability of precipitation enhance the chances of drought.

Desert

South of the steppe lands, principally in the Trans-Volga area (the land east of the Volga River in southern USSR), aridity increases until true *desert* conditions exist. Precipitation in the desert is generally less than 10 inches (25 centimeters) a year, and very hot dry conditions prevail in the summer. Winters are cold. The vegetation of the desert consists mostly of clumps of *grass and xerophytic plants* that can store moisture; there are large expanses of bare earth, rock, and sand. In widely scattered oases and along the few rivers flowing through the desert, a rich plant life is supported on alluvial soils. Large parts of the desert are used for *livestock grazing.* Crop cultivation is limited to areas watered by streams or by irrigation projects.

The Subtropical South

Two small but important regions are the *humid subtropical region* along the east coast of the Black Sea and the *dry summer subtropical region* along the south coast of the Crimean Peninsula. The Crimean Mountains help protect the narrow coastal region of the peninsula from severe cold winds of the north. The mild climate and moisture from the Black Sea contribute to a *varied agriculture,* including nuts, fruits, and vineyards. The Crimea is a famous *resort area.* Old palatial residences of the wealthy, including the tsars' former palace at Livadia, have been turned into rest and vacation hotels for Soviet workers and bureaucrats.

Much of the USSR's southern border is mountainous. The mountains of Central Asia remain the home of pastoralists who graze their sheep in the mountains during the summer and winter them in the valleys. Many of the once-fortressed towns on the mountains have been abandoned and are now in ruins. (TASS from Sovfoto)

The humid subtropical area is also a favorite *vacation spot.* This area is renowned for its specialized agriculture, of which tea and citrus fruits are most important crops. The moisture-laden winds from the sea bring large amounts of precipitation. The soils are quite fertile and support a luxuriant vegetation.

Mountain Areas

The climate, soils, and vegetation of the *mountain regions* are diverse, reflecting the location of the mountains, their local relief, and—most important—altitude. Many valleys, foothills, and mountain meadows of the mountain systems in central Asia and in the Caucasus support relatively large populations on productive agricultural lands.

POPULATION

With a population of approximately 272 million, the USSR is the *third most populous state* in the world. The size and nature of the population are important to Soviet economic development. An understanding of the most significant population characteristics contributes to our appreciation of the economy's spatial distribution as well as its contemporary and possible future economic and political problems.

Ethnic Groups

There are more than *a hundred different ethnic groups* in the Soviet Union. These peoples vary widely in their language, history, religious tradition, physical characteristics, and geographic distribution. The presence of different ethnic groups within a sovereign state frequently is *a threat* to the unity and even the existence of the state. The Soviet Union, however, has maintained a tightly controlled state with apparently little conflict among its diverse ethnic groups.

To minimize potential ethnic conflict, the USSR not only uses the all-pervasive power of the central state but also employs a governmental structure that allows *a modicum of ethnic rights.* These rights vary depending in part on the group's size, culture, and economy. Among the rights are: use of native languages in schools, courts, places of business, newspapers, and books; maintenance of ethnic customs (with severe limitation on religious practices); and for fifty-three of the groups a politically recognized territory. The laws, customs,

and acts of the ethnic groups cannot, however, conflict with the dictates of the central government. The central government, which is synonymous with the national Communist Party, is controlled by the dominant ethnic group in the country—the Russians. Stalin defined the nature of the government of the Soviet Union as *"national in form, socialist in content."*

Although ethnic groups are granted protection of some of their customs and traditions, a process of *cultural and economic integration* has continued throughout the Soviet period. Russians have migrated into other ethnic areas and in many are now the major population group. Russian is a required language for students.

Political Territorial Units

Highest of the political territorial units of the USSR is the *Soviet Socialist Republic (SSR).* According to the Soviet Constitution, the USSR is "a federal state formed on the basis of a voluntary union of equal Soviet Socialist Republics," but the union is anything but voluntary. It was formed by Bolshevik force. Among the many rights the constitution bestows upon the SSR is the right of secession; however, in reality no SSR is allowed to secede, for such a move would indicate "bourgeois nationalist tendencies." Rough guidelines for the formation of a socialist republic include a population of more than 1 million, location on the periphery of the country, and a majority population of the ethnic group for which the republic is named. The *fifteen republics* represent large ethnic groups (Table 12–1 and Figure 12–2). They do not, however, represent the fifteen largest population groups.

Thirty-eight other ethnic groups have an *administrative unit of lesser status.* The peoples of these units have various rights of language and custom but are subordinate to the laws of the SSR in which they are located as well as to Moscow. Where there are no significant ethnic groups, the territory is divided into administrative districts.

Ethnic Composition

Ethnic composition by *language groups* is shown in Figure 12–3. The eastern Slavs (Russians, Ukrainians, and Belorussians) belong to the *Indo-European language family* and are the largest group in the USSR, 72% of the country's population; the *Russians* alone comprise 52% of the total population. The Russians predominate in central and northern European Russia, Siberia, and Kazakh-

TABLE 12–1 The Soviet Socialist Republics (SSR)

	Common Name	Capital
Union of Soviet Socialist Republics (USSR)	Soviet Union	Moscow
1. Russian Soviet Federated Socialist Republic (RSFSR)	Russia	Moscow
2. Estonian Soviet Socialist Republic	Estonia	Tallin
3. Latvian Soviet Socialist Republic	Latvia	Riga
4. Lithuanian Soviet Socialist Republic	Lithuania	Vilnius
5. Belorussian Soviet Socialist Republic	Belorussia	Minsk
6. Ukrainian Soviet Socialist Republic	Ukraine	Kiev
7. Moldavian Soviet Socialist Republic	Moldavia	Kishinev
8. Georgian Soviet Socialist Republic	Georgia	Tbilisi
9. Armenian Soviet Socialist Republic	Armenia	Yerevan
10. Azerbaijan Soviet Socialist Republic	Azerbaijan	Baku
11. Kazakh Soviet Socialist Republic	Kazakhstan	Alma Ata
12. Turkmen Soviet Socialist Republic	Turkmenia	Ashkhabad
13. Uzbek Soviet Socialist Republic	Uzbekistan	Tashkent
14. Kirgiz Soviet Socialist Republic	Kirgizia	Frunze
15. Tadzhik Soviet Socialist Republic	Tadzhikstan	Dushanbe

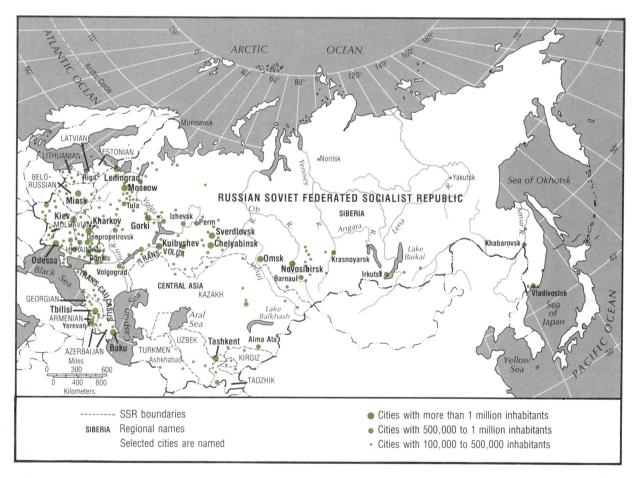

FIGURE 12–2 The Soviet Socialist Republics and Cities with More Than 100,000 Inhabitants.
Politically the Soviet Union is divided into Socialist Republics, most of which are in the west. Note also that most of the nation's major cities are in the western one-quarter of the country.

193

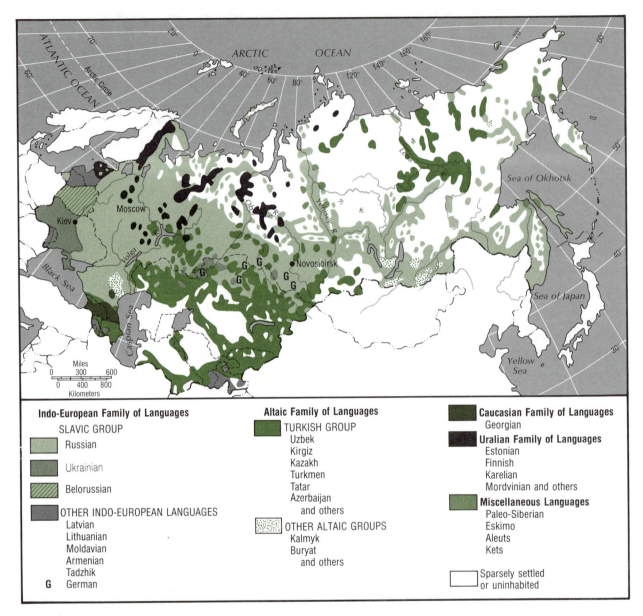

FIGURE 12–3 Major Language Groups of the USSR.
Society in the Soviet Union is often falsely characterized as monolithic. This map of
languages shows that there is considerable cultural diversity.

stan, but they are distributed throughout the USSR and are important even as minority groups, especially in cities.

Other ethnic groups representing the Indo-European family of languages include the Latvians and Lithuanians in the Baltic region, the Armenians in the Caucasus, and the Tadzhiks (an Iranian people) in central Asia.

There are almost 40 million people speaking *Altaic languages*. These groups have an *Islamic heritage* and are found principally in Central Asia, the

middle Volga Valley, and the Caucasus. The *Uralian family* of languages is represented by a number of groups found in northern European Russia and in western Siberia. Most prominent of this family are the Estonians and their close relations, the Finns and Karelians.

The *Jewish people* form the sixteenth-largest ethnic group in the USSR. They are predominantly urban and live mostly in the European Soviet Union. Moscow's attempt to establish a Soviet homeland for the Jews in the far east, the Jewish

Autonomous District, has not been successful in attracting the Jewish population. Less than 7% of the population of the district is Jewish.

Other groups in the nationality patchwork are the numerous Caucasian peoples in the Caucasus, of whom the Georgians are most notable. There are also Mongols, Koreans, and the many small indigenous tribes of Siberia, such as the Eskimos.

Ethnic Demographic Differences

The Slavs, Baltic peoples, and other groups of modern European culture have passed into that stage of the *demographic transformation* in which birth rates, death rates, and rates of natural growth are low. In contrast, other ethnic groups with a more traditional and agricultural life-style maintain higher birth rates, which, with low death rates, result in faster population growth.

Ethnic groups in Central Asia, the Caucasus, and Siberia have a higher rate of population growth than the Slavs and other European groups. In 1959, the Eastern Slavs comprised about 76% of the country's population, whereas in 1979 they made up 72%. The proportion of Russians declined from 55% to 52% during the same period. At the present rates of population growth the *Russians will be a minority by the year 2000.* This fact in itself does not mean that the Russian grip on the country will be lessened; it does mean, however, that the handful at the top of the Communist hierarchy must be prepared to deal with the challenge of ethnic discontent. Anti-Russian riots and demonstrations have occurred in Central Asia and the Baltic States; there are numerous underground national movements in existence, and reports of widespread discrimination and maltreatment of minorities in the armed forces all indicate that although the security of the state is not threatened, *ethnic tensions* will further test the system.

Demographic Characteristics

From the early part of the twentieth century to the present, the Soviet population has passed from a stage of high birth and death rates to a stage of *low birth and death rates.* In 1913 the birth rate was 4.5%, in 1959 it was 2.5%, and now it is 1.8%. The death rate for this same period went from 2.9% to 1.0%. The resultant rates of natural increase were 1.6% in 1913, 1.7% in 1959, and 0.8% now. Life expectancy in European Russia at the end of the nineteenth century was thirty-one

years for men and thirty-three years for women. Life expectancy now is sixty-four years for men and seventy-four years for women. These figures compare with a life expectancy of sixty-nine years for men and almost seventy-seven years for women in the United States.

Figure 12–4 shows the distribution of the Soviet population by sex and age in 1970. This *population pyramid* is a valuable tool to demonstrate some Soviet demographic characteristics. One striking characteristic of the Soviet's post–World War II population is the *imbalance in the male and female population.* For the entire country there are 87.6 males for every 100 females (a population of 95–99 males per 100 females is considered normal). It will take about two generations before a normal ratio is attained. The abnormal ratio between males and females is confined to the older age groups and reflects not only the longer life expectancy of women but also the destructive effects, on men in particular, of war, revolution, and the collectivization process of the 1930s.

The shortage of males has required the use of *female labor* in many occupations, including heavy labor. Furthermore, *Communist theory advocates the use of women* in the labor force as a means to increase production and for greater equality. The availability of labor is also affected by variations in

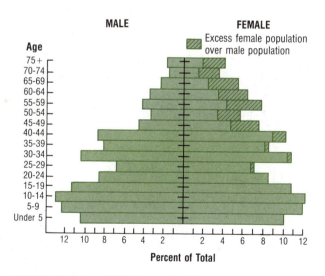

FIGURE 12–4 USSR Population Pyramid.
A population pyramid is a useful device to illustrate certain demographic characteristics. The Soviet Union has a youthful population, but note that the number of children under five years is smaller than the immediately older age groups. The effects of war and revolution are also apparent in older age groups where the female population is significantly larger than the male.

195

the birth rates. During World War II, the birth rate dropped sharply; therefore, severe labor shortages appeared in the late 1950s and early 1960s. To remedy this deficit, the Soviet government adopted a number of policies, such as shortening the period of secondary schooling by a year and requiring students to interrupt their studies to work where manpower deficiencies existed.

During the 1960s, the number of young workers entering the labor force steadily increased; however, the lower birth rates in the 1960s generated further *labor shortages* from the late 1970s. The decline in the birth rate in the 1960s resulted primarily because the fewer women born during the war years were entering the prime childbearing years. Other factors that affect the birth rate include the high proportion of women in the labor force, severe shortages of housing, and the availability of abortions and various forms of contraception.

Although the Soviet Union theoretically accepts the Marxist position regarding population growth, discussed in Chapter 2, it does not have a clearly defined pronatal policy. Bachelors and childless couples pay an additional income tax, and awards and monetary stipends are given to mothers with four or more children. Such payments are small and are not a real incentive for large families. On the other hand, the availability of government-provided contraceptive devices and abortions, together with numerous economic factors, encourages small families. The declining birth rates and prospective labor shortages have stimulated governmental discussion of a more forceful pronatal policy.

Distribution of Population

The Soviet *population is unevenly distributed* (Figure 12–5). In the western Ukraine, rural population densities are higher than 250 per square mile (100 per square kilometer), whereas large tracts of tundra and desert are essentially unpopulated. Almost three-quarters of the population live in the European USSR, which contains about one-quarter of the national area. In the European USSR, areas

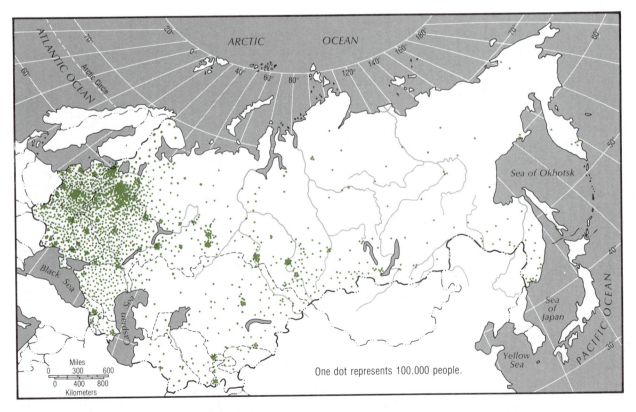

FIGURE 12–5 Distribution of Soviet Population.
Population distribution in the USSR is uneven. Most of the people live in the European west. Much of the north and east is sparsely settled.

196

with densities of at least 25 people per square mile (10 per square kilometer) are mostly south of the 60th parallel. The areas of *highest population densities* are in the southwestern Ukraine, in the lowlands of the Trans-Caucasus, in the vicinity of industrial regions and large cities such as the eastern Ukraine, the Moscow area, and the middle Volga lands.

Outside the European part of the country, the heaviest population concentrations are found in the foothills and valleys of the central Asian mountains, along principal rivers, and in irrigated areas. Scattered islands with population densities exceeding 25 per square mile (10 per square kilometer) are found in the southern portions of Siberia. Here the Trans-Siberian Railway has been an important contributing element in agricultural and industrial development. Throughout most of the taiga, tundra, and desert zones, densities are 2 or less per square mile (1 per square kilometer).

Urbanization

Despite increasing industrialization and commercialization in tsarist Russia, only 18% of the population lived in cities in 1913 (Table 12–2). It was not until the Soviet Union embarked on an all-out program of *industrialization* in the late 1920s that urban growth became significant. By 1940 the urban population had more than doubled to 33% of the total population.

World War II temporarily slowed the pace of urban growth. The return to peacetime conditions stimulated city growth. By 1961 as many people

TABLE 12–2 Growth of Urban Population in the Soviet Union, 1897–1981

Year	Urban Population (millions)	Rural Population (millions)	Percent Urban
1897	18	106	15
1913	29	131	18
1940	63	131	33
1959	100	109	48
1966	108	108	50
1970	136	106	56
1981	169	98	63

SOURCES: USSR Central Statistical Committee, *Narodnoye Khozyaystvo SSSR Za 60 let* (Moscow: Statistika, 1977); and USSR Central Statistical Committee, *Narodnoye Khozyaystvo SSSR v 1980 2* (Moscow: Finansyi Statistika, 1981).

lived in cities as in rural areas in the Soviet Union. Since 1913 the *urban population has increased almost sixfold.* The rural population, on the other hand, was 33 million less.

The *distribution of cities,* like that of the population as a whole, is uneven throughout the Soviet Union. The map (Figure 12–2) of large cities (100,000 or more) shows that the greater number lie in the European Soviet Union. The distributional pattern of these cities forms a wedge with a broad base stretching from the Baltic Sea to the Black Sea. In Siberia large cities are concentrated primarily along or near the Trans-Siberian Railway. Of the eighteen cities with more than 1 million people, all but three (Novosibirsk, Tashkent, and Omsk) are in the European part of the country. The population and urban pattern clearly suggest the dominant role the western sections of the USSR play in the agricultural and industrial life of the country.

197

13

The USSR: Economic Activity

The Soviet Union's most significant achievement during the first seventy years of existence is clearly its *rapid industrial development.* Industrialization has progressed to the point that the USSR has emerged as *a major world power* second only to the United States. The Soviet Union points with great pride to its success in working toward the attainment of its two principal goals: (1) surpassing the United States in the production of goods, thereby becoming the world's greatest economic power, and (2) raising the material level of life in the USSR to the highest levels. Past accomplishments give the Soviet leaders confidence of success in achieving anticipated growth. They believe they will prove their system superior to any of the capitalist varieties. At a time when so many poor nations are searching for ways to raise their own standard of living and attain a higher level of modernization, the Soviet economic system offers *an alternative* to the Western capitalistic system that some poor nations have adopted.

The Soviets contend that *the key* to their economic development lies in the application of the "truths" of social and economic development expounded in *Marxist-Leninist philosophy.* This body of ideas bestows on a handful of people the power to control and plan virtually every segment of the economy through a large bureaucracy.

PLANNED DEVELOPMENT

The state planning agency *(Gosplan),* following Communist Party directives, works out detailed plans. A plan is devised for a five-year period and for each separate year of the period. The complexities of these vast *economic blueprints* invariably lead to numerous problems. Frequently, overoptimistic production goals have had to be reduced. Because the targets have usually been expressed in quantity of production, the Soviet economy has been faced with perennial problems involving the *waste of resources,* the *inefficient use of labor,* and *poor-quality goods.*

Despite these and many other shortcomings, the USSR has achieved notable success in its overall economic growth. This growth, however, has largely been the result of the *development of heavy industry.* Sectors of the economy such as agriculture and light industry (notably consumer goods) have suffered from low rates of investment. By using Soviet official statistics this point can be substantiated. The statistics indicate that if one assumes the values of producer goods, consumer goods, and agricultural production in 1940 equal to 1, then present values would be 29 for producer goods, 11 for consumer goods, and only 2.4 for agricultural goods.

The *Soviet economy,* its planning, and its administration are neither a "chaotic mess" nor a model of perfection. The Soviet leaders, like their counterparts in the satellite states of Eastern Europe, are concerned with increasing economic efficiency and reducing waste. Unlike the governments of Hungary and Czechoslovakia, which have permitted some decentralization, Soviet leaders are not willing to loosen their hold on the economy in order to allow greater decision making on the farm or in the factory and mine. Nevertheless, concerned over *decreasing rates of economic growth* that have characterized the 1960s, 1970s, and 1980s, the Soviet leaders are searching for ways to increase the efficacy of their economic planning and administration. They are attempting to accomplish this by developing better and more *scientific means* of directing the economy, for example, by the use of mathematical methods and computers. The government has actually *increased* the central control over industry in order to better meet plan objectives. It is the Soviet *planner's goal* to reduce the waste of materials and increase the productivity of the economy by expressing production goals not in quantitative terms but in relation to *costs and profitability of production.* The present leadership has also increased investment in the production of consumer items and agricultural production. These sectors of the economy, however, still lag far behind heavy industry.

A significant *change in Soviet economic policy* is an increased willingness to forsake a traditional policy of *autarky* (economic self-sufficiency) in favor of increased cooperation with Western industrial nations. This change involves not only *expanded trade* such as grain purchases but also exchange of technical information and even Western *investments* in developing new industries where the Soviets are reluctant to commit their capital.

The Soviet Union has made rapid industrial growth by emphasizing heavy industries such as iron and steel making and machinery. Here is a synthetic rubber plant in which hydrocarbons from petroleum are the principal raw material. Production of synthetic rubber lessens Soviet dependence on imported natural rubber. (Novosti from Sovfoto)

INDUSTRIAL RESOURCES

A critical factor contributing to Soviet economic development is the diverse and rich *natural resource base.* The Soviet Union ranks among the world's leaders in reserves of oil, natural gas, coal, iron ore, timber, copper, and chromium. The application of the Soviet model of development to another state requires adjustment for any deficiency of that state's industrial resources compared to the Soviet Union's.

Much of the Soviet Union's *resource potential lies in remote sections* of the country, accessible to the market areas only at tremendous cost. This high cost results from expensive long hauls by railway; it may even be necessary to build transport facilities to the site of the deposits. Many deposits are located in areas of harsh environmental conditions. It is difficult to attract workers to these sites. Furthermore, the costs of building the required housing and transportation facilities and the mining operations themselves may push the expenditures beyond practical economical levels.

Energy Sources

The most significant development in the Soviet fuel industry in recent decades has been the *growth of oil and gas production.* Soviet petroleum production has more than quadrupled since 1960, gas production has increased ten times while the output of coal has grown by only 30%, and the production of wood as a fuel has declined by a third. Today the Soviet Union is the world's largest producer of both petroleum and coal, and it vies with the United States as the number one supplier of natural gas.

Approximately one-half of Soviet oil comes from Siberia, and virtually all of this Siberian oil comes from the *West Siberian fields* (Figure 13–1). Most of the growth in Soviet petroleum production since the early 1970s has come from these extensive deposits. Production is hampered by the harshness of the environment, especially *permafrost,* which complicates drilling and the laying of pipelines. West Siberian oil moves by a growing *pipeline network* which also carries the petroleum of the country's second most important producing area, the *Volga-Urals fields,* eastward into eastern Siberia and westward to major industrial and urban centers of the European USSR and Eastern Europe.

Most notable among the remaining oil-producing areas of the USSR are several fields in the vicinity of the *Caspian Sea* and *scattered deposits* in the Ukraine and northern European Russia. Recently discovered deposits along the upper reaches of the Lena River in eastern Siberia will help to provide for future petroleum needs.

Natural gas has become increasingly important in the Soviet Union's energy program. Estimates give the USSR as much as 40% of the world's gas reserves, and the Soviet government has placed a high priority on the development of these reserves. With increased output of natural gas the Soviets will be able to *substitute gas for petroleum* in heating and thus save the scarcer oil for transport and industrial uses. Also, expanded production will mean more *gas for export* by pipeline to Eastern and Western Europe. Sales in the latter area are especially important, for they will provide the Soviets with Western currency for the purchase of much-needed Western technology. Several United States companies have expressed interest in the Siberian fields and have proposed financial and technical assistance in exchange for gas. Japan has also indicated an interest in Siberian gas, but these deposits are especially important to Western Europe, where they are expected to account for as much as 40% of West European natural gas consumption in the near future.

Approximately one-third of Soviet natural gas comes from the fields of *western Siberia.* Other important fields are in the Ukraine, northern Caucasus, Central Asia, and southern Urals.

Not only is the Soviet Union the *leading coal producer* in the world, but its coal reserves (about one-third of the world's proven reserves), rank among the world's largest. It is estimated that, at present rates of consumption, Soviet coal will last for almost eight thousand years. More than 93% of these *reserves* are located in Soviet Asia, the majority in Siberia. Coal production, however, is concentrated in the western USSR. Soviet Europe (including the Urals) accounts for about 60% of the extracted coal. The *Donets Basin,* in the southern steppes, alone produces 35% of total Soviet output. The *Kuznetsk Basin* is the Soviet's second major source of coal. This West Siberian deposit meets local needs, and coal is shipped as far west as the Urals.

The *lower-grade fuels* (peat and oil shale) are locally significant as energy sources for small thermal electric stations. With the increasing availability of natural gas and petroleum, these less efficient fuels are losing their comparative cost advantages.

Most electricity in the Soviet Union is generated

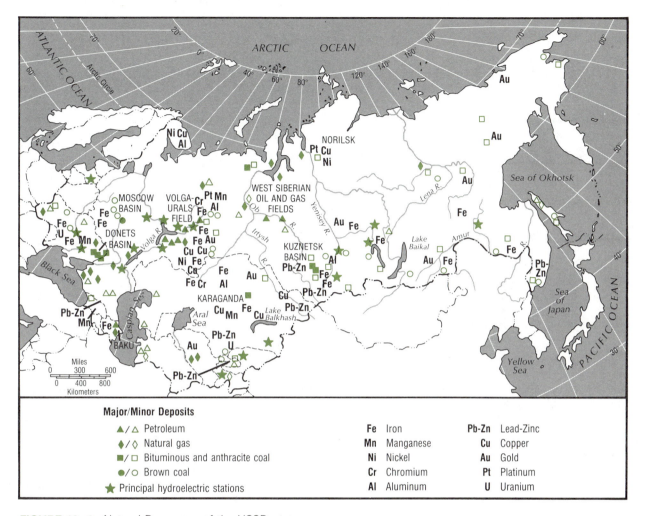

FIGURE 13–1 Natural Resources of the USSR.
The Soviet Union is rich in minerals. High-quality coal in the Donets and Kuznetsk basins and near Karaganda, hydroelectric power along the Volga River, and oil and gas production around the Caspian Sea and in the Volga-Urals field provide abundant energy. Most of the nation's metals are found in the west and along the southern fringe of the country.

by *thermal stations* utilizing coal, peat, oil, and gas. *Atomic power plants* produce only a very small share of the country's electricity; however, Soviet plans are eventually to produce as much as 25% of all electricity by nuclear power. Only about 15% of all electricity generated in the USSR comes from hydroelectric stations, in spite of the fact that the hydroelectric potential, estimated at twice that of the United States, is enormous. A major handicap is that about 70% of this potential lies within Siberia, far from the centers of demand. Consequently, the Soviets have concentrated on thermal and atomic stations within the consuming areas rather than continued development of the remote Siberian hydroelectric resources.

The European Soviet Union's *hydroelectric potential* is more fully developed, particularly the Volga and its tributary, the Kama. The construction of hydroelectric stations and their necessary reservoirs has converted these rivers into a string of large lakes. The hydroelectric stations supply electricity to the Volga cities, the Urals, and the Moscow area. The Caucasus and Central Asian mountains have also seen expansion of hydroelectric capacity since 1960.

Metallic Ores

Complementing the energy resources is an ample and diverse base of *metallic resources*. However, tungsten, tin, and aluminum resources, especially bauxite, are in short supply. Soviet *iron ore reserves* (according to Soviet estimates) are the largest in the world, representing 40% of all known reserves. The most important of the producing iron ore deposits are in the west, whereas the majority of the known reserves are in Siberia and Kazakhstan. Approximately 50% of the iron ore extracted in the Soviet Union comes from the *Krivoy Rog deposits* in the Ukraine. These ores supply not only the country's largest concentration of iron and steel works in the Ukraine but also other factories in western USSR and in Eastern Europe. The second major area of iron ore mining is in the *Urals.*

The Soviet Union possesses a wide array of other mineral ores. The *ferroalloy metals* such as manganese, nickel, and chromium are found in ample quantities. Most important of the deposits are those accessible to the principal steel-producing centers. The Soviets are second to the United States in *copper* production and claim that their reserves are the largest in the world. The Soviet Union is also a major producer of lead, zinc, gold, silver, and platinum.

INDUSTRIAL REGIONS

In a *command-type economy* such as the Soviet Union's, planning for regional distribution and development of industries is a major task. The difficulty is compounded by the nation's large area, unequal distribution of natural resources, and the diversity of ethnic groups at different levels of economic development.

From the start of the industrialization program, the Soviets have attempted to *disperse their industrial production* beyond the limits of the European part of the country for ideological as well as strategic reasons. The goal of rapid industrialization was to build "*socialism in one country,*" to use Stalin's phrase. The Soviet Union would then be able to withstand any attack from the Western capitalist powers, which Stalin believed to be imminent. The development of industrial production in the militarily more defensible areas, the eastern portions of the country, seemed advisable. With much fanfare, the Soviets embarked on *expansion of the Urals industrial base* and the development of the *Kuznetsk metallurgical base.*

World War II was the greatest stimulus to increase industrial production in the east. The Ger-

mans occupied an area of the European Soviet Union that included 40% of the population and the bulk of heavy industry (62% of coal production and 58% of steel production). From 1940 to 1943, for example, industrial production increased 3.4 times in western Siberia. During the post–World War II period, the west was reconstructed and regained its dominant position in industrial production, though at a level slightly below previous times.

Although Soviet economic planning policies emphasize the *balanced distribution of economic activities,* including the development of the less economically advanced areas, they also *stress production close to raw materials and markets* to minimize transportation costs, the development of *specialized forms of production* in areas best suited for them, and creation within each region of *adequate production* to meet the basic needs of the population.

Governmental *regional development policy* seems to favor increased industrial investment in the west, particularly in medium-sized cities in areas with large labor supplies. Siberian development is slated to continue at a rate slightly higher than the national average. The high cost of attracting and sustaining labor in the east, however, means that development there will be more technological than labor intensive.

There are *six areas of prime industrial significance:* the Center (Moscow area), the Ukraine, Leningrad, Mid-Volga, Urals, and Siberia (Figure 13–2).

The Center

The Center, with the Soviet Union's most populous and largest industrial city, *Moscow,* owes its industrial prominence to a number of factors including a *large market,* an ample supply of *labor* both trained and unskilled, and excellent transport *accessibility* to all parts of the country. Electric lines from Volga power stations and gas and oil pipelines from the Ukraine, north Caucasus, the Volga-Urals field, Central Asia, and western Siberia supply the Center with important sources of *energy.* All of these factors assure this area's continued prominence in industrial production, although its share of total industrial output may slowly decline.

The *industrial resource base of the Center is weak.* Energy resources of brown coal and timber are inferior. There are a small iron ore deposit, phosphorus for fertilizer, and some building mate-

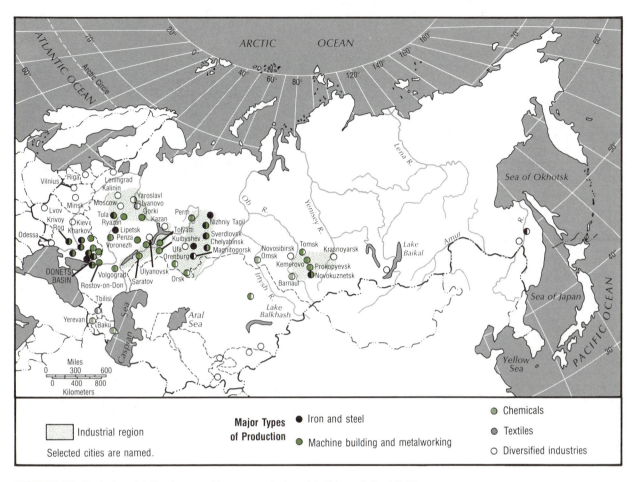

FIGURE 13–2 Industrial Regions and Important Industrial Cities of the USSR. Industry in the Soviet Union is concentrated in five regions. Except for the Moscow area, these regions are also areas with abundant local energy supplies. Industry is oriented toward capital goods, although in recent years consumer-goods production has been increased.

rials. Textile manufacturing is prominent. Around Moscow, 30% of all industrial workers are employed in the manufacture of linen, cotton, wool, and silk fabrics. Other major industries include metals, machine construction, engineering industries, chemicals, food processing, and woodworking.

Ukraine Industrial District

The Ukraine industrial district is the *principal heavy manufacturing area*. The availability of coal, iron ore, and ferroalloy metals facilitates major *iron and steel* production including various forms of steel, *heavy machine construction,* and the *coal-based chemical industry*. The largest concentration of cities within this industrial area is on the coalfields of the Donets Basin.

The industrial resource base of the Ukraine extends beyond coal and iron ore. Energy resources are supplemented by gas fields to the north and by the gas and oil fields of the northern Caucasus. The *base of raw materials* also includes such minerals as salts, potash, mercury, and brown coal. Furthermore, the *high productivity of agriculture* within the district and throughout the southern steppes of western European Russia has stimulated the development of extensive food-processing industries and the production of agricultural equipment.

The two largest cities of the Ukraine, Kiev and Kharkov, are located on the northern margins of the Ukrainian manufacturing area. In addition to its political function as capital of the Ukraine, Kiev is a diversified industrial city. Machinery, textiles, and food processing are major industries in both

203

these cities. To the southwest is the diversified industrial port and shipbuilding city of Odessa.

Leningrad

Leningrad, the second largest city in the country, occupies a position similar to that of Moscow. It lies in a region with a *deficient resource base.* Aside from hydroelectric stations, local energy resources are limited to peat deposits and oil shales. Leningrad industries include machine tools, equipment for hydroelectric plants, and shipbuilding.

Mid-Volga Area

The industrial strength of the *Mid-Volga industrial area* rests primarily on the *extensive energy resources.* Not only is one of the major petroleum-producing fields of the Soviet Union (Volga-Urals field) found here, but there are important gas fields and surplus hydroelectric power generated by some of the country's large dams on the Volga River. During World War II this area experienced rapid industrial and population growth. It was located east of the front for security but was readily accessible to areas of need in the west. Since the late 1950s industrial production has increased in

the Volga area at a rate greater than that of the Soviet Union as a whole.

The Mid-Volga area has several advantages in addition to its fuels and hydroelectric power. It is *connected by the Volga River* and its tributaries to large areas of the western Soviet Union. The Volga River system is the Soviet Union's major water route, carrying over 60% of all the freight transported by river.

Accessibility by water, rail, and pipeline not only has stimulated the development of the Mid-Volga's energy resources but also has allowed the expansion of industrial activities as a whole. One major example is the large automotive plant built as a joint venture with Fiat, the Italian firm, at Tolyatti. The plant is to produce 600,000 cars a year, 48% of the planned yearly Soviet automobile output.

The Urals

In terms of overall industrial production, the *Urals* ranks third behind the Center and Ukrainian areas. Urals industry depends chiefly on the *rich and varied local mineral deposits.* In addition to its important iron and steel industry, the area is known for smelting and refining of copper, zinc refining, and

The Volzhsky motor works at Tolyatti built jointly with the Italian Fiat company. The plant was completed in 1972 and incorporates assembly-line techniques. (TASS from USSR Consulate General)

the production of alumina and refining of aluminum.

Sverdlovsk, a *major railroad center,* is the largest of the Urals cities. Here machine construction, especially for the mining industry, dominates; ferrous metallurgy is also a prominent industry.

Siberia

The *Siberian manufacturing region* lies between the Ob and Yenisey rivers in western Siberia. The rich *coal deposits* of the Kuznetsk Basin and the *hydroelectric power* of the region are major reasons for the development of industries in this area. The region includes the complex of *metallurgical industries* in the Kuznetsk Basin and several cities outside the basin, including Novosibirsk, the largest city in all of Siberia. Novosibirsk is a *major transport center* located on the Trans-Siberian railway and the Ob River. It is also a major industrial city producing metallurgical products, machines, foods, and textiles.

AGRICULTURE

Soviet agriculture has not experienced the same successes as industry. From the late 1920s to the early 1950s, agricultural output barely kept pace with population growth. Since the early 1950s, however, production of agricultural goods (both crops and livestock) has grown almost three times faster than population. Despite this recent growth in agricultural output, Soviet *farm production still lags behind that of the United States.* For every American farm worker, the Soviet Union has eight. About 23% of the Soviet labor force is engaged in agriculture compared to 2% in the United States. The area sown to crops is 73% greater in the USSR than in the United States, yet Soviet crop production is equal to only 80% of that of the United States. Overall, the average American farmer is ten times more productive than his Soviet counterpart. Whereas the Soviet agricultural worker grows enough food to feed eight of his fellow citizens, the United States farmer grows enough food to feed fifty-two Americans.

The reasons why Soviet agriculture has developed less successfully than industry are found in part in the *extreme environmental conditions* in the country, and in part in the *organization of agriculture* that has emphasized the development of industry.

Position of Agriculture and Soviet Development

The decision to embark on an all-out program of rapid industrialization in the late 1920s required the *collectivization of agriculture* for several reasons. The major factors in the decision to collectivize were political as well as economic:

1 The peasant class represented a capitalist or latent capitalist element that was ideologically unacceptable to the regime.
2 It would be more efficient to control the peasantry grouped in large farms rather than scattered in smaller units.
3 By forcing the peasants into large collectives, agricultural prices and wages could be controlled at low levels to allow accumulation of capital for industrial expansion.
4 This control would facilitate the flow of foodstuffs to the cities to feed the growing industrial labor force.
5 Large-scale units could be mechanized to increase productivity of agriculture and also to free labor for the growing industrial activities.

War raged in the countryside. The better-off peasants and opponents of the collectivization process were exiled, imprisoned, or murdered. Livestock herds were decimated as many peasants slaughtered their animals rather than surrender them to the new collectives. By 1940, however, virtually all peasant households were part of the collective agricultural economy.

Two forms of farm organization emerged: the collective farm, or *kolkhoz,* and the state farm, or *sovkhoz.* The collective farm is a group of workers responsible for seeing that state production quotas are met. After the quotas are fulfilled and the needs of the collective farm are satisfied (capital for farm repairs, taxes, and seed for the next season), the remaining crop is divided among the workers as their share of the profit. The peasants are, therefore, residual claimants to the farms' production. The state farm *(sovkhoz)* workers, on the other hand, are paid a set wage, and the total costs of the operation are underwritten by the state. Needless to say, the cost of operation for the state was greater for the *sovkhoz* than for the *kolkhoz.* The *kolkhoz* system was greatly favored by the Stalin regime; however, *kolkhoz* efficiency suffered because of the government's fixed lower prices for agricultural products. The capital thus

205

provided for the operation of the *kolkhoz* was insufficient and failed to provide work incentives. The state's investment in fertilizers, machinery, and other necessary technological improvements was woefully inadequate. All these factors contributed to stagnant agricultural production through the early 1950s.

The critical difference between survival and starvation was the *private sector of Soviet agriculture;* collective and state farm workers and some industrial workers were permitted to raise products on plots of 0.5 to 1.25 acres (0.2 to 0.5 hectares). In 1953 private plots made up only about 4% of the cultivated land in the country but produced 72% of the potatoes, 48% of the vegetables, 52% of the meat, 67% of the milk, and 84% of the eggs. The majority of the collective farmers' incomes was generated by sales of products from these private plots in the free market. Today the private plots produce one-quarter of all foodstuffs and 30% of all meat and poultry on less than 3% of the country's farmland.

The "Permanent Crisis"

The present leadership of the Soviet Union, like Leonid Brezhnev and Nikita Khrushchev before them, inherited the agricultural problem which has been called by some Western critics the Soviet Union's *"permanent crisis."*

The *dire situation of agriculture* at the death of Stalin forced his successors to turn their attention to the needs of the agricultural economy. A program of expansion of cultivated lands in the dry steppes of the Trans-Volga region resulted initially in a significant increase in agricultural production. This *virgin-and-idle-lands program* brought 116 million acres (47 million hectares) of new land under cultivation mostly in western Siberia and northern Kazakhstan. In these new lands precipitation ranges from 16 inches (41 centimeters) in the north to 9 inches (23 centimeters) in the south, so they are *a marginal farming area.* Anticipated production levels have not been attained, and actual production varies from year to year. Nevertheless, the increased wheat area has freed land in the moister areas of European Russia for other crops.

During the Brezhnev regime the Soviets adopted a more *tolerant attitude toward the private plot,* further increased prices on purchases from the collectives, and introduced a guaranteed minimum wage for the collective farm worker. Prices on machinery and fertilizers were reduced, and

the state recognized the right of collective farms to participate more in the planning procedure. The last half of the 1960s was marked by notable success in Soviet agricultural production. The output of agricultural products increased by 23% from 1965 to 1971.

The need of the Soviet government to import large quantities of grain underscores the fact that Soviet agriculture is still beset by *numerous problems.* Despite increased wages, the average collective farmer makes less per year than the average industrial worker or even the average state farm employee. Soviet agriculture still suffers from inadequate mechanical equipment and deficient storage and transportation facilities. The more productive young workers leave the farms because of the low salaries, restricted opportunities for advancement, and a scarcity of amenities that are available in an urban life. Furthermore, central decision making in the Soviet system contributes to interference by party bureaucrats and inefficient use of agricultural resources.

In 1950, Soviet agricultural output was about 60% of that of the United States. Now Soviet farm production equals 80% of American production but must feed a population 38 million larger. Since the early 1970s, Soviet agriculture plans have called for production levels equaling those of the United States. Prospects for Soviet agriculture to meet this goal even in the foreseeable future remain remote.

Agricultural Regions

Soviet agriculture suffers from the *institutional restraints* of its organization and from physical *environmental handicaps.* It is, of course, possible that improved strains of plants, land amelioration programs, or even domes with controlled environments will someday reduce nature's unfavorable aspects. It is also true that even under similar environmental conditions in the United States and Canada, agricultural productivity surpasses that of the Soviet Union. Although the land area of the Soviet Union is two and one-half times that of the United States, its area suitable for crop cultivation is only one-third greater. Only about 11% of the Soviet territory is *arable.*

Figure 13–3 shows generalized *zones of agricultural use* of the Soviet Union. The areas with no agriculture or widely scattered small farms clearly occupy the majority of the Soviet lands. Areas with too short a growing season and poor soils cover

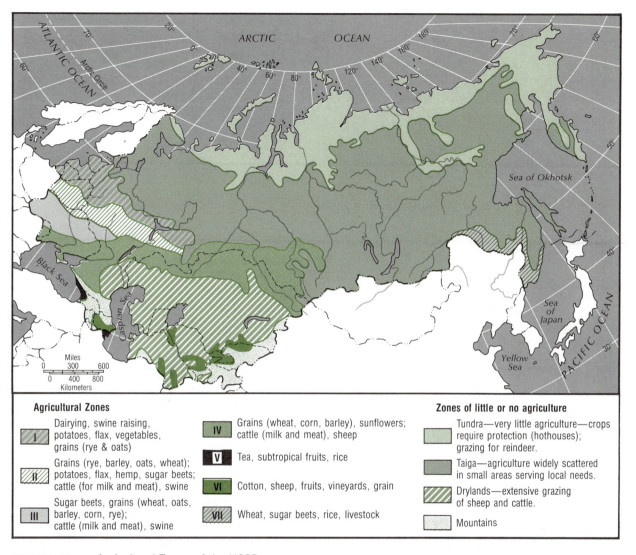

Agricultural Zones

	Dairying, swine raising, potatoes, flax, vegetables, grains (rye & oats)
I	

	Grains (rye, barley, oats, wheat); potatoes, flax, hemp, sugar beets; cattle (for milk and meat), swine
II	

	Sugar beets, grains (wheat, oats, barley, corn, rye); cattle (milk and meat), swine
III	

IV	Grains (wheat, corn, barley), sunflowers; cattle (milk and meat), sheep
V	Tea, subtropical fruits, rice
VI	Cotton, sheep, fruits, vineyards, grain
VII	Wheat, sugar beets, rice, livestock

Zones of little or no agriculture

	Tundra—very little agriculture—crops require protection (hothouses); grazing for reindeer.
	Taiga—agriculture widely scattered in small areas serving local needs.
	Drylands—extensive grazing of sheep and cattle.
	Mountains

FIGURE 13–3 Agricultural Zones of the USSR.
Agriculture in the Soviet Union is confined largely to the southwest quarter of the nation. Since World War II attempts have been made to expand the area of crop production into the dry lands east of the Caspian Sea and into western Siberia.

the northern European Soviet Union and virtually all of Siberia except the south. Also, the bulk of Central Asia is desert. To these generally nonagricultural zones must be added the scattered mountains of the west, central Asia, and Siberia. The remaining area comprises the agricultural zones of the USSR, but even among these lands are areas too cool, too moist, or too dry, which would be judged marginal in the United States.

The agricultural zones indicated in Figure 13–3 are found mainly south of the 60° N parallel in the European part of the country and south of the 57° N parallel in Siberia. In the south the agricultural area is limited by high *evapotranspiration rates.*

Zone I, bordering the southern limit of the taiga forest from the Baltic to the Urals, is an area of *scattered agricultural development.* Within this zone *dairying and swine production* are a major activity. *Flax and hardier grains* like rye and oats do well in the short moist summers. *Potatoes,* long a staple of the European Soviet diet, are a principal crop; a variety of other vegetables are also raised in this zone.

In *zone II* the frequency of *cropped land increases,* and *grains* (rye, oats, barley, and wheat)

207

Wheat planting in the dry steppes of central Asia. About 122 million acres (47 million hectares) of this type of land have been put to cultivation in the virgin land program fostered by former Soviet Premier Khrushchev. (Novosti from Sovfoto)

with *potatoes, flax, and hemp* are characteristic. Dual-purpose (milk and meat) *cattle* are extensively raised.

Sugar beets are the most important industrial crop in *zone III.* Higher temperatures than in the north balanced by adequate precipitation favor sugar-beet cultivation, together with *grains* (wheat, corn, barley, rye, and oats) and *potatoes.*

The subhumid steppe, *zone IV,* has fertile *chernozemic soils* and is the most important *wheat-growing region* in the Soviet Union. Winter wheat is grown largely in the European section, while spring wheat dominates the east. *Corn* has become an increasingly important crop in the west, and plantings of this crop have expanded into the drier steppe region of the east. *Sunflowers* for oil are another major crop, particularly in the western sections. A variety of grains and other crops (sugar beets, potatoes, flax, and barley), along with *milk and meat livestock* and sheep raising, round out the agriculture of this region.

Three additional regions are noted for their specialized crop production. One is the Trans-Caucasus *(zone V),* where *citrus crops* and *tea* thrive in the well-watered, protected western portion. In the drier eastern Trans-Caucasus area, tea is grown along with *cotton* and *rice* in irrigated regions. *Zone VI* is the principal *cotton-growing* area of the Soviet Union in the irrigated valleys of Central Asia and the eastern Caucasus. *Rice,* the staple

food of Central Asia, is also widely grown in the irrigated areas along the rivers and in the far eastern *zone VII.*

THE USSR IN THE COMING YEARS

At present, the USSR's position as *a major world power* is unshakable. The development of Soviet economic power has indeed been impressive. Many claim this growth has occurred not because of but in spite of the political system of the country. The almost single-minded drive toward *industrial and military development,* the abundance of *raw materials,* and the use of *Western technology* have enabled the country to achieve high levels of economic growth. Progress has been costly. *Natural resources have been wasted* and *labor inefficiently* used. Most important, personal liberty and material well-being have been *sacrificed* for the state's developmental plans.

The Soviet Union is faced with a number of *problems,* the first of which is a *slowing of economic growth.* Efforts to remedy this difficulty have focused largely on reducing waste and developing a more efficiently run economy, not on fundamental organizational changes.

The shortages and consequent high prices of raw materials, particularly energy resources, that have threatened the Western industrial states have had less effect on the Soviet Union. The rich natural resource base of the Soviet Union gives it an enviable position. Not only is the USSR's own economic growth provided for, but it can become a more important supplier of raw materials to deficient industrial states. This fact cannot help but enhance the USSR's political and economic position in the world. To maximize this potential, however, the Soviet Union needs Western technological assistance to exploit its far-flung resources. Thus there occurred in the early 1970s what some observers viewed as a major turning point in Soviet economic strategy: its *expanded commercial relations* with the United States and other states of the industrialized West. The full impact of these new trade agreements may be felt in the near future. There still remain, however, strong suspicions that the Soviet objective is to expand its economic power for political and military purposes.

Agriculture remains the major bottleneck in the Soviet economy. Despite significant improvements in production during the last decade, Soviet agriculture fails to provide the quantity and quality of foodstuffs promised by the government. Frequent crop failures require imports of food from abroad.

The ability of the USSR to increase its output of foodstuffs is important not only for its own well-being but also for a food-deficient world as a whole.

FURTHER READINGS

Eastern Europe

Two excellent and thorough, although somewhat dated, studies of Eastern Europe are N. J. G. Pounds, *Eastern Europe* (Chicago: Aldine, 1969), and Roy E. H. Mellor, *Eastern Europe: A Geography of the Comecon Countries* (New York: Columbia University Press, 1975).

DORNBERG, JOHN, *East Europe: A Communist Kaleidoscope* (New York: Dial Press, 1980). A survey of historical, political, and social developments in Eastern Europe (excluding Albania) directed to the general reader.

FISCHER-GALATI, STEPHEN (ed.), *Eastern Europe in the 1980's* (Boulder, Colo.: Westview Press, 1981). A collection of articles by East European experts surveying recent industrial, agricultural, political, cultural, and educational developments in the East European countries.

KOSTANICK, HUEY L. (ed.), *Population and Migration Trends in Eastern Europe* (Boulder, Colo.: Westview Press, 1978). An interdisciplinary study on population in Eastern Europe.

SHOUP, PAUL S., *The East European and Soviet Data Handbook: Political, Social, and Developmental Indicators, 1945–1975* (New York: Columbia University Press, 1981). A collection of demographic, ethnic, economic, and social data that provides a convenient source of information for the person interested in postwar changes in East European and Soviet society and life.

TURNOCK, DAVID, *Eastern Europe* (Boulder, Colo.: Westview Press, 1978). The best study available for an overview of East European industry, industrial resources, regions, development, and east-west economic relations.

The USSR

The best and most comprehensive texts on the Soviet Union are two books by Paul E. Lydolph. His *Geography of the U.S.S.R.* (Elkhart Lake, Wis.: Misty Valley Publishing, 1979) is an excellent topical treatment of Soviet geography with an emphasis on the economic aspects of the USSR. The second volume, also titled *Geography of the U.S.S.R.,* 3rd ed. (New York: Wiley, 1977), organizes Soviet geography (also with an economic stress) according to the nineteen major economic regions of the country.

BERG, L. S., *Natural Regions of the USSR* (New York: Macmillan, 1950). The classic study of Soviet landscape zones by an eminent Soviet geographer.

DAVIS, R. W. (ed.), *The Soviet Union* (London: Allen and Unwin, 1978). An interdisciplinary examination of contemporary Soviet literature, foreign policy, economics, politics, and other topics.

GOLDMAN, MARSHALL I., *The Enigma of Soviet Petroleum: Half Empty or Half Full?* (London: Allen and Unwin, 1980). Examines the all-important role that petroleum plays in Soviet economic life and discusses production problems and the controversial issue of the potential of Soviet oil output and trade.

———, *U.S.S.R. in Crisis: The Failure of an Economic System* (New York: W. W. Norton, 1983). Reviews the USSR's economic strategy for development and the reasons why after early success the system has faltered.

MATHIESON, R. S., *The Soviet Union: An Economic Geography* (New York: Barnes and Noble/Harper and Row, 1975). An explanation of the economic geography of the USSR in the post–World War II period to 1973.

NOVE, ALEC, *The Soviet Economic System* (London: Allen and Unwin, 1977). An excellent and exhaustive study on Soviet economic institutions, problems, and reforms.

PARKER, W. H., *A Historical Geography of Russia* (Chicago: Aldine, 1968). The historical geographical development of the Russian Empire and the Soviet state.

SCHERER, JOHN C. (ed.), *U.S.S.R. Facts and Figures Annual,* Vol. 5 (Gulf Breeze, Fla.: Academic International Press, 1981). A most helpful resource for the person looking for a wide variety of data and information on the Soviet Union. Covers such topics as government and party personnel, political developments, demographic, economic, social, and cultural affairs.

U.S. CONGRESS, JOINT ECONOMIC COMMITTEE, *Soviet Economy in a Time of Change* (Washington, D.C.: Government Printing Office, 1979). A two-volume, comprehensive collection of scholarly articles on the past, present, and future performance of the

Soviet economy. More than seventy experts on the Soviet Union contributed their efforts to produce this valuable resource for anyone wanting to understand Soviet economic life.

WESSON, ROBERT (ed.), *The Soviet Union: Looking to the 1980's* (Stanford, Calif.: Hoover Institution Press, 1980). Focuses on challenges and possible changes in economic, political, foreign, and nationality policies of the Soviet Union.

WHITING, ALLEN S., *Siberian Development and East Asia* (Stanford, Calif.: Stanford University Press, 1981). The economic development of East Siberia and the strategic implications of these changes vis-à-vis China, Japan, and the United States.

V

Japan and Australia/ New Zealand

Jack F. Williams

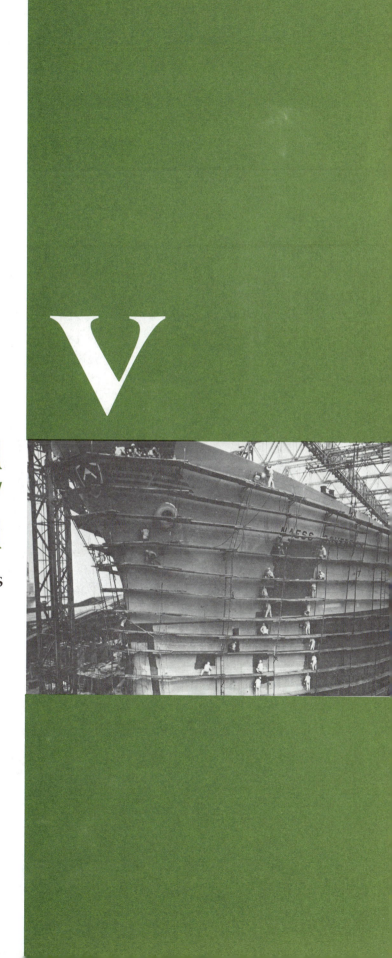

14

Japan: Industrial Wonder of Asia

To date, Japan is the only non-Western country to have entered the relatively exclusive club of rich nations. By whatever standards one wishes to use, Japan is a rich nation. The mere fact that no other non-Western nation has yet reached so high a stage of affluence—the fifth or "high mass consumption" stage in Rostow's theory—makes the Japanese experience unique in the history of the modern world. It is all the more exceptional when viewed in relation to the resource foundation of Japan.

RESOURCES

The Physical Base

Japan is often used as an example to disprove the old *theory of environmental determinism* because the Japanese have succeeded seemingly in spite of the natural environment and poor resource endowment. Yet a closer analysis reveals that the situation is more complex.

Location and Insularity

Japan's unique role in East Asian civilization can be attributed in part to the nation's *relative isolation* off the east coast of Asia. The country consists of four main islands (Hokkaido, Honshu, Shikoku, and Kyushu), plus many lesser islands, that stretch in an arc about 1400 miles (2253 kilometers) long from around 45° N to 31° N (not counting the Ryukyu Islands chain, which continues in another arc southward to near 24° N); see Figure 14–1. The Japanese islands are separated from the Korean peninsula by the Tsushima Strait, which is only 115 miles (185 kilometers) wide. Moreover, in premodern times, there was still the mountainous Korean peninsula to traverse before arriving at the great culture center of China.

The *advantages of Japan's location* were several. On the one hand, Japan had natural protection and was never successfully invaded in its his-

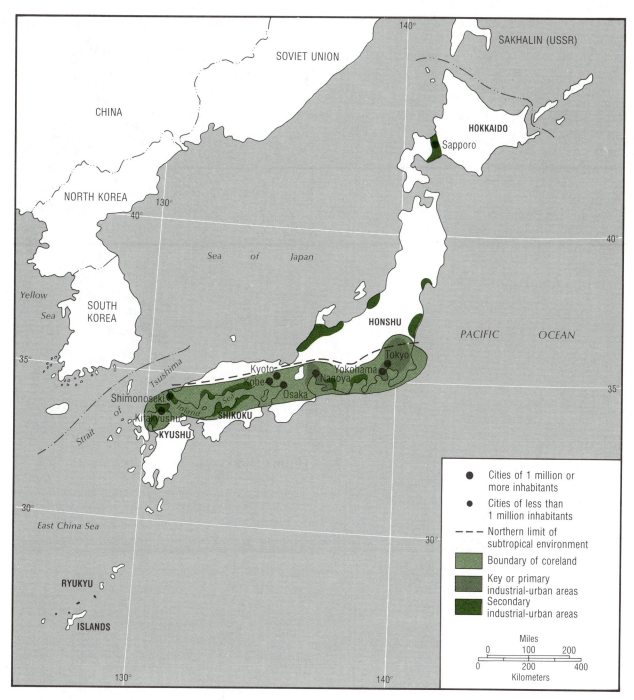

FIGURE 14–1 Japan: General Locations and Coreland.
Japan is an island nation and has often been compared to the United Kingdom. Historically, Japan's insular location provided a natural barrier to outside forces; in the modern world, it contributes to accessibility since ocean transport is the cheapest way to move goods.

tory. On the other hand, the relative proximity to China enabled Japan to adopt many aspects of Chinese culture on a largely voluntary and selective basis. The Japanese were able to adapt these innovations to their situation and create a truly unique and brilliant culture of their own.

Japan's proximity to China in modern times has contributed to the increasingly important economic relationship between the two nations. That relationship is based on a classic situation of *economic complementarity*. Japan buys raw materials, especially oil and coal, from China and is helping China develop its mining sector. China, in turn, buys manufactured goods and whole industrial plants from Japan.

A Temperate Land

Japan's long latitudinal sweep within the temperate zone, combined with insularity, has benefited the country in terms of its climate, which is roughly comparable to that of the east coast of Anglo-America from New England to northern Florida (Figure 14–2). The *maritime location* means that Japan has no real dry season. There is sufficient rainfall throughout the year for crop growth.

A *significant dividing line* is roughly along the latitude 37° N. South of this line double-cropping is possible, but north of it the winters are usually too long to permit double-cropping. This basic fact has been reflected in Japan's historical development. Settlement north of the line, in northern Honshu and Hokkaido, came much later than in central and southern Japan. Even today most of Japan's population is found south of 37° N. The northern lands remain a much less developed frontier region.

A Crowded Land

Japan is *a small country* when compared with the other great powers of the world. With a mere 143,000 square miles (370,370 square kilometers), it is slightly smaller than California. Japan's land problem stems from too many people on too little land; the population/land ratio is high.

Compounding the problem is the *rugged terrain*. The islands are actually the summits of immense submarine ridges thrust up from the floor of the Pacific Ocean. The island chain is but one part of the unstable orogenic (mountain-making) zone that encircles the Pacific Ocean. There are hundreds of volcanoes scattered along the archipelago, and earthquakes are a common occurrence.

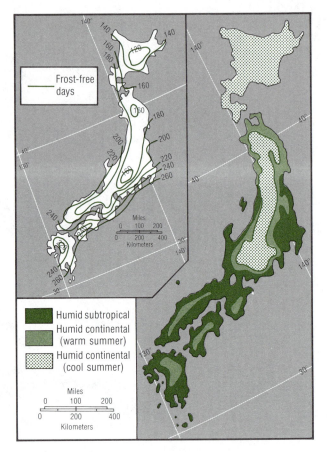

FIGURE 14–2 Climatic Regions and Frost-Free Days in Japan.
Japan's climate is similar to that of the eastern United States except that Japan experiences more moderate weather. The surrounding seas cause lower temperatures in the summer and warmer temperatures in the winter. Rainfall is more abundant. Much of Japan has a long growing season.

In this geologic setting, low, *level lands are in short supply.* What level land does exist is found in narrow river valleys and alluvial coastal plains separated from one another by stretches of rugged hills. Thus Japan's more than 119 million people are actually concentrated in a land area slightly smaller than the state of Indiana. The population per square mile of arable land is one of the highest in the world (see Table 24–1). Japan is thus one of the most crowded of human landscapes (Figure 14–3).

The Japanese have been able to support such a *dense population* because quite early in their history they adopted an *intensive form of irrigated agriculture* from China. This agriculture, almost like gardening, produces relatively high yields per land

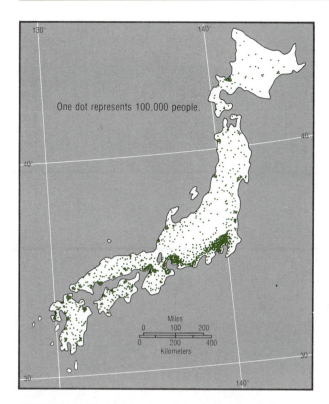

FIGURE 14–3 Distribution of Japanese Population. Japan has one of the highest population densities in the world. Densities are greatest on the plains of eastern Honshu and around the Inland Sea. To help feed its large population Japan turned to the sea and to international trade.

One dot represents 100,000 people.

unit, and can be found in many parts of Asia. The chief distinction of Japan's variant of this agricultural system is that the Japanese, in modern times, have developed productivity to a degree that has not been matched by any other Asian country.

A Maritime Nation

In premodern times, land communication was difficult in Japan. Thus the surrounding seas provided links among the islands and along the coast, as well as contact with the outside world. Most of Japan's population is still located close to the sea. Not surprisingly, the sea has always played an important role in Japan's national life.

The rugged coastline, with countless bays and inlets, the most important of which is the great Inland Sea between Honshu and Shikoku, made *fishing a major activity* of the Japanese early in their history. The Japanese continue to obtain a major share of protein for their diet from fish and other sea products. In modern times, the fishing industry has become a global enterprise, with vessels roaming the four corners of the world in search of seafood for Japan's large population. The Japanese moved into *shipbuilding* early in their national life, but especially with the industrialization effort at the close of the nineteenth century. Japan became *one of the major maritime powers* of the world by the 1920s, a position it has held to this day. In

Japan is a maritime nation. Fishing is a traditional occupation, and fish and other sea products are important elements in the Japanese diet. Today their fishing industry is among the largest, with Japanese vessels working the sea's resources throughout the world. (Consulate General of Japan, New York)

In the post-World War II period, Japan has become the world's principal shipbuilder. Nearly all the materials needed must be imported, yet well-organized operations with highly skilled labor have overcome the shortage of local raw materials in this labor-intensive activity. (Consulate General of Japan, New York)

fact, in the post–World War II period, Japan surged ahead as the leading shipbuilder in the world, far outdistancing any of its rivals.

Mineral Resources

Japan is practically *devoid of significant mineral deposits.* As a result, the Japanese have had to rely primarily on imported raw materials to supply their industrial development. Among the important industrial raw materials the country is self-sufficient only in limestone and sulfur. The nation's limited supply of coal is inadequate to meet domestic industrial and residential needs.

The demand for mineral raw materials has mushroomed since World War II. These now account for nearly 70% of Japan's total imports, and a major share of Japanese investment overseas is for securing raw materials. Hence, *foreign trade* assumes a special importance for Japan.

The magnitude of this *dependency on overseas sources* is best illustrated by the case of petroleum. Oil is Japan's largest single import commodity, accounting for one-third of total imports. By the early 1980s, Japan was annually importing more than 250 million tons of petroleum at an annual cost of $60 billion in 1981. Japan is the second largest importer of petroleum after the United States. Oil accounts for two-thirds of Japan's total energy consumption, and nearly 70% of the oil comes from the Middle East.

To remedy this *dangerous dependency situation,* Japan has embarked on a strict energy conservation and diversification program, promoting decreased use of oil and increased use of other energy forms (Figure 14–4). It is hoped that by 1990 oil will account for only half of the country's energy needs, with sharp increases for coal, natural gas, and nuclear power.

In balance, the Japanese have responded to their physical environment by channeling their national development to take maximum advantage of their opportunities and by overcoming environmental deficiencies. Yet much of this progress might not have been possible without the remarkable characteristics of the Japanese as a people and culture.

The Human Resource Base

Japan's development experience is a powerful demonstration of the fact that, in the final analysis, any nation's development is largely determined not by its physical resource base but by the quality of its *human resources.* Compared to most other nations, the Japanese have a *high degree of racial and cultural homogeneity.* This is one of the country's great strengths because it has helped

216

Japan is dependent on imports of raw materials and food products and exports of manufactured goods. Major port facilities such as those of Yokohama have been built to further trade. The accumulation of materials at port sites has stimulated manufacturing in the local area. (Consulate General of Japan, New York)

foster a sense of national identity and cohesiveness that has fueled Japan's modernization in the last hundred years.

The Emergence of the "Japanese"

The Japanese achieved their *distinctive physical and linguistic identity* 2000 years ago. The Japanese language is similar to Korean and the Altaic languages of northern Asia. Although it continues to use a large number of Chinese characters in the written form, the Japanese language is actually very different from Chinese and was one of the factors that helped Japan to preserve its cultural distinctiveness.

Although the Japanese are a homogeneous group today, many racial strains have been blended into the people over a long period. *Neo-*

lithic peoples inhabited the islands for several millennia. These were hunters, fishermen, and gatherers, and had only partial Mongoloid racial features. The *Ainu*, another important element in the creation of the modern Japanese, were a proto-Caucasian people who inhabited much of northern Japan and are still found in small numbers on the island of Hokkaido. Other traits may have been acquired from *peoples of Southeast Asia* (Figure 14–5).

The real beginnings of the Japanese as a distinctive people and culture can be traced, however, to southwest Japan in the centuries just before the Christian Era. At that time a culture known as Yayoi developed in north Kyushu, spreading eastward into the Inland Sea area by around 200 B.C. *Yayoi culture* evolved from Mongoloid peoples who migrated from the mainland under the push of the

217

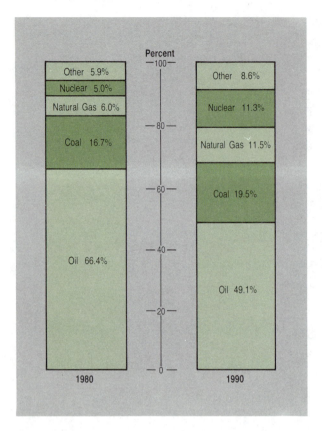

Percent

Other 5.9%
Nuclear 5.0%
Natural Gas 6.0%
Coal 16.7%
Oil 66.4%

1980

Other 8.6%
Nuclear 11.3%
Natural Gas 11.5%
Coal 19.5%
Oil 49.1%

1990

FIGURE 14–4 Japan's Changing Energy Sources. Japan has few local energy resources. All of the petroleum and much of the coal used in Japan must be imported. Industry is the principal user of energy in Japan. By 1990 it is expected that Japan's dependence on oil will be somewhat reduced.

expanding *Han Chinese.* These peoples brought with them paddy-rice agriculture and the use of bronze and iron.

Southwest Japan was a logical place for *Japan's culture hearth.* Aside from the fact that it was close to Korea and China, the climate there was much milder than places farther north in the islands. Moreover, the sheltered coastal fringes of the Inland Sea area provided reasonably good habitats for these peoples to establish themselves (Figure 14–5).

By the third century A.D., *Yayoi expansion* resulted in the shifting of the center of their culture from north Kyushu to the Yamato Plain at the eastern end of the Inland Sea, near present-day Nara and Kyoto. This area remained the focus of Japanese culture for more than fifteen hundred years. By the sixth century, the Japanese had begun to push northward. As they reached the northern half

of Honshu, they encountered the culturally very different Ainu. The clash of cultures and competition for land resulted in military struggles along the frontier with the Ainu, most of who retreated to the island of Hokkaido. In northern Honshu the Japanese found a forested environment with a much cooler climate and few directly usable land resources. Hence, the major development of Japan as a nation took place south of about 37° N.

During the *Tokugawa period* (1615–1867), the focus of power shifted farther eastward to Tokyo (then called Edo) in the Kanto Plain, where it has remained ever since. *Japan's core area* today covers essentially the area in which the nation's foundations and early growth were laid. This core is a belt, approximately between 34° N and 36° N, that extends from Tokyo on the east to Shimonoseki at the western end of Honshu and encompasses the Inland Sea. The major share of Japan's

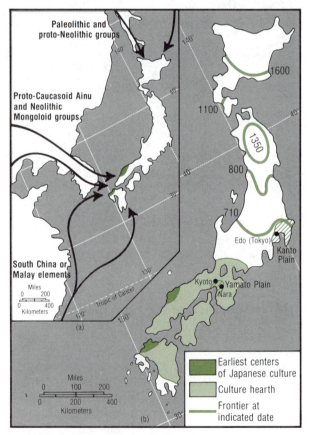

FIGURE 14–5 Origin and Spread of Japanese Culture. Many of Japan's culture characteristics originated in China. Japan's insular location, however, fostered the development of a distinctive culture.

population, cities, industry, and modern economy is concentrated in this zone (see Figure 14–1).

National Character and Social Attitudes

Much has been written and said about the *character and attitudes* of the Japanese people. As is true of most peoples of the world, the Japanese are extremely complex. Nevertheless, a few traits can be singled out as having had particular impact in shaping the destiny of Japan. Most of these characteristics have evolved from centuries of national development. Some were not firmly stamped on the national character, however, until the Tokugawa period.

Perhaps the single most important characteristic to help explain the developments of the last hundred years is the remarkable *adaptability* of the Japanese. Just as they adapted to their peculiar natural environment, they have been willing to borrow culture and technology from others when they clearly saw the superiority of something foreign. Examples are their adoption of aspects of Chinese culture in the premodern era and their adoption of Western technology in the modern era. Certainly contributive to their adaptability have been the innate intelligence of the Japanese and their deep *respect for education.* In the course of the last century, the Japanese have evolved from the master imitators of Western *technology to technological innovators* of the first rank. Also important to Japan's success in development has been the great capacity of the Japanese for *thrift and hard work,* qualities undoubtedly instilled by the relatively difficult natural environment in which they had to create a nation.

Underpinning these qualities have been the basic *homogeneity and unity* of the Japanese already referred to, although political unity has been a more recent attribute. Out of that homogeneity grew a deep and unshakable consciousness of being Japanese, a clear recognition that they are unique.

The *role of the individual* in Japanese society has also been critical. In Japan, *the group is all-dominant.* Everything in life including work, study, and play is done largely within a group context. Moreover, *society is extremely hierarchical,* with clearly defined social classes and modes of behavior. Yet there is social mobility; it is possible to move upward in society through education, hard work, and intelligence.

The Japanese have long had a strong sense of *personal self-discipline* and of *duty* to country and society, together with respect for authority and willingness to obey orders. Japan's government and business leaders were thus able to set goals for the country's development and to count on the total commitment of the people to these goals. The defeat in World War II perhaps weakened some of these qualities, but the Japanese remain today without question one of the most vigorous and purposeful societies in the world.

MODERN JAPAN: THE TRANSFORMATION

In the more than a century that followed the overthrow of the feudal Tokugawa rule in 1867, Japan was physically transformed almost beyond recognition, yet some of the transformation was superficial. Many of the basically feudal characteristics of Japanese society were retained. Although the social transformation has been more rapid since the end of World War II, Japan still has a *dichotomous nature,* a curious mixture of the *traditional and the Western.*

In a manner of speaking, Japan has gone through *three transformations* since 1868. The first consisted of the early period of modernization and industrialization in the late nineteenth century, followed by the move to heavy industry and militarization that reached its peak with World War II. The second transformation was the reconstruction and return to international power in the postwar period. The third transformation, now in its initial stages, involves the search for new directions in a world of increasing scarcity of raw materials and discontent with past growth strategies.

The Meiji Period and Its Aftermath

The Japanese experience illustrates one kind of cooperative and balanced relationship between the government and people useful in the economic development of a nation. The Japanese adopted a *pragmatic approach* to transforming Japan step by step. Within a mere fifty years, the leaders created a sound and modernized economy, and Japan achieved a position of national security and international equality. This position was attained in part by the government's providing a proper environment for development. The government also took a direct role in industrial development by pioneering many industrial fields and by encouraging businessmen to move into new and risky ventures. The government helped fund many ventures by providing private entrepreneurs with aid and privileges. Active cooperation between *government and big business* worked well in Japanese society and continues to be a basic characteristic of Japan's economy.

The role of the people was equally important in making a success of the transformation. Thousands of individual Japanese responded eagerly to the new economic opportunities. In the long run, it was this *private initiative* operating within the context of Japanese society that produced the bulk of Japan's economic modernization.

A critical aspect of this national response, however, was the emphasis in the first two decades of the Meiji period on *development of the traditional areas* of the economy: agriculture, commerce, and cottage industry. The Japanese did not attempt to build modern industry immediately and directly on a weak local economy, as many developing nations in the post–World War II period have mistakenly tried to do.

Growth of Industry and Empire

The *modern industries* deemed important by the government were those on which military power depended. Hence the government led the way in developing shipbuilding, munitions, iron and steel, and modern communications. At the same time, as the need for raw-material imports grew, export industries were strongly encouraged, particularly silk and textiles.

In a quantitative sense, *Japan's takeoff period* really did not begin until after the Russo-Japanese War of 1905. Between 1900 and the late 1930s, the production of *manufactured goods* increased more than twelvefold. *Export trade* grew twentyfold in the same period; manufactured goods accounted for most of the increase. The Japanese excelled at producing *inexpensive light industrial and consumer goods* cheaper than many other countries. It was a production approach that continued to serve the Japanese well. Foreign markets, however, played a less important role in this export strategy than is commonly believed. Japan's economic growth in the early twentieth century was largely self-generated.

Two decisive events that shaped the course of Japan's development in the twentieth century were the victories over China in 1895 and over Russia in 1905. Success in these wars started Japan on *a course of imperial conquest* that ended in the disaster of World War II. These two wars also greatly expanded Japan's territorial control to include Taiwan, Korea, and parts of northern and coastal China, including Manchuria.

Growth of Population

A significant change that accompanied the modernization of Japan after 1868 was an upsurge in *population growth*. During the latter half of Tokugawa rule, Japan's population had stabilized at about 33 million. For the period from 1868 to 1940, however, Japan provides the classic illustration of the interaction of economic and demographic factors. As industrialization and urbanization proceeded, birth and death rates both declined. Population began to increase, but the rate averaged only about 1.5% a year up to 1940. Still, that growth rate was sufficient to double the population to just over 73 million by 1940.

At the close of Tokugawa rule, three-fourths of Japan's population consisted of peasants, and about four-fifths of the labor force was engaged in agriculture, forestry, and fishing. Since 1868 the *rural population has declined* steadily in proportion to the urban population.

Rise from the Ashes of War

When Japan surrendered in August 1945, the nation was prostrate. Destruction from war had been catastrophic, especially in urban and industrial areas. The nation was stripped of its empire and consisted only of the main archipelago. The future of the 72 million Japanese seemed bleak indeed. Yet within a decade Japan was thriving; most of the physical destruction of the war had been erased, and the nation was well on its way to regaining a position of economic might in Asia and in the world. This revival is attributable in substantial part to the resilience of the Japanese people.

The "Cold War" in the late 1940s, the fall of China to the Communists in 1949, and the Korean War in 1950 led the United States to return the reins of government to the Japanese much faster than might otherwise have been the case. The United States needed a strong Japanese ally and played a critical role in rebuilding Japan.

The American Role

American assistance took several forms. *Financial aid* consisted of billions of dollars in foodstuffs, military procurement orders during the Korean War (1950–53), and other aid. Of even greater long-range benefit was the *policy of relatively open doors* for Japanese exports to the United States. By 1970 the United States was buying about one-third of Japan's total exports. This huge market was and still is of great importance to the Japanese. Another benefit was military protection by the United States, which enabled Japan to spend less than 1% of its GNP annually for military expenses. Also beneficial to Japan was *American technology.* With their own industry nearly leveled to the ground by the war, the Japanese bought American technology at attractive prices and revitalized their industry. This approach gave Japanese industry a competitive advantage over many other countries, including the United States, whose plants and technology were much older and could not be replaced so easily.

Development Strategy of the Japanese

The *policies of the Japanese government* and businessmen were probably even more important than the American role. The close cooperation between government and business continued and even grew. *Cooperation* was particularly important in financing the modernization process. Banking credit was backed by the government and made *heavy capital investment* possible. The economy was geared to a *high-growth-rate strategy* that relied on large increases in productivity to provide surpluses to pay back capital debts.

The Japanese government, with great skill, has guided the economy much as in a socialist state. *Growth industries* are identified and supported with generous assistance of many kinds. As Japan's economy has grown during the last century, industry has gradually progressed from labor-intensive light industrial production to *capital-intensive heavy industry.* In the 1970s the official government stimulus shifted more to such industries as automobiles, precision tools, and computer electronics, an indication of the great sophistication of contemporary Japanese industry.

On the other hand, *inefficient or non-growth-potential industries* are dealt with ruthlessly. The attitude is that uncompetitive industries should be forced to the wall by governmental financial practices. Some of the resources of those uncompetitive industries can thereby be freed for more efficient enterprises. Transfer of resources from less to more efficient sectors is a key ingredient of economic progress. Japan's vigorous pursuit of this transfer is in marked contrast to many other countries.

At the same time, Japan has been *protectionist* in the postwar period. High *tariff barriers* have been raised against thousands of products from foreign countries. Foreign investment in Japan is very restricted. The rationale for a protectionist policy was that Japan's economy was too weak to withstand uncontrolled imports and foreign investment. This position was generally accepted by the United States and other nations in the 1950s and 1960s. Because of events that took place during the early 1970s, however, Japan has been forced to start lowering its protective barriers.

The *Zaibatsu*

The *corporate structure* also has played a key role in Japan's development. When any traditional country industrializes, there are shortages of capital, skilled labor, and technical resources. To obtain rapid growth, resources must be concentrated. Since there was no model of socialist development in the late nineteenth and early twentieth centuries for Japan to follow, it was logical for concentration of resources to fall into private hands. Hence, the *zaibatsu,* or "financial cliques," emerged out of the close relationship between government and business. By the 1920s, the *zaibatsu* controlled a large part of the nation's economy.

The *zaibatsu* were very efficient. Certainly they provided the *entrepreneurial strength* that led to the modernization process. Moreover, the importance of foreign markets and raw materials helped keep *zaibatsu* prices competitive. Although much of the nation's wealth became concentrated among a few immensely rich and powerful families, enough *profits from industry filtered down* to establish a substantial and growing Japanese middle class.

Efforts by the United States occupation authorities to break up the *zaibatsu* were not very successful. Many have reemerged in the last two decades. These cliques, joined by many other giant corporations outside the *zaibatsu* system, have played a vital role in Japan's second modern period of development.

The Double Structure of the Japanese Economy

The resurrection of the *zaibatsu* has not surprised the Japanese. This attitude is related to the peculiar *double structure* that has long characterized the economy. Basically, the structure consists of a handful of *giant combines* and thousands of *tiny workshops* with only a relatively few medium-sized firms. This structure had fully emerged by the 1930s and was in part the result of a "split technology." The leaders of modern industry followed Western technology, while the owners of small shops were rooted in the traditional ways of old Japan.

The *giant modernized companies,* because of large outlays for advanced techniques, have succeeded in greatly increasing the *productivity of their labor forces.* The *medium and small firms* have relatively little capital outlay and rely on *cheap labor* to make their products competitive. The relationship between the two levels of the economic hierarchy is close. The larger companies *job out* substantial parts of their production to the smaller firms because it is cheaper to do it that way.

The employees in the larger firms have reaped the greatest benefits from Japan's growth. A *paternalistic relationship* between workers and management is very strong in these larger businesses. Workers tend to stay with a firm for life and to identify with a particular company rather than a skill. If a person's skill becomes obsolete, the company provides *retraining* with no loss in pay. Unions thus do not resist new technology. Employers in turn have great freedom to shift workers from one job to another and can invest huge sums for training without worrying that the employees will leave the company. Labor mobility remains extremely low. In return for their employees' absolute loyalty and hard work, the large companies reward them with *lifetime security,* relatively modest salaries, and generous fringe benefits. These workers have a level of living comparable to those of workers in many Western countries. The majority of workers in the small firms have not shared as much in the modernization process. Their job security and fringe benefits are far poorer.

Population Stabilization

The *rising standard of living* of the Japanese in the postwar period can also be attributed to the gradual stabilization of population growth. In 1948 the government passed the *Eugenics Protection Law* legalizing abortion for economic as well as medical reasons. Public and private efforts were also made to spread birth-control practices. Government propaganda stressed the advantages of a small family. Many other factors, including more years of higher education, later marriages, and the two-child family, also contributed to a steady decline in the rate of population growth.

Although the total population is now over 119 million, the annual growth rate has fallen to around 0.7%, a rate typical of countries in the latter phase of stage III of the *demographic transformation.* The population is projected to stabilize at about 132 million by the year 2020.

Growth of Urbanization

One of the most dramatic developments of the postwar era, and one with profound consequences for Japan, has been the rapid increase in *urbanization* and its concentration in a small fraction of the country. Farm population as a percentage of total population declined from 85% in the early Meiji period to about 50% in 1945. It is now less than 20%. The farmers and rural people have migrated to the cities.

Today more than 50 million people live in the *three great metropolitan regions* centered on Tokyo, Nagoya, and Osaka. Tokyo already has more than 26 million, and Nagoya more than 9 million. Tokyo is believed to be the largest urban area in the world in population. A major share of the remaining 68 million Japanese are found in other parts of the core region, between Shimonoseki and Tokyo. Movement into these zones has been phenomenal in the last two decades. If migration and natural population growth rates were to continue, by the year 2000, 80 million Japanese would be located in the *"Tokaido Megalopolis"* (named after the ancient Tokaido highway) stretching between Tokyo and Osaka.

The *reasons for urbanization* in Japan are much the same as in other countries. However, certainly an important additional factor was the desire of Japan's industrialists and government to concentrate industry in coastal centers. In this way, industry could take advantage of economies of scale.[1] Heavy industries, such as iron/steel and petrochemicals, that rely on large volumes of bulky imported raw materials, could be located right at the

---

[1]"Economies of scale" means it is usually cheaper to produce goods on a large scale in a few factories than on a smaller scale in many small factories.

point of unloading from ships, to reduce costs further. Moreover, large planned industrial zones, called *kombinats,* were developed to permit industries that rely upon each other for basic inputs, such as petroleum refining and petrochemicals, to be located next to each other. All of these measures have contributed to the remarkably high productivity of Japanese industry.

The Consequences of "Japan, Inc."

Japan today is faced with a host of pressing *problems* that are the consequence of the development policy pursued since World War II. This policy, which emphasized *growth of the GNP at the expense of social welfare,* succeeded in catapulting Japan to the forefront of the industrial nations. The price, however, was high.

Regional Imbalances

Japan exhibits sharp *regional imbalances* in the distribution of population and levels of economic development. As urbanization and industrialization grew in the core area, another kind of double structure occurred. There was an increasing concentration of the population and modern economy on the outward or Pacific side of the nation at the expense of the inner side, which looks toward the Sea of Japan. This resulted in an increasingly marked geographical division between a modern industrialized, urbanized, densely populated Japan and a "backwoods" Japan that has remained underdeveloped and not much changed from the rural Japan of many decades ago.

Urban Ills

As assortment of *urban ills* has been another byproduct of economic growth. The term *kogai,* meaning "environmental disruption," is much on the minds of millions of people in the core region. *Kogai* has taken many forms. One form stems from the decision in the postwar period to opt for an automotive society. Starting in the early 1960s, the automobile was vigorously promoted. Rapid increases in per capita income in the 1960s led to an explosion in demand for cars and other vehicles, and Japan is now the second largest automobile consumer in the world. Heavy traffic is worsened by the haphazard way in which the cities have grown.

The Japanese have not neglected public trans-portation, however. In fact, they have been among the pioneers in development of high-speed express trains and electrified railways, such as the famous Bullet Express between Tokyo, Osaka, and other cities. The major cities also have excellent subways and bus systems, but because of the large populations in the urban areas, the demand for public transportation far exceeds capacity.

Many *social problems* have also arisen in the cities. Some of these problems are physical and tangible. These include the high price of housing, the extremely crowded living conditions, and the lack of modern sanitary facilities. Some of the bleakest industrial slums in the world are found in Japan's cities.

Other problems are less tangible but no less important. Many stem from the vast social changes that have swept Japan since World War II. Included are the *breakdown of the family,* the increasing *independence of children* from parental authority, the rising desire of young married couples to live alone away from parents and relatives, and juvenile delinquency and crime in general (still much below the levels of the developed Western countries). Other changes are reflected in the *increased freedom of women* in a society where women's liberation has been slow to develop, the trend toward pursuit of happiness, the weakening of the Spartan work ethic, and the neglect of old people in a society where social security has always been provided by the family, not the government. This last problem will become especially severe by the year 2015 when the proportion of people over age 65 will come to 18.5% of the total population (Figure 14–6). Most of these problems are aspects of industrialization and urbanization found in any country, but in Japan they have taken on their own special nature.

Pollution of the Environment

Of all Japan's problems, one of the best known is *environmental pollution.* Until very recently, Japan had done the least of any major industrial nation to protect its natural environment from the effects of *uncontrolled industrial development.* The seriousness of the problem was finally recognized by the government in the 1970s. Every city and prefecture in Japan has been affected to some degree. Most serious have been air pollution and water pollution. Reported cases of the effects of air and water pollution of human health are widely believed to be merely the tip of the iceberg. Much less eas-

FIGURE 14–6 Japan's Changing Population Pyramid.
Japan's population age structure is expected to change. From 1950 to 2015 the percentage of the population under 15 years will decline while the old-age population will grow.

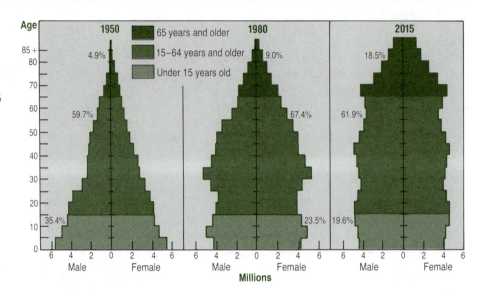

Rural Problems

Rural areas have not escaped the effects of modernization either. The *farmer's natural conservatism* was reinforced in the 1940s by the land reform program that awarded tenant farmers the *small plots* they had tilled for generations. These small holders, farming on an average of only 2.5 acres (1 hectare), have been loyal supporters of the postwar conservative governments. In return, the government has offered the farmers the highest subsidies and support prices for their rice crop of any Asian country. These policies have contributed to the high food prices urban consumers pay. Yet farming by itself is not profitable enough to sustain a farm family. Only 13% of the farm households derive all their income from farming today. More than 60% of farm households earn the major share of their income from nonfarm sources. Agriculture has reached the point where further gains in productivity can be achieved only by removing the *marginal farmer* from the land and consolidating landholdings into cooperatives or larger private farms. Progress along these lines has been slight because of the reluctance of farmers to part with their land, unless they are fortunate enough to be in the path of urban-industrial expansion.

Japan's food situation has been further complicated by *changes in diets*. Food habits have been partly westernized, especially in large cities. As a result, imports of foodstuffs have risen dramatically in the last two decades. Japan now imports nearly 30% of its food supply, and the proportion of self-sufficiency is steadily declining. Major agricultural imports are wheat, soybeans, barley, corn, and sugar. The government has ambitious plans to increase domestic food production and to reduce imports. It remains to be seen, however, how successful the program will be unless significant changes are made in other aspects of Japan's economy.

Foreign Trade and Aid: International Ill Will

Another problem emerging from Japan's success has been a rising tide of *antagonism toward Japan's foreign trade, aid, and investment policies*. Japan's exports still account for only slightly more than 10% of the GNP, a smaller figure than for many countries of Western Europe. The total volume of exports, however, has risen enormously because of the high growth rates of the economy. Because Japanese products have been marketed at relatively low prices and are generally of high quality, they have sold extremely well.

Nowhere is this *trade advantage* more apparent than in the United States, which rapidly developed an enormous trade deficit with Japan by the late 1960s. The imbalance soared to more than $18 billion in 1981 and was a major factor in the overall trade deficit for the United States in the early 1980s. It is sometimes said that the *United States has become an economic colony* of Japan. The United States supplies Japan with foodstuffs and industrial raw materials and in return buys vast amounts of manufactured goods. Cries for protectionist policies in the United States have led to strained relations with Japan.

Japanese agriculture is largely small-plot horticulture with rice as a major crop. Mechanization is confined to small machines such as garden tillers and the rice-threshing machine shown here. (Consulate General of Japan, New York)

Similar complaints are raised by *other major trading partners of Japan,* particularly in Southeast Asia. There the problem is complicated by the memories of wartime experiences that have been revived by the sometimes undiplomatic behavior of Japanese businessmen and tourists who have flooded the region. In many respects, Japan's *economy has become a global one.* Not only does Japan buy raw materials and sell finished products, but also it has set up overseas operations, sometimes on a *joint-venture basis* with other industrialists. Labor costs have continued to rise in Japan. It has been found more profitable to move labor-intensive industry, such as assembly plants for electronic goods, to foreign areas where labor costs are much lower. Although many benefits accrue to these countries, *Japanese investments have met with a mixed response.* Critics contend that Japan has succeeded in creating economically what it failed to create militarily in World War II— the *Greater East Asia Co-Prosperity Sphere.* That is, Japan is the headquarters of an Asian eco-

nomic system dependent on Japanese capital and leadership. The other Asian countries provide the cheap labor, raw materials, and markets for industrial manufactures in a neocolonial relationship. Although the Japanese vehemently deny this allegation, there is no escaping the fact that the economies of the countries of East and Southeast Asia are increasingly tied to Japan. Even in the realm of foreign aid, Japan has gained more brickbats than praise.

Correcting the Mistakes of the Past

As Japan approaches the decade of the 1990s, many people believe that Japan will have to *change the path of development* it has followed since World War II. The international oil crises of the 1970s, as well as the growing cries for trade barriers and protectionism in the major markets of the United States and Western Europe, have alerted the Japanese to the need for change and reform.

225

Remodeling the Japanese Archipelago

It remains unclear, however, just how sincere Japan's government and leaders are in their professed desires to correct the problems created by an expansionist economy over the last two decades. For example, *various proposals* to redress the very unbalanced regional development of Japan, such as the visionary plan proposed by the Kakuei Tanaka government in the early 1970s, have been aired publicly. These plans aim to relieve the congestion in the Tokaido megalopolis and to end the economic stagnation of some rural areas. The ultimate benefits, theoretically, would be a major redistribution of population and modern industry, a reduction of pollution in existing industrial areas, and considerable improvement in the quality of life.

Serious problems have arisen in connection with such plans. Land speculation, concerns about spreading pollution to unspoiled parts of the country, and political opposition, as well as other factors, have derailed most of these plans. Nonetheless, at least the Japanese are seriously thinking about *alternatives to regional development* in their crowded country.

Other Approaches

Even without the plan for remodeling the Japanese archipelago, Japan appears to have begun *shifting its priorities.* The economy is becoming increasingly *service oriented,* like that of the United States. It is hoped that by deemphasizing the industries most responsible for pollution and exports, Japanese discontent and international concern about the country's aggressive trading practices will be lessened. Some leaders believe that the old route to economic growth may no longer suffice. For one thing, the heavy industries have reached capacities exceeding domestic and foreign demand, and surplus production is no longer easily disposed of, as in the iron and steel industry. The shipbuilding industry has had to retrench drastically. Likewise, the markets for automobiles are approaching saturation. Moreover, if Japan does not *lessen its emphasis on heavy industry,* it could face severe shortages of raw materials and increased dependency on supplier nations. Industrial workers are already in short supply because of past rapid economic growth. They are likely to decrease further in numbers as the country's population pyramid continues to change, with a declining number in the productive age group of 15–64 years. Workers are also demanding higher wages.

Japan, in short, is now going through *a transition* to a still more mature economy, perhaps *a sixth stage* beyond Rostow's theory. This stage is more strongly oriented toward services, with a reduced demand on the world's resources and a less lopsided trade balance. A multibillion-dollar pollution control program was put into effect in 1971 when the Japan Environment Agency was established with attention focused initially on the Tokyo area. Another law passed in 1970 was designed to assist more than a thousand areas in Japan where a sharp fall in population in the last decades has made the maintenance of a normal life difficult for the residents. These are but a few examples of the seeking of new directions in a world vastly different from the one of the early 1950s.

The Japanese Experience as a Model

In reviewing the last century of Japan's development experience, one important question remains unanswered. Can Japan's experience be used as *a model for other developing countries?* Certainly the Japanese experience has been unique in the sense that Japan will always have the distinction of being the first non-Western country to have achieved a high level of development. Japan, however, is almost certain to be joined by other countries within a few decades. Obviously, every country has its own peculiar mix of population and resources, and each country's experience will be unique. Yet the basic factors that have been the backbone of Japan's development could be viewed as essential to any country's development strategy. These factors include national unity, a sense of purpose, strong and effective government, hard work, thrift, and a sense of pride. Without these basic requisites, it is questionable how successful any development policies could be.

From another point of view, it might be asked whether or not other countries ought to model their development after that of Japan. One is reluctant to recommend that a country follow directly in Japan's footsteps to the tune of promoting *a high-growth-rate strategy* with minimal concern for social welfare and the natural environment. Yet, to a disturbing degree, that is exactly what many developing countries in Asia and elsewhere are doing. The Japanese model is being intensively studied and emulated around the world, in both rich and poor countries.

15

Australia and New Zealand: Isolation and Space

The "secret" of development of Australia and New Zealand lies in the successful *transplantation of Western society and economy* to virgin territories that, owing to accidents of geography and history, had been largely unknown and untouched by the peoples of Asia.

Australia and New Zealand share *many common characteristics.* Both were founded in the late eighteenth century as British colonies of white settlement. Both have large land areas in proportion to their population, high standards of living, and much closer ties with the United Kingdom and the United States than with most of the Asian countries. Both Australia and New Zealand developed in their early periods as sort of *"supermarkets" for Britain;* that is, they provided many of the foodstuffs Britain and the British Empire needed. That function is still important but no longer dominant. New Zealand remains the more pastoral and agriculturally based of the two countries, with its high level of living dependent on abundant production of dairy products, meat, wool, and other animal products. Australia's wealth is more diversified, with rich deposits of minerals, coal, and natural gas; a bountiful agricultural basket of meat, dairy products, wool, wheat, and sugar; and, increasingly, industrial manufactures.

Australia and New Zealand *depend on trade* with the industrialized nations to maintain their high standards of living. For many decades, this trade was directed primarily toward Great Britain and the British Commonwealth. Because of tariff and other trade privileges, it was profitable for Australia and New Zealand to market their products thousands of miles away while largely ignoring the nearer but poorer Asian market. Since World War II, however, and particularly since Britain entered the European Common Market, the overseas relations of Australia and New Zealand have undergone a metamorphosis. Ties with the British have gradually weakened, whereas those with the United States, Japan, and to a lesser extent other countries of Asia, especially Southeast Asia, have assumed new importance.

AUSTRALIA

A Vast and Arid Continent

Much of Australia's development experience is related to the continent's physical environment, particularly its *isolation, vastness, aridity,* and *topography.* With nearly 3 million square miles (7.8 million square kilometers), including the offshore island of Tasmania, Australia extends 2400 miles (3862 kilometers) from Cape York in the north at 11° S to the southern tip of Tasmania at 44° S and about 2500 miles (4023 kilometers) from east to west (Figure 15–1). Although this land area is approximately equal to that of the lower forty-eight states of the United States, Australia's population is far smaller, just over 15 million. On the basis of population, Australia is actually one of the smaller countries of the world.

Australia is *sparsely populated,* and its people are concentrated in a relatively small part of the continent because most of its territory is too dry to be usable (Figures 15–1 and 15–2). Only 11% of the area gets more than 40 inches (102 centimeters) of rain per year, while two-thirds has less than 20 inches (51 centimeters).

Four major natural regions are distinguished on the basis of climate and relief (Figure 15–2). The *core region* of Australia is the humid eastern highlands, which extend in a belt 400–600 miles (644–966 kilometers) wide along the east coast. The narrow and fragmented coastal plains along the base of the highlands are the only part of Australia not subject to recurrent drought. Most of Australia's population, major cities, agriculture, and modern industrial economy are concentrated in this coastal fringe.

The *three other natural* regions have various disadvantages for human settlement, and the land

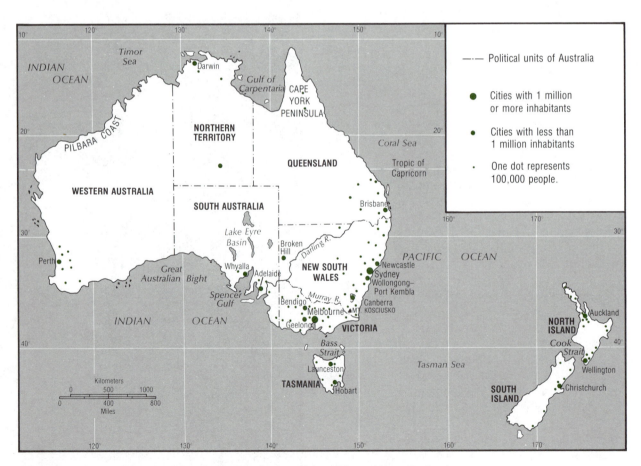

FIGURE 15–1 Locations and Population Distribution of Australia and New Zealand. Australia and New Zealand are two European-culture outposts in Asia. Both are sparsely populated and trade more with European nations than with their Asian neighbors.

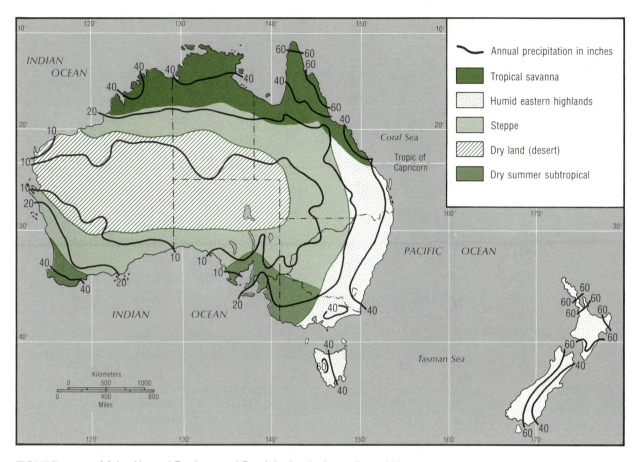

FIGURE 15–2 Major Natural Regions and Precipitation in Australia and New Zealand.
Much of Australia is little used because water is scarce. A large area of Australia receives less than 20 inches of precipitation yearly. New Zealand is much better watered.

use is confined largely to mining and livestock raising. Population density is very light in these regions. Along the northern fringe of Australia are the tropical savannas, where the monsoon climate—six months of heavy rain followed by six months of almost total dryness—makes settlement and agriculture extremely difficult. In the southwestern corner of Australia and around Spencer Gulf the climate is Mediterranean, or dry summer subtropical. These areas have the second major concentration of population, particularly around the cities of Perth and Adelaide, but the total population is still sparse. The huge interior of Australia is desert surrounded by a broad fringe of semiarid grassland (steppe) that is transitional to the more humid coastal areas. These *dry areas* cover more than half the continent. The western half of Australia is in fact a vast plateau of ancient rocks with a general elevation of only 1000–1600 feet (305–488 me-

ters). The few isolated mountain ranges are too low to influence the climate significantly or to supply perennial streams for irrigation.

Shortage of Arable Land

Australia actually has a remarkably small amount of *arable land* for such a large landmass. Fully one-third of the continent in the arid interior region has been found totally unusable for agricultural purposes, including livestock raising. Another 40–42% of the land area to the north, east, and west of the arid interior receives only enough rain to support cattle and sheep. The remaining land area, about 25%, receives sufficient rainfall to support agriculture, but rough terrain and poor soils further reduce the truly arable area to no more than about 8%. Much of the arable land is already used for *sheep and cattle raising and dairy farming*. In real-

229

The Darling River near Broken Hill is only a chain of waterholes during most of the year. The line of gum trees marks the river's course and forms the only break in a semi-arid grazing zone. (Australian Information Service)

ity, less than 2% of Australia's total land area is devoted to crop cultivation at the present time, an indication of the very *extensive*[1] *use of the land resources* of the continent.

Settlement and Population Growth

The long delay in the discovery and settlement of Australia was due to many factors, including the vastness of the Pacific Ocean, the direction of prevailing winds and currents, and lack of any sign that the continent possessed resources worth having. Until 1788, Australia was inhabited only by *Aborigines,* numbering perhaps 300,000. These dark Negrito-type peoples of complex origin had been in Australia for thousands of years living a very primitive existence as hunters-gatherers. In 1770, Captain James Cook became the first known European to reach the east coast of Australia, the one part of the continent that appeared suitable for settlement.

Exploration and settlement by adventurers, emancipists (convicts who had served out their sentences), and others continued into the nineteenth century. Immigration from Britain was encouraged by land grants in the developing continent, but the total population remained very small.

One of the greatest stimuli to development and immigration was the *gold rush of the 1850s,* which brought large numbers of prospectors and settlers. In 1901 the six Australian colonies—New South Wales, Victoria, South Australia, Queensland, Western Australia, and Tasmania—were federated into the Commonwealth of Australia.

One of the most important developments of the nineteenth century was the implementation of the "White Australia" policy as the first trickle of nonwhite immigrants set foot on the continent. Successive Australian governments have recognized the dangers of a relatively small white population controlling such a large land area so close to the overpopulated regions of Asia. Immigration into Australia thus has been a major concern of governmental policy. In general, this policy has been characterized by three main features:

1 Alternating support for and opposition to large-scale immigration, closely linked to periods of domestic prosperity and depression.
2 Strong preference for immigrants of British origin.
3 Exclusion of nonwhite immigrants, with only a few exceptions, from the late nineteenth century until the 1970s.

Britons predominated in the immigration pattern until World War II and were aided by the Australian

---

[1] An "extensive" agricultural activity is one in which limited amounts of labor and capital per unit area are expended on a relatively large area, as, for example, wheat farming.

The Murray River area in New South Wales is being developed for irrigation agriculture. Much of the area, however, is still used for dryland cropping and livestock raising. (Australian Information Service)

government. After World War II the government changed its policy and accepted other European and Anglo-American immigrants as long as they were white. These new immigrants have helped create an increasingly distinct "Australian" character for the country to replace the dominantly British type of Australian society. Australia continues to cultivate immigration with the aim of maintaining net annual migration as nearly as possible at 1% of the total population.

The White Australia policy, officially termed *Restricted Immigration Policy,* was quietly shelved in 1973, as Australia's focus shifted away from Europe. Replacing the old policy is one that now selects immigrants on the basis of education, job skills needed by Australia, and potential for adapting to life in Australia (meaning primarily English-language ability). For example, in 1980–81, Australia's population grew by 1.64%, or around 240,000, of which just over half came from immigrants, with Asians constituting one-quarter of the immigrants. In spite of this liberalization in immigration policy, Australia remains essentially a white bastion in the Asian Pacific world.

An Urbanized Society

An unusual characteristic of Australia, for a country with so much land and so few people, is the *high degree of urbanization.* About two-thirds of the

population live in cities with more than 100,000 residents. Around 40% of the people live in the two great metropolitan areas of Sydney and Melbourne, each with populations of about 3 million. There are several reasons for this pattern. Although the production of agricultural products is still very important to the economy, these activities are extensive and employ few people. Moreover, especially since World War II, Australia has encouraged *industrialization* as a means of supporting its population, providing more home-produced armaments for defense, and securing greater economic stability from a more *diversified and self-sufficient economy.*

A further characteristic of Australia's urban population is that all of the five largest cities (in decreasing size, Sydney, Melbourne, Brisbane, Adelaide, and Perth) are *seaports,* and each is the capital of one of the five mainland states of the commonwealth. This uniqueness is due to the fact that much of the country's production is exported by sea, and much internal trade is also conducted by coastal steamer. Moreover, before federation in 1901, each state built its own rail system focusing on its chief port.

The Bases of Australia's Economy

Australia's *high standard of living* can be attributed to a small population and a reasonably well-devel-

231

Adelaide, one of Australia's five largest cities, is the capital of South Australia, a major port, and a road and railway center. (Australian Tourist Commission)

Only a small part of Australia's land area is cropped. Wheat is the most important crop and is cultivated on large farms using mechanized equipment as illustrated by these combines. (Australian Information Service)

oped and diversified export economy dependent on production of agricultural, mineral, and industrial goods. This *production trilogy* provides a solid base on which to build a prosperous economy.

From initial settling of the country to recent times, *agriculture* was the real mainstay of the economy, generating about 80% of export income in the early postwar years. Although the total value of farm exports has increased, their relative share of export trade has plunged significantly, to only about 45% currently, mainly because of the dramatic increase in mineral exports. Agriculture is dominated today by sheep, cattle, and wheat, extensive forms of agriculture well suited to Australia's environment (Figure 15–3). *Sheep ranching* became the first mainstay of the economy in the nineteenth century and provided wool for Britain's textile industry. By 1850, Australia was already the world's largest supplier of wool, a distinction it still holds, although wool now accounts for just under 10% of total exports. Since World War II mutton has been exported in increasing quantities. Sheep ranches, or "stations," are usually quite large, some encompassing thousands of acres, and motor vehicles and airplanes are important equipment on them.

Generally, cattle have been relegated to those areas not suitable for sheep, and the *cattle industry* remains secondary in importance. The dairy industry has also seen strong growth since World War II. *Dairy farming* is confined largely to the eastern and southeastern coastal fringes. Development

of refrigerated shipping has enabled Australia to supply northern markets with meat and dairy products. Much of the increased demand for wool, beef, and dairy products has come from the countries of East and Southeast Asia, especially Japan, where rising standards of living are changing consumption habits.

Wheat has also benefited from modern technology. The introduction of mechanization in the twentieth century permitted the extensive cultivation of the crop, and some 60% of Australia's total cropland is now devoted to wheat. Like Canada and the United States, Australia has become one of the great breadbaskets of the world.

Australia also produces many other crops and is virtually *self-sufficient in foodstuffs*. Among these, sugar is one of the more important. Sugarcane is grown along the northeastern coastal fringe. Australia annually produces more than 3 million tons of sugar. Most is exported to Japan and other Asian markets, making Australia one of the largest participants in world sugar trade. Other important crops include a wide variety of temperate and tropical fruits for both domestic consumption and export markets.

Major constraints against further growth in agricultural exports are tariffs within the European Community (EEC) that exclude Australian meat, butter, grain, fruit, and sugar from Britain and other European markets. Moreover, the EEC also *dumps its agricultural surpluses* in other markets where Australia is competing. Hence the growth

Sheep raising became Australia's first economic mainstay, and wool still accounts for about one-quarter of the nation's exports by value. (Australian Tourist Commission)

potential for agriculture is not as great as for mining.

In terms of *mineral resources,* Australia is a veritable cornucopia (Figure 15–3). The continent is among the most favorably endowed in this regard of any major world region. Moreover, the resource base is not just in one commodity, such as petroleum in the Middle East, with the severe limitations that such a narrow dependency can produce. Rather, in Australia the resource *base is both varied and large* in volume. Australia sits upon enormous reserves of coal, uranium, iron ore, alumina, natural gas, lead, zinc, and other minerals. The mining sector expanded at remarkable rates through the 1960s and 1970s, so that Australia now accounts for a fifth of world coal exports, 30% of bauxite, and 22% of alumina. Although many countries buy these mineral products, Japan has quickly developed as the major market, especially for coal and iron ore for its steel industry. Moreover, Australia now ranks third (after the United States and Indonesia) in Japanese overseas investment, with Japanese capital playing a major role in development of Australia's mineral wealth. Petroleum is Australia's only major mineral deficiency. Domestic production is inadequate to meet demand, and thus petroleum imports annually constituted 12–14% of total imports by the early 1980s.

Industry remains the weakest link in Australia's economy. Industry developed so far has been primarily import-substitution geared to consumer goods, plus partial processing of mineral and agricultural products, and the modest beginnings of some heavy industry, such as iron and steel and automobiles. Industry remains *protected by tariff barriers,* however, because the relatively small domestic market makes most production uncompeti-

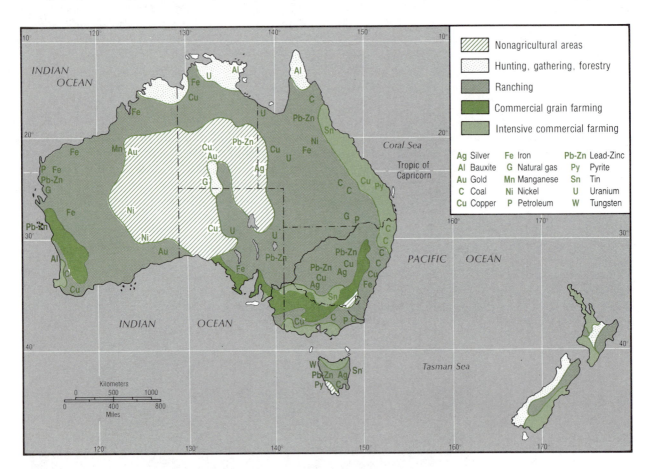

FIGURE 15–3 Rural Land Use and Mineral Resources of Australia and New Zealand.

Australia and New Zealand have extensive forms of agriculture. Ranching dominates much of the area of each country. Other extensive activities are mechanized commercial grain farming in Australia and forestry in New Zealand.

234

tive with foreign imports. Like Great Britain, Australia is also plagued by extreme trade unionism that tends to stifle productivity improvements.

Manufacturing is concentrated in the state capitals, which have three-fourths of all factory workers. Each of the capitals has attracted a dominant share of the state's industrial activities because of the availability of markets, labor, fuel, business and government contacts, and access to overseas and internal transportation systems. Much of the remaining industrial activity is located at a few large provincial centers on or near the coast, such as Wollongong and Newcastle. The leading industrial state is New South Wales, centered on Sydney; Victoria state is second, focused on Melbourne.

Thus, although there have been significant *sectoral shifts* in Australia's economy in recent decades, the country is still essentially *dependent on the export of primary or semiprocessed primary products* in exchange for manufactured goods from the rich world, with the United States and Japan together accounting for 35–40% of both total imports and exports. The United States dominates in sales to Australia, and Japan dominates in purchases from Australia. A truly fundamental change in this pattern must await a number of developments: further increases in Australia's population, greater efficiency in Australian industry, further development of the continent's resources, and greater trade ties with the countries of Asia. None of these developments is likely to occur quickly, and thus Australia will remain for the foreseeable future in the unusual position of a rich nation whose wealth is not derived from industry.

NEW ZEALAND

New Zealand, lying more than 1000 miles (1609 kilometers) southeast of Australia, consists of two main islands—North Island, with the smaller area but two-thirds of the population, and South Island—plus a number of lesser islands. The country is located entirely in the temperate zone from about 34°S to 47°S. Like the Japanese islands, New Zealand is a section of a *circum-Pacific mobile crustal belt* and forms the crest of a giant earth fold that rises sharply from the ocean floor.

South Island consists of nearly three-fourths mountainous terrain, dominated by the Southern Alps rising to elevations above 10,000 feet (3048 meters). North Island is less rugged, but many

peaks still exceed 5000 feet (1524 meters). New Zealand has a *humid temperate climate* commonly known as marine west coast, with mild summers and winters, but in the highlands weather conditions are severe enough for glaciers to form.

A Pastoral Economy

Settlement of the islands is confined largely to the fringing lowlands around the periphery of North Island and along the drier east and south coasts of South Island. A large section of the country is relatively unproductive, although a *tourist industry* is being developed in the mountains.

The Dutch explorer Abel Janszoon Tasman was the first European known to have sighted the islands in 1642, but it was not until Cook arrived in 1769 that exploration and settlement really got under way. Hence the development of New Zealand closely parallels in time that of Australia. The climate is ideal for growing grass and *raising livestock,* and New Zealand, like Australia, has specialized in this activity from the very founding of the country in the eighteenth century. Some two hundred years later, New Zealand's economy is still extremely dependent on specialized production of animal and dairy products.

The country has one of the *highest proportions of livestock in relation to human population* in the world, a ratio of 25:1. Pastoral industries completely dominate exports, and because of the country's small population and high standard of living, New Zealand is *a world leader in per capita trade.* New Zealand is among the top two or three exporters in the world of mutton, lamb, butter, cheese, preserved milk, wool, and beef. In exchange for huge volumes of these products, New Zealand receives most of its manufactured goods and considerable quantities of food. The 3% of the land area that is cropped is devoted in large part to the growing of animal feeds.

The Need for Industry and Diversification

With such a heavy dependency on trade and a very *narrow economic base,* New Zealand is far more vulnerable to the vagaries of world economic conditions than Australia. Not surprisingly, its per capita income is only about half that of Australia's.

Attempts at *diversification,* primarily by industrialization, have not been very successful. For one thing, New Zealand does not have the rich mineral

Livestock raising, especially sheep, is the backbone of New Zealand's economy. Only about 3% of the land area is cropped, mostly for animal feeds. (Consulate General of New Zealand, New York)

resources that Australia has. Furthermore, the *local market is too small* and dispersed. Large-scale production and efficient marketing are thus restrained. The *cost of skilled labor* is also high, and competition from overseas producers, such as Japan and the United States, can be severe, as Australia has learned. Most of the present manufacturing industries are *high-cost producers* surviving under tariff protection.

In spite of the predominantly agricultural economy, most of New Zealand's small population of just over 3 million live in cities, just as in Australia's case. The chief distinction between the two countries is that New Zealand's cities are much smaller, the largest being Auckland, with 800,000 people, on the North Island. Other major cities are Wellington, the capital, also on North Island, and Christchurch and Dunedin on South Island.

FURTHER READINGS

Japan

ASSOCIATION OF JAPANESE GEOGRAPHERS (eds.), *Geography of Japan* (Tokyo: Teikoku-Shoin, 1980). A collection of writings about Japan's geography by members of the Association of Japanese Geographers. A somewhat specialized but very useful supplement to Trewartha's book.

HALL, ROBERT B., *Japan: Industrial Power in Asia,* 2nd ed. (New York: Van Nostrand, 1976). A brief introductory survey emphasizing modernization, industrialization, and urbanization.

KORNHAUSER, DAVID, *Urban Japan: Its Foundations and Growth* (New York: Longman, 1976). An interesting account of the formation of the Japanese urban and rural landscapes viewed in both their historical setting and changing present.

PATRICK, HUGH (ed.), *Japanese Industrialization and Its Social Consequences* (Berkeley: University of California Press, 1976). A series of articles focusing on

Japan's manufacturing labor force from economic, cultural, and political points of view.

REISCHAUER, EDWIN O., *The United States and Japan* (New York: Viking Press, 1965). Reischauer's classic work on Japan illustrating the history of Japanese-American relations. An excellent book for the beginning student.

TREWARTHA, GLENN T., *Japan: A Geography* (Madison: University of Wisconsin Press, 1965). The basic geography of Japan, still useful, especially for the treatment of the physical landscape.

Australia and New Zealand

BLAINEY, G., *The Tyranny of Distance* (New York: St. Martin's Press, 1966). An interesting analysis of Australia's vastness and isolation and their effects on development.

CONDLIFFE, J. B., *The Economic Outlook for New Zealand* (New York: Praeger, 1969). The general economic picture and some possible future trends as seen by a well-known economist.

MEINIG, D. W., *On the Margins of the Good Earth: The South Australian Wheat Frontier, 1869–1884* (Skokie, Ill.: Rand McNally, 1962). An important study on the early period of wheat farming and grazing on the frontier of the steppe.

ROBINSON, K. W., *Australia, New Zealand, and the Southwest Pacific,* 3rd ed. (London: University of London Press, 1974). A fine overview of the whole region of Oceania, but especially Australia.

ROWLAND, D. T., ''Theories of Urbanization in Australia,'' *Geographical Review,* 67 (1977), 167–76. A review of several theories that have been offered to explain the dominance of Australia's metropolitan centers.

SPATE, O. H. K., *Australia* (New York: Praeger, 1968). Still a classic general study of Australia.

Latin America

Don R. Hoy

16

Latin America: Foundations and Processes of Change

When Christopher Columbus reported his famous "New World Discovery" in 1492, a chain of events began that transformed the western hemisphere. Age-old cultures, some of high attainment levels, were shattered almost overnight. European nations established colonies and introduced their value systems and their political, economic, and social organizations. All mother countries applied to their colonies the *mercantile philosophy,* which, along with usurpation of native-held lands and enslavement of Indians, formed the basic foundation of the colonial economy. Latin America felt the brunt of the European conquest a century or more before the European presence was significant in Anglo-America.

THE CONQUEST

At the time of the Spanish and Portuguese conquests, some *75–100 million Indians* inhabited Latin America. Most lived within the densely settled realms of *four high-culture groups:* Aztec, Maya, Chibcha, and Inca (Figure 16–1). These groups had well-developed social, political, and economic organizations, although the Mayan civilization was declining and had lost some of its internal cohesiveness. Agriculture formed the livelihood base, and advanced land-management techniques were common. Irrigation, land drainage, fertilization, terracing, and crop and land rotation were widely practiced. Crops used by these and other groups within Latin America apparently had been largely developed independently of the outside world. The more well-known *cultigens* included maize (corn), manioc, sweet and white potatoes, pineapple, cacao, tomato, avocado, cotton, tobacco, and peanuts. In addition, there were a number of plants, such as beans and squash varieties, that were also cultivated in many other parts of the world. *Few domestic animals* were found in Latin America. The dog was ubiquitous, and the Incas had the guinea pig as a pet

FIGURE 16–1 Pre-Columbian Indian High Cultures of Latin America.
At the time of European discovery, Latin America was occupied by nearly 100 million
Indians. Most of the pre-Columbian population lived in four high-culture clusters: Aztec, Maya, Chibcha, and Inca. The rest of the Indian population was widely scattered and had simple or primitive means of production.

and food source, the llama as a beast of burden,
and the alpaca as a source of fiber.

Outside the regions of high culture lived numerous small, scattered, and loosely organized
groups ranging in economic attainment from total
dependency on hunting, fishing, and gathering to
simple agriculture.

Areas with large, *high-culture Indian groups*

were especially attractive to the European conquerors. These groups possessed the elements
needed to satisfy the conquerors: accumulated
stores of *gold and silver*, abundant and well-
trained *labor supply* for work in the fields and
mines, and a large population for the church to
convert to Catholicism. Smaller Indian groups had
neither the wealth nor the work skills required by

the Europeans. Moreover, the *diseases* brought with the conquest severely depleted all Indian groups, but in the high-culture areas a sufficient labor residue remained. Smallpox, measles, typhus, and probably yellow fever and malaria were among the diseases introduced from Europe and Africa.

After the first flush of conquest, and faced with the decimation of the Indian labor force, the Spanish imported *African slaves*. Later the Portuguese, and in the seventeenth century the English and French, followed Spain's example. In the Antilles the descendants of these slaves make up a large part of the present population. The use of African slaves on the mainland of Latin America was neither very successful nor widespread, since the slaves could escape to the backlands, areas the Spanish and Portuguese did not effectively control.

The early Portuguese and Spanish rarely brought their families. Indeed most Europeans apparently intended to return to their European homes. *Miscegenation* between Spanish and Portuguese men and Indian and African women was widespread and produced two new racial groups: the *mestizo* (part European, part Indian) and the mulatto (part European, part African). The mestizo became the more numerous. These racial groups varied in their distribution and have given rise to distinct differences among the various parts of the region.

Early in the colonization of Latin America, a number of Old World crops and animals were introduced that eventually gained wide acceptance. These contributions *expanded the resource base*. For example, *sheep, goats, and cattle* could forage on alpine meadows and in grassland areas that were marginal for Indian crops. *Wheat,* a plant that can grow in relatively dry areas, extended the cultivable area. Imported *cotton* varieties outproduced native cotton, and the latter type all but disappeared. *Sugarcane* and *bananas,* raised in theretofore sparsely populated areas, and coffee have become major export crops throughout tropical America. *Rice* and a number of vegetables diversified the subsistence farmers' diet. *Pigs and chickens* were readily accepted on small farms where they scavenged for food. Even the grape and olive were adopted in the more temperate areas. The *breadfruit tree,* brought from Polynesia by the famous Captain Cook, became an important food source for slaves in the Antilles. Finally, the *horse,* mule, and donkey proved superior to man and the llama as beasts of burden and enabled outlying areas to carry on trade with the population centers.

Equally important was the introduction of *iron and steel tools,* which were far superior to the wooden and stone implements of the Indians. The digging stick gave way to the hoe and plow, and the stone and obsidian ax to the machete. Digging tools provided a means to expand mining, and the cart made transport cheaper and more efficient.

The Europeans brought a completely *new lifestyle* to Latin America. Landownership was essentially unknown to the communal Indians. Although used for decoration, gold and silver had no economic value to Indians. Towns built along European lines far surpassed the limited pre-Columbian market and religious centers. Trade and manufacturing were greatly expanded over the barter, short-distance marketing, and handicrafts to which the Indians were accustomed. Catholicism was imposed, and Indian ways of worship were destroyed. European languages became the official tongues, and those unable to speak one of them were destined to servitude. The Europeans imposed their will on all others and were quite successful in establishing their value systems and cultural institutions.

THE COLONIAL PERIOD

Evolution of a *dual economy* began early in the colonial period. Production for export to mother countries was largely the function of *large landholdings* controlled by Europeans. In the colonies of northern European nations (British, French, and Dutch), the *colonial plantation* was typical. African slaves were introduced into the Antilles as laborers. Management was carried out by Europeans, although the owners often lived abroad. Large blocks of the best land were set aside for these plantations, and the entire economy and life in general revolved around the plantation. In less accessible areas and on poor-quality land, crops for the local market were raised on small properties.

In Spanish and Portuguese colonies, the large landholdings were the *hacienda,* the *estáncia,* and the *fazenda* (the first two terms are Spanish and the third Portuguese). The hacienda and fazenda usually refer to land units devoted to crops and animals, whereas the estáncia normally refers to a livestock operation—that is, cattle and horses. These large landholdings were not so export-oriented as the plantation, since neither Spain nor Portugal was a large enough market to accept the vast productive potential of its colonies.

Also, in contrast to the plantation, labor was obtained from the local Indian supply, and the owners usually lived on the property or in a nearby city. In less accessible areas and on poor lands, small Indian subsistence economies and age-old practices persisted.

Especially in Spanish-controlled areas, the *patron system* and debt peonage were used to assure a continuing supply of Indian and mestizo laborers. The patron system is both an economic and a social system. The patron, the large landowner, functioned both as an employer and a "godfather." Indeed, he was literally a godfather to many of the peons' (workers') children. He practiced a *paternalistic policy* toward his peons and in return gained their allegiance. In all things the patron's word was law, and few thought to dispute his prerogatives. The patron-peon relationship thus made the patron almost a feudal lord and the peons serfs.

Debt peonage further reinforced the patron system. Under the peonage system, the patron's power over his workers was given legal sanction. Wages paid to the peons were minimal, but housing and garden plots, and often some food, were provided by the patron. For other needs the patron maintained a store that granted credit to the workers. Over time, the peons usually got increasingly in debt, and these debts were inherited by the workers' heirs. Under debt peonage, no worker could leave his patron without approval. The patron, therefore, had a legal weapon to use against dissatisfied peons. Should a peon pay off his debts, there were still strong social ties causing him to remain with his patron. The hacienda, fazenda, or estáncia was a cultural institution as much as an economic organization. The peon's family and friends lived on the hacienda, and the peon usually felt a strong attachment to his patron. Neighboring patrons, in any event, rarely accepted a peon from another hacienda, and few alternative employment opportunities existed.

Mercantilism

All mother countries attempted with varying degrees of success to operate their colonies according to the *mercantilistic philosophy*. A colony could trade only with its mother country; intercolonial trade was severely restricted. As a result few linkages existed among colonies, even of the same mother country, and there was little feeling of commonality. Rather, competition was encour-

aged. The effect of this mother-country policy is evident in modern Latin America where nations of common cultural heritage are often at odds with one another.

The mercantilistic policy also strongly affected *economic activities*. No colony was permitted to produce anything in competition with the mother country. Manufacturing especially was curtailed. For Spanish and Portuguese colonies, this policy meant that sheep raising and wheat, olive, and grape growing were suppressed. The environmental realm of the British, French, and Dutch colonies was so different from that of northern Europe that the agricultural economies of colonies and mother countries were inherently complementary rather than competitive. In the early nineteenth century, however, competition did develop between the *sugarcane* of the colonies and *sugar beets* of northern Europe. Since the colonial sugarcane industry was already firmly established and no alternative activities could be found, the industry was permitted to continue, but tariffs on imported cane sugar were increased to the advantage of local beet sugar producers.

Social Organization

Society in colonial Latin America was organized almost on the caste system; that is, a person could rarely rise above the station in which he was born. In the northern European colonies, the African *slaves* formed the great lower class and had few privileges. Above them were the *freedmen,* who usually were mulattos. Mulattos were the craftsmen and gang bosses. Some had been educated and functioned as secretaries and accountants. Just above the mulatto were a small number of *Europeans of limited wealth* who lived as small farmers and competed with the mulattos for jobs on the plantations; this group has almost disappeared in the Antilles. At the top of the social order were the *Europeans of wealth.* They were the government officials, major merchants, plantation owners, and managers.

In the Spanish and Portuguese colonies, a similar *stratification based on race* prevailed. At the top was the small class of *Europeans* born in the mother country who held all the major governmental, military, and religious posts. Just below them were the *creoles* (Europeans born in the New World), about whom life in the colonial cities revolved and who owned large rural landholdings. The creoles generally were the wealthy class but prohibited by royal decree from holding the most sensitive government positions. It was as *landed aristocracy* that the creoles obtained their wealth and power. The creole measured wealth not only in terms of money but more important in landownership. This attitude toward land was fundamental to colonial Latin America and is only somewhat diminished today. Below the creole was the *mestizo*, who on ranches became the *vaquero* (cowboy) and on farms the overseer. In the army he was the noncommissioned officer, and occasionally he was the village priest. In town he was the salesclerk and craftsman. Finally came the *Indian*, who performed the various jobs requiring manual labor (farming, mining) and who often farmed a small plot of land for his own needs. In the high-culture areas, many Indians were able to maintain their communal lands but were forced to work a certain number of days for local patrons. Others had their lands taken from them and became peons under a patron.

PROCESSES OF CHANGE

For the majority of the population, freedom from colonial rule in the early nineteenth century did little to change life-styles. Throughout much of the nineteenth century the new nations experienced the *growing pains of self-rule*. Internal disunity led to stormy periods of near anarchy and strong, personalized, dictatorial rule, to battles over the location of the national capital, and to border conflicts. The *seeds of change* were planted, however, and the processes that are now modifying Latin America had their origin in the nineteenth and earlier centuries.

Population Growth

For most of Latin America's history, population growth has been slow and birth and death rates have been high. In a few nations (Argentina, Brazil, and Chile), large-scale *immigration* from Italy, Germany, Spain, and other European countries in the second half of the nineteenth century added a new dimension to the population. Yet Latin America did not equal its pre-Columbian population until about 1920.

Today Latin America's rate of growth is one of the highest in the world. By the year 2000 the population is expected to be ten times the 1900 level (Figure 16–2). All parts of Latin America are pre-dicted to grow substantially during the rest of the century, but tropical South America and mainland Middle America will show the greatest increases. A *fast-growing population* is one reason for Latin America's status as a poor or less developed region.

By far the most important factor in population growth is the *natural (internal) growth rate;* although pertinent in the past, immigration is not important today. Two characteristics of population are responsible for the rapid population increase: a *high birth rate* and a *declining death rate,* particularly in infant mortality. Traditionally, high birth rates prevailed and were necessary to maintain a viable population, since death rates were equally high. During this century and especially since about 1920, the death rate has dropped dramatically (Figure 16–3).

Although the population explosion is a factor in the *low per capita income,* it is also a reflection of *greater productivity.* Most people in Latin America do eat enough calories and protein, and most have a relatively long life expectancy. On the other hand, few acute observers would deny that heavy population pressures exist in some areas—for example, the Antilles and areas of pre-Columbian high civilizations.

Urbanization

Latin America's population is dynamic not only from the point of view of growth but also in urbanization. *Migration* from rural to urban areas and from small villages to large towns is progressing rapidly. At present about 60% of Latin America's population live and work in towns and cities. By 2000, 75% are expected to be urban dwellers, a figure comparable to that projected for the rich or developed world.

Figures 16–4 and 16–5 show the expected urban and rural populations for Latin America and its principal subdivisions. In the space of forty years, 1960–2000, Latin America's urban population is expected to increase from 100 million to 500 million, one-half of this number in tropical South America, principally Brazil. The rural population is also expected to increase but at a much slower rate; in southern South America it is projected to decrease.

Urbanism is more than living in a town or city. It also means a distinctive life-style, different economic activities, labor specialization, and economic interdependence. These characteristics are

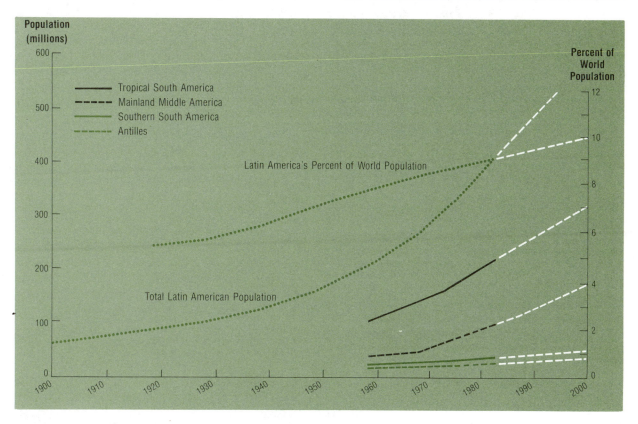

FIGURE 16–2 Latin American Population Growth, 1900–2000.
Latin America's rate of population growth is one of the most rapid in the world. By the year 2000 the region will contain about 600 million people—10% of the world's inhabitants.

SOURCES: Latin American Center, *Statistical Abstract of Latin America* (Los Angeles: University of California, 1982); *World Population Situation in 1970* (New York: United Nations, 1970); and *Economic and Social Progress in Latin America* (Washington, D.C.: Inter-American Development Bank, 1982).

reflected in changing social and political attitudes. A person born and raised in a rural area, especially of a subsistence background, finds the city an almost unfathomable complexity; yet nearly one-half of the urban dwellers are migrants. Few successfully make a direct farm-to-city transition. It seems more likely that movement from farm to regional town to large city is more common, often taking at least two generations. Moreover, a selective process occurs in the migration with the more educated, wealthy, and economically motivated participating in the movement. In many Latin American areas, the remaining rural population is almost devoid of a class of incipient entrepreneurs.

Economic Development

During the twentieth century and especially since 1950, a number of economic developments have altered Latin America's economic structure. Over-

all, the economy has grown rapidly, although the high rate of population growth has diminished per capita values. The economy has become *diversified* with manufacturing and tertiary activities providing more and more jobs and contributing greatly to the growth of the gross national product. *Foreign investment,* largely from the United States, has increased in spite of a growing spirit of nationalism and actual nationalization of some foreign corporations.

The gross national product for most Latin American nations has grown substantially. Much of this growth has resulted from *increased productivity,* especially in manufacturing and service industries. In fact, some authorities place Latin America in an intermediate category between the rich and the poor nations. Per capita GNP for Latin American nations is not a good measure of income for most of the population, since wealth is concentrated in the upper-class stratum and increasingly in urban

245

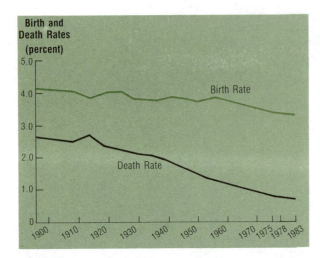

FIGURE 16–3 Latin American Birth and Death Rates. One reason why Latin America's population is increasing rapidly is that the birth rate remains high while the death rate has decreased significantly. Perhaps Latin America is in stage 2 of demographic transformation.
SOURCE: *Economic and Social Progress in Latin America* (Washington, D.C.: Inter-American Development Bank, 1975, 1982).

areas. For much of the lower-class and rural population, incomes have remained low. Moreover, inflation (loss of buying power) has reduced real gains in income. For nations such as Venezuela, total and per capita gains result largely from governmental petroleum revenues which go to the population only indirectly.

Agriculture remains the basic source of livelihood for much of Latin America's population, but other activities have increasing importance. Agriculture's share of the labor has declined dramatically from 70% in 1940 to 46% in 1960 to 40% in 1980 (Figure 16–6). The total number of agricultural workers has actually increased during this time and the result is *severe population pressure* in some rural areas. In contrast to agriculture, *manufacturing* has increased both relatively and in absolute number. In 1960, 18% of the labor force were in the secondary sector, and by 1980 some 28% of the labor force were part of the manufacturing sector. Employment in *tertiary activities* (services and construction) remained essentially unchanged from 1960 to 1980.

The *Latin American economy* has grown substantially. In the early 1960s, Latin America's GNP increased at a rate of slightly more than 5% annually. During the decade of the late 1960s to late 1970s, the growth rate averaged almost 7% yearly. More recently the growth rate has slowed

to about 5.4%. This recent decline in GNP growth rate results from worldwide inflation, high energy costs, and limited demand for Latin America's basic exports (raw materials and foodstuffs).

Growth rates differ among sectors of the economy. Agriculture's growth was about 4% in 1960 and 2.6% in 1980. Manufacturing's annual growth rate has fluctuated between 5.5% and 6.0% during the decades of the 1960s and 1970s, whereas tertiary activities have continually grown from a 5.3% annual rate in the early 1960s to 7% in the early 1980s.

There is also a significant *difference in per capita productivity* among the various economic sectors. Workers in agricultural activities in 1960 were only half as productive as workers in general; 46% of the labor force accounted for only 20% of the GNP. By 1980 agricultural workers were relatively even less productive. In contrast, secondary and tertiary workers had productivity ratings significantly higher than average. If we assume income

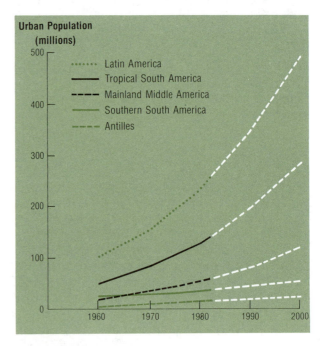

FIGURE 16–4 Latin American Urban Population Projected to 2000.
Latin American population growth is mainly in the large cities. From 1960 to 2000 the urban population should increase from 100 million to 500 million people. Urban population growth comes from both natural increase in the cities and migration from the countryside to the cities.
SOURCES: *World Population Situation in 1970* (New York: United Nations, 1970); and *Economic and Social Progress in Latin America* (Washington, D.C.: Inter-American Development Bank, 1982).

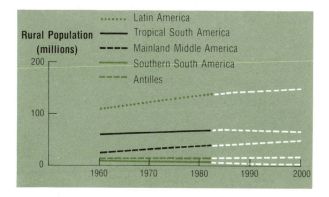

FIGURE 16-5 Latin American Rural Population Projected to 2000.
Traditionally Latin America has been a rural region. During the remainder of this century, however, the rural population is expected to increase only slightly.
SOURCES: *World Population Situation in 1970* (New York: United Nations, 1970); and *Economic and Social Progress in Latin America* (Washington, D.C.: Inter-American Development Bank, 1982).

is related to productivity, then the urban worker has a much greater income than a rural worker. It is little wonder that urbanization is increasing rapidly.

Foreign Investment

Independence of former Spanish and Portuguese colonies opened the door to economic interests in the United States and northwestern European nations. *Britain* early invested heavily in banking, railways and utilities, mining, meat-packing plants, and stock-raising operations. These activities provided the British Isles not only with income but also with products for the home market. In the twentieth century, British and other European investments have declined as United States corporations and more recently those of Japan have moved aggressively into Latin America.

Latin America has been particularly *attractive to United States companies*. From the time of the *Monroe Doctrine*, the United States has considered Latin America within its sphere of influence. United States government policy has protected private investment and has assured a degree of security from attempts to control foreign operations. Only since about 1960 have Latin American nations begun to exert a significant influence over foreign investments within their territories.

At the turn of the century, *United States private investment* amounted to $320 million, about 47% of all its foreign investments. Most of the investments were in mining, railways, and crops for export and were concentrated mainly in Mexico. The Spanish-American War peace treaty gave the United States Puerto Rico and a "protectorate" over Cuba. Investment in the Cuban sugar industry was rapid. By 1914, on the eve of World War I, Mexico and Cuba were by far the most important investment areas. After World War I the area of investment expanded; Cuba and Mexico remained the most important investment areas, but petroleum in Venezuela and Colombia and mining in other South American nations received substantial

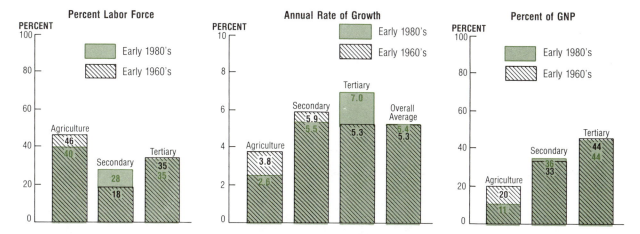

FIGURE 16-6 Sectoral Economic Characteristics of Latin America.
The character of economic activity is changing in Latin America. Primary activities (mainly agriculture) are decreasing in importance while urban-oriented activities are increasing.
SOURCE: *Economic and Social Progress in Latin America* (Washington, D.C.: Inter-American Development Bank, 1982).

247

investor-company interest. Although private investment in Latin America continued to increase, its share of total United States foreign investment declined as companies expanded into other parts of the world. The depression of the 1930s and World War II caused a drop in foreign investment. By 1943, the heart of the war period, investment in Latin America had declined significantly. After World War II, private investment entered Latin America at an unparalleled pace. In the period since the war, mining, manufacturing, and service activities have received most of the investment interest. The sharp decline in the percent of all United States foreign investment in Latin America reflects increased investment activity in Europe and elsewhere (Table 16–1).

Foreign investment is a *mixed blessing* to Latin America. Corporations have provided badly needed capital and equipment, managerial and technical skills, and transport and marketing facilities. These assets have produced jobs, tapped heretofore unused resources, and created infrastructural improvements in the form of roads, banking, communications, and energy. In Central America, companies opened up little-used coastal tropical rainy areas for export banana production. Thousands of jobs were created not only directly in banana operations but also to service the industry and workers. Transport facilities (railways, roads, and ports) were built that all could use. Revenue was generated and taxes were paid to the government for use in other parts of each nation. In Venezuela, petroleum companies mapped, explored, and partly tamed parts of the Maracaibo Basin and have contributed greatly to Venezuela's large per capita gross national product. Cuba was probably most greatly affected by foreign investment. Within the space of a few years after the Spanish-American War, the island was the scene of feverish sugarcane development. New lands were

opened, processing plants built, and transport systems established. Later, tourism provided jobs in Havana and other resorts. These are but a few examples of the benefits of foreign investments.

The other side of the story is not so rosy. Much of the wealth (profits) generated by foreign investment has been "exported" to the companies' home nations. Petroleum and other mining operations have traditionally sent 50% or more of their profits home. The same situation existed in the sugar industry of Cuba. In some countries, the private corporations were so strong they influenced national elections and local governmental policy. Banana companies in Central America are alleged to have "bought" politicians and may have sponsored revolutions to overthrow governments. In 1973 the International Telephone and Telegraph Company was accused of trying to prevent Salvador Allende's election as president of Chile and, when that failed, of attempting to overthrow his government. Where investment was particularly great—for example, in Cuba—the economy became heavily oriented to the export market and the country became an *"economic colony"* of the investor country.

The position of foreign investment in Latin America is changing. Most Latin American nations have reassessed their policies toward foreign corporations and have moved, or are moving, to exert *greater measures of control*. Some forms of control are designed to integrate these traditionally export industries with local economic developments. Mexico, for example, encourages the processing of products before export by variable taxation. Raw materials (lead and zinc bars) are taxed at a higher rate than finished products (batteries). Mexico and other nations also have lower taxes for companies with majority ownership by nationals. Many countries require employment of nationals at all levels of management; technical and managerial skills are consequently nurtured. Still other nations limit the amount of profits that can be taken out of the country.

Finally, there is a growing trend for partial government ownership and *nationalization* of foreign corporations. Ecuador is in partnership with petroleum companies developing its eastern oil fields. Chile, before nationalization of its copper industry, held 51% of the stock. Nationalization is becoming common. Between 1959 and 1961, Cuba nationalized its sugar industry and just about every other externally owned business. Chile and Peru in the early 1970s nationalized most of the foreign mining

TABLE 16–1 United States Investments in Latin America

Year	Value (millions of dollars)	Percent of Total Foreign Investments
1900	320	47
1914	1,600	46
1930	5,200	33
1943	3,400	34
1950	4,400	37
1981	39,900	17

SOURCE: *Survey of Current Business* (Washington, D.C.: Department of Commerce, various issues).

enterprises; additionally, Peru expropriated some of the larger corporate-owned agricultural lands. In contrast to other nations, Venezuela has proceeded slowly, gradually gaining control of petroleum operations and eventually nationalizing the industry completely (1976).

Agrarian Reform

The percent of GNP (Figure 16–6) clearly demonstrates that agricultural development has not kept pace with other sectors of the economy. There is a growing *rural-urban dichotomy* as many agriculturists continue traditional farming practices emphasizing hand tools, while in the city modern technology is rapidly replacing craftsmen and artisans. In recent years nearly every Latin American nation has announced an *agrarian reform policy* by which the fundamental structure of the agricultural economy may be changed. These reforms have two objectives:

1 Rural modernization, basically to improve overall productivity, especially yields per unit of land and per worker.
2 "Social justice," which is interpreted differently from country to country, but in general means the opportunity for each worker to own land, to enjoy greater income, and to participate in the benefits of modern society such as electricity, schools, and medical care.

Most of Latin America's rural problems are traceable to the colonial system of land tenure and social organization. From almost the first day of European conquest, a process of *land consolidation* began by which the conquerors and those who came later acquired large landholdings *(haciendas, fazendas,* and *estáncias).* In time the countryside came to suffer from both latifundia and minifundia. *Latifundia* describes the large landholdings held by a small percentage of the population, *minifundia* the small landholdings held by most of the people—the small farmers. The owners of large estates usually had the most productive land and used it inefficiently. In contrast, most minifundias were on poorer land and were not large enough to provide full-time employment. The small farmer was constantly faced with day-to-day problems of inadequate food supply and lacked the capital, knowledge, and opportunity to apply advances in technology even if he had learned that such technology existed.

Over time, as the *rural population grew,* the pressures on the land increased; but for the small farmer there were *few alternative occupations.* Through inheritance, the already small land parcels were further subdivided until they could be divided no more. Some of the younger generation had to migrate. Some went to sparsely settled areas, but they were often owned by a large landowner, were isolated, and if in the tropical lowlands were disease ridden. Others, in search of work, went to the large landowners, but oversupply of workers meant low wages even if jobs were available. More fortunate were those who found employment in mines or, in the late nineteenth century, on foreign-owned railway construction and on coastal plantations. In the twentieth century, urban development has provided an additional outlet for rural overpopulation.

Agrarian reform has been heralded as the answer to the plight of the rural population. Most reform programs have social as well as economic aspects. Sometimes these two objectives are in conflict. This conflict is seen in the most widespread and popular reform, *land redistribution.* If economic aspects were the sole determining factor in improving agricultural productivity, land redistribution would probably play a minor role. Instead, small land parcels would be consolidated into far larger units so that mechanization and capital could be accumulated. Also, it is easier to spread technological knowledge among a small group of entrepreneurs than to a large mass of almost illiterate peons. Indeed, where land redistribution has been accomplished, productivity has often declined. Socially, however, land redistribution is most important. There is an egalitarian belief that the poor have a right to land control. Moreover, it is hoped that ownership of land may facilitate other changes that in the long run will be advantageous. In any event, political and social pressures demand land redistribution. Failure to redistribute land could lead to political unrest and armed revolts. Indeed, armed revolution has already occurred in Mexico, Bolivia, and Peru.

The reform movement is changing the relationship between the large landowner and his workers. The patron system, in which the employer provides housing, a plot of land, and other noncash allowances, is giving way to a *wage system.* Debt peonage and indentured work arrangements have been largely eliminated, and *unionization* of farm labor is increasingly important. For many workers the destruction of the patron-peon relationship has resulted in loss of security. The obligations of the owner to care for his sick and old workers no longer exist. Indeed, the necessity to pay wages has encouraged many landowners to mechanize their operations, reducing the need for manual labor.

Improvement of the infrastructure is another aspect of agrarian reform. Almost everywhere *farm-to-market roads* are being constructed, and new roads are pushing back the agricultural frontier. New and improved roads reduce the cost of moving goods; the farmer gains access to a larger market area, and in return products become available to him at lower cost. In some countries—for example, Mexico and Peru—*irrigation projects* have increased the arable land. In others, land *drainage* has brought new land under cultivation. Nearly everywhere *rural education* programs have been instituted.

Most reform programs are designed for gradual implementation. Yet so slow has been the pace of agrarian reform that in many countries it exists more in name than in reality. Technical problems such as surveying property lines and building roads, the high cost of reform, and the resistance of the landed gentry have all acted to retard agrarian reform. Where reform policy has been implemented, it is often no more than land redistribution.

Land for redistribution has come from three sources:

1 Some lands have been expropriated from foreign ownership; in many cases, however, these lands are located away from the areas of heavy population pressure.
2 Some of the locally owned large landholdings have been nationalized and given to the workers. In Peru, Bolivia, Mexico, and more recently Chile, land for redistribution has come from the local gentry.
3 Most countries have an agricultural frontier beyond which lies unused land (see Figure 17–4).

Colonization of the sparsely settled frontier areas is supported by the landed aristocracy, who see it as a safety valve for surplus populations; they hope by encouraging colonization to hold onto their lands. The peons look upon the frontier as a place of opportunity. Governments favor colonization because it satisfies some of the demand for agrarian reform without economic or social unrest and because development of the sparsely settled areas can contribute to the economic growth of the nation and may lead to the use of nonagricultural resources.

Colonization has some disadvantages. It is expensive. All kinds of infrastructure must be provided: roads, schools, towns, and banks. Often the area colonized has an environment different from what the settlers previously experienced. Crops and land management techniques that are unfamiliar to the settler may be required. Most colonization zones are in the rainy tropics. Unless carefully controlled, endemic diseases can spread rapidly. The malaria-carrying mosquito, once very susceptible to DDT, is now increasingly resistant to the chemical. Some authorities believe destruction of the covering forest may lead to serious and irreparable damage to the soil.

Nationalism and Multinational Integration

Nationalism (the allegiance to the nation-state) and the move toward multinational economic integration are, in some respects, complementary forces. In other aspects, however, these two forces are contradictory. For almost two decades the multinational integration movement held center stage. In recent years, however, nationalistic fervor, disunity within nations and among neighboring countries, and changing allegiances have paralyzed many of the economic integration efforts.

A sense of *national identity* has been missing in most Latin American countries. *Cultural pluralism* and a diffuse and elusive national allegiance remain regional characteristics. In some countries, however, nationalism has asserted itself. Mexico, since the 1910–17 revolution, has successfully pursued a policy of national identification and unification. Argentina, during Juan Perón's first tenure as chief of government and during the 1982 Falkland Island war with the United Kingdom, experienced a swell of nationalistic fervor for a time. Since about 1950, every country in Latin America has sought to subordinate local and provincial loyalties to national interests. In many cases they have been at least partly successful.

Nationalism takes several forms. In Mexico, a determined blending of both Spanish and Indian cultures has formed the basis of a nationalistic philosophy. A similar sentiment has characterized a segment of Peruvian and Bolivian political thought. In Brazil, national unity is fostered by a frontier spirit, the challenge to tame the west. Movement of Brazil's capital from Rio de Janeiro to interior

Brasília is symptomatic of frontier nationalism. In Cuba, nationalism is fostered by intensive indoctrination programs and economic and political reorientation.

Nationalistic attitudes are heightened in many countries by real or imagined threats from the outside and the need for internal unity. Often the United States is characterized as an adversary, and the international communist movement is considered by some to pose a threat to national security. Recently the question of territorial control of offshore resources has become a rallying point for national interests. Many Latin American nations have extended their territorial sea limits from 12 miles (19 kilometers) to 200 miles (322 kilometers) from the shore. Finally, nationalism is expressed in the desire for greater control over national resources. Foreign ownership of the natural resource endowment is being attacked, and a conscientious effort is being made to make sure that foreign corporations cannot exert much influence.

One goal of nationalism is *self-sufficiency*. Few, if any, nations today can be completely self-sufficient and still provide the material benefits demanded by their citizens. A balance is usually struck between the desire for higher standards of living facilitated by trade and the desire for independence of external ties. Industrialization, particularly heavy manufacturing, is often part of the spirit of nationalism.

The European Economic Community (the Common Market) has demonstrated that *regional specialization* can foster rapid economic development. Tariff barriers among member nations have been abolished, allowing the easy movement of finished products and raw materials among all member nations. Each nation can concentrate on producing those things for which it has an advantage and send its surplus to other member nations in exchange for goods it cannot readily produce. The success enjoyed by the European Common Market is viewed with great interest in Latin America. The various governments reason that a similar organization might spur development, yet the desire of each country to supply its own needs limits the application of regional specialization. Three multinational economic unions have been formed: the Latin American Free Trade Association (LAFTA), the Central American Common Market (CACM), and the Caribbean Common Market (Caricom).

The *Latin American Free Trade Association* was formed in 1961 and grew to eleven members: Argentina, Bolivia, Brazil, Chile, Colombia, Ecuador, Mexico, Paraguay, Peru, Uruguay, and Venezuela. LAFTA's basic goal has been to reduce tariffs among member nations, but progress has been slow. LAFTA operations were suspended in 1982, and the future of the association is clouded. Trade among member nations has increased but not to the degree hoped. A subgroup, the Andean Group, has been formed and may continue to play a role in economic integration. Formed in 1969, it comprises Bolivia, Colombia, Ecuador, Peru, and Venezuela. The members are committed to a multidimensional development program. Not only are there scheduled tariff reductions, but also a common position is sought on several other aspects of development.

The *Central American Common Market (CACM)* was established in 1960 and includes Guatemala, El Salvador, Honduras, Nicaragua, and Costa Rica. Patterned after the European Common Market, CACM has reduced tariffs among member nations and set up a common tariff structure on imports from other nations. The Central American Common Market nations, along with the United States, have funded the Central American Bank for Economic Integration to provide capital for a wide variety of development projects; transport and industrialization receive particular attention. Other treaties have been signed on common educational curricula and texts, defense, and future formation of common economic and political policies. Like LAFTA, however, the Central American Common Market is moribund. The 1969 war between El Salvador and Honduras, the 1979 revolution in Nicaragua that changed the country's economic and political orientation, and the continued internal conflict in El Salvador and Guatemala have effectively halted the CACM integration movement.

The *Caribbean Common Market (Caricom)* was established in the late 1960s and is composed of former and currently associated British territories of the Caribbean. Members are Antigua, Barbados, Dominica, Grenada, Guyana, Jamaica, Montserrat, St. Kitts–Nevis–Anguilla, St. Lucia, St. Vincent, and Trinidad and Tobago. To date, a number of policy statements have been agreed upon and tariffs among members almost abolished. Aside from petroleum products from Trinidad, however, little trade occurs among members.

Last, the nations that are the major exporters of bananas, coffee, and cacao have banded together in an attempt to control production and prices. By controlling the amount exported from each country, the producing nations can set the selling price. A recent example of this idea is the Organization of Petroleum Exporting Countries (OPEC), which has greatly increased oil prices throughout the world. Producers of tropical export crops point out that prices for their exports have not changed much in the last thirty years, but the finished products they import have doubled or tripled in price. These tropical export crop organizations have members not only in Latin America but also in Africa.

SUMMARY

Latin America is in ferment. Many of the processes of change have become important only in the last thirty-five years, and the rate of change is accelerating. Change always brings uncertainty and confusion. As old systems are challenged, individual and group relationships are disrupted. It is not clear what characteristics the new relations will have. The peon freed of patron control must make decisions he never before faced. Foreign corporations must reassess their investment opportunities. Nationalists must weigh the relative advantages and disadvantages of foreign investment in their country and how much interdependence they should permit in the multinational economic integration movement. Fundamental to these changes are population growth and urbanization-industrialization. How extensive these changes are and what direction they will take are questions that we can only ponder.

17

Latin American Regions: The North

We now look more closely at Latin America's regions in order to appreciate differences from one place to another and to understand some of the development problems and potentials each part possesses. For our purposes Latin America is divided into six regions: mainland Middle America, the Antilles, northern South America, Andean South America, southern South America, and Brazil.

MAINLAND MIDDLE AMERICA

Mainland Middle America includes Mexico and the Central American nations of Guatemala, Belize, El Salvador, Honduras, Nicaragua, Costa Rica, and Panama (Figure 17–1). All of these nations except Belize were under Spanish dominion in colonial times. All except Belize and Panama have been independent for more than 150 years. All have rapidly growing populations.

Diversity characterizes many aspects of the region. Racial groups range from the descendants of Aztec and Mayan cultures in Mexico and Guatemala to whites in Costa Rica and blacks of European culture along the Caribbean coast. Many Indians continue the ways of their ancestors little changed four centuries after European contact. *Cultural pluralism* is one inhibiting factor in economic development. *Economic diversity* is exemplified by the mestizo farmer who cultivates a small hillside patch of maize, beans, and squash with hand tools, while a short distance away his children live in the city, wear European clothes, work in a factory, and buy their food. Physically, mainland Middle America varies from the steppe and desert environments of northern Mexico to the rainy tropics of southern Central America. Most of the region's population live in the highlands, although in this century other environments have witnessed an influx of people (Figure 17–2).

FIGURE 17-1 Nations, Cities, and Major Highways of Latin America.

Mexico

Mexico was one of the first Latin American nations to undergo many of the *processes of change* described in Chapter 16. Agrarian reform was instituted in the 1920s. Large amounts of foreign investment have been received, and Mexico early began a program of control. In some industries, such as the railways, the government bought out private companies and now operates them as national utilities. *Mexicanization* (majority ownership of corporations by Mexicans) rather than nationali-

zation, however, has been the dominant trend. Industrialization and urbanization have progressed rapidly, and secondary and tertiary activities now produce the bulk of Mexico's wealth. Mexico has reached Rostow's "takeoff" stage of development (Chapter 4).

In the area of *social development,* Mexico has made some dramatic advances. For more than fifty years, education has been a principal concern of the government. More than 10% of the government's revenue has gone to education in recent

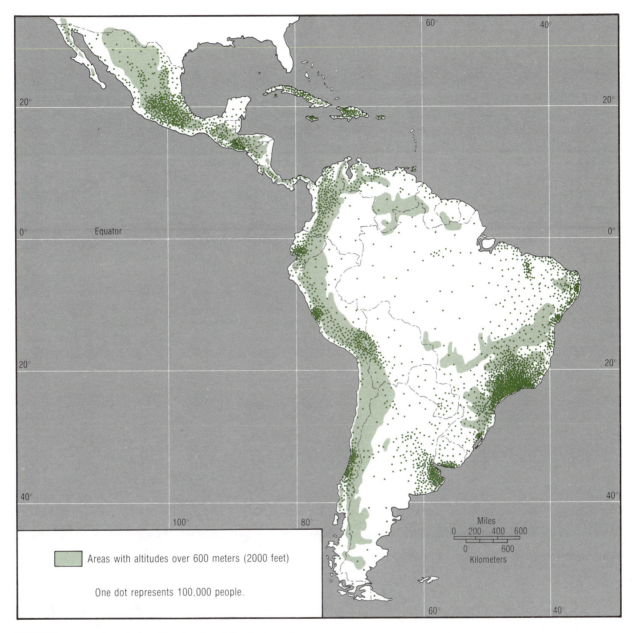

FIGURE 17–2   Population Distribution in Latin America.
The distribution of population in Latin America is oriented toward the fringes of the land. Almost 90% of the population is located within 200 miles of the coast. Some densely settled areas are situated in the highlands formerly occupied by pre-Columbian high cultures. Other densely populated areas are the Antilles, the northeast coast of Brazil, the Rio de Janeiro–São Paulo area, and the Argentine Pampas.

years. At present, 78% of Mexico's people are considered literate; in 1950, 56% were so classed. Another 8% of the government's expenditures is for public health. Improved sanitation, including potable water and sewage facilities, and pre- and postnatal care are widespread and partly explain the decline in death rates.

In spite of a continued expansion of economic activity (6–8% yearly since 1960), *increases in population have diluted economic gains* on a per capita base. Mexico's population has grown rapidly, especially since about 1930; it has doubled in the last twenty-five years. By 1990 population is projected at 97.6 million. The rate of population growth has also increased and is now one of the world's highest.

Mexico's rapid rate of economic and social change has its roots in the late nineteenth and early twentieth centuries. *Porfirio Díaz* came to power in 1876 and for thirty-four years ruled autocratically. Under the Díaz government, efficiency improved, and economic development was stimulated. *Foreign investments,* particularly from the United States, were encouraged. Mining and land-ownership laws were changed to benefit the foreign investor and the local aristocracy. Railway construction and public works projects provided the *infrastructure* to increase production and commerce. Díaz's policies favored the upper class, but the lot of the lower classes worsened. Indian communal lands and small properties were usurped and formed into large haciendas. *Debt peonage* was common, and education was reserved for the elite. By 1910, *latifundia* were highly developed, and the wealth of the nation was concentrated in the hands of only 3–5% of the population.

From 1910 to 1917, Mexico was racked by *revolution,* and Díaz was finally overthrown. More important, however, the revolution marked a fundamental change in national direction. The victorious forces committed themselves to a new national constitution that provided for universal suffrage and education, restriction of foreign and church ownership of property, a minimum-wage law, arbitration of labor disputes, and agrarian reform. Succeeding governments have adhered to these concepts.

National Unity

Basic to Mexico's development was the *creation of national unity.* The large unassimilated mass of Indians, half of the population in 1910, were looked upon as inferior by those of European culture. The postrevolutionary governments sought unity by investing heavily in basic education and an improved transport and communications network and by extolling the virtues of Indian heritage. Mexican artists and writers of the 1920–40 period revered the dignity of manual labor and Indian ancestry. A large degree of national unity and allegiance is now an accomplished fact.

Agrarian Reform

Agrarian reform has been the keystone of Mexico's economic and social development programs. On the eve of the 1910 revolution, 95% of Mexico's rural families were landless. Land for the landless was one battle cry of revolution, and in the succeeding years *land redistribution* became a central government policy. Fundamental to the agrarian reform movement was the *ejido* system. Under the *ejido* system, tenure rights for land expropriated from *haciendas* were given to villages. In turn, the village elders allotted land to individuals. The allotted land was not privately owned, but the occupant had the right of use. The user could improve the land, build on it, work it, and will the right to use it to his heirs. This land, however, could not be mortgaged or sold. In a few areas, especially in irrigated areas, it was not feasible to subdivide an already operating farm. In these cases the *ejido* was operated as a unit with members sharing in the work and profits. At present, nearly half of the population live and work on *ejidos.*

In recent years the agrarian reform movement has done more than redistribute land under the *ejido* system. Since 1950 various *colonization* projects have been started in the rainy tropical areas of the Gulf of Mexico and Pacific coasts. In these projects private landholdings are encouraged. In northern and central Mexico, numerous *irrigation projects* have brought large areas of desert and steppe environment into agricultural production. Agricultural credit, farm-to-market roads, improved technology in the form of seed, fertilizers, and pesticides, water supply and sewage disposal, and housing are all part of the movement. In the early 1970s some *ejidos* were adapted for tourist use or small-scale manufacturing.

Not all *hacienda* land has been expropriated and turned into *ejidos.* In both northern and southern Mexico, some *large land units remain.* Even where *ejidos* have been organized, the *hacienda* owner was permitted to retain at least 100 hectares (247 acres) and often more, so private property is still a very important part of Mexican life.

Agriculture

Mexico's numerous physical environments facilitate the production of *many different crops* (Figures 17–3 and 17–4). Governmental policy also supports crop diversity. The *Aztec crop trilogy* of maize, beans, and squash remains the principal land use for much of central Mexico and provides the bulk of the rural population's diet. These cultigens are also important subsistence crops in other parts of Mexico. Commercial *agriculture varies regionally.* Wheat and barley are important land uses

In northern Mexico irrigation projects have opened formerly desert areas to crop production. *Ejidos* in this part of Mexico are often operated as single units rather than in small individual plots as characteristic elsewhere. (United Nations)

in central Mexico. In the northern part of the Central Plateau region, cotton is grown on irrigated land and is one of the nation's leading exports. Elsewhere on the Central Plateau most of the land is used for cattle. Along the Gulf Coast, rice, citrus, bananas, and sugarcane along with cattle raising are major commercial activities. On the eastern flanks of the Sierra Madre Oriental and in the Chiapas Highlands, coffee is a major export crop.

The Yucatán Peninsula, around Mérida, is one of the world's leading suppliers of henequen, a hard fiber used in bagging, carpeting, and twine. Much of the peninsula and adjacent southern Mexico is still a frontier area.

Mining

During the Díaz period, foreign investment in *mining* was stimulated. Old mines were revitalized, and new mines were developed. Since the revolution, mining has continued to play an important economic role, and ownership by national companies has been encouraged. Today, Mexico is of *world importance* as a producer of lead, zinc, silver, sulfur, copper, and a number of minor minerals. Coal (near Sabinas) and iron ore deposits (near Chihuahua) support a growing iron and steel industry (Monterrey, Monclova, and Mexico City) that supplies most of the national demand (Figure 17–5, page 260). During the Díaz period, Mexico became a major producer of petroleum for export. After nationalization in 1938 production greatly decreased but in succeeding years has regained its former level, making the nation self-sufficient. In the late 1970s vast oil reserves were discovered, and large quantities of oil are exported to finance governmental social and economic development projects.

Industrialization and Urbanization

Much of Mexico's rapid economic growth is attributable to *industrialization*. As in most other na-

FIGURE 17–3 Major Physiographic Regions of Latin America.
Latin America is composed of many different physiographic regions. Each region possesses a distinctive environment that provides different opportunities and problems for development.

tions, Mexico's manufacturing was initially based on processing of commodities such as textiles, milling, and furniture making. *Cottage or household manufacturing* was particularly important. Although these activities are still important, processing has moved from the home to the factory, where machines have replaced hand labor. More complex forms of manufacturing made their ap-

pearance around 1940. Today, in and around the large cities of Mexico City, Guadalajara, Monterrey, Ciudad Juárez, Puebla, and León, many *modern factories* produce a multitude of consumer and capital goods. Some examples of *consumer goods* are automobiles (assembled), pharmaceutical products, refrigerators, washing machines, and television sets. *Capital goods* include iron and

Agricultural frontiers

::: Livestock raising

Sedentary subsistence agriculture

Mixed subsistence-commercial agriculture

Shifting subsistence agriculture (much land unused)

Middle-latitude forests (little used)

Commercial agriculture

C Coffee
Co Cotton
B Bananas
S Sugarcane
W Wheat

Miles
0 200 400 600

0 600
Kilometers

FIGURE 17–4 Predominant Rural Land Use in Latin America.
Rural land use in Latin America depends partly on the natural endowment of an area and partly on the area's location relative to markets. Much of the tropical interior of South America is beyond the agricultural frontier. Parts of the tropical lowlands of Mexico and Central America are also a frontier area.

steel bars and sheets, chemicals, cement, machinery, and fertilizers. The possession of many raw materials, adequate energy resources, and an abundant labor supply are prime factors in the rise of manufacturing. Manufacturing now contributes more to the GNP than agriculture, although more people are employed in agriculture.

Urbanization has progressed rapidly. Greater Mexico City now has a population of more than 16 million and is growing at a rate of 3.5% annually. By the year 2000, Mexico City's population will be 32 million, according to projections. Guadalajara and Monterrey are each over 1 million and growing faster than Mexico City. Along the United States–

FIGURE 17–5 Principal Mineral-Producing Areas of Latin America.
Mineral production is an important activity in many parts of Latin America. Petroleum
production in Ecuador, Venezuela, and Mexico is of world importance. Brazil has
many deposits of iron and ferroalloys. Bauxite, the ore from which aluminum is obtained, is mined in Jamaica, Surinam, and Guyana.

Mexican border the cities of Ciudad Juárez, Tijuana, and Mexicali have more than 250,000 inhabitants and are growing about 5% annually. Many of these cities are now facing *urban problems* similar to those found in the industrialized world. Mexico City, for example, has problems of water supply, sewage disposal, and smog.

With urbanization have come *economic interdependence* and service activities. *Service activities* are the most rapidly growing sector of the economy. Service activities also have grown in response to increasing *tourism*. Tourists, largely from the United States, each year spend over $1 billion. Acapulco is heavily oriented to tourism, as are many areas along the highways from Mexico City north to the border.

Present Status

Mexico is one of the most advanced of all Latin American nations. The development process began more than fifty years ago. The population has been increasingly unified and educated. A variety of natural resources and their integrated use have led to diversity of production and multiple exports; Tampico and Veracruz are the nation's principal ports. In contrast to most other Latin American nations, no one or two products dominate the export list. The economy is becoming increasingly commercialized. A well-developed infrastructure, largely in the north and central sections, has encouraged commercialization. An estimated 40% of the people, mainly urban dwellers, live in reasonable comfort and form a growing middle class. Yet the economic crisis of the late 1970s and early 1980s, during which the peso went from twenty per dollar to more than a hundred per dollar and hundreds of thousands of Mexicans migrated to the United States, indicates a fragile economy.

Today, there are *three Mexicos:*

1 A large group of poor peons in rural areas who live in the old ways.
2 Middle- and upper-class citizens in the urban areas who live like other members of a modern, industrialized society.
3 The transitional group of villagers and rural migrants who live in urban shantytowns and are caught between the traditional and modern worlds.

Central America

Like Mexico, Central America is an area of *rapid population increase.* Nearly 25 million people inhabit the isthmus, and it is projected that by 1990 the population will grow to almost 30 million. Unlike Mexico, Central America has *limited mineral wealth,* and manufacturing is important only in the larger cities; agriculture is the principal source of employment. *Small size,* both in area and in number of inhabitants, is a limiting factor in development. Moreover, agrarian reform and modernization of the economy are only weakly supported. In Rostow's economic growth stages, Central America is somewhere between the first stage (tradi-

tional society) and the second (preconditions for "takeoff").

Central American Common Market

Ever since independence from Spain, there have been occasional attempts to reunite Central America. The most recent unification move is the *Central American Common Market (CACM)* of which Guatemala, El Salvador, Honduras, Nicaragua, and Costa Rica are members. Today the CACM is in *suspension.* Quarrels among member countries and insurrection within nations have brought the effective operation of the union to a halt.

To many, *CACM is a logical step* to overcome the difficulties of each nation's small size. They reason that many economic activities are precluded by the limited markets of the individual countries. By *elimination of tariffs* among member nations, manufacturing plants can be constructed to serve all members. There is a minimum number of customers (a *threshold*) for every activity. Consequently, the more customers there are, the greater the number of possible activities. Looking back to the discussion on resources in Chapter 2, we can see that by expanding the market, *resources may be created.* Tariff elimination can also promote *regional specialization.* Each nation no longer needs to strive for self-sufficiency but can produce those things for which it has an advantage and trade for whatever it cannot readily produce itself. Theoretically, the common market should stimulate economic activity and spur growth, sorely needed in the light of a rapid rate of population growth.

From a practical standpoint, however, the Central American Common Market faces *problems.* One problem is the *attitude of the people.* Few have a supranational point of view of Central America as a unit or see economic integration as a useful goal. Many are highly *nationalistic* and place their own country's welfare above the goals of the common market. Governmental policies are often *protectionist* and in conflict with common-market goals. Other Central Americans have a still more *restricted view* and pay primary allegiance to their local community. The large Indian population of Guatemala represents this latter view, and similar feelings are found in many outlying areas of all Central American nations. A second problem is that the economies of the member nations are more *competitive than complementary.* Only a few products make up the bulk of the exports for each country, and those exports are generally the same. Central American nations have little surplus to trade among themselves. Third, *rivalry* among nations has greatly hampered common-market development. Guatemala and Costa Rica, the two strongest members, have continued to vie for leadership. Internal political dissent and revolutionary activity in Nicaragua, El Salvador, and Guatemala have weakened the market still further.

The Countries of Central America

Guatemala. *Guatemala* is the largest of the Central American nations, but it is also one of the poorest. The *northern half* is sparsely populated and forms the agricultural frontier. In this area colonization projects have been established, and roads connecting the region to the rest of the country have been constructed. The *highlands* that parallel the Pacific Ocean contain the bulk of the population. In the western part of the highlands live most of the Indian population who follow traditional ways of life and still possess many customs of their Mayan ancestors. In and around Guatemala City, mestizos and whites follow European life-styles. Guatemala City (850,000) is the only truly modern city of the nation and is the social, economic, industrial, educational, and political center. Along the *southern coast* and slopes of the adjacent mountains lies the main commercial agricultural zone. Cotton, sugar, coffee, and cattle are widely grown and raised.

El Salvador. *El Salvador* is the *most densely populated* of all Central American nations. The effects of *heavy population pressure* are seen in eroded hillsides, the flow of rural migrants to shantytowns in the principal cities of San Salvador (350,000) and Santa Ana (100,000), the migration of Salvadorans to less densely settled areas in neighboring Guatemala and Honduras, and the continued unrest and often armed conflict between government forces and those demanding sweeping change. In contrast to other Central American nations, El Salvador does not have an agricultural frontier. Increased agricultural production must involve higher yields per unit of land rather than expansion of cultivated areas. The traditional landholding patterns have not been broken, and little has been done to improve agricultural efficiency. Rather, emphasis is on industrialization. El Salvador, however, does not possess

many attributes for manufacturing. No minerals of significance are known; power supply must be imported; the only local raw materials are those of agriculture; and the labor supply, although abundant and cheap, is not skilled.

Honduras. *Honduras,* with an area six times that of El Salvador, has a population of only 4.1 million. Some parts of Honduras are sparsely settled. Outside the major cities of Tegucigalpa and San Pedro Sula where services and limited manufacturing provide employment opportunities, *agriculture* is almost the sole occupation. In the *highlands,* subsistence agriculture, livestock raising, and commercial coffee growing are the dominant activities. Foreign-sponsored banana plantations along the *coast* around San Pedro Sula contribute much of the nation's exports. Aside from these plantations, the rural scene is characterized by low productivity, poor infrastructure, uneven land distribution and use, and few credit facilities. Except for San Pedro Sula, which is a thriving, growing city, other urban places, including Tegucigalpa, present an impression of *stagnation.*

Nicaragua. *Nicaragua* has had a history of *political instability.* For a time in the nineteenth century the British tried to control part of the Caribbean coast. Later an American adventurer briefly captured the capital. In the early part of the twentieth century internal conflict between conservatives and liberals contributed to internal instability. Finally, in 1937, the Somoza family gained control of the government and held power until overthrown in 1979 by the leftist *Sandinistas.* The Sandinistas have reoriented the nation away from close political and economic ties with the United States to the Soviet bloc.

Nicaragua is basically an agricultural nation with most of the production coming from the lowland area near Managua. Much of eastern Nicaragua is sparsely populated and underlain by sandy, unproductive soils. Banana plantations, so common along the east coast of the isthmus, have never been important in Nicaragua.

Costa Rica. *Costa Rica* differs from other Central American nations in several aspects. More than any other Central American nation, Costa Rica has developed a spirit of *national unity.* Nearly 90% of the population is *literate;* the *infrastructure* is well developed; politically the government is democratic and forward-looking with a history of *stabil-*

ity. Agriculture employs 30% of the labor force, and large landholdings exist. Yet productivity is relatively high, and small- and medium-sized owner-operated farms are widespread. Within Costa Rica's small area, there is considerable *regional specialization.* In the highlands around San José high-quality coffee is grown; on the fringe of the highlands dairying is important. Along the coast are banana plantations and cacao, and in the northwest beef cattle are raised. *Manufacturing,* limited but expanding, centers in and around San José. As with other mainland Middle American nations, population growth is rapid. In Costa Rica many rural people are moving into the rainy tropics and to the San José area. Costa Rica presents a picture of progress, yet the high energy costs and a high population-growth rate spell the need for continued economic expansion.

Panama. *Panama* owes its existence and its *economic viability* to the Panama Canal. The canal built by the United States on land leased from Panama is being turned over to the Panamanians under a treaty signed in 1978; they are to assume full control in 1999. The canal divides Panama. The *eastern part* of the nation is little developed. The *western part* contains numerous banana plantations along the coast and beef cattle, rice, and staple food crops in the interior. Tertiary activities in and along the Canal Zone contribute more than 55% of the total GNP and are the most rapidly growing sector of the economy. Panama City on the Pacific Coast and Colón on the Caribbean Sea are the major urban centers. Both cities cater to tourists and transit passengers by offering duty-free goods.

The area in and around the Panama Canal is truly a *"crossroads of the world."* English and Spanish are spoken by a majority of the population, and United States currency is used everywhere. Since the canal acts as a funnel, finished products and raw materials can be brought together in Panama for transshipment elsewhere and for processing. Manufacturing, however, is only slightly developed.

Belize. *Belize,* a former British colony, gained independence in 1982, but its viability is in *question.* Its population numbers only about 160,000, and much of its 8866 square miles (22,963 square kilometers) is little used. For many years the nation's principal exports were wood and other forest products. Within the past thirty years, however, sugar and citrus fruit have come to dominate the export market. The English influence on the predominantly black population is evidenced in language

The Panama Canal divides Panama into two parts. The change of control of the Canal Zone and canal from United States jurisdiction to Panama is under way. The canal provides the major focus of Panama's activities.

and Protestant religion. The city of Belize is the principal urban center.

The Antilles

The Antilles differ from mainland Middle America in three important ways. First, except for Haiti, *export production dominates* the economy of all islands. The Antilles, more than any other part of Latin America, are dependent upon trade with the industrialized world, not only to sell their products but also to obtain the goods they need. Inflation, recession, and prosperity in the industrial nations are quickly reflected in the Antilles. Although many islands of the Antilles are politically independent, they remain in essence economic colonies. Second, the *pre-Columbian Indian population was destroyed* early in the European colonization period and has left little imprint on the present landscape. African slaves were introduced as a labor supply, and their descendants now form the principal population element in the Lesser Antilles (the small islands between Puerto Rico and Trinidad), Jamaica, and Haiti, and a significant minority in Trinidad/Tobago, Cuba, Dominican Republic, and Puerto Rico. East Indians also form an important segment of Trinidadian society. All of the populations of the Antilles except Haiti's have a European culture base; Haiti has a mixture of African and Eu-

ropean cultures. Third, the Antilles are at the mercy of *hurricanes*. These strong tropical storms begin in the Atlantic somewhere east of the Lesser Antilles, move generally westward increasing in strength, and once in the Caribbean Sea generally curve northward through the Greater Antilles (Cuba, Hispaniola, Jamaica, and Puerto Rico). From June to December, the danger of hurricanes is very real. Many of the rural inhabitants build small, inexpensive homes and accept a nearly total loss if caught in the path of a hurricane. Crops grown in this area of mainly tropical wet and dry environment are usually those that can bend with the wind (sugarcane) or can be quickly brought back into production (bananas).

Cuba

At the beginning of the twentieth century, Cuba was a poor colonial remnant of the once great Spanish Empire. For more than thirty years sporadic revolts had disrupted the economy and had made much of the population destitute. Land had little value, and much of it was used for cattle ranching or only occasionally cultivated. *United States "protectorate"* status quickly changed Cuba. Foreign investors bought large sections of the level and fertile land and transformed Cuba into the *world's leading exporter* of cane sugar. Cuba's

Tourism is a major industry in the West Indies. The beaches are considered among the finest in the world. Warm temperatures throughout the year and nearness to the United States are additional advantages for the West Indian tourist industry. (United Press International)

proximity to the United States also fostered *tourism,* and Havana became a major vacation spot. From 1900 to 1959, Cuba was an *economic "colony"* of the United States. On the island a well-developed infrastructure was built, sanitation facilities were greatly expanded, and per capita income was substantially increased. Much of the wealth produced, however, was exported as corporate profits, and substantial *disparity in income* between the rural poor and the urban upper class existed.

In 1959 the insurgent army of *Fidel Castro,* with much popular support, gained control of the country. From that time Cuba has undergone a *social and economic reorientation.* After a break in diplomatic relations with the United States, Castro turned to the Soviet Union for assistance and trade. Foreign investments were nationalized; most locally owned property was confiscated. The economy was redirected along *Marxist lines,* with economic and social objectives set by a central planning council. Today its goals are to increase export crop production, mainly sugar, and expand agricultural production for the domestic market to make the nation more self-sufficient. The council also supervises construction of rural and urban housing and expansion of electricity and potable water for Cuba's poor. Schooling is emphasized for both young and old.

Cuba's Marxist orientation has been of *questionable success.* Certainly present-day Cuba provides *better social services* to the population. Yet *economically Cuba is faltering* and a large percentage of Cuba's population (greater than 10%) have escaped or emigrated from the island. Many of the migrants are from the skilled and entrepreneurial class and now reside in southern Florida.

Puerto Rico

Until the 1940s, Puerto Rico had progressed little since becoming a dependency of the United States. Unlike Cuba, with its large area of level land and excellent soils for sugarcane, Puerto Rico received United States corporate attention only along the level fringe of the island. Interior Puerto Rico is hilly and mountainous.

In the post–World War II period, Puerto Rico began a period of continuing economic growth through a program called *Operation Bootstraps,* a three-pronged development plan. The first prong was *industrialization.* Like other Caribbean islands Puerto Rico had few raw materials other than those of agriculture, few power sources, a small market, and a cheap but unskilled labor supply. Industry was attracted by tax exemptions, government training programs for labor, and a strenuous governmental advertising effort. So successful was the industrialization movement that by 1956 manufacturing produced more wealth than agriculture. Much of the manufacturing is centered on San Juan.

Agricultural improvement was the second prong of the program. Rural experiment stations were established that, with the help of soil conservation agents, introduced better land-management practices. Dairying and truck gardening, new kinds of land use, competed with the traditional crops of coffee in the highlands and sugarcane in the lowlands. In more recent years, rural development has centered on improving housing and bringing water, electricity, and schooling to the farm family.

The third prong was expansion of *tourism.* Again the government played a major role by sponsoring hotel construction and by an extensive promotion campaign. In 1953, 118,000 persons visited the island; by 1963, tourists numbered nearly 500,000; and recently, the figure has grown to 1 million. Much of Puerto Rico's tourist trade has come from the United States.

Hispaniola—Dominican Republic and Haiti

Haiti occupies the western third of the island of Hispaniola, and the Dominican Republic the eastern two-thirds. There are *marked contrasts* between these nations that share the same physical environment. The statistical comparisons found in the Appendix give only a partial picture.

Haiti, with its *Afro-French culture* but almost entirely *black* population, is a country of extreme *poverty* and has an aura of *decay.* The hoe and *machete* are the only tools on its small *subsistence* farm plots. Deterioration is evidenced by the pot-holed roads and abandoned irrigation systems, erratic plumbing and water supply in the city of Port-au-Prince, the erosion of hillsides, and the decline of the population's social organization.

The *Dominican Republic* has a *Hispanic flavor;* and although there are areas of hoe and machete subsistence agriculture, there are also large farms with plows and tractors. Road transportation is especially good, and the *infrastructure* is improving. The Dominican Republic is a poor country but gives an *impression of progress and change.*

265

NORTHERN SOUTH AMERICA

Along the Caribbean fringe of South America, the three Guianas (French Guiana, Surinam, and Guyana), Venezuela, and Colombia make up an amorphous but large Latin American region. All were at one time part of the Spanish Viceroyalty of New Grenada, but the Guiana area was never effectively occupied by the Spanish and subsequently was colonized by the British, Dutch, and French.

Venezuela

Venezuela has one of the highest per capita GNPs in Latin America, much of which comes from the *petroleum* industry. Oil and oil products account for more than 90% of the country's exports, for 25% of the GNP, and for more than 60% of the government's revenue. Petroleum and development are synonymous terms in Venezuela.

The *modern oil industry* of Venezuela began in the early part of this century when the dictator Juan Vicente Gómez encouraged *foreign petroleum companies* to develop the nation's reserves. The petroleum fields of the Llanos and the Maracaibo Basin became major sources of oil entering international trade. For many years Venezuela exported crude oil to refineries on the nearby Dutch islands of Aruba and Curaçao, in the United States, and in northwestern Europe. Since World War II, *refineries* have been built in Venezuela, the *by-products* providing raw materials for tires,

This modern steel mill built in the 1960s is part of the Venezuelan interior development program. The view is northeast across the Orinoco River and the Venezuelan Llanos. (Wide World Photos)

synthetic fibers, medicines, and a host of other industries. In addition, the government, in partnership with the petroleum companies, has obtained a greater share of the industry's profits. In 1976, Venezuela *nationalized* the industry. Nationalization is the final step in Venezuela's move to regain control over its single most important economic activity.

Iron ore along the northern fringe of the Guiana Highlands has also contributed substantial exports for Venezuela. Originally developed by United States corporations, the operations were nationalized in 1975. The ores were located in a little-developed part of Venezuela and required *large capital investments* and an *extensive infrastructure*. Once developed, the area received additional investments from the Venezuelan government with the goal of creating a national center of heavy industry. Hydroelectric power and imported coal are now used to make iron and steel and aluminum. *Ciudad Guayana* (population over 100,000) is the center of this ambitious development program.

For many years Venezuela has had a policy of "sowing the oil." Governmental profits derived from petroleum and iron ore have been reinvested to stimulate other sectors of the economy and to provide a better level of living for the population. Funds have been used for highway construction and farm-to-market roads, low-cost housing and education, agricultural improvement and colonization, and industrialization. By these means the wealth produced by mining is allocated to the pop-

Petroleum provides much of Venezuela's export earnings. Developed by foreign corporations, largely those of the United States, the oil industry was nationalized in 1976. Shown here are storage tanks and wells of the Lake Maracaibo field. (United Nations/V. Bibic)

266

ulation; the mining industry itself employs only about 3% of the labor force.

Venezuela is one of the most *urbanized* nations of Latin America; nearly 80% of the population is considered urban. Of Venezuela's 18 million people, 2.4 million live in the metropolitan area of Caracas; the next four largest centers are Maracaibo (650,000), Valencia (370,000), Barquisimeto (335,000), and Maracay (225,000).

Most of Venezuela's population is located in the Cordillera Mérida highlands. Away from the urban centers, small-plot agriculture is characteristic. Population density is high. South of the highlands stretches the broad, flat, and sparsely settled Llanos, devoted mainly to cattle raising. Along the highland-Llanos fringe, however, some agricultural colonization is taking place. Much of the region south of the Llanos, except for the Ciudad Guayana area, is outside of effective government control.

Colombia

Colombia's 27.7 million citizens are distributed in several *separated population clusters* in the western third of the nation. The eastern two-thirds, sparsely inhabited and little developed, is a continuation of the Llanos and a small section of the Amazon Basin. In the west the great Andean mountain system is divided into three parallel ranges: the Cordilleras Oriental, Central, and Occidental. In the *Cordillera Oriental* are several intermontane basins in which the pre-Columbian high-culture Chibchas lived. Many of the early Spanish colonists settled in the same area, and Bogotá (3 million) eventually became the largest city and focal point of the nation; Bucaramanga (320,000) is similarly situated. In the *Cordillera Central,* other settlers established the town of Medellín (2.2 million) and moved southward along the mountainside. Along the *Caribbean coast,* population grew around the ports of Barranquilla (850,000), Cartagena

(620,000), and Santa Marta (150,000) and expanded inland. Some of this expansion extended up the *Cauca Valley* and centered on cities such as Pereira (250,000) and Cali (1.2 million).

The various *population clusters are poorly interconnected;* east-west movement is particularly difficult. The Cauca and Magdalena rivers provide a somewhat better line of communication in a north-south direction, but even along these routes transport is difficult. Limited interconnection among population clusters has meant *strong regional loyalties,* lack of economic integration, and political factionalism.

In spite of numerous large cities, Colombia is *basically rural.* Coffee provides about 60% of the nation's exports, and almost 30% of the labor force is engaged in agricultural pursuits. Efforts to diversify the agrarian sector and lessen the poverty of most rural inhabitants have not been very successful. As a consequence, migration rates to the city and unrest among the rural population are high.

Colombia's seven largest *cities* have about 33% of the nation's population. Within the cities is produced a full range of manufactured products from textiles and foods to metalworking and chemicals. Metalworking is facilitated by a steel plant located in the Cordillera Oriental at Beléncito. Medellín has had a tradition of textile processing, and in recent years several other kinds of industries have located in the area. Cali in 1912 had but 27,000 inhabitants; in 1950, 200,000; and today, more than 1 million. Much of Cali's growth has come from industrialization by both national and foreign corporations.

Colombia has a diversity of natural environments in which the growth of a wide variety of crops is feasible. Moreover, coal, iron ore, and petroleum are, if not abundant, more than adequate for further development. The lack of a viable political organization and an adequate infrastructure has hampered economic modernization attempts.

18

Latin American Regions: The South

Three final regions complete the broad realm of Latin America: Andean South America, southern South America, and Brazil—a single nation with a larger area than any of the other five regions.

ANDEAN SOUTH AMERICA

Andean South America includes Ecuador, Peru, and Bolivia. These nations have several characteristics in common and face similar development problems. All are in Rostow's stage 1 of development (traditional society), although Peru is nearing stage 2 (preconditions for "takeoff"). All are predominantly *rural;* half of the labor force in Bolivia and Ecuador is engaged in agriculture. Peru has 38% of its labor force in agriculture. Manufacturing accounts for 10–20% of the labor force in each country. More significantly, all three nations have *plural societies* with the descendants of the Incas making up a large part of the population, although ineffectively integrated into the national economy. There is a great gulf between the rural Indian and the Western-oriented urban dweller. Cultural pluralism is also evidenced in a *dual economy.* There is a sharp line dividing economic activities for local consumption and those for the world market.

Ecuador

Ecuador has long been a nation beset with social and economic problems. During the *colonial period,* Ecuador was largely neglected. The few Spanish who did settle there gained their livelihood from the sweat of Indian farm workers on *large estates* in the highland basins of the Andes. The Spaniards themselves lived in Quito, the capital and former Inca center, and in smaller regional centers. For many of the Indians, European conquest had little impact except for introducing new crops and animals. In the Oriente, east of the Andes, the fierce Amazon Basin Indians and a paucity of resources of interest to the Spanish kept the area free from colonial administrative control. To the west, the

Pacific coast and the Guayas Lowlands were largely neglected (see Figure 17–3). Part of the area was disease ridden (malaria and yellow fever), and part was quite dry.

The *Pacific coast,* especially the fertile Guayas Lowlands, is Ecuador's major center of growth. Guayaquil (900,000) is now the nation's largest city and functions as the *principal port and manufacturing center;* it is growing at a rate of 6% yearly. Control measures have freed many of the coast's rainy tropical areas from disease, and since 1940 a vigorous *road-building* program has opened large areas for settlement and commercial agriculture. Ecuador is the world's leading *exporter of bananas,* all of which come from the coast. *Fishing* in adjacent Pacific waters has contributed substantially to the nation's wealth. Ecuador claims a 200-mile (322-kilometer) territorial control of the bordering Pacific, a reflection of the importance of fish and of *budding nationalism. Natural gas* resources offshore of Guayaquil are now being developed. Along the coast there is the aura of *rapid economic growth.*

In contrast, the *Andean segment* is a stagnant area little changed since colonial times. Population is confined to several *intermontane basins* and surrounding hillsides. Quito (600,000) is the urban center of the region and serves as the national capital, the home of the landed aristocracy, and a regional service center. *Agriculture,* largely for local consumption, is the major economic activity of the area. The *best lands* are in large farm units; these estates specialize in dairying and maize. The *poorer lands* are divided into thousands of small Indian subsistence plots with wheat and barley cultivated on dry lands, potatoes raised at higher elevations, and sheep pastured on the alpine meadows. The Andes of Ecuador present a classic example of *minifundia* and *latifundia.*

The *Oriente* is a sparsely settled area inhabited largely by Amazon Indian groups such as the Jivaro who possess few modern techniques of production. In past times, the Oriente has received temporary influxes of people from the Andes in search of balsa wood, rubber, and cinchona (quinine used for malaria). Each time, however, the gathering of these *wild products* has lost out to commercial production elsewhere or to synthetic substitutes. In 1967 a major *oil* field was discovered in the Oriente, and Ecuador has become a leading oil exporter. Nearly 500,000 barrels per day can be shipped by pipeline to a Pacific terminal.

Ecuador's 8.5 million inhabitants are about evenly divided between Andean Indian farmers and Spanish and mestizos who live in the cities and on the Pacific coast. These *two groups* differ greatly in their cultural outlooks and economic well-being. There is a minimum of communication between them. The *Indian* is oriented to the *local group* rather than to the nation, is *illiterate,* and has an income far less than the national average. *Prospects* for economic gain are dim, for *population pressure* has become increasingly intense. The Indian's worldly possessions are few: a one-room hut, a sleeping mat, and the clothes on his back. The *city dweller,* of Spanish descent or a mestizo, has a more worldly outlook, is *nationalistic,* and has an income much greater than the national average. Most urban residents also are *literate,* have adopted *Western dress and ways of life,* and have alternative economic opportunities. These two groups present the nation with a problem of *cultural dualism.*

Bananas, oil, and fish provide Ecuador with valuable *exports,* but national development is hampered not only by cultural dualism but also by *isolation.* The Andes are a formidable *barrier* to internal transport. *Few all-weather roads* connect the intermontane basins of the highlands or the highlands to the coast. Roads are almost nonexistent in the Oriente. Ecuador as a nation is isolated by its *west-coast location.* Only with the opening of the Panama Canal have the markets of eastern Anglo-America and Europe become available to Ecuador, but even so the nation is off the major shipping lanes. Since about 1950, with the rise of Japan as a major trading nation and population growth along the United States Pacific coast, Ecuador's access to markets has improved. Japan particularly has become an important trading partner with Ecuador and other west-coast South American nations.

Peru

Like Ecuador, Peru has three major areal units: the coast, the Andes, and the east. Peru, however, has *a more diversified economy.* Much effort has been made to incorporate the Amazon lowlands into the national system, and in recent years Peru has pursued a *policy of national unity.* In many respects, Peru is in the forefront of Andean development.

The *Peruvian coast* is a narrow ribbon of desert. Settlement is confined to some forty rivers that rise in the Andes and cross the coastal zone to the

Shantytowns are common sights in large Latin American cities. This one is in Lima, Peru. Rapid population growth and urbanization have led to an urban population explosion. Many of the urban poor must use whatever material they can find to build their homes. (United Nations)

sea. Most of the nation's *commercial crops* are grown in the *river oases,* cotton and sugarcane being the most important export crops (see Figure 17–4). Formerly these export crops were grown on large estates, many owned by foreign corporations, but in the late 1960s, most estates were nationalized and formed into worker cooperatives. The oases near the Lima-Callao area (population 3.7 million) are oriented to truck gardening for the urban market. The southern oases have a subsistence economy, but some grapes and olives are grown for the national market and food crops for the nearby regional center of Arequipa (250,000). Since 1950 the government has begun many projects to supply greater amounts of water to expand the irrigated area.

The coast is endowed with some significant *mineral deposits.* In the north *petroleum* has been pumped since the late nineteenth century. In the south are *copper and iron ore.* The iron ore is shipped to Chimbote where it is processed into iron and steel sheets, rods, and ingots.

Lima and its port, Callao, form the nation's social, political, and economic focal point. While preserving many of its colonial traditions, the city has become increasingly industrialized. Historic buildings on narrow streets contrast with modern architecture and broad avenues. Around the city's fringe and also on formerly vacant land near the city center are *shantytowns* populated by the urban poor and migrants from the countryside and smaller regional towns. The Lima-Callao area is experiencing a population growth rate of nearly 6% yearly, about double the national growth rate.

As in Ecuador, the *Peruvian Andes* are inhabited mainly by *peasant Indians* of Inca descent, the coastal area by Europeans and mestizos. Population in the Andes is dense, and many peasants *mi-*

grate both temporarily and permanently to the coast and to a lesser degree to the east in search of work. *Land tenure* in the highlands remains much the same as in colonial times. Large estates control the *best lands*. On the *poorer lands* and in isolated pockets, many small farms, often on communally owned land, dot the landscape, and the *poverty-stricken inhabitants* live much as in pre-Columbian times. The Peruvian government has attempted to break the landed aristocracy's hold on the land and peasants. Some of the large estates have been broken up and allotted to the peasants who worked them. These actions are intended to further a larger policy of *national unification* by integrating the Indian into the national Spanish-mestizo culture.

The Andes of Peru are also noted for extensive *mineral deposits* (see Figure 17–5). Coal is mined in several locations along the western flank of the Andes. The most famous mining district is Cerro de Pasco, an old colonial silver center that has been redeveloped to produce copper, silver, gold, lead, zinc, and bismuth.

Peru has made repeated attempts to tame its large eastern lands of the *Amazon Basin.* For more than a hundred years the nation has sponsored *colonization* projects in the area, and these attempts have greatly intensified in the last forty years. *Roads* have been built across the Andes into the eastern plains, and along each road have come highland peasants in search of new economic opportunities. Illustrative of the interest in the east is Peru's proposed *Marginal Highway* projected to extend along the eastern flank of the Andes (Figure 18–1). Much of the east, however, remains sparsely populated. Iquitos is its urban center and Peru's major port on the Amazon River. The existence of *petroleum* deposits has been known for some time, but production is limited by inadequate transport. Active exploration for oil continues, and hopes remain high that vast reserves will be found.

In many respects Peru's development problems, like those of Ecuador and Bolivia, are similar to the problems formerly encountered by Mexico. In recent years the Peruvian government has been playing an active role in directing change, both economic and social. Many of Peru's policies mirror those instituted much earlier in Mexico: seizure of political power from the landed gentry, unification of the society, an emphasis on education, and agrarian reform.

SOUTHERN SOUTH AMERICA

Southern South America is composed of Chile, Argentina, Uruguay, and Paraguay. The first three nations share several common features that separate them from surrounding areas. Chile, Argentina, and Uruguay have relatively *high per capita GNPs.* Each is highly *urbanized,* has a well-defined *middle class,* and has a very *high literacy rate.* The nation-state idea is well established, and the populations are *culturally unified.* Measured by most standard indicators, these three nations fall in or near the rich side of the poor-rich continuum of the world's nations. This alignment with rich nations is accentuated further by demographic features. A review of birth, death, and growth rates shows that they have followed the course of the *demographic transformation* model (Chapter 2) and are now nearing the model's final stage. Birth and death rates are low. According to Rostow's stages of economic development, Argentina is probably in stage 4 (the drive to maturity); Rostow placed Argentina at the beginning of stage 3 (the "takeoff") in 1935. Chile and Uruguay fall within stage 3.

Paraguay is included within the southern South American region because its Guarani Indian population has intermarried with the Spanish immigrants to form a unified mestizo society with European cultural values; Paraguay's connection with the outside world is largely through Argentina. Moreover, the physical environment within which Paraguayans live is an extension of the environment of adjacent Argentina. Finally, Paraguay, along with Argentina, Uruguay, and Brazil, is part of a regional development program of the Río de la Plata watershed.

Chile

Chile has achieved *cultural unity* in spite of its long and narrow shape. Central Chile, between 30° S and 42° S, is the heart of the nation, and it is within this area that a *cohesive society* was formed. To the north is the bone-dry Atacama Desert, sparsely populated but possessed of nitrates, copper, and iron that provide most of the nation's exports. To the south is a very humid, rough, and rugged area, also sparsely inhabited but with abundant forest and waterpower potential. In the far south some coal and petroleum are found; on the windy and wet tip of the continent sheep are raised for wool, which is exported.

It is within *central Chile* that most of the popu-

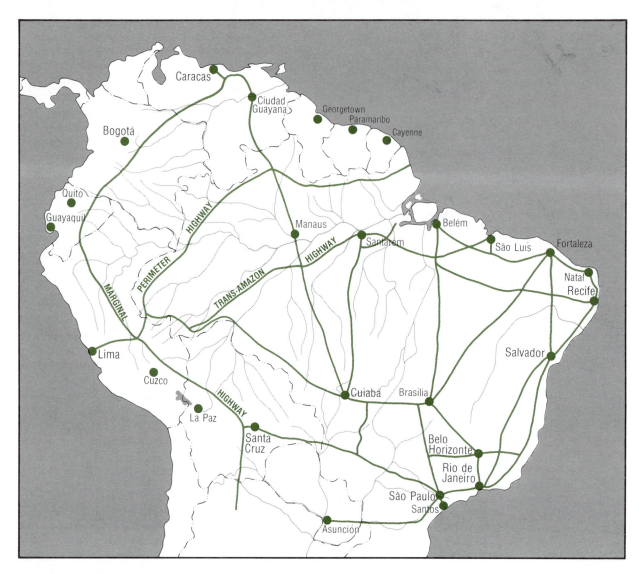

FIGURE 18–1 Proposed Amazon Basin Highway System and Connections to Other Regions.

A major development project is under way in the Brazilian part of the Amazon Basin. As part of this development effort an extensive highway system is planned. Work on some of these roads, such as the Trans-Amazon Highway, is well along. The other nations having lands in the Amazon Basin have proposed the construction of the Marginal Highway, but this highway remains only an idea.

lation lives and where most of the industry and agricultural lands are located. Chile is an urban country (82%), and most of the cities, including Santíago, Valparaíso, Concepción, and Temuco, are in the middle section.

Chile's unified society is of recent vintage. In colonial times the economy revolved around rural estates with a patron-peon organization. The patron's hold was not strong, however, for *population pressure was low,* and to the north, south, and

east across the Andes in Argentina were frontier areas. From 1879 to 1883, Chile successfully pursued a *war* against Peru and Bolivia over *nitrate deposits* in the northern Atacama area that united the Chilean population in a common cause. Nitrates for fertilizer and explosives became the nation's major export, and mining it provided an alternative economic opportunity for the peons. Later, copper mining provided more jobs. Early in the twentieth century, especially during World War I when goods from the industrialized world were difficult to obtain, *industrialization* for the local market began. In Chile the landed aristocracy invested in manufacturing activities and services more readily than did their counterparts elsewhere in Andean South America. The *alternative economic opportunities* in the cities provided another and increasingly important outlet for the rural farmer.

Cultural unity, national allegiance, valuable export products, lack of heavy rural population pressure, and diversification of the economy by manufacturing and service largely account for Chile's development. In addition, about 94% of the population is *literate*. *Political stability* has been characteristic throughout most of Chile's history. In recent years political power has passed to *urban-oriented parties* that campaigned for economic and social reform. These reforms have two main lines. The first is greater *control of foreign investment,* especially the mining industries that provide most of the nation's exports. The second is *land redistribution* in the rural sector.

In 1970, *Salvador Allende* became the first self-acknowledged Marxist president of Chile. Allende was elected by a minority of the electorate with barely over one-third of the vote in a three-man race. The Allende government attempted to reorganize Chile's economic and social structure very rapidly. In the process the mining industry was nationalized, land redistribution was accelerated,

and most banks and the communication media were taken over by the government. A true *socialist economy* was in the offing. This restructuring was not without problems and serious opposition. Allende almost bankrupted the nation and alienated a large part of the population. Congress repeatedly attempted to rescind Allende's actions; strikes of protest were common among workers and in the middle class. In 1973 the protest was brought to a head, and the *military revolted.* Allende was killed, and a military government was formed to rule the nation. Today, Chile is a politically *divided nation with a disrupted economy.*

Argentina

Argentina, in its early independence period, had to *struggle for national identity.* In colonial times Argentina lay on the fringe of the Spanish empire in Latin America. The most developed areas were along the Andean flanks in the west and northwest, and commerce was directed through Lima. The small fort at Buenos Aires and the surrounding scattered cattle and horse *estáncias* were but outposts to defend against Portuguese penetration southward. With independence came a period of *internal conflict,* and a *reorientation* from the west to the east began.

Argentina's *modern development* began about 1860 with the *settlement of the Pampas,* the

Copper is an important export of Chile. Developed by foreign (largely United States) corporations, the copper industry was nationalized in the 1970s. Here copper ingots are being processed for both export and local needs. (United Press International Photo)

broad, flat, and fertile grasslands surrounding Buenos Aires. Before 1860 the Pampas had been divided into *large ranches* devoted to cattle and horse raising. The cattle were valuable only for hides and tallow, since there was no nearby market for beef. In many respects the Pampas was like the American West with cowboys (gauchos) and nomadic Plains Indians. About 1860 *immigrants* from Italy, Spain, Germany, Switzerland, Britain, and other European countries began coming to the Pampas in large numbers; by 1930 some 6.3 million immigrants had come to Argentina, largely through Buenos Aires. Some stayed in Buenos Aires, but many went into the country where they worked as tenants on the estáncias. The normal *owner-tenant agreement* was one of work for land. The tenant raised alfalfa, built barbed-wire fences, and sank water wells powered by windmills; in exchange he could cultivate some of the estáncia's land on which he raised cereal grains, especially wheat, and garden crops for his family's needs. After cultivating the land for five to seven years, the tenant planted alfalfa or pasture grasses and moved on to another piece of land. Under this system many tenants were able to *accumulate wealth,* which they used to buy land. In time the Pampas came to be one of the most agriculturally productive areas in the world, and by the early 1900s, the value of crops exceeded livestock. Livestock, however, remains very important, and Argentine beef is exported around the world.

Today, Argentina is one of the most *developed nations* in Latin America. Its per capita GNP rivals that of the rich world. Its population is both *literate* (94%) and *urban* (86%), and population growth is a moderate 1.5%. The nation's 29.1 million people are European and have a well-developed sense of national unity. The recent Falkland (Malvinas) Island war with the United Kingdom over ownership of the islands is a reflection of this nationalistic identity. *Political strife* is now rampant, however, over the direction of the nation's future. Buenos Aires and the Pampas form the country's *heartland* and are by far the nation's most important area. Greater Buenos Aires (Buenos Aires and its satellite cities) has a population of 8.7 million people, making it one of the largest urbanized areas in the world. A well-integrated *road and railway system* radiates outward from the city, and goods funnel through the city's port destined for the world market. From the *Pampas* come beef, maize, and wheat, which are exported largely to the developed world. *Imports* are mainly manufactured goods from the United States and Western Europe. The development of an internal transport network, steam-powered ocean vessels, and refrigeration were vitally important to the economic advancement of Argentina and particularly the Pampas. Without these *innovations,* the export sector of Argentina's economy would be severely hampered.

Most of Argentina's *manufacturing and service activities* are in Buenos Aires; some 45% of the nation's industrial labor force is located there and about 50% of all those engaged in tertiary activities. From 1946 to 1955 the government regulated and redirected the national economy toward manufacturing under a broad *policy of creating national self-sufficiency.* Metal fabricating, petrochemicals, and plastics came to vie with the more traditional agricultural processing industries of meatpacking, milling, and textiles. Government investments in manufacturing, an increasing bureaucracy, and numerous social projects such as housing and water supply were paid for by taxes on the agricultural sectors. Eventually the tax burden became too great, and the entire *economy foundered.*

Several *other regions* provide products that are *complementary* to the economy of the Pampas. In *the west* near the Andes the old cities of Córdoba, Mendoza, and Tucumán have become important growth centers. The west, a dry area, has an agricultural base with the dry lands devoted to cattle raising and irrigated zones to sugarcane and grapes for the national market. In the extreme *northwest* are a few small mines of asbestos and several metals. Some petroleum is produced from a giant structural trough located just east of the Andes. To the north of the Pampas is a low, humid area between the Paraná and Uruguay rivers. The location of this *area between rivers* has created problems of access, particularly since the rivers are wide and subject to great variation in water flow. The Río de la Plata watershed development program in which Argentina, Paraguay, Uruguay, and Brazil are participants may make the Argentine "Mesopotamia" a more productive area, but already the area is noted for its tea, maize, flax, cattle, and sheep. In the south is *Patagonia,* a sparsely populated, dry, windswept plateau, most of which is devoted to sheep raising, but in the lower, well-watered valleys are irrigated alfalfa fields, cattle raising, vineyards, and mid-latitude fruit. In this century some mineral resources have been exploited. Petroleum is found in a trough to the east of the Andes and in the south near the coast. Also in the south near the coast are small quantities of coal and iron ore that are shipped to Buenos Aires for iron and steel processing.

Argentina's outlying regions are *primarily product suppliers* for the Pampas and its center, Buenos Aires. Much of the Pampas' population is supported by these regions, and their development is reflected in the improved well-being of the nation's heartland.

BRAZIL

Brazil is by far the largest of all Latin American nations. Indeed, it is the fifth largest in the world in terms of area and *sixth in population*. Of the seven most populated countries, Brazil's rate of population growth, 2.3%, is the highest. If the present rate continues to the year 2000, Brazil's population will increase to 187 million.

Brazil's *large area* is underlain by a diversity of geologic formations. Some of these structures contain rich and extensive *ore deposits*. The early Portuguese settlers discovered gold and diamonds and established mines to extract them. The more important minerals for manufacturing such as iron ore, bauxite, and ferroalloys are only now being extensively developed. It appears that Brazil holds vast quantities of many minerals.

There are several different *physical environments* within which many different crops are grown. In the north is the Amazon Basin with its rainy tropical climate and luxuriant rain forest. Just to the south is a large area of tropical wet and dry climate. Some of the wet and dry area is located at elevations above 2000 feet (610 meters) with moderate temperatures that permit a combination of mid-latitude and tropical crops to grow. Only the eastern part of the tropical wet and dry area is extensively used; the western part, like the Amazon Basin, is beyond the agricultural frontier. Southern Brazil has a mid-latitude environment and is almost fully settled; mid-latitude crops and animals dominate the agricultural scene.

Sleeping Giant

Brazil has been called the *sleeping giant,* for in spite of its size and resource base, it has never attained the status of a world power. Apparently the giant is now stirring; not only is the population growing rapidly, but so is the economy. The total GNP is growing at the rate of 8% yearly, the per capita at 5.6%. *Industrialization* has contributed much to the nation's economic growth; manufacturing accounts for 29% of the GNP and is growing at the rate of nearly 8% yearly. Brazil's *industry is diverse* and has many branches that use high levels of technology and automation. Examples of modern factory products include electrical equipment such as motors and generators, transport equipment such as cars, trucks, and ships, and iron and steel production. Iron and steel, the basis of modern manufacturing, have grown rapidly; steel production in 1970 was 5.4 million tons and is expected soon to climb to 20 million tons.

Brazil is *increasingly urbanized.* At present, 67% of the people are urban dwellers. Six urban areas have more than 1 million inhabitants: São Paulo (8.4 million), Rio de Janeiro (7.3 million), Belo Horizonte (1.3 million), Recife (1.2 million), Salvador (1.2 million), and Pôrto Alegre (1.2 million). There are many other cities with more than 100,000 people. The growth rate for most urban centers is between 4.5% and 6.0%; Brasília, the new capital, is growing at the rate of 10%. With urbanization have come the problems of traffic congestion, deficiencies in public services, and air pollution.

The *rural sector* of the economy has grown at a slower pace, 3–5% yearly. About 40% of the population is engaged in agriculture and supplies low-cost foodstuffs and industrial raw materials. Moreover, it is agriculture that provides the bulk of the exports: coffee, cotton, and sugar. Manufactured goods and minerals account for only 17% of all exports. The rural sector contributes 20–25% of the GNP and is receiving much attention in an attempt to improve production rates.

One example of Brazil's drive toward development is its *push to occupy its territory effectively* and to tap an enlarged resource base. In 1960 the national capital was moved from Rio de Janeiro to Brasília on the agricultural frontier in the Brazilian Highlands. This move stimulated the western push of settlement. More recently has come the planned occupation of the Amazon Basin with an extensive road-building program. Road building is opening both the highlands and Amazon Basin to an influx of settlers.

Boom and Bust Cycles

One reason for Brazil's limited development has been the *exploitative nature* of the economy. From Portuguese colonization to the present, emphasis has been on derivation of maximum wealth in minimal time without regard to long-term, stable growth. Several cycles of *boom and bust economic activity* have been experienced.

Sugarcane in the Northeast was the first commodity of importance. Large plantations with African slaves yielded great profits to the owners in the sixteenth and seventeenth centuries. For a long time, the Northeast was Brazil's most important region. The sugar industry eventually collapsed under stiff competition from other parts of Latin America, accelerated by the inability or lack of initiative to apply improved levels of technology used elsewhere. The second cycle was the push inland to the Brazilian Highlands north of Rio de Janeiro where *gold and diamond deposits* were discovered. The exploitation of these minerals led to partial settlement in the interior. Moreover, the discoveries encouraged other colonists to come to Brazil. Yet after the many surface ores were removed, there was an exodus from the area.

The third cycle occurred after 1850 with the search for areas suitable for *coffee cultivation.* It is around São Paulo that coffee production reached its zenith; it is there too that the famous *terra roxa* soil exists that is ideal for coffee. Until early in the twentieth century, coffee commanded a high price, and its cultivation was expanded into areas environmentally marginal for its growth. When prices dropped, these areas were abandoned.

Coffee culture put Brazil ''on the map,'' and for many years Brazil exported more coffee than all other nations combined. By the 1910s, world production exceeded demand, and prices rapidly dropped. The Brazilian government tried to protect its most important export by buying the harvest and holding back coffee from the market until an acceptable price was attained. Other Latin American nations, notably Colombia, undercut Brazil and took a large part of the market. Since then, Brazil has repeated its attempt to *control coffee prices* but each time eventually has lost out to other countries. In recent years other coffee-producing nations have joined Brazil in trying to limit production and control the price. Such endeavors, called *valorization,* can only work when producing nations are in full cooperation.

The rise of coffee in Brazil illustrates a method of land development that requires little money and is similar to the *tenant-owner agreement* used in the Argentine Pampas. When coffee production began its great expansion, thousands of immigrants from Europe entered Brazil; most went to the southern half of the country as tenants on large *fazendas* (farms) whose owners were rich in land but poor in money. Between the owner and tenant, a working agreement was made that involved little or no cash. The tenant leased a portion of the *fazenda* (often uncleared land) on which he planted coffee bushes provided by the owner. For a period of years the tenant tended the young plants and cultivated his own crops, both subsistence and commercial, between rows of coffee. After five to seven years, the coffee bushes began to bear fruit, and the tenant moved on to another plot. In this manner an entire fazenda could be put to coffee with little capital; once harvesting began, cash-wage employment was available to the tenant and his family.

Land of Contrasts

Today, Brazil is a *nation of contrasts.* The western half of the country is sparsely populated, but the east and along the coast are densely settled. There is both a traditional and modern Brazil. *Traditional Brazil* is the rural and agricultural sector with the controlling landowner group of European extraction and the workers of European, African, and Indian stock. The workers are poor, often illiterate, and cultivate their small farms or work on large fazendas using hand tools. *Modern Brazil* is urban and industrialized. Living levels are high, literacy is nearly universal, and machines have replaced muscles. In the south, modern Brazil includes mechanized grain-farming areas. Almost all modern Brazilians are of European origin. The difference between traditional and modern Brazil is striking.

Regions of Brazil

A third contrast is *regional difference.* There are significant variations in population and in economic activity among the six regions of Brazil shown in Figure 18–2.

São Paulo

The state of *São Paulo* is Brazil's *most modern and productive region.* Per capita income is the national average, and nearly a third of the national GNP is produced there. São Paulo accounts for almost two-thirds of the nation's *industrial* output, producing practically all of Brazil's motor vehicles and leading the nation in the manufacture of textiles, cement, shoes, paper products, soluble coffee, pharmaceuticals, and electrical goods. Petroleum

FIGURE 18–2 Brazil's Regions.
Brazil is one of the largest nations in the world. The regions along the coast are largely settled, but the interior regions (Central West and North) are part of Brazil's frontier. Ambitious projects are under way to develop these interior regions.

refining, petrochemicals, and steel production are important.

São Paulo state is also the leading region in *agricultural production*. Coffee, beef, sugar, cotton, peanuts, truck crops, and rice are the principal farm products. Tractors and fertilizers are widely used.

Urbanization has progressed rapidly. The city of greater São Paulo has grown from a backland trade center to one of the world's largest urban areas. Growth has been rapid, particularly in the last half century; its population increased from 600,000 in 1935 to 4.5 million in 1960 and 8.1 million in 1970, and it is projected to be more than 24 million by 2000. São Paulo city alone accounts for 55% of the nation's manufacturing with almost 25,000 factories. It is the nation's leading *financial center* with the largest number of banks and the biggest stock exchange. Rapid growth has not come without problems. Many *migrants* from the countryside and other regions have been attracted to the city; most are illiterate and unskilled and live in great poverty. Only a small part of this group has access to the city's water supply and sewage network. Transport, housing, public health, and schools are gen-

erally inadequate. The World Health Organization considers São Paulo's *air pollution* level near the danger point. Other cities, including the port of Santos, have experienced similar growth and accompanying problems.

The South

Southern Brazil has its own *distinctive flavor*. In the nineteenth century large numbers of *German and Italian farmers* settled in the area, and many retained their national identity well into the twentieth century. German and Italian architecture is still noticeable. *Agriculture* is the basis of livelihood, but several cities of over 100,000, and growing at rates of about 4% yearly, are located on or near the coast; Pôrto Alegre (1.2 million) and Curitiba (600,000) are the largest. Most of the region falls within the *well-settled* part of Brazil, but the western section is still part of a *pioneer-fringe*. Transport facilities have been greatly expanded and have contributed much to the region's prosperity. Exploitative agriculture is less prevalent in the South, perhaps because of the migrant farmers who have settled there. Cattle ranching and *mechanized wheat farming* characterize part of the agricultural zone. Coffee raising reaches its southern physical limit in this region, since freezing temperatures penetrate from the south almost to the northern limit of the region. Rice, maize, and hogs are raised in the more traditional manner; hogs are associated especially with German settlements. Some Italian communities are noted for their grapes and wines.

Urban areas are basically regional service centers, but in the larger cities, such as Pôrto Alegre, industrialization based on processing farm and ranch products provides employment to a sizable part of the labor force. Milling, meat packing, tanning, textiles, and breweries are typical types of manufacturing.

Mineral resources of the South are limited. Small deposits of low-grade coal are mined to provide energy in the region and for shipment to the East and to São Paulo.

The East

The East is a region of *contrast between modern and traditional Brazil*. Large modern urban centers are in stark contrast to some of the nearby rural areas little changed in the last hundred years. The East is basically the *hinterland (tributary area)* of Rio

277

de Janeiro, the former capital and second city of the nation. The region has experienced several *waves of exploitation,* first sugarcane, then gold and diamonds, coffee, rice, and citrus fruit. Following each boom period was a time of population decrease and reversion to a grazing economy.

Rio de Janeiro is the focal point of the East. Rio was founded early in the colonization movement as a defensive position, and its excellent natural harbor and location near the inland gold fields gave the city an early advantage. When the capital was moved from the Northeast to Rio de Janeiro, its functions increased. As long as Brazil was a coastal nation, Rio continued to be a national focal point. Access to the interior, however, was difficult, since just behind the narrow coastal plain is a steep escarpment. The advent of the motor vehicle and paved roads has partly offset this natural disadvantage, but other cities located inland have taken some of Rio's former trade area. Rio itself faces a severe shortage of level land suitable for urban development.

The hinterland of Rio de Janeiro has both an *agricultural and mining base.* Most of the land is used for grazing; cropland is devoted to the cash crops of sugarcane, coffee, rice, and citrus, but much land is used for subsistence crops. *Land rotation* is a common farming practice; after cropping a plot for a few years, soil erosion and depletion make further cultivation uneconomical. The land is then allowed to ''rest,'' and another plot is put to crops. Part of the East is one of the most mineralized areas in the world, the *Mineral Triangle.* Only within this century has large-scale mining been developed, and further development is in the offing. Gold, diamonds, and a number of precious and semiprecious gems have been mined for centuries. *Industrial minerals* such as manganese, chromium, molybdenum, nickel, and tungsten are of growing importance. *Iron ore,* however, is the most important mineral. The amount of iron ore reserves is unknown but is certainly one of the largest in the world. A conservative estimate is 15 billion tons of easily recoverable high-grade ore.

Volta Redonda is Brazil's iron- and steel-making center. Located close to the iron ore with nearby limestone and water and the São Paulo and Rio de Janiero markets, the site is almost ideal. Coal, the one vital ingredient that is lacking, is brought in from the south and imported from the United States. The Volta Redonda plants began production in 1946, and since then have increased in capacity several times. Today, Brazil is the leading steel producer in South America and is expected to become the world's tenth-largest steel maker.

The Northeast

The *Northeast* was once the center of Brazilian culture, the location of the capital, and the most developed part of the nation. This favored position was based on sugarcane cultivation under the plantation system with African slaves and on the Northeast's location as the nearest part of Brazil to Portugal. Proximity was advantageous during the days of sailing ships but lost importance with the advent of steamships. Sugarcane is still grown in the region, but it is no longer a prosperous industry. Today, the Northeast is *poverty-stricken,* and per capita income is less than half the national figure. Thousands of people have migrated to other parts of Brazil, adding to the number of unskilled laborers searching for work in the cities farther south and in the new capital of Brasília.

Population pressure is great and livelihood precarious even in the best of times. Over the years, settlement has moved inland away from the moist coast into a drier environment. In the backlands, hillsides are cultivated with *minimum regard for conservation,* and pastures are overgrazed. Both practices have resulted in soil destruction and rapid water runoff from the land. The naturally *semiarid environment* has become still more moisture deficient because of these human actions. Droughts are frequent, and when they occur, there is a large movement of people from the interior to the already overcrowded coast and to other parts of the nation. Yet so strong is the tie to the Northeast that many persons forced to leave return when they can.

The Northeast has two cities with more than 1 million people, Recife and Salvador, and one other with more than 500,000, Fortaleza. These port cities function as service centers. *Manufacturing is limited.*

Since about 1960 a number of government-financed projects have been directed to the Northeast, and a regional development corporation has been established. Dams for hydroelectric power and irrigation are being built and roads improved and extended. Industries locating in the Northeast receive special tax rebates. None of these efforts have yet become effective.

Central West

The *Central West* region is part of Brazil's *Sertão,* the backlands where a frontier spirit and life prevail. Many parts of the region remain essentially unknown, most is *sparsely settled,* and much remains outside effective control of the government. The new capital of *Brasília* was constructed on the eastern edge of the Central West to symbolize and encourage the occupation of the backlands. Roads have been built to connect Brasília with other parts of the nation and westward to new lands. Along the eastern fringe of the region, some commercial agriculture has evolved, but elsewhere subsistence agriculture, grazing, small-scale lumbering, and mining are carried on.

In the *pioneer zone,* the frontier is gradually moving westward as new lands are opened for agriculture. Brazil believes its future lies in the west. Pioneering is at best a risky venture, but the farmers have discovered a *relationship* between vegetation and quality of land for cropping. The existence of this relationship is especially important, since there are few weather stations and soil surveys to provide more scientific data. Six vegetative types are recognized. In descending order of quality for agriculture, they are as follows:

1 Mata da primeira classe (first-class forest).
2 Mata da segunda classe (second-class forest).
3 Campo cerrado (mixed grassland and woodland).
4 Cerradão (scrub woodland).
5 Campo sujo (grassland with scattered trees).
6 Campo limpo (grassland).

Mata vegetation is regarded as good-quality land suitable for cropping. *Campo sujo, campo limpo,* and *cerradão* are considered generally unusable for crops, although mechanized cropping might be feasible. Generally, however, these types of vegetation are used only for grazing. Use of areas covered by *campo cerrado vegetation holds the key* to occupation of the Central West, since this vegetative type covers about 75% of the region. *Campo cerrado* soils are droughty and lacking in plant nutrients. Extensive mechanized agriculture emphasizing wheat and soybeans has had some success in the eastern part, and alfalfa, a deep-rooted plant, offers another possibility. The main use of *campo cerrado* areas, however, is for grazing.

The Amazon Basin

In recent years much effort has been exerted in taming the *Amazon Basin.* Although the basin has long been considered an area of great potential and occupies more than 40% of the nation's area, it has only 4% of the population. Only along the Amazon River and its principal tributaries have people of European culture been able to gain a foothold. Manaus (400,000) near the center of the basin and Belém (700,000) near the Amazon mouth are the major urban places. Away from the rivers live an unknown number of Indians who, although touched by Western society, still cling to their traditional ways; they gain their livelihood by shifting subsistence agriculture, emphasizing manioc, maize, and beans, and supplement their diet by hunting, fishing, and gathering.

In the nineteenth century, many parts of the Amazon were explored for *rubber* trees (*Hevea brasiliensis*), the only significant source of natural rubber in the world. Tappers extracted the latex, smoked it into balls, and sold the balls to traders

The effective settlement and use of the Amazon Basin have long been a dream of many Brazilians. This dream is now being realized through costly and ambitious governmental development programs. The key to these programs is the Trans-Amazon Highway shown here. (United Press International Photo)

279

for shipment to the United States and Europe. Rubber first gained importance with the development of the *vulcanization process,* but the demand greatly increased in the latter part of the century with the pneumatic bicycle tire and shortly afterward the automobile. Rubber prices skyrocketed, but still rubber was gathered only from the wild tree of Brazil. Finally, some seeds were smuggled to England, and these seeds became the foundation of the rubber plantations of Malaya, Sumatra, and elsewhere. Once plantation rubber came on the market, the price dropped, and wild-rubber gathering greatly decreased in importance.

Starting about 1950, two small attempts at taming part of the Amazon Basin have met with some success. The first has been the introduction of *Japanese farmers* who have been able to produce rice, jute (a hard fiber), and black pepper on a commercial basis. The second has been the *mining* of manganese ore in the extreme northeastern part of the basin. Many other minerals are known to exist in the basin but have yet to be mined. Of these, the search for petroleum has been the most extensive. Although oil has been discovered, deposits extensive enough for development have not been found.

More recently the Brazilian government has initiated a most ambitious development program for the Amazon Basin. In 1970 work on the *Trans-Amazon Highway* began and heralded the start of a *massive road system* throughout the basin and connection of the region to other parts of the nation (Figure 18–1). With the development of the road system comes the expectation, already partly realized, that migrants from the Northeast and elsewhere will settle the area. Plans call for government-sponsored construction of urban areas, lumber and pulp plants, hydroelectric plants, and schools. Governmental officials and most Brazilians look upon the Amazon Basin *development program* as vitally necessary to assure continued economic growth of the nation and a better life for its citizens.

This view, however, has been challenged by some ecologists who maintain that although short-term economic growth may result, *disastrous results* may occur over a long time. Many ecologists believe that destruction of the tropical forest may substantially change the area's climate, that drier and warmer conditions will prevail. Some estimates are that the forest could be cleared in less than eighty years. They also point out that many of the soils of the basin are low in plant nutrients and require careful handling to prevent their destruction. Whether or not the ecologists are right, Brazil is committed to Amazon Basin development.

FURTHER READINGS

The *ejido* system of Mexico has received much attention. A recent study is Elinore M. Barrett, ''Colonization of the Santo Domingo Valley,'' *Annals, Association of American Geographers,* 64 (1974), 34–53.

A number of books provide a comprehensive geographical survey of Latin America, for example: Harold Blakemore and Clifford T. Smith (eds.), *Latin America: Geographical Perspectives* (London: Methuen, 1974); Alice Taylor (ed.), *Focus on South America* (New York: Praeger, 1973); and Robert West and John P. Augelli, *Middle America: Its Land and Peoples,* 2nd ed. (Englewood Cliffs, N.J.: Prentice-Hall, 1976).

An interesting and informative account of the flow of crops, animals, and disease into and out of Latin America and the effects of the exchange is Alfred W. Crosby, Jr., *The Columbian Exchange: Biological and Cultural Consequences of 1492* (Westport, Conn.: Greenwood Press, 1972).

The Brazilian attempts to develop the Amazon Basin have received much attention. Two short but useful articles on this subject are William N. Denevan, ''Development and Imminent Demise of the Amazon Rain Forest,'' *Professional Geographer,* 25 (1973), 130–35, and Martin T. Katzman, ''Paradoxes of Amazonian Development in a Resource-Starved World,'' *Journal of Developing Areas,* 10 (1976), 445–60. Nigel J. H. Smith, *Rainforest Corridors: The Trans-Amazon Colonization Scheme* (Berkeley: University of California Press, 1982), provides a recent assessment of tropical rainforest use.

Statistical data on Latin American nations covering such subjects as demography, labor, health, education, economic activity, foreign trade, and national accounts are readily available in Latin American Center, *Statistical Abstract of Latin America* (Los Angeles: University of California, annual). The role of foreign investment in Latin America and the current status of multinational corporations in Latin America is reviewed by Paul E. Sigmund, *Multinationals in Latin America: The Politics of Nationalism* (Madison: University of Wisconsin Press, 1980).

The colonial antecedents of Latin American land use are described by David L. Clausen and Raymond E. Crist, ''Evolution of Land-Use Patterns and Agricultural

280

Systems,'' *Mountain Research and Development, 2* (1982), 265–72.

The problems of adopting new techniques are illustrated by David L. Clausen and Don R. Hoy, ''Nealtican: A Mexican Community that the Green Revolution Bypassed,'' *American Journal of Economics and Sociology,* 38 (1979), 371–87.

The effects of oil development in Mexico are analyzed by Jesus A. Velasco, *Impacts of Mexican Oil Policy on Economic and Political Development* (Lexington, Mass.: Lexington Books, 1982).

Shantytowns are a familiar sight in urban Latin America. These settlements are examined by Susan Lobo, *A House of My Own: Social Organization in the Squatter Settlements of Lima, Peru* (Tucson: University of Arizona Press, 1982).

VII

Africa South of the Sahara

Leonard Berry/
Douglas L. Johnson

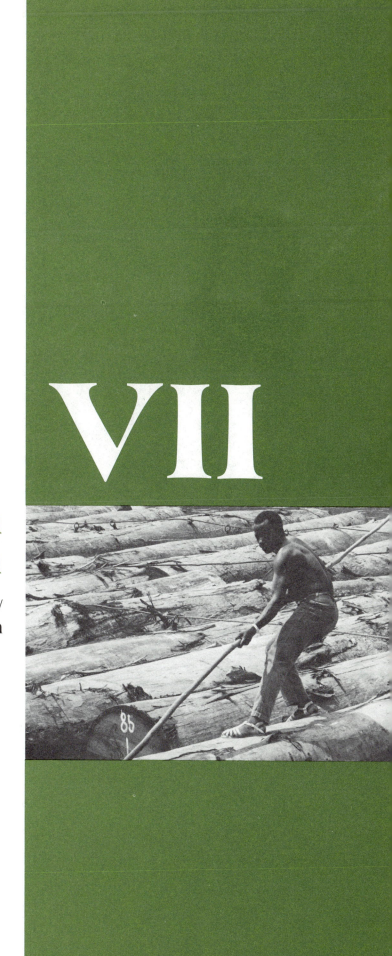

19

Africa:
East and West

AFRICA SOUTH OF THE SAHARA: AN OVERVIEW

Africa south of the Sahara, or Subsaharan Africa, is a culture world dominated by black African culture and political systems (Figure 19–1). To this broad world region with its distinctive cultures has been added the influence of European settlement and colonialism. Although political colonialism is now a thing of the past, its imprint persists in many ways, including European trading partners and European languages. The northern tier of African states south of the Sahara has long witnessed a strong Arab influence. Here black African culture and Arab culture have mixed and fused over a broad transition zone extending from the middle of the Saharan Desert to the more humid areas along the desert's southern border. For this reason the boundary between Arab North Africa and black Subsaharan Africa is drawn through the Sahara. Countries to the north of the boundary focus on the Mediterranean Sea and are clearly Arab. Countries south of the boundary exhibit both an Arab and black heritage, and contact is generally to the south.

Africa south of the Sahara is one of the poorest parts of the world. Two-thirds of the least developed nations (a United Nations designation for nations with a very low GNP per capita) are found here. The region is rich in cultural and historical tradition, although diversity is also a trademark of the cultural scene.

Economic well-being, environment, mineral wealth, religion and culture, and patterns of colonial and postcolonial change are the most important factors. Oil and other minerals are important to the Union of South Africa, Zimbabwe, Nigeria, and Gabon. South Africa is the only country in the region with a fully developed industrial and service sector, although Zimbabwe and Nigeria have well-diversified economies. Environment in the area ranges from little-disturbed tropical rain forest occupying much of central Zaire to the wide ex-

FIGURE 19–1 Africa South of the Sahara: General Locations and Subregions. Africa south of the Sahara comprises many nations with diverse cultures. Traditional and modern cultural and economic systems are often found within a small area.

panses of very dry desert forming most of northern Mali and Sudan. The diversity of livelihood systems of the region reflects the Arab, African, and European culture systems which interact here.

Finally, most of the countries of the region have gained independence since 1960. They bring into their new political states colonial heritages of administration, language, and external linkage that have complicated their development and their interaction with each other.

This chapter deals with the diverse range of states in East and West Africa. Chapter 20 will focus on the central and southern African states. No sharp dividing lines separates these two areas, but a range of historical, political, and economic circumstances make for a convenient division.

AFRICA EAST AND WEST

The countries of East and West Africa (Figure 19–2) fall into *two major groups*. Some are members of the world's *least developed* nations. Others are in a middle group of nations that are poor but have a *more diversified resource base* and have prospects for a higher rate of economic growth. This chapter will look first at the problems of the least developed countries of East and West Africa. It will then turn its attention to nations at an intermediate stage of economic development, as well as those few states with sufficiently diversified resources to possess prospects for rapid economic growth.

The Least Developed Countries

The least developed nations are characterized by smallness in area or population, illiteracy, limited wealth, significant environmental problems, a low percentage of their labor force in manufacturing, a colonial heritage, and, in many cases, isolation. Although lower per capita GNP is a distinctive characteristic of all least developed countries, the causes of limited economic development levels are varied. East and West Africa have more nations classified as least developed than any other part of the world. These nations are Benin, Burundi, Cape Verde, Chad, Djibouti, Ethiopia, Gambia, Guinea-Bissau, Mali, Niger, Rwanda, Sierra Leone, Somalia, Togo, Uganda, and Upper Volta.

These countries are divided into three regional groups: a West African Sahelian group that includes Mali, Niger, Chad, Upper Volta, and Cape Verde; a West African humid area group comprising Benin, Sierra Leone, Gambia, Guinea-Bissau, and Togo; and a more diverse East African group composed of Djibouti, Ethiopia, Uganda, Somalia, Rwanda, and Burundi (Figure 19–3). We shall begin by discussing themes common to all least developed countries, and then explore the specific differences that characterize each regional group.

Environment

Most of the least developed nations are vulnerable to fluctuating weather conditions. These fluctuations make for uncertainty in terms of both basic food supplies and export earnings. Since agricultural products account for a large proportion of the total economy, *weather fluctuations* can have a significant impact on the overall economic situation. For example, during the Sahelian drought of the 1970s, the GNP of Mali dropped considerably.

Environment-related problems such as the lack of good drinking water, insect infestation, and disease also take their toll on the health and vigor of the inhabitants. Although malaria is the most widely known of tropical diseases, yellow fever, schistosomiasis (bilharzia), and onchocerciases (river blindness) are the most prevalent in Africa (Figure 19–4). Animal diseases, such as infections carried by the tsetse fly and east coast fever, cause large areas of land to be removed from certain kinds of economic activity.

Among environmental problems of the least developed countries is the absence of ecosystem diversity. *Uniform environmental conditions* often result in a narrow range of crops and other agricultural activities. Of the countries considered in this section only Ethiopia shows a wide range of diversity of ecosystems. In Ethiopia the great altitudinal differences between one part of the country and another result in major differences in temperature, precipitation, soils, and vegetation. There are thus a variety of natural ecosystems and a considerable range in crops and land potential. Despite this diversity, Ethiopia's social, economic, and ecological problems have kept per capita GNP on a par with the poorest of the least developed nations.

The other countries are characterized by much more homogeneous environmental systems, mainly of two types: (1) arid or semiarid and (2) humid. Both of these ecosystem types have advantages and disadvantages. When a whole country is occupied mainly by one ecosystem, however, the lack of diversity places severe restraints on the

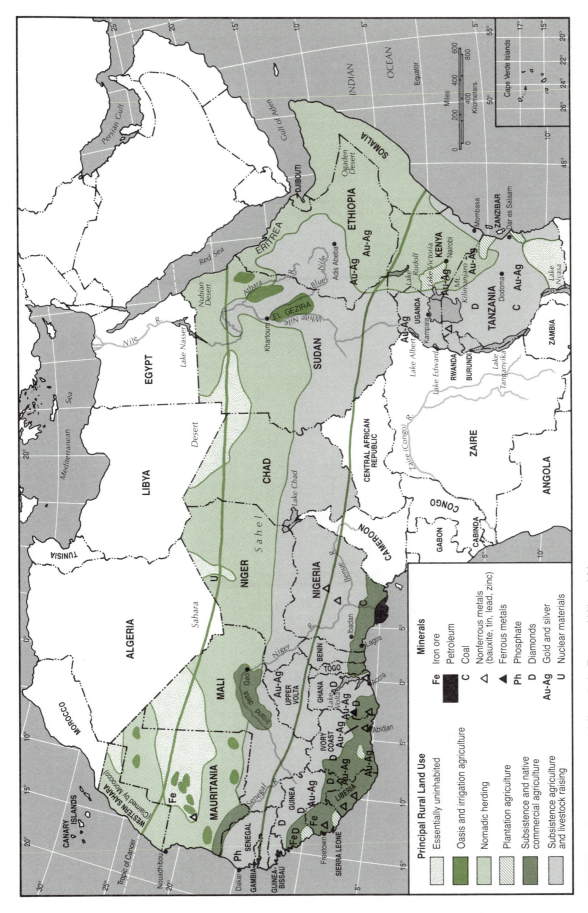

FIGURE 19–2 Land Use and Minerals in East and West Africa.
Many different forms of land use characterize East and West Africa ranging from traditional nomadic herding to modern plantations. A number of different minerals, such as petroleum in Nigeria, are important exports.

Principal Rural Land Use

- Essentially uninhabited
- Oasis and irrigation agriculture
- Nomadic herding
- Plantation agriculture
- Subsistence and native commercial agriculture
- Subsistence agriculture and livestock raising

Minerals

Fe	Iron ore
C	Petroleum
△	Nonferrous metals (bauxite, tin, lead, zinc)
▲	Ferrous metals
Ph	Phosphate
D	Diamonds
Au-Ag	Gold and silver
U	Nuclear materials

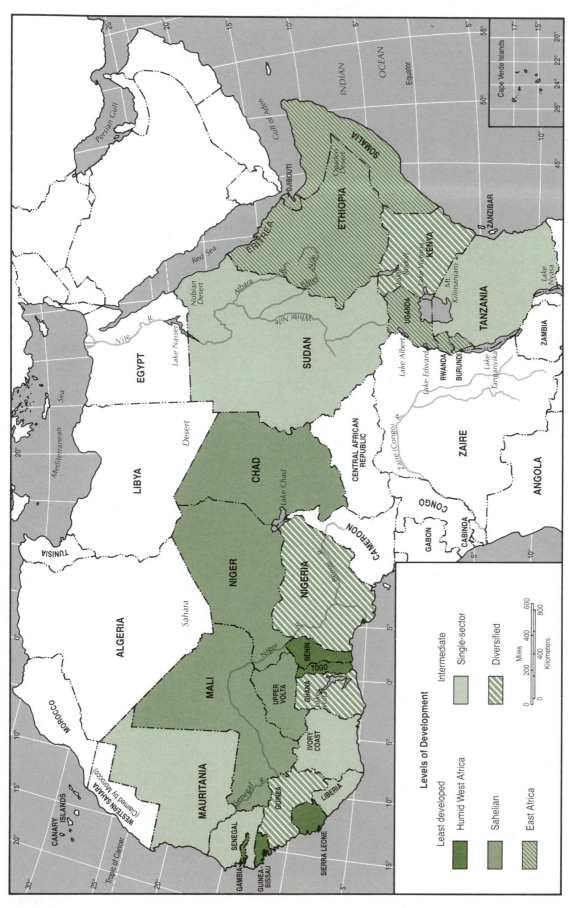

FIGURE 19–3 Africa East and West: Levels of Development.
Many East and West African countries are among the poorest in the world, yet others have important actual or potential growth sectors in which development is possible.

Levels of Development

Least developed
- Humid West Africa
- Sahelian
- East Africa

Intermediate
- Single-sector
- Diversified

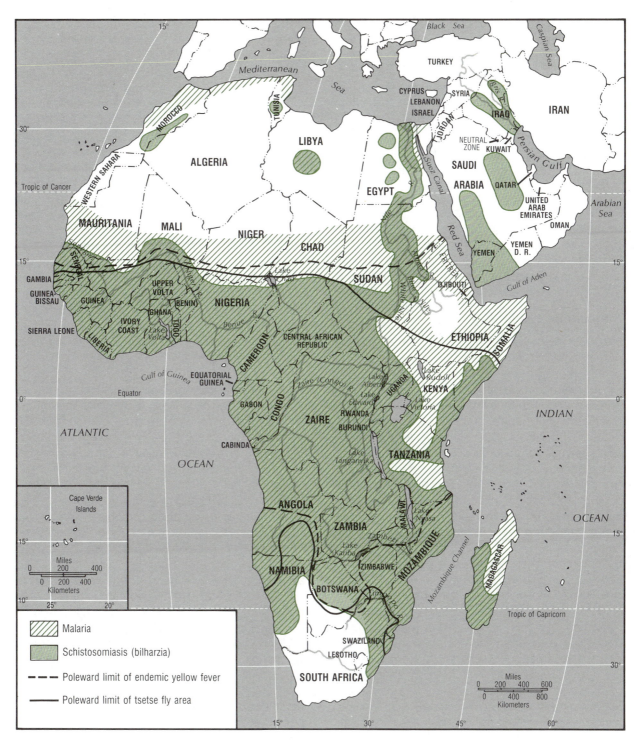

FIGURE 19–4 Disease-Infested Areas in Africa.

Disease is a major inhibitor of development in some parts of Africa. Malaria, schistosomiasis, and yellow fever are endemic over much of the area. The tsetse fly carries trypanosomes that infect both people (sleeping sickness) and livestock.

SOURCES: Ernst Rodenwaldt, *World Atlas of Epidemic Diseases* (Hamburg: Fale, 1952–61); and *Atlas of the Distribution of Disease* (New York: American Geographical Society, 1950–55).

range of agricultural and pastoral life-support systems that can be economically established.

The problems of single-ecosystem countries have been dramatically highlighted when those ecosystems are placed under stress—for example, during the recent *Sahelian drought.* The whole of the countries of Niger, Mali, and Chad were affected by drought, and international help was essential to the survival of many of their people and for support of the economy. Neighboring Nigeria and, to a lesser extent, Upper Volta were only affected in part by the drought, and resources were reallocated from the unaffected areas of these countries. Nigeria was particularly fortunate because its humid coastal zones escaped the drought. Niger was less environmentally diverse and suffered disproportionately.

Subsistence Activities

The least developed nations are grouped together primarily because of their *low levels of economic development.* There are common characteristics of *poverty* and *simple economic structures.* The economic systems of these countries have a high proportion of their population involved in subsistence farming. They also frequently contain a small enclave of more advanced agricultural activities in the form of foreign-owned farms, state farms, or plantations, and they may have limited industrial sectors. In addition, relatively low levels of international trade reduce the potential for additional income. There are few societies anywhere that are completely removed from the cash exchange economy, yet there are many countries in which a large portion of the inhabitants work first to feed themselves and their families and only then to earn cash for the extras in their lives.

Goods that are sold—cattle, maize, cotton, peanuts, and the like—find their way into local and perhaps national and international markets. If the crop is grown specifically for export, government or cooperative marketing arrangements are made for the orderly movement of goods from local to national markets. However, for small peasant farmers located far from good communications or population centers, selling their crops is almost as big a problem as growing them.

Modern Agricultural Enclaves

The development of *cash crops* was encouraged by the colonial powers mainly to serve overseas markets. Two main methods were utilized. The first was *native smallholder production* of such crops as peanuts, cotton, and cacao. The second was the establishment of *plantation* enterprises both for these crops and others such as rubber, pineapple, and tea that were thought to need large-scale investment and experienced management. In some countries, expatriate farmers were encouraged to set up private plantations. In others, governments initiated state farms or joint ventures with overseas companies. In semiarid areas attempts have been made to introduce *large-scale ranching* to indicate possibilities for future animal husbandry in these areas. Also involved, however, is the separation of much land from the traditional pastoral economy, the investment of considerable capital, including imported animal stock, and considerable management skill.

The Ivory Coast is a good example of a country that has been able to rise above least developed nation status on the basis of its agricultural development. Indeed, until recently the Ivory Coast has been something of an economic miracle. For the decade 1965–75, an economic growth rate of more than 7% per year gave the Ivory Coast an economy more expansive than many in Europe. Nearly 1.5 million workers from the surrounding countries have flocked there to share in the boom. Perhaps the most remarkable feature of Ivory Coast development is its nearly total reliance on growth in the economy's agricultural sector. Coffee, timber, and cacao account for 80% of all exports, with such specialty crops as pineapples, peanuts, sugarcane, bananas, and rubber comprising most of the remaining earnings. Unlike the small industrial sector where management and ownership are concentrated in French hands, most of the Ivory Coast's cash crops are produced by peasant farmers. This widely diffused growth plus great political stability have created a climate conducive to large amounts of foreign investment to complement government expenditures on ports, roads, and other infrastructural improvements. Although the immediate future of the Ivory Coast's economy appears bright, there are problems that have significantly slowed its rate of growth, especially the country's vulnerability to price fluctuations on the world market, deteriorating commodity prices, and the global economic recession of the early 1980s.

Industrial Sector

In the least developed countries, the *industrial sector is small.* Industries are typically concerned with *first-stage industrial processing* of mineral and agricultural products, such as peanut shelling or cotton ginning. Small-scale production of cement, cheap household goods, beer, soft drinks, shoes, plastic goods, and textiles, and heavier industry including construction materials and products comprise the bulk of the consumer-oriented industrial establishment. Since the internal market has low purchasing power, and many countries in this group are small, market-oriented industry is limited.

Links with the International Trading Network

Table 19–1 illustrates the *pattern of trade* for selected countries of the least developed group. The total volume of trade in relation to population size is low. These countries still have largely self-sufficient societies. Typically, *exports* consist of primary agricultural or mineral products, often in commodities for which there are many alternative sources. Generally, these commodities have shown only slowly increasing price levels over the last two decades, and some, despite inflation, have shown a drop in value. On the other hand, the small volume of *imports* typically includes oil, which is vital for industry and communication, together with *machinery,* specialized goods of all kinds, and fertilizer. Most imports have shown steeply increasing price levels over the past two decades, especially since 1970.

The Sahelian Least Developed Countries

The Sahelian group of least developed countries are illustrative of the problems of development in arid and semiarid areas. All except Cape Verde lie on the south margin of the Sahara. A large part of each country is desert and much of the rest is semiarid rangeland. Although productivity is low, total available grazing resources are immense. These resources have been quite effectively used by *nomadic* or seminomadic peoples in flexible systems of land use.

Nomadism, however, is associated with a closely knit social and economic system. This system, effective in its use of environment, is often difficult to fit into the fabric of new nation-states. Governments tend to favor settlement of nomadic communities. Taxes, security, ease of administration, and a modern national image are all contributing factors in this desire to settle the nomad. The flexible nomadic system has almost entirely broken down in the face of the combined effects of the severe drought that began in 1969 and the impact of political and economic pressures.

Problems arose not only because of the *drought* but also because of other *changes affecting nomadic people.* Growth in numbers of livestock, changes in the amount of land available because of increased settled agriculture, new national boundaries, and the *effect of modern technology* (new wells, medicines, and roads) have played a part in intensifying the effect of the Sahelian drought. *A basic issue* in the use of the resources of such dry areas is the way in which new patterns of life emerging in the post-drought years will be adapted both to the ecological needs of the environment and the socioeconomic needs of the people.

The southern parts of the Sahelian countries have greater amounts of rainfall, and dryland agriculture complements production from irrigated land along the rivers. These agropastoral regimes

TABLE 19–1 Trade Patterns of Selected Least Developed Nations in Africa

Country	Major Sources of Imports	Percent of Total Imports by Value	Major Destinations of Exports	Percent of Total Exports by Value
Chad	France	45	France	72
	Japan	22	West Germany	7
	West Germany	10	Netherlands	6
Ethiopia	United States	55	Italy	20
	West Germany	9	United States	18
	Italy	9	Japan	18
Uganda	United States	26	United Kingdom	31
	United Kingdom	23	Japan	16
	Japan	19	West Germany	12
Upper Volta	Japan	34	France	80
	France	27	West Germany	8
	Italy	19		

provide a wider range of survival options when faced with drought. The combined production from animal husbandry and agriculture, which make up most of the economy of the Sahelian countries, provides but a modest level of living for the peoples of the area. When drought and attendant problems have reduced one of these components, considerable outside assistance has been needed.

West African Humid Least Developed Countries

The West African humid least developed countries present a different set of environmental problems. The natural vegetation, particularly in lowlands, is a thick forest with several layers of vegetation. This *biomass* (a term used to include all plant and animal life in an ecosystem) when cleared provides a good environment for the production of root crops like yams and cassava and has potential for cash crops such as palm oil, cacao, and rubber. Clearing is hard work, and soils are fertile only locally. The most common soil type in this zone is the deep-red *latosolic,* which deteriorates quite rapidly under continual cultivation. Most of the nutrients of the humid tropical ecosystems are locked in the vegetation mass itself. Unless the natural vegetation can be replaced with a similar cultivated system, nutrients are lost through soil erosion.

Traditional *shifting cultivation* provides a mechanism whereby land is allowed to regenerate after a few years use while a new plot of land is cleared. When plenty of land is available, shifting cultivation seems a reasonable economic use of labor and land from the point of view of the traditional cultivator. Clearing is effected economically by cutting and burning the smaller trees and litter. The ash provides a short-lived fertilizer, and the small size of the plots prevents major soil erosion problems. After a few years, the vegetation is allowed to reestablish itself, and a new nutrient-rich plot is cleared. This system, however, is impossible to sustain when land is in short supply. As population grows, land formerly under shifting cultivation comes into permanent use with consequent problems of soil fertility and soil erosion. The small countries in West Africa have yet to evolve productive small-farm systems that are alternatives to either shifting cultivation or plantation agriculture. In countries such as Ivory Coast, *plantation-type farms* have been successful in preserving soil and maintaining production levels, but only because of capital inputs in the form of fertilizers and machinery that are far beyond the capability of the small peasant farmer.

A further problem of the humid tropics is that *animal husbandry,* apart from pigs and chickens, is difficult. Cattle suffer from a variety of diseases in the wetter areas and are plagued by the *tsetse fly* in drier savanna conditions. The result is a gener-

One traditional means of raising water for irrigation is the ''Persian wheel.'' Buckets are strung in series around a wheel. As the wheel turns, the buckets dip into the well and then raise the water for distribution in irrigation ditches. (United Nations)

ally lower level of human nutrition in some of the humid areas than in the arid zones where animal products are used much more.

These conditions are characteristic of each of the West African countries and result in a generally low level of production. Diversity is brought about by small-scale mineral development. Sierra Leone has scattered diamond deposits, and 35% of Togo's exports are phosphates. These additional resources, however, have not significantly changed the overall economy of the countries concerned.

East African Least Developed Countries

The East African countries have less uniform characteristics. Rwanda and Burundi are small, densely populated, isolated countries located mostly in humid uplands. Ethiopia is a large country with one of the largest ranges of climate and environmental conditions in Africa, including humid uplands and lowland arid plains. The northern half of Uganda is a humid, fertile plateau and mountain area, while arid conditions prevail in the east. Somalia is a country with only a small area of rain-fed agriculture supplemented by irrigated areas in the south.

The 4.6 million Somalis are mostly a nation of pastoralists, although some commercial banana production takes place under irrigation in the south. The other East African countries similarly are predominantly agricultural, with all of them growing and exporting coffee. Ethiopia, with a population of more than 30 million, is one of the largest countries in Africa. However, it has not yet been able to translate this size advantage into a productive and growing economy. Adis Abeba with 1.2 million inhabitants is one of Africa's largest cities and the headquarters of the *Organization of African Unity* (OAU) and other pan-African bodies. A combination of infrastructure deficiencies, archaic social and political forms, internal separatist movements, and conflicts with Somalia have retarded Ethiopia's economic development.

The East African countries, like many of the least developed states, all share the characteristic of *isolation.* The need to move goods through a neighboring country to reach the coast often means not only that long land distances have to be covered but also that there are *problems of external control* over national trade routes. Uganda relies completely on Kenya for access to the sea because routes through Zaire, Sudan, Tanzania, or Ethiopia are so difficult that they do not present realistic alternatives. Ethiopia, although it has a long coastline, has poor access to the sea. The coast has few anchorages, and the central government is plagued by a large-scale revolt supported by the Muslim population of coastal Eritrea. Landlocked Rwanda and Burundi have long land routes through Uganda and Kenya, through Tanzania, or through Zaire. The East African least developed countries are plagued by low production levels and by poor trade routes for their exports. Equally damaging are the added costs that long and difficult transportation routes place on imports.

INTERMEDIATE NATIONS

Two types of intermediate countries are identifiable. The first set of countries is dominated by a *single economic sector.* These include Mauritania, Ivory Coast, Tanzania, and Sudan. Most are grounded in agriculture, although Mauritania is heavily dependent on a single mineral resource, iron ore. The second set of countries has *more diversity* in their countries' national economies, some on the basis of a single mineral resource with an agricultural base, others on several mineral resources and an agricultural base. These include Ghana, Nigeria, Guinea, and Kenya.

Countries with Dominantly Single Sector Resources

In some respects, nation-states are not unlike ecosystems. Complex ecosystems are more stable and have less risk of destruction than simple ecosystems. Nation-states with a diverse economy are likewise less vulnerable than those whose economy depends heavily on only one or two products. *Vulnerability* increases when the country must compete with many others in the world market. States possessing diverse environments and a number of developed economic components stand a good chance to provide prosperous and stable conditions for their citizens. A few countries have an abundant single resource of such value that it overcomes all other difficulties.

Problems Associated with a One-Item Resource Base

Many countries have development prospects in one sector of the economy. Ivory Coast, Tanzania, and the Sudan are examples of countries nearly totally reliant upon *agriculture* not only as a prime source of *daily subsistence* but also as a generator of *foreign exchange.* In contrast, Mauritania is an agricultural and pastoral community that struggles to maintain a precarious existence, although mineral exploitation offers the best hope for future economic growth.

Dependence on only one sector to stimulate growth may cause serious problems. Some of these problems are internal; others are produced by the operation of international markets. Internally, concentration upon agricultural development often takes the form of cash-crop production for overseas sales as a means of gaining foreign exchange. Agricultural developments of this type often place great *strain on local socioeconomic systems.* Frequently, capital generated from commercial farming benefits only a small section of the agricultural community. *Inequitable distribution of wealth* in turn may create social tensions where none existed before or may encourage a drift of population away from economically less advanced rural regions toward the seemingly greater economic opportunities of urban areas. Commercial agriculture may also *disrupt traditional agricultural patterns* by encouraging a removal of land from subsistence food production or by encroaching upon fallow or pasture land. The growth of agricultural settlement on government irrigation schemes has in some cases encroached upon land formerly an essential part of a pastoral system.

Externally, concentration of development in the commercial agricultural sector, particularly when only one or two crops are exported, may make a country extremely vulnerable to *price fluctuations* on the world market. Cotton in the Sudan, cacao in Ghana and the Ivory Coast, and coffee and cloves in Tanzania are all major foreign exchange earners. As long as commodity prices are high, the economy of these countries prospers. When prices fall, however, it can mean a major economic setback. Year-to-year *weather conditions* can also result in considerable differences in production and unforeseeable fluctuations in the national income.

The economies of almost all of the countries considered have provided *a narrow base* upon which to attempt to develop *manufacturing.* Agricultural processing industries frequently remain small in size, and most products are either consumed fresh or shipped in an unprocessed state to overseas markets. Even where a local mineral resource such as Mauritania's iron ore is developed, it is exported overseas after only minimal processing.

Alternative Development Strategies

There are great difficulties when a country attempts to raise personal income significantly from an agricultural base. The governments of one-resource countries have tried various solutions to foster economic growth. These alternative strategies may be ideological in nature, such as Tanzania's emphasis on instilling a spirit of *ujamaa* (socialist cooperation) and pride in the people. There may be attempts to increase the diversity of the natural ecosystem by the *application of technology.* Giant irrigation schemes such as Sudan's Gezira are examples of an attempt to create totally new agricultural environments to spur increased productivity. Still other strategies call for the creation of new sources of *revenue,* for example, from a tourist industry based on mild climate and an attractive waterfront or from the development of known mineral resources. Let us consider four such strategies: tourism in the Ivory Coast; irrigation technology in Sudan; mineral development in Mauretania; and self-help in Tanzania. Each strategy has its good points and its bad, its record of successes and failures.

Tourism.   All countries in the intermediate group gain some income from tourism, but the Ivory Coast has placed special emphasis on the *tourist industry* as a means of accelerating economic growth. Tanzania has important tourist potential but has hesitated to commit resources to this sector. The Tanzanians fear that tourism would benefit mainly foreign companies and would encourage "servant" attitudes among local people. For the Ivory Coast there have been few such inhibitions.

To reduce the risk of sudden market variations for its agricultural exports, the Ivory Coast has searched for *alternative methods* for extending its economic boom. Investment in the tourist industry has been adopted as one means of rapid future economic growth.

Large sums of money are being expended to increase the number of hotel rooms, increase accessibility to interior locations of cultural and environmental interest, and *improve tourist facilities.* More than ten thousand hotel rooms now exist in the country, at least half in and around the capital, Abidjan. *Visitor-nights* were over 1 million in 1981.[1] Lured by long, sandy, palm-lined beaches, luxurious hotels, exotic game parks, and excursion air fares, tourists are expected to spend lavishly in Abidjan and several other target zones along the west coast and in the interior.

The tourist alternative is not necessarily a satisfactory solution. The *culture conflict* inherent in a tourist industry threatens local values and often accounts for the hostility that may accompany a massive tourist influx. Thus, although tourism may help diversify a local economy, it is *a mixed blessing.* Furthermore, because the tourist's ability to spend is linked directly to the health of the global economy, it may expose the tourist-oriented economy to many of the same vicissitudes that affect a cash-crop concentration.

Application of Technology. Use of *technology* to overpower the environment and increase production is a *Western strategy.* Diffusion of Western technology throughout the world has made it possible for underdeveloped countries to emulate the activities of the industrialized nations. For example, in the arid zones, large-scale agricultural development projects based on dam-fed irrigation systems have been a common pattern of development. The Sudan has adopted this strategy to promote growth in its agricultural sectors as well as provide hydroelectric power for projected industrial opportunities.

Sudan has used *irrigation technology* over a long period. Begun in 1925 with the construction of the Sennar Dam on the Blue Nile, the first stage of development brought irrigation to more than 1 million acres (405,000 hectares). Much of this newly irrigated land lies in the *Gezira,* a triangular area between the White and Blue Niles. Formerly a semiarid plain used primarily by pastoral nomads, much of the Gezira has been converted into *intensive and productive agricultural land.* The primary product in the Gezira region is *long-staple cotton;* this variety is used to make fine cloth, and its export is the major foreign exchange earner for the country.

The Gezira is not only Sudan's major cash-crop region, but it is also an important source of *food*

for *internal consumption.* Half of the country's wheat is produced in the Gezira, and substantial amounts of other crops are raised in the region's coordinated and integrated irrigation system. Its importance has led to major expansions. The Manaquil Scheme has extended irrigation to an additional 800,000 acres (324,000 hectares), and the ar-Rusayris Dam has added another 300,000 acres (121,000 hectares). Much of this new cropland is planted in *short-staple cotton* in an attempt to avoid excessive dependence on the specialty cotton market and to provide an easier-to-process fiber that can meet the competition of artificial fibers. The Khashm al-Qirbah Dam on the Atbara River also irrigates 740,000 acres (300,000 hectares) of cotton and food crops in another development project. These projects employ relatively *simple technology and construction* and are able to rely on an efficient and cheap gravity-flow water system. Moreover, in each scheme the tenant farmers participate in the local development board's decisions.

A Mineral Alternative to Agropastoralism. Mauritania is an interesting comparison to other nations in the intermediate group, for its one major activity suitable for rapid economic growth is its mining sector. Although most of Mauritania's small population (1.7 million) is actively employed in agriculture, only 1% of the country's land is arable. Most of the land is concentrated in the extreme south in the Senegal River valley, and the remainder of the country is composed of the *low productivity ecosystems* of the semiarid Sahelian grassland and the Sahara Desert. Except for the desert oases in the center and north that produce dates, and the millet zones along the Senegal, most of Mauritania is suitable only for livestock raising. It is no accident that *nomadic pastoralism* continues among 70% of Mauritania's population. Although efforts to improve all sectors of the traditional agropastoral economy are being made, the low productivity of the land presents many problems. Drought such as that which has blighted crops and decimated herds in West Africa's Sahelian zone is the most important of the environmental problems. The risk of repeated severe drought is the prime reason why agricultural opportunities for major economic growth are limited.

From an economic development perspective, Mauritania's situation would be truly desperate were it not for the existence of several *rich mineral*

---

[1]Visitor-nights is a tourist-industry measure of the number of beds occupied by tourists in a given statistical period (generally a year). Since tourists usually remain in a country for more than one night, the figure drastically overstates the total number of tourists actually visiting a country.

Water is critical for development in arid countries in East and West Africa. The above dam construction site on the Blue Nile (Sudan) is one of many developed during the post-World War II years. Note the harsh landscape in the background. (United Nations)

deposits. These deposits permit hope that the pitfalls of a one-resource agricultural environment can be avoided to at least some degree. The existence of rich *iron ore* deposits in northern Mauritania has been known since 1935. Only since 1960 has exploitation of these iron reserves, in many cases with a metal content of 60 percent, been possible. A great deal of *infrastructure,* including construction of a railroad to move the ore to the coast and port facilities at Nouadhibou to speed handling, had to be built before major exports became possible. Exports began in 1963 and rose rapidly. *Copper* deposits also promise to generate substantial revenue, and a search for petroleum is under way.

As a result of royalties and taxes generated by the mining industry, Mauritania's foreign trade balance has improved from one of small deficits to a modest net favorable balance. This favorable balance is not without its problems, for it has proved difficult to distribute the newly found wealth, modest though it is, equitably among Mauritania's citizens. Today some 15% of the population, largely those with managerial and professional skills or positions in the government bureaucracy, control 65% of the country's wage income. Such *inequities* are bound to have future political and social repercussions.

Ujamaa in Tanzania. Like all other countries in the intermediate group, Tanzania derives most of its resources from *the agricultural sector.* Farming and animal husbandry are the activities of the bulk of the population. Exports of agricultural commodities such as coffee, cotton, cloves, and sisal are the country's major foreign exchange earners. All of these crops are affected by fluctuations in world-market prices, cloves having been especially hard hit by a steady decline in price over the last

296

decade. This situation has sparked *efforts to diversify the economy* of Zanzibar and nearby islands, where most of the cloves are grown, by introducing cattle ranching and new crops such as rice, cacao, and coffee.

Tanzania is unusual with respect to *the strategy* selected to deal with the development problems. Tanzania has not adopted many aspects of Western Technology. Instead, emphasis has been placed on developing the motivation and aspiration of the people in ways that seem authentically Tanzanian and African. *Ujamaa* is a philosophy of pride in self, of commitment to self-improvement, and of development of the individual, region, and nation along communal and cooperative lines. This type of *African socialism* is seen by Tanzanians as preserving the best features of the African tribal, group-conscious social system. Involving the people at grass-roots level in the development process has produced generally good results.

Originally the government experimented with building a series of new villages with modern amenities in an effort to promote rural development. The experiment was not successful as costs were high, and many Tanzanians were reluctant to move to the new settlements. Emphasis shifted to introducing development projects into existing villages, the so-called *ujamaa* villages. Local people were encouraged and often exhorted to undertake *communal projects* in return for government assistance. Cooperative agricultural activities have been promoted with marketing and credit arrangements handled by the cooperative. *Government-controlled price structures* are also important in encouraging farmers to produce crops in the amounts required. For example, a small upward change in the price-support structure for rice (an alternative crop) encouraged a spurt of rice cultivation as farmers rapidly adjusted to the improved economic returns.

Great efforts are also being made to utilize technology that is *appropriate to the local area*. Tanzania has stressed improvements in its existing *peasant technology*—for example, producing stronger and more efficient ox plows rather than introducing tractors and other machinery. With foreign exchange reserves declining, this policy also conserves badly needed cash. The break with the colonial era is also symbolized by Tanzania's decision to *move its capital* from the coastal city of Dar es Salaam, the former center of colonial administration, to a site at Dodoma in the central part of the

country. The cost of such a move is tremendous, both in terms of abandonment of existing facilities and infrastructure along the coast and in creating a new city in the water-deficient central interior.

Countries with Diversified Resources

Four countries in the area have more diversified resources combining agricultural, forestry, and mineral resources. These are Nigeria, Guinea, Ghana, and Kenya. Guinea has the world's second largest bauxite deposits together with good agricultural potential. Ghana is a major cacao exporter and has used its surplus capital to spur industrialization and diversify its agricultural base. Nigeria and Kenya serve as examples of countries with diversified resource bases.

Nigeria

Nigeria is economically different from all the other countries of the region. Nigeria has gained its *different economic status* through the discovery of oil and gas in the southern part of the country. The massive income generated in the latter part of the 1970s slackened considerably in the early 1980s, and it has proved difficult to invest the oil income successfully in other parts of the economy.

Nigeria, with a population of 84 million and a per capita GNP of $873 billion, has the largest economy in Africa south of the Sahara except for South Africa. Much economic growth has occurred since 1968 because of the *discovery and exploitation of oil* in the Niger delta area. In 1968 exports were valued at $321 million with oil accounting for only $58 million. In 1972, before the abrupt rise in oil prices, exports were $2.2 billion with oil representing $1.8 billion of the total. In 1974 oil revenues soared to $7 billion. This growth has *revolutionized the whole economy* of a country that was recovering from the effects of a long and devastating civil war when oil began to be important.

Before the oil boom the *main staples* of the export trade were cacao, peanuts, and tin. Nigeria also has a great variety of *indigenous production systems*. Cattle and peanuts are dominant in the north, and palm oil, rubber, timber, and cacao in the south. Agriculture in Nigeria, however, is reported to be in an indifferent state; output in 1971 was only about 36% of the 1960 level. Perhaps the major factor accounting for the decline in agriculture has been the social and economic differential between towns and surrounding countryside. Wages, standards of health, and *diversity of opportunity* have all become greater in the city, and young people are reluctant to stay on farms. With all data indicating that the country will maintain a rapid growth of population, increased food production is mandatory.

Oil extraction is already making a major difference in some aspects of internal investment in the country. Nigeria has always placed a high priority on investment in *higher education;* there are five universities and many *technical colleges.* The country has a variety of research institutions of national and international character. These are signs that the oil money is being used to reinforce and broaden some of these institutions.

Some of the oil money, as in other newly rich countries, is being used to improve Nigeria's power in the West African region. Nigeria has engaged in such *humanitarian* and *prestige-enhancing projects* as contributing to drought relief in Mali and sponsorship of the 1975 All-Africa Games. Efforts to improve *regional commercial ties* have also emerged. Nigeria has invested heavily in Guinea's iron ore mines in the expectation that much of the ore can be used in its own planned iron and steel industry. Transit taxes have been removed on goods crossing the border of some of its neighbors in an effort to promote regional trade. There are additional efforts to promote regional development through the Lake Chad Basin Commission. In contrast to these positive regional initiatives, in early 1982, Nigeria expelled many guest workers from neighboring countries when its own economy lost momentum.

Nigeria has long had a *varied industrial sector.* Little modern industry exists, however, and most of the industrial sector is made up of thousands of small units working at the craft or local level. This sector continues to flourish and dominate many parts of the country. In recent years investment in Western-type industry and services has been heavy. As export earnings grew, imports of food, machinery, and equipment grew phenomenally even as the import of manufactured goods also exploded. The rapid increase in world oil prices helped pay for this import expansion. Attention, however, has turned recently toward investing oil revenue in *infrastructural improvements,* such as paved roads and new railway equipment, and in

large-scale industrial enterprises. Attention now focuses on increasing oil-refinery capacity and beginning an iron and steel industry. If the new oil wealth can be invested in successful projects of this type, dependence on a high volume of imports can be somewhat reduced.

Kenya

Kenya is a country that does not fit easily into any simple classification of countries by degree of economic development. In some respects Kenya should be regarded as a least developed country with an economy based on agriculture and with a relatively low per capita GNP. In other respects it can be regarded as having more diversified resources because of its important service-center role in East Africa and Africa as a whole.

Kenya's relatively small area of humid upland is highly productive, with coffee, tea, and fruit as the major export crops. Flowers, vegetables, and fruit are airfreighted to Europe in a modern world production system. However, much of the rest of Kenya consists of arid and semiarid areas where grain production and livestock raising are the main enterprises. Although some meat is exported, the arid part of the nation relies heavily on subsistence activities.

Kenya, in terms of physical resources, depends very much on its agricultural base, one that is coming under pressure from a rural population that is growing at the rapid rate of 4% a year. However, Nairobi's central location and history of expatriate capital investment have encouraged its evolution as a regional and continental center and as a small industrial center for East Africa. Tourism has been one form of investment, and the game parks and beaches of the country continue to provide attractive vacation packages.

Tourism has been helped by the development of Nairobi as a center of the African air network. In addition, Nairobi has been chosen as the headquarters location of a number of international and regional agencies and as an important conference center.

Industry in Kenya includes food processing, particularly dairy products and tea and coffee. Light manufacturing also exists, but the major industrial foreign exchange comes from the export of petroleum products from the refinery at Mombasa. In some years, refined petroleum is the top exchange earner for Kenya.

20

Central and Southern Africa

Central and Southern Africa is the second major region of Subsaharan Africa (Figure 20–1). Botswana, Cameroon, Lesotho, Madagascar, Malawi, and Swaziland, like their counterparts in East and West Africa, are among the world's least developed nations. South Africa, in contrast, is relatively rich. The other nations are intermediate in economic well-being.

COLONIAL LEGACY

Central and Southern Africa is a region where the stamp of Belgian, Portuguese, and British *colonialism* is strong. The colonial history is still being written in the case of Namibia. Zimbabwe achieved majority rule in 1979, and Mozambique gained independence from Portugal a few years earlier. Most of the others gained independence during the 1960s.

Unlike much of East and West Africa, the *transition to independence* in Central and Southern Africa has not been peaceful for a number of the countries. In Zaire the rapid exodus of the Belgians in 1960 left a country with few trained local people except at the lowest levels of government. Not surprisingly, the following years saw many problems in the life of this large and diverse country including internal strife, the death of a president under suspicious circumstances, and eventually a long period of military rule. In Angola and Mozambique, independence from Portugal was gained only after extended revolutionary struggle, and in both countries there remains armed dissent against the government, a relic of the factionalism of the independence movement. Similarly, Zimbabwe, newly created from Rhodesia, won independence with majority rule only after a long war against the white minority government. The aftermath is continual *internal strife* between opposing groups of the revolutionary forces. In contrast,

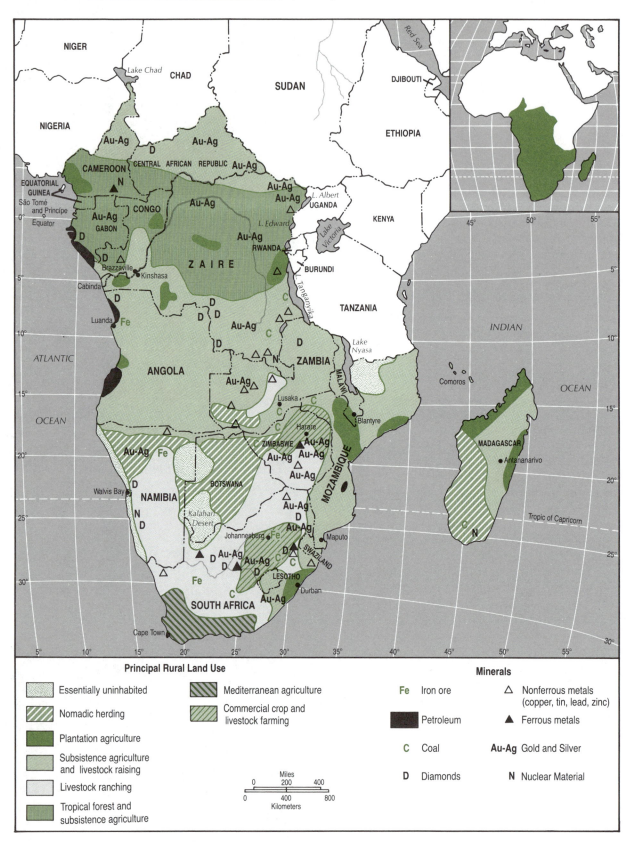

FIGURE 20–1 Land Use and Minerals in Central and Southern Africa.
Central and Southern Africa exhibit many forms of land use. Some, such as herding,
support only a few people per square mile, but others are intensive and provide a
livelihood for many. The mineral industry supports few people directly but provides
valuable foreign exchange, especially in Zaire and South Africa.

Cameroon, Zambia, Botswana, Malawi, Swaziland, and Lesotho gained independence peacefully and have not experienced armed internal conflict.

RESOURCE BASE

Minerals

The countries of Central and Southern Africa tend to be better endowed with a broader *range of resources* than those of East and West Africa. Gabon, the Central Africa Republic, Equatorial Guinea, and Zaire have considerable timber resources. Zaire, Zambia, and Zimbabwe are important for copper production, and Zaire has a number of other mineral resources. Angola and Gabon have oil and gas and several metallic mineral resources. Botswana, once a country with few apparent mineral resources, has gained economically from the discovery and mining of diamonds and other minerals. Swaziland and Namibia also have mineral wealth. Finally, the Republic of South Africa is one of the world's major producers of a whole range of minerals including gold, diamonds, and coal. It is this great abundance of *mineral resources* (not shared by Malawi and Lesotho) that has helped many of the nations of the region to achieve "intermediate" economic status.

Agricultural Base

Despite the importance of mineral production, *agriculture remains the backbone* of most economies in the region and is the most important source of employment. The economy of Malawi is almost entirely based on agriculture. Botswana relies on cattle and other livestock as a major source of income for most of its people. Mozambique and Angola both have diversified and potentially productive agricultural sectors. Zimbabwe has a highly productive agricultural economy, and the Republic of South Africa has one of the world's more diversified farming systems. Lesotho, Namibia, and Zambia have small agricultural economies. Zaire's once-flourishing agricultural sector is currently in disarray.

PHYSICAL ENVIRONMENT

The diversity of the region is perhaps best expressed in its *great range of physical conditions.* The Zaire River basin is the locale of most of Africa's tropical rain forest, nurtured by its year-round rainfall. The river reflects the climatic situation in its high regular flow that provides Zaire with a large but only partly utilized hydropower. Angola, Gabon, Mozambique, Cameroon, and the Central African Republic also have large areas of rain forest.

In contrast, Namibia and Botswana are some of the driest parts of the continent with less than 2 inches (5 centimeters) of annual precipitation. In between lie areas of plateau Africa with moderate (30–45 inches, 76–114 centimeters), seasonally distributed rainfall and with mediocre-to-good potential for agriculture. The moist uplands of Zimbabwe and Malawi are ideal for a wide range of crops, including tea and coffee. The lowland areas of Zimbabwe and Zambia are more suited for maize (corn) and animal raising and have varying degrees of productivity.

The remaining physical zones are found mostly within the Republic of South Africa. There climates include coastal Mediterranean conditions and inland subhumid to semiarid plateaus with a range of grain crops, including wheat and maize, and wide areas in which livestock raising predominates.

ECONOMY

The *economic systems* of Central and Southern Africa are *oriented in three directions:* (1) southward to South Africa; (2) outward to the ocean and Europe and North America; and (3) outward to the Soviet Union.

For most of the countries in Southern Africa, the dominant orientation of their economy is southward to the Republic of South Africa. The sheer *economic weight of South Africa* plus the orientation of infrastructure toward it almost ensures this domination even though political will might prefer another set of links. The enclosed or semi-enclosed countries of Lesotho, Swaziland, and Botswana have especially close ties with South Africa, and Namibia is currently being administered by the Republic. In addition, Zimbabwe, Malawi, Mozambique, and to a lesser degree Zambia all retain close economic ties with South Africa. Zaire, Gabon, Zambia, Zimbabwe, and Angola all have export economies which are closely linked with Europe and North America, mostly based on natural resources. Angola and Mozambique trade with the Soviet Union, but they also have strong export and import ties with Europe and North America.

301

SUBREGIONS

Central and Southern Africa are subdivided into *four subregions.* West Central Africa includes Cameroon, the Central African Republic, Gabon, Congo, Equatorial Guinea, and Zaire. Central Africa is composed of Angola, Zambia, Malawi, and Mozambique. Southern Africa includes Botswana, Lesotho, Namibia, South Africa, Swaziland, and Zimbabwe. The islands consist of the Comoros, Madagascar, Mauritius, and the Seychelles. Case studies are presented for each of these subregions.

West Central Africa

Two countries provide a sample picture of West Central Africa: both Gabon and Zaire are intermediate in economic status. Gabon is one of Africa's main petroleum product exporters and also possesses a timber and mining industry. It's strong economy, shared by a population of less than 1 million, results in a high (and inflated in terms of the wealth of most people) per capita GNP of more than $3900. Zaire with its diversified mining industry is a large country of 31.8 million people and a per capita GNP of only $225. Potentially a wealthy country, Zaire's economy is almost bankrupt—the result of a combination of low international· prices for Zaire's minerals, mismanagement and corruption, and the basic difficulties of leading a country of this type to a more sustained pattern of economic development.

Gabon: A Timber, Petroleum, and Mining Economy

Gabon is a small country where subsistence agriculture has been replaced as the basic form of livelihood first by the *timber industry* and then by mining and petroleum production. In the last twenty years Gabon's GNP has increased from $167 million to nearly $3 billion.

In terms of climate, soils, and general environmental resources, there is a moderate long-term potential for production from the land. At present this potential is only being realized in terms of lumber. In the 1970s less than 0.5% of the land was under cultivation. Of this, two-thirds was for subsistence crops and the remainder was planted with cacao, coffee, and palm oil for export. The proportion of land under cultivation is slowly declining as urban opportunities encourage people to move to the towns. One result of this situation is a high import bill for foodstuffs. Forests cover 56 million acres (23 million hectares), or 25% of the country,

Gabon's economy differs from those of its neighbors. Subsistence agriculture is of minor importance, and lumbering and mining are major sources of livelihood. (United Nations)

and account for more than half of the total exports. Timber is exploited by large European companies and increasingly by small Gabon enterprises. A great advantage of timber production is that processing industries can be established without massive capital investment. Timber processing accounts for more than 50% of Gabon's industry and for much local employment.

In the 1960s *mining* became increasingly important, and in the 1970s and 1980s petroleum and natural gas have become economically dominant. Manganese, uranium, iron ore, and gold provide a diversified and expanding mining economy.

Gabon has great potential, but major future constraints exist in the human resources of the country. Wealth is being generated rapidly from mining, yet many of the rural population still live at low economic levels.

Zaire: Mineral and Hydroelectric Power as a Development Focus

Zaire is a *multiple-mineral-resource nation* with considerable agricultural potential. Zaire has relied heavily on copper as the main basis of export earnings. Copper accounts for 60% of export earnings in most years. Cobalt and diamonds provide important mining income and rank with coffee and palm oil in export values. Zaire is the world's largest producer of industrial diamonds. In addition, there is a range of other as yet little-developed mineral resources including silver, iron ore, manganese, and possibly oil.

Zaire has considerable *agricultural potential,* although agriculture is still a struggling sector. Coffee, cacao, tea, cotton, and tobacco are produced for export, but with the increasing rate of urban population growth in Kinshasa (currently over 2.5 million) and the relatively high degree of employment in mining within the country, food imports continue to grow.

Zaire has the long-term advantage of one of the world's great *hydroelectric power* sites at Inga near the mouth of the Zaire River (formerly the Congo). Initially 300 megawatts were developed, and in 1976 another 1000 were added. This energy resource is one key to Zaire's development prospects, increased by the growing price of other energy sources.

A number of *industrial projects* are being considered to use hydroelectric power, including a steel complex, a caustic soda factory, a polyvinyl (PVC) factory, a nitrogen fertilizer factory, and an aluminum smelter. The steel complex is already under way, and the caustic soda and PVC plants are in the planning stage. Power will also be used for the copper-producing area in Shaba (Katanga).

Zaire, with a large urban population and major mineral and power resources, may well become a nation with a major *manufacturing* sector. Manufacturing is one conscious focus of government development plans. Much of the industry will be capital intensive, and there is a great need for massive training of skilled workers and an upgrading of employment possibilities in the rural and service sectors. A long period of economic modernization extending through the 1980s will probably be necessary for sustained economic growth and fulfillment of Zaire's dream to be an industrial nation.

Central Africa

Zambia: The Copper State

Zambia is a landlocked country and like other Central African countries has strong ties both with the north and the south. Its mineral industry, especially copper, has placed it firmly in the intermediate countries in economic status.

The problems of Zambia are a good example of one kind of situation related to *mineral development.* Zambia, with a population of 6.2 million and a per capita GNP of $586 is a moderately rich country compared with others in Africa. It had a favorable balance of trade until the sharp decline in copper prices at the end of the 1970s. Still, twenty years after independence, copper accounts for more than 90% of foreign exchange earnings. It is directly responsible for 20% of government revenue and contributes additional indirect sums. Cobalt, zinc, and tobacco are other minor export earners. Zambian towns appear on an average to have as high a standard of living as anywhere in Africa and higher than some European countries. The countryside, however, is a different matter. Zambia has considerable agricultural potential yet imports 40% of its food.

There are many reasons why Zambia is *food deficient.* For some years after independence, little investment occurred in agriculture, and the proportion of people on the land dropped from 70% in 1963 to near 55%. The higher standards of living in the towns compared with the countryside are a major factor in a rural-to-urban migration. Employment opportunities in the cities are not growing as fast as the urban growth rate. Large ranges in wealth among working adults occur in the towns, and greater *discrepancies in standards of living* occur between town and country. Ironically, much wealth is being lost because there is not yet a sufficient reservoir of trained technical and managerial workers to run the copper mines and other industries. Zambia thus has an unemployment problem and a need to employ many foreigners at high salaries.

Efforts to solve the *employment problem* are beginning. Although some Zambians welcome skilled foreigners who are ''valuable to the economy,'' others are worried about the present labor structure in economic, social, and political terms. Through the University of Zambia and formal technical-education programs, the Zambian government is attempting to create a reservoir of skilled manpower quickly. Training is relatively easily acquired; job experience takes much longer.

In agriculture the position is changing. One success story is sugar, where there was steady improvement from 1960, when all needs were met by imports, to 1974, when Zambia became self-sufficient. Prospects in the future are good for cattle. If World Bank plans materialize, Zambia could become a major exporter of meat and maize—for which Zaire is a large potential market—and of fruit, particularly pineapples and mangoes.

Zambia has defined a *policy of diversification* of resources and greater spread of wealth. It will take some years for this policy to become fully effective.

Southern Africa

The countries of Southern Africa have one thing in common: their *strong economic ties* with the Republic of South Africa. Lesotho is physically surrounded by South Africa, while Swaziland, backed against Mozambique, is cut off on three sides by the Republic. Both countries are economically dependent on South Africa. Namibia is currently being administered by the Republic of South Africa pending long-drawn-out negotiations over its eventual independence. Zimbabwe and Botswana have close economic, commercial, and infrastructure links; nearly all goods entering and leaving Zimbabwe and Botswana pass through South Africa. For all of these countries the tension between economic reality and the political desire of the nations to distance themselves from South Africa is a major and continuing problem.

Three southern African countries are profiled: Zimbabwe, Botswana, and South Africa. Zimbabwe is one of the most economically advanced of the intermediate countries in Africa. Botswana is classified as least developed, though its mineral wealth has greatly increased its national GNP in the last decade. South Africa is one of the most developed nations in Africa and stands as a bastion, a white power amid growing black nationalism.

Zimbabwe

Zimbabwe, formerly Rhodesia, is a newly independent country (1980). The country is small, covering an area of approximately 150,804 square miles (390,582 square kilometers). The population of 8.4 million is made up of roughly 200,000 Europeans and 8.2 million Africans.

The *economy* of Zimbabwe is more diversified than those of most African countries with mining accounting for more than 50% of foreign exchange earnings. *Manufacturing* produces about 25% of the GNP. A well-developed *commercial agricultural* sector contributes to foreign exchange as well as

provides most of the nation's food supply. Zimbabwe is a leading exporter of agricultural products.

As the second-largest producer of chrome in the world, Zimbabwe has *strategic importance*. Its landlocked situation and reliance on transport through South Africa, however, are a disadvantage. The two nearest seaports are in Mozambique, but one is operated by South Africans and the other is often crippled by guerrilla attacks. Most importantly, Zimbabwe depends on the South African government for oil-importing facilities.

Mounting *oil costs* hinder development, and the search for new sources of energy is a high priority of the government. Electric power is being developed by the joint Zimbabwe-Zambia Kariba Dam on the Zambezi River; thermal power based on local coal deposits will improve Zimbabwe's energy status still further. Moreover, there is a continuing search for nonpetroleum fuels: biogas, ethanol from sugar, wind, and solar power.

Although the majority of Zimbabwe's population is directly dependent on agriculture for a living, a small minority of some five thousand European farmers and some two hundred African farmers are responsible for the highly developed *commercial sector*. Most African farmers are subsistence oriented and confined to specific communal land areas. Communal lands are a consequence of the country's colonial history. A *division of land* between a minority of Europeans and the majority of Africans took place in 1930. White farmers were allotted roughly half of the land, including the most fertile upland areas, and the native lands with a communal form of tenure were created out of the remainder, with a small area reserved for African freehold farms. The communal lands suffer from heavy *population pressure* that has deleterious effects on land resources.

The *land question* is central to the reconstruction of the Zimbabwean economy for two important reasons. First, land is the basis of livelihood for most of the population, so that farming improvements are the key to increased standards of living. Second, *environmental deterioration* in the areas of peasant farming is a mounting concern. Reports detail the extension of cropland at the expense of both grazing and woodland in the communal lands, though not all have the same degree of problems of soil erosion, deforestation, and overgrazing.

Zimbabwe is a vitally important country in southern Africa. It is important economically because of its strong mining and agricultural economy. It is important politically in terms of the working out of its own internal problems and also in terms of its relationship with South Africa and with other parts of black Africa.

Botswana

Botswana is a country that has experienced a *dramatic growth* in GNP as a result of the development of a thriving mineral industry based on diamonds and coal. During the 1970s the mineral extraction industry grew at an average annual rate of almost 17%.

Before mineral development, Botswana was dominated by *pastoral activities,* because only a small part of the southern margin of the country has enough rainfall for sustained agriculture. Cattle were and are the dominant agricultural resource, and beef exports, mostly to the Republic of South Africa, constitute the most important agricultural product. With a population of about 900,000, Botswana has few significant towns, and the capital, Gaborone, is situated very close to the border with South Africa, a measure of the close economic dependence on that country.

The *mineral industry* has led to the establishment of a number of small towns, but the country is essentially one of scattered small villages. Male migration to work in the mines of South Africa has long been an important facet of economic life for

The central business district of Johannesburg, the largest city of South Africa. Johannesburg is a major industrial and commercial center with gold mining nearby. (United Nations/Jerry Frank)

families in Botswana, and a period of work in South Africa is still part of the life experience of many of the Botswana men. The growth of an indigenous industry and a political concern for greater economic independence have reduced these linkages.

South Africa

South Africa is a large country in terms of area, 471,000 square miles (1,220,000 square kilometers) and population (30.2 million), but most of all it is large in terms of its economy. Its per capita income of about $2500 gives it the largest GNP of any country south of the Sahara. South Africa's *economic structure* is different from most African countries. Nearly 50% of the people live in towns; only 30% of the work force is in agriculture; and the remainder are in industry and services. These are characteristics of a country with a *well-developed, integrated economy.*

Demography also makes the Republic of South Africa different from other African countries. There are about 4.5 million people of European descent, mainly of Dutch and English origin though with an important German component. About 2.5 million people are of Asian (mostly Indian) or mixed background. The majority of the population (more than 22 million) are of African origin. The white population originated from initial settlements as early as the 1650s, and they regard South Africa as their homeland and heritage as much as black Africans do. The domination of both the government and the economy by the white population and the long-established policy of *apartheid,* or separation between the two groups, have led to a major conflict within South Africa and between South Africa and much of the rest of the world.

South Africa is often in the news because many countries do not approve of the official apartheid policies of the government. South Africa is also in the news because it is a leader in the production of a number of important minerals (Figure 20–2). Its great economic strength may well be a major reason why other countries are unable to influence internal South African policies significantly.

South Africa is the world's leading producer of gold, antimony, platinum, and diamonds and the second-leading producer of uranium, chromium, manganese, vermiculite, and vanadium. It is also important for its reserves of copper and nickel. Moreover, South Africa has more than 90% of the continent's known coal deposits and these deposits enable the nation to meet most internal energy needs and still export coal to Zimbabwe and Zambia. The *mineral sector* is the leading edge in South Africa's economic growth.

South Africa's southerly location also provides a strong *comparative advantage* in the production of fruit and vegetables for northern hemisphere markets. In addition, sheep and cattle raising and production of several mid-latitude crops serve local needs and provide some foreign exchange. Potentially at least, South Africa is a very prosperous country. Paradoxically this prosperity is beginning to cut away the country's archaic social structures. At present, the nation is not effectively using a large part of the available human resources, but there is some hope that more liberal economic and social views will prevail.

The Islands

The Indian Ocean islands can be regarded as separate from or linked with the mainland. Of the four main island groups, Madagascar is perhaps the most closely linked with the mainland, even though it does not quite regard itself as African. All four nation-states are regarded as being of major strategic significance as outpost control points in the Indian Ocean.

Madagascar

Madagascar is a *poor nation* with a per capita GNP of $330. It is mainly an *agricultural country.* A population of 9.5 million puts the little-known island republic well into the middle range of the African countries in terms of size. The island has a wide range of climates as different parts of the island come under the influence of moist or dry air streams. Most of the rainfall is brought by northeast winds, so that the northern and eastern parts of the island are much wetter than the semiarid southwest. Rainfall totals range from 25 to 100 inches (64 to 254 centimeters) annually.

This *wide range of environmental conditions* is reflected in the large numbers of different food crops. Rice is the most important, but there are more than fifty other significant food crops that, together with rice, utilize more than two-thirds of the cultivated land. Coffee, beef, and cotton are the major exports but French government aid is an important factor in maintaining economic viability.

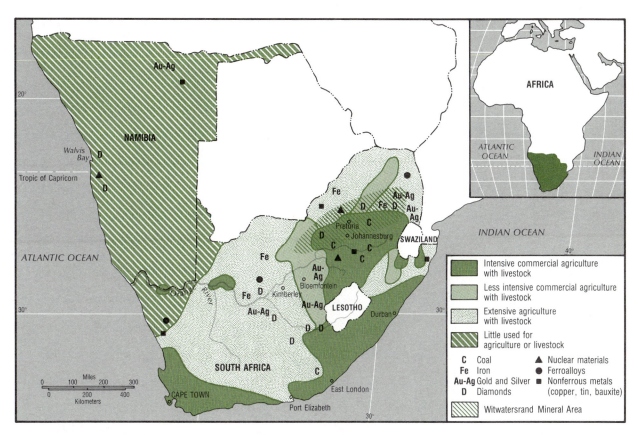

FIGURE 20–2 South Africa's Agricultural Systems and Mineral Deposits.
South Africa is one of the most developed nations in Africa. The country possesses
varied ecosystems and a diversity of minerals. South Africa has been able to establish an integrated economy based on a diverse resource base.

FURTHER READINGS

ALLAN, WILLIAM, *The African Husbandman* (Edin-
burgh: Oliver and Boyd, 1965). A classical and pro-
vocative statement that is a valuable introduction to
African subsistence production systems.

CARR, CLAUDIA J., *Pastoralism in Crisis: The Dasa-
netch and Their Ethiopian Lands* (Chicago: University
of Chicago, Department of Geography, Research
Paper No. 180, 1977). Approached from a holistic
perspective that integrates physical and social environments, this cultural ecological case study insight-
fully details the changes affecting the pastoral mode
of production.

DAVIDSON, BASIL (ed.), *The African Past: Chronicles
from Ancient to Modern Times* (Boston: Little,
Brown, 1964). A well-written general account that
presents many aspects of African civilization.

GROVE, A. T., *Africa,* 3rd ed. (New York: Oxford University Press, 1978). A comprehensive, literate,
and valuable regional text.

MONOD, THÉODORE (ed.), *Pastoralism in Tropical Africa* (London: International African Institute/Oxford
University Press, 1975). One of the few volumes to
combine under one cover the results of both Franco-
phone and Anglophone researchers.

HANCE, WILLIAM A., *The Geography of Modern Africa,* rev. ed. (New York: Columbia University Press,
1975). An excellent comprehensive introduction to
the entire continent.

————, *African Economic Development* (New York:
Praeger, 1967). A detailed examination of development schemes and mineral resources using a case-
study approach; it is an admirable complement to
the preceding more general text.

NORCLIFF, GLEN, and TOM PINFOLD (eds.), *Planning African Development* (Boulder, Colo.: Westview
Press; London: Croom Helm, 1981). Although the
eight essays consider only Kenya, the issues raised
are those encountered throughout the region.

OMINDE, S. H., and C. N. EJIOGU (eds.), *Population Growth and Economic Development in Africa* (London: Heinemann; New York: Population Council, 1972). A valuable, comprehensive compendium treating African demography.

PRATT, D. J., and GWYNNE, M. D. (eds.), *Rangeland Management and Ecology in East Africa* (Melbourne, Fla.: Krieger, 1977). The best scholarly statement on the topic that is also accessible to the lay person.

VIII

Middle East
and
North Africa

Douglas L. Johnson/
Leonard Berry

21

The Mediterranean Crescent

NORTH AFRICA AND THE MIDDLE EAST: COMMON THEMES

Arab Character

The Middle East and North Africa together constitutes a distinctive region (Figure 21–1). This distinctiveness is a product of many common features. The principal feature of the region is its Arab character. Arab tradition is represented by a shared spoken language, Arabic, by a common cultural tradition and pattern of historical development, and by participation in a common awareness of being Arab. Approximately 120 million people share the Arab culture and tradition. There are, however, some non-Arab ethnic groups, such as Turks, Kurds, and Persians in the Middle East and Berbers in North Africa, who maintain different cultural and linguistic patterns yet remain integral components of the region.

Religion

All of these groups share a common religion, Islam. It is the spread of this religion, first by the Arabs and later by other Islamized groups, that accounts for the dominance of Arabic culture. Because Islam's sacred scripture, the Koran, is recorded in Arabic, this language has acquired a high status. Many who today consider themselves to be Arabs are the descendants of other ethnic groups who gradually adopted Arab culture. Of course, many peoples accepted Islam without abandoning their indigenous culture. Some groups, such as the Copts of Egypt or the Maronites of Lebanon, adopted Arabic speech and culture without converting to Islam. Still other people, such as Armenians and Jews, remain faithful to their ancestral religious and cultural traditions. The spread of Islam south of the Sahara into zones of black African culture resulted in still greater diversity of speech and ethnicity. Islam also spread eastward into Pakistan, Bangladesh, Indonesia, and the Philippines. The result is a complex mosaic of languages, cul-

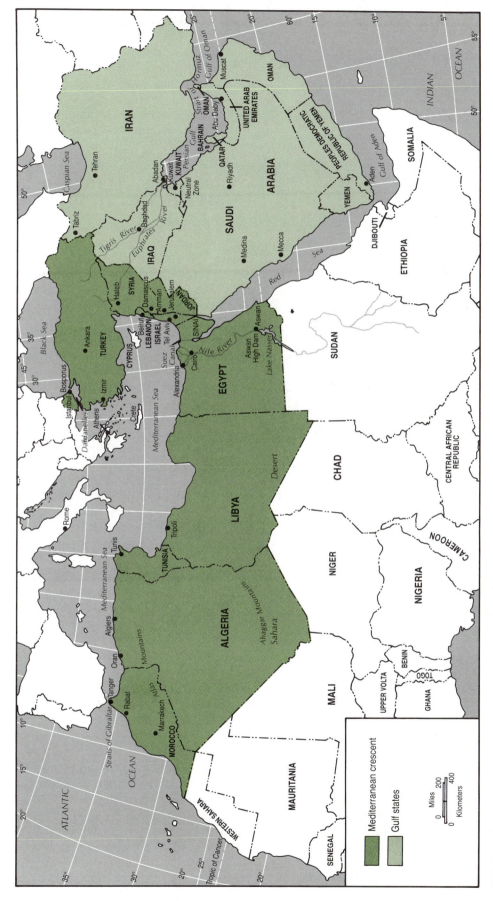

FIGURE 21–1 North Africa and the Middle East: General Locations.

tures, and religions, although the core elements of that mosaic remain the Arab people and the religion of Islam.

Crossroads Location

The Middle East and North Africa is also a bridge between Europe and Africa. With Europe, the Middle East shares many common scientific and literary traditions, and much of the ancient culture of Greece and Rome was preserved by Islamic scholars. Trade across the Sahara and overland between the Mediterranean and the Red Sea or Persian Gulf linked European economies with the Orient. Less benignly, European colonization in the nineteenth and twentieth centuries had a disruptive impact on Middle Eastern society. The conflicts engendered by this contact continue to disturb the region.

Aridity

Aridity is a salient feature of the region. Ribbons of cultivation along the region's few rivers, as well as islands of higher agricultural potential where mountain ranges capture more moisture, relieve the arid monotony. Few countries have more than 50% of their land surface useful for farming, forestry, or pasture. Many, such as Egypt and Libya, have only 3% of their territory usable for agriculture. The impact of aridity both constrains agricultural possibilities and encourages intensification.

Diversity

Any region that stretches from the Atlantic Ocean to Afghanistan can be expected to contain a great deal of diversity. In the Middle East and North Africa, no contrast is greater than between those countries with rich oil resources and those without. Morocco, Syria, Jordan, and Yemen are handicapped in their efforts to modernize and industrialize because they lack petroleum. In contrast, the extremely high per capita incomes of Kuwait, the United Arab Emirates, and Qatar are made possible by oil revenues. Other elements of diversity include relatively industrialized states such as Turkey versus preindustrial societies such as Yemen; sectarian fragmentation within Islam, Christianity, and Judaism; large, diverse economies (Turkey, Egypt) compared to less developed ones; large states (Algeria, Sudan) versus small (Bahrain, Abu Dhabi, Lebanon); and traditional value systems in conflict with modernist and secular schools of

thought. These themes are pursued in this and the following chapter. The remainder of this chapter will focus on the North African and Middle Eastern states that fringe on the Mediterranean Sea. Chapter 22 will focus on the Middle Eastern states that are oriented toward the Persian Gulf.

OVERALL FEATURES OF THE MEDITERRANEAN CRESCENT

Stretching in a crescent from the Atlantic coast of northern Africa to the eastern Mediterranean, the coastal zones of the Middle East are distinctive and distinguishable from the Islamic core areas of the Arabian Peninsula and the Persian Gulf (Figure 21–2). Several features of this coastal region set it apart from the Middle East core.

First, these coastal zones are an *interface environment,* a zone of contact between the Middle East and Europe. Indeed, the Mediterranean Sea has more often served as a link among the lands around its shores than as a barrier. This link has *ancient roots.* Greeks settled eastern Libya in the seventh century B.C. Carthage, in modern Tunisia, began life as a Phoenician trading colony. The Romans unified all of the Mediterranean Basin into one political unit, and in the process promoted trade, cultural contact, and population exchanges. Muslims, Jews, and Christians look to Jerusalem as a common focus for all or important parts of their religious experience. The Mediterranean has served as a permeable membrane through which have filtered the religious, artistic, literary, and scientific accomplishments of the peoples dwelling on or near its shores. In many instances, *cultural contact* has been associated with political conflict and conquest. Alexander the Great overthrew the Persian Empire and extended Greek culture deep into southwest Asia. Islamic expansion into Spain and the Balkans permanently influenced the culture and political organization of the southern peripheries of Europe. More recently, European colonial expansion in North Africa and the Levant has left a patina of European culture, a plethora of psychic scars, and a patchwork of political entities that pose problems for present policy makers. Contemporary labor migration from Turkey and North Africa brings a decidedly Middle Eastern flavor to many of Europe's metropolitan areas. As *a crossroads* of contact between east and west and as a seedbed of *cross-cultural hybridization,* the coastal Mediterranean crescent has acquired a character-

The recent increase in oil prices has brought great wealth to petroleum-exporting nations. Some of the North African and Middle Eastern nations are major oil producers. Active exploration continues, as shown here in Libya. (Photo Library, Mobil Oil Corporation)

istic diversity and complexity, an intricate, small-scale, mosaic-like texture both fascinating and perplexing.

This intimacy of contact has spawned *a variety of images and myths* in the West. Two are particularly prominent. One is associated with the *cities* of the coastal crescent that serve as contact points between Near Eastern and Western culture. Often they are pictured as sleazy and decadent, as a mélange of *casbah* and palace in which high life and low life, sophisticated culture and abject poverty are inextricably linked. In contrast stands the heroic image of the nomadic desert warrior. The practitioner of an austere, simple, openhanded lifestyle, the *nomad* depends on mobility and careful animal husbandry to survive in an inhospitable environment. The elegant simplicity of life in the des-

ert, together with its apparent rejection of crass materialism in favor of such values as freedom, self-reliance, and independence, appealed to Victorian romanticism. That most Middle Easterners were and are farmers rather than nomads and live in villages rather than in large cities has done little to dim the images launched in literature and perpetuated in movies and television.

A second feature of the coastal crescent is its contrasting *lowland centers of political and economic power* and its uplands, where political, ethnic, and religious dissidents have congregated for survival. The Levantine coast is a perfect example of the use of mountainous *uplands as refuge areas.* Alawites inhabit the Syrian coast, Maronite Christians dominate Mt. Lebanon, and Shiites and Druze are scattered across south Lebanon, south-

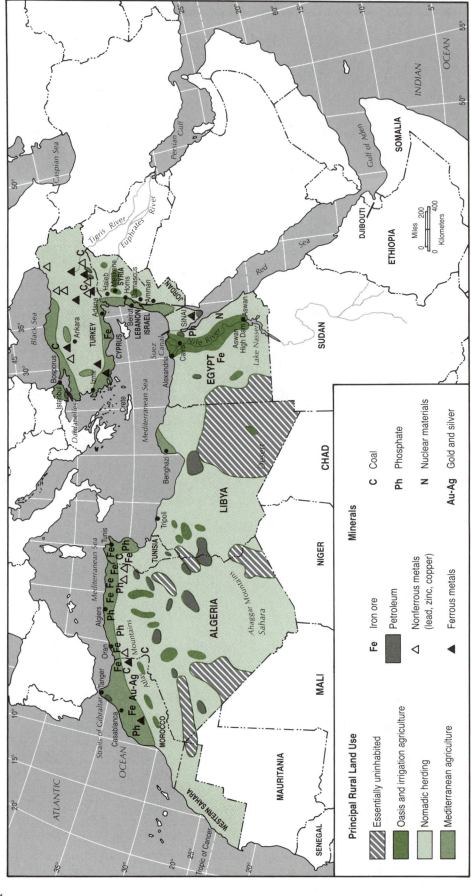

FIGURE 21–2 Land Use in the Mediterranean Crescent. Much of North Africa and the Middle East is too dry to support much economic activity. The region's population is concentrated in the more humid coastal sections and along the major rivers.

ern Syria, and northeastern Israel. In North Africa, Berber speech and culture survive in southern Tunisia and the mountain massifs of the Algerian and Moroccan Atlas. Habitually these mountain tribesmen have been at odds with the governments of the coastal cities. It is no accident that a main center of resistance to colonial regimes was found in such upland areas. The continuing presence of regions that are only partially integrated into the political, economic, and social life of crescent countries markedly complicates their internal unity.

As physical systems, these *mountain massifs* also dominate the landscapes of the Mediterranean crescent. With the exception of Egypt, which derives its water supply from the mountains of Ethiopia, the traditional livelihoods of the region are enormously influenced by mountain ranges. The precipitation that falls in the highlands supported the *forests,* now largely removed, that supplied the fleets of the Mediterranean states. As *catchment areas* for the perennial rivers of the region, these mountain zones are of critical importance to the agricultural prosperity of the lowlands.

The coast has also been the base from which the struggle between *farmer and herder* began. Many parts of the coast and the adjacent interior can be used for either nomadic pastoralism or sedentary agriculture. At times, people have shifted from one livelihood to another in response to political and environmental conditions. The forty years of wilderness existence experienced by the Israelites in their flight from Egypt and the decision by Lot to concentrate on herding and move into drier areas reflect the interchangeability of the practitioners of each livelihood system. In history, farmers moved into drier zones when security was guaranteed and environmental conditions were suitable. The power base of the farmer always was the urban-oriented coastal zone. When this power base weakened, settlement retreated and the nomad pressed into potentially cultivatable zones. The fluctuating fortunes of herder and farmer in controlling the land resources of the drier interior margins of the coast is a recurring theme in the settlement history of the region.

A further characteristic of the coastal zone is its

In recent years Tunisia has built a number of hotels and initiated other projects to attract increased tourist traffic. Shown here is a modern hotel where the government sponsors training of hotel personnel. At the right is the restored fortress of Monastir. (United Nations/B. Graham)

diversity of resources. Not only can pastoralism and extensive agriculture be practiced, but the region's position on the margins of several *tectonic plates* also makes *mineral opportunities* available. While these resources are generally limited and of greater importance historically and internally than they are internationally, some minerals are significant. Morocco is the world's largest producer of internationally traded phosphate; Turkey is a major source of rare but critical minerals such as molybdenum; Libya supplies a large amount of high-grade oil to Europe; and Cyprus was once an important source of copper for the Mediterranean world.

Location of minerals and agricultural products in proximity to the coast made trading easy. The Mediterranean has facilitated the exchange of products between the crescent and Europe and favored a *mercantile and maritime tradition* in the coastal ports. The more cosmopolitan and diversified character of these sites and their outward orientation sets them apart from less diverse parts of the Middle East.

It is in these coastal settings that many of the earliest *cultural and political traditions* emerged. These centers on the Nile, along the Bosporus, in the floodplain of the Orontes, and in northern Tunisia provide a depth and a richness to the local scene rarely found elsewhere. There is a continuity to the area's culture that increases the variety of the coastal zone. Few traditions have died out completely, and many fragments of the past survive to the present. Feuds with an ancient origin as well as the remnants and memories of past glories vie for attention.

Finally, of no less significance is the contemporary *pattern of interaction* between the Mediterranean crescent countries and Europe. This interaction builds on past patterns. Workers migrate in large numbers to overseas jobs. They tend to move to the major metropolitan areas of the former colonial rulers in part because language and cultural problems are minimized. Some adopt permanent or quasi-permanent residences. All send money home to their families. Even in time of economic difficulty in the host country, jobs held by the migrants continue to provide economic opportunities for them because native workers scorn work the migrants will accept.

Coastal states are *maximizing their resource base* by expanding the *employment opportunities* available to their citizens. For those people of the coast with technical skills and language compatibility with the rest of the Arab world, the flow is often away from Europe toward the opportunities present in the oil economies of the Persian Gulf. But whether it is selling skills or labor, the migrants from the coastal states are following a *time-honored pattern.* They are converting their talents and the *locational advantage* of a crossroads situation into economic opportunites that are less available to their more isolated inland cousins.

INTEGRATED ECONOMIES

Three Mediterranean crescent countries, Algeria, Egypt, and Turkey, are discussed in this chapter. Although each country possesses unique features, they have many basic similarities. Many of the remaining states of the Mediterranean rim are much smaller and are described in the following section. A few, such as Morocco, Tunisia, and Libya, share numerous features with the larger crescent states, and they are treated as variants on a common theme in this chapter.

Features of Similarity

First, one important common feature of Algeria, Egypt, and Turkey, in comparison with other Middle Eastern countries, is that they are relatively *self-supporting.* This self-sufficiency comes from a combination of three factors: (1) an *important industrial sector* that processes raw materials; (2) a diversified and productive *agricultural base;* and (3) a developed *service sector.* Coupled with a *large population* that includes a significant proportion of well-educated individuals, this diverse natural and human resource base is the foundation for the development of a balanced, integrated economy.

Second, these countries have a high proportion of their population living in urban areas. This *urban emphasis* is the product of a long, rich history of urban life. In these countries are found several of the region's largest cities: Cairo (5.1 million), Istanbul (3.9 million), and Algiers (2.2 million). Unlike many other countries where one large *primate city* dominates the national landscape, several urban centers have attained major size and importance: Ankara (2.6 million), Izmir (1.7 million), and Adana (1.0 million) in Turkey, Alexandria (2.3 million) in Egypt, and Oran (633,000) in Algeria. Egypt with 45.9 million, Turkey with 49.2 million, and Algeria with 20.7 million people all have large populations. These cities serve complex *functions.* They are administrative, industrial, cultural, trading, and, often, religious centers. Their economic role is enhanced because the economies of which they are a part are large, well-integrated, and relatively powerful.

Third, each country has a *rich agricultural base.* This base is essential both to feed the local population and to earn foreign exchange in overseas markets. The wines, dates, and olives of Algeria, cotton of Egypt, and citrus of Turkey are important export commodities. The domestic food-processing industries that agriculture supports are significant focuses of development. Nonetheless, in all three countries (but most particularly in Egypt where the pressure of population on agricultural resources is especially severe) maintaining and increasing agricultural productivity is a continuous struggle.

Fourth, a significant feature of each country's economic diversity is its appreciable *mineral wealth.* Algerian oil and natural gas are important world resources. The coal, iron ore, and chrome of Turkey are less abundant but provide the base for the largest heavy industry complex in the Middle East. Egypt's oil and phosphates meet internal needs and, together with hydroelectric power from the Aswan Dam, are important bases for economic growth.

Finally, environmentally each country is characterized by a large area of *semiarid and arid territory* of little value to the agricultural economy. Agriculture thus tends to be concentrated in limited areas and to be more intensive than in many other countries in Africa and the Middle East. Egypt with the narrow fertile strip of the Nile valley and delta is the extreme example.

Features of Diversity

The differences that distinguish Algeria, Egypt, and Turkey from each other are largely due to cultural heritage, historical development, and the impact of colonialism. Turkey and Egypt are states with a long, uninterrupted cultural heritage and a proud tradition of political and economic independence. In contrast, Algeria is a country in which more than a century of cultural conflict and discontinuity, terminated by a lengthy struggle for independence, interrupted indigenous patterns of life and development.

Both Turkey and Egypt share a common interest in their *cultural richness* and past achievements. This historical dimension gives an added impetus

to their pursuit of economic development. The richness of *Egyptian civilization* was a wonder in Greco-Roman times, and the splendor of its scholarly institutions increased after the Arab conquest in A.D. 640. At al-Azhar University and other religious centers in Cairo, Arab scholars were largely responsible for preserving the writings of the classical world and transmitting them to the Christian West. At a period when the Dark Ages characterized Western Europe, the acme of scholarship and invention was reached in the universities of the Middle East. Even when political independence was lost, first to Ottoman Turkish domination and then to indirect British political and military control, a spirit of *national independence survived.*

The Turks as the heirs of the *Byzantine Empire* assumed the mantle of empire in the eastern Mediterranean. So successful were their military campaigns that Vienna nearly fell to their armies in 1529, and only when outstripped by the industrial development of the West did the Ottoman Empire lose its internal cohesion. Even then, Turkey retained sufficient vitality and creative drive to renew itself and escape colonial rule. Turkey's post–World War I leader, Kemal Ataturk, took his country down a path of modernization, industrialization, and territorial reorganization that left Turkey politically and militarily stronger, and ethnically more homogeneous, than it had been in the last days of the Ottoman Empire.

Algeria

Whereas Turkey and Egypt were able to stave off many aspects of *colonialism,* or to avoid colonial control altogether, Algeria has been the scene of *major cultural conflict.* This conflict began with the establishment of a French protectorate in 1830. Large-scale migration of Europeans to Algeria followed, and tremendous pressures were placed on local cultural institutions. Today, despite an end to French control, Gallic civilization maintains a very visible presence in Algeria. This French-Arab conflict mirrors the long opposition of Berber culture, protected by its rugged mountain retreats, to a succession of Phoenecian, Roman, and Arab invaders. It is perhaps ironic that guerrilla opposition to the French has done much to reduce the differences between Arab and Berber in Algeria and speed the process of Berber assimilation into the mainstream of Arab culture.

Algeria has some similarities to other oil states. It has a rapidly increasing income from *oil* and is using this new wealth as the basis for economic growth. Like Egypt, however, Algeria has gained a position in the world far above that which might be suggested by its size and economic status. One reason for this position is Algeria's long struggle for independence from French colonial domination. Algeria sympathizes with and is respected by many Third World revolutionary groups. Another reason is the strong and expanding economy that Algeria is building. This economic strength is made possible by the good mixture of resources available to the nation.

Algeria has the most diverse and integrated economy of the four North African countries. With Libya, Algeria shares a wealth of *petrochemical resources*. Like its near neighbors, Morocco and Tunisia, Algeria possesses rich *agricultural resources* in the northern and coastal sections of the country. The combination of minerals (iron ore, phosphates, lead, zinc, and mercury) with petroleum and farming provides a breadth and richness to the resource base that is lacking elsewhere in North Africa.

Although 80% of the country is classified as too dry or too steep to be productive for crops and pasture (Table 21–1), the *desert* portions of Algeria

are vital to the national economy. Here *small oases* produce high-quality dates for export. The major resources in the desert, however, are *oil and natural gas* (Table 21–2). These two commodities produce 35% of the country's gross national product. Algeria has been a pioneer in the production of liquefied natural gas (LNG) for export to the industrialized states. Half of Algeria's LNG is shipped to the United States. To lessen this dependence on the United States, a pipeline to Italy is under construction. Most of the revenue earned by the sale of oil and natural gas is invested in the northern part of the country where the bulk of the population and productive agricultural land is located.

This *northern region* is relatively *fertile and well watered*. Once the granary of the Roman Empire, the area today grows wheat and other cereal crops, especially in the drier plateau country between the Saharan Atlas and the coastal mountain ranges. The coastal climate is ideal for citrus, vines, olives, and other fruits and vegetables, and European markets are close at hand. Outside the commercial agricultural zone, many peasants till poor soils for a low subsistence level of life; however, many people have left these subsistence areas and moved to the coastal cities where there

Bedouins from the south of Algeria graze their camels and sheep. Traditional ways of life are rapidly disappearing in many North African and Middle Eastern countries. (United Nations)

TABLE 21-1 Land Use in Algeria, Egypt, and Turkey, 1971 and 1980

	Total Land Area (thousand hectares)	Arable		Permanent Crops		Permanent Pasture		Forest		Other	
		Thousand Hectares	Percent of Total Area	Thousand Hectares	Percent of Total Area	Thousand Hectares	Percent of Total Area	Thousand Hectares	Percent of Total Area	Thousand Hectares	Percent of Total Area
Algeria											
1971	238,174	6,287	3	569	<1	37,761	16	2,424	1	191,133	80
1980	238,174	6,875	3	634	<1	36,321	15	4,384	2	189,960	80
Egypt											
1971	99,545	2,728	3	115	<1	0	0	2	<1	96,700	97
1980	99,545	2,700	3	155	<1	0	0	2	<1	96,688	97
Turkey											
1971	77,076	25,001	33	2,578	3	10,800	14	20,170	26	18,528	24
1980	77,076	25,354	33	3,125	4	9,700	13	20,199	26	18,698	24

SOURCE: Food and Agriculture Organization, *Production Yearbook* (Rome, 1982).

319

TABLE 21-2 Oil Production in Mediterranean Crescent Countries

Country	Total 1979 (thousand barrels)	Total 1980 (thousand barrels)	Percent Change 1979 to 1980
Algeria	445,300	347,600	−22
Egypt	193,672	213,798	+10
Libya	756,975	638,750	−16
Morocco	141	104	−26
Syria	59,500	62,952	+6
Tunisia	41,830	42,561	+0.02
Turkey	20,140	16,449	−18

SOURCES: M.-A. Nicod, "North Africa," *Bulletin of the American Association of Petroleum Geologists,* 65, No. 10 (October 1981), 2047–75; Darwin O. Hemer, John E. Mason, and Gregory C. Hatch, "Middle East," *Bulletin of the American Association of Petroleum Geologists,* 65, No. 10 (October 1981), 2134–61; and M.-A. Nicod, "Petroleum Developments in North Africa in 1979," *Bulletin of the American Association of Petroleum Geologists,* 64, No. 11 (November 1980), 1739–75.

are greater economic and social opportunites. Also, many Algerians, as well as Tunisians and Moroccans, have migrated to France. Their labor is important to the French economy, and their remittances sustain the family members left behind.

The basis for *industrial growth* exists in Algeria. Oil and gas resources are backed by phosphates and major iron ore deposits. Together with lesser quantities of coal, lead, and zinc, these resources form a solid base from which to promote industrialization. Skills and infrastructure are well enough developed to allow the major development initiative to remain firmly in Algerian hands.

A distinctive feature of Algeria is the single-minded *political strength* and acumen of its leaders. This feature has enabled the government to demand and get major *internal sacrifices* in return for the promise of *long-term benefits.* Luxury imports are strictly curtailed, and some 40% of the GNP is currently being reinvested in the economy. Such allocation of investment has enabled great headway to be made on a major iron and steel plant, a large plastics complex, and many small consumer industries. Algeria's recent yearly industrial growth rate of 7–10% is likely to continue unless a world oversupply of both oil and gas reduces the funds available for investment.

Algeria has been able to take a *leadership role* in the Arab group of countries, the Organization of Petroleum Exporting Countries, and the Third World in general. This position of influence in the Third World is matched by respect in both capitalist and socialist countries. As a result, Algeria was able to serve as middleman between the United States and Iran to help settle the hostage crisis that destroyed relations between those two countries in 1979–80. Algeria also supports those revolutions and wars of national liberation that it views

as legitimate. This position is a consequence of Algeria's own struggle for national independence against French colonialism, and explains why Algeria has so actively assisted the Palestinian Liberation Organization. There is good reason to predict that economic growth and political struggle will continue to be characteristic of Algeria.

Egypt

Egypt is a *country of contrasts.* Cairo and Alexandria are two of the largest cities in Africa, but peasants in Egypt live about the same as their ancestors did a thousand years ago. The Egyptian *economy is diversified,* with a wide range of basic and processing industries and many large and small consumer-oriented enterprises. Yet population grows at a rate of 3% per year and makes economic development *a constant struggle* to stay abreast of population increase. Egypt has the largest population of any country in the Arab world, yet only 3% of its land surface is arable. The country is a *center of Muslim culture* and tradition and is a *powerful political force.* For example, it manufactures most of the world's Arabic language films, and its colleges enroll 500,000 students, many of whom come from other Arab countries. At the same time, employment opportunities for college-educated Egyptians are limited and many must seek work elsewhere. More than 2.5 million Egyptians of all skill levels find employment in other Arab countries. This employment pattern increases Egypt's influence outside its borders. At the same time, the decline of pan-Arab nationalism and Egypt's support for the Camp David agreements, which established peaceful relations with Israel, have isolated Egypt politically from other Arab states, at least temporarily.

Egypt achieves its position of influence and potential power on a *limited resource base. Agricultural resources* are constrained by a lack of water and suitable soils. Except for a thin strip of land along the Mediterranean coast, there is very little rain in Egypt. Instead, cultivation is concentrated in the valley and delta of the Nile River where adequate water and good soils coincide. Only in a few small oases is significant cultivation possible outside the Nile Valley. Alfalfa, cotton, rice, maize, and wheat are the main crops. Of these, cotton is the most valuable export crop and it is also the basis for a significant and growing textile industry (see Table 21–3). *Land use is very intense,* and given the current socioeconomic situation, major breakthroughs to higher levels of production are unlikely. An attempt to achieve a breakthrough in agricultural production was made when a high dam was constructed at Aswan. *The Aswan High Dam* created a huge lake. This water represents the flood that annually inundated Egypt and provided a natural supply of irrigation water and soil-enriching silt to the fields. The Egyptians expected to *increase agricultural production in two ways.* First, releasing the stored water slowly made it possible to raise two, and sometimes three, crops a year on the same land. Second, surplus water could also be channeled into the desert to develop areas that were without rainfall.

Both good and bad *economic and ecological effects* have occurred since the completion of the dam. Desert soils have proven more difficult to manage than anticipated, and many development areas are afflicted with salinization (crop-growth-inhibiting salt accumulations in the soil). The spread of year-round cropping has expanded the habitat of *schistosomiasis,* a debilitating parasitic disease with a complicated life cycle that affects a large percentage of the workers engaged in irrigated agriculture. Moreover, even with water saved by the Aswan High Dam, Egypt faces serious water shortages in the foreseeable future. This impending *water deficit* encourages efforts to conserve water, to develop new water sources in Sudan, Egypt's neighbor to the south, and to achieve closer political and economic ties with Sudan. In contrast, *hydroelectricity* produced at the Aswan High Dam has promoted industrial growth, and a fertilizer industry has emerged to restore fertility to soils now deprived of Nile silt. Equally important, the dam has been *an important symbol* of Egypt's determination to achieve development. Thus the dam has had a psychological effect that may be as significant as its economic impact.

Much of North Africa and the Middle East is dry and sparsely populated. Where water is found it is associated with dense concentrations of people. Here is a satellite photograph showing the delta of the Nile in the immediate foreground. On either side of the delta is desert. Just above the delta is the Suez Canal connecting the Red Sea on the right with the Mediterranean Sea on the left. Across the Suez is the Sinai Peninsula. (NASA)

TABLE 21–3 Employment in Manufacturing in Egypt

Industry	Employed 1971 (thousands)	Employed 1976 (thousands)	Percentage Change 1971 to 1976
Food processing	107	118	+10.3
Textiles/footwear	240	291	+21.3
Wood products	8	9	+12.5
Paper products/printing	22	27	+22.7
Chemicals/petroleum/plastics	51	64	+25.5
Nonmetallic minerals (glass, etc.)	27	36	+33.3
Basic metal industries	16	37	+131.3
Fabricated metal products, machinery, and equipment	76	72.4	−4.7
Other	1	1	0
Total	548	655.4	+19.6

SOURCE: International Labour Office, *Year Book of Labor Statistics* (Geneva, 1981).

Despite efforts to increase agricultural productivity, Egypt's *food imports* now exceed $8 billion annually. This amount is a significant proportion of Egypt's trade deficit and increases the incentive to achieve agricultural growth. *A new effort* to promote growth is reflected in plans that integrate agricultural, industrial, and urban development in northeastern Egypt. These plans envisage the reconstruction and improvement of areas along the *Suez Canal and in the Sinai* that have been regained from Israeli control. If funds can be found for these projects, the Egyptians hope these areas will play the same role in the 1980s that the Aswan High Dam did in the 1960s and 1970s.

Industry is diversified and growing steadily. Between 1971 and 1976, the last year for which data are available, Egyptian industry expanded its labor force by 19.6% (Table 21–3). These are important gains, but even with this record of growth, there are *major constraints* on rapid expansion. Food processing, textiles, and petrochemicals remain the major industries. These industries reflect the important role of agriculture in the overall economy, the central position that cotton occupies with respect to the textile industry, and the significance of oil and phosphate resources. Most other industries, such as metal processing, are based on imported materials. Not reflected in the statistics in Table 21–3, however, is *small-scale manufacturing* organized on a traditional, household scale. Frequently engaging too few employees to be included in official statistics, these craft enterprises make an important contribution to the national economy.

Turkey: Unrealized Potential

The end of World War I witnessed the demise of *Ottoman Turkey* and the emergence during the next decades of a new, smaller, but more homogeneous Turkish state. Led by *Kemal Atatürk,* the forces of modernization cut the ties to the past. In this process *the army* played a crucial role and served as an integrating institution of national regeneration to replace the discredited Ottoman bureaucracy. Many old customs were abandoned, a secular state was created, and great impetus was given to industrial and agricultural development.

Turkey is unusual among Middle East states in two ways. First, it possesses sufficient *mineral resources* (coal and iron) to develop its own *heavy industry* without having to import basic materials. Second, it has more *usable land* as a percentage of its total territory than any other country in the region. These advantages, combined with virtual *self-sufficiency in petroleum,* help create the most powerful and integrated economy in the region.

Much of Turkey's development is centered on *industry.* More than 85% of the capital invested in development is absorbed by industrial projects. Until the global economic recession of the early 1980s, Turkey's industrialization program made rapid progress. Concentration on basic *metallurgy and textiles* stimulated the development of Turkey's primary resource base. By 1976, Turkish industry employed 737,900 workers, and manufacturing and mineral extraction generated 25% of the country's GNP. This record of significant *industrial growth* is genuine and impressive.

Nevertheless, rapid *population growth* throughout the 1960s and 1970s outpaced the ability of Turkish industry to absorb new workers. As a result, large numbers of Turks sought *employment abroad,* a pattern of labor migration that replicates the experience of the North African countries. The bulk of this migrant labor flowed to West Germany, Turkey's traditional European ally, and the *remittances* of this labor force were important forces for social and economic change.

The problems of sporadic industrial growth emphasize the continued importance of the *agricultural sector.* Most of Turkey's industrial products are consumed at home, and only rare minerals, such as chromite, meerschaum, and manganese, enter world trade. In contrast, three-quarters of all *exports* are agricultural products. Of these, cotton, tobacco, fruit, and nuts are the most important. Much of the *commercial crop production* is concentrated in coastal districts that have a mild dry summer subtropical climate, rich soils, gentle slopes, adequate moisture, good potential for irrigation, and relatively easy access to good transportation facilities. During the last fifteen years, these *coastal districts* have experienced considerable intensification in their agricultural activities. Because most of Turkey's farmland is held in private ownership, continued agricultural growth depends on recognition by farmers of the positive economic benefits that follow from a change in focus from subsistence to commercial activities. This *change in orientation* is being achieved in many areas as dry farm operations are converted to *irrigation,* more fertilizers are employed, and more high-yielding commercial crops are planted. These changes are assisted by liberal credit policies, government extension services, and support from cooperative societies in which Turkish farmers participate with increasing frequency.

These progressive trends in the coastal, commercial districts have not been matched elsewhere in the country. Although much of *interior Turkey* is suitable for *cereal cultivation* and *animal husbandry,* there are serious constraints on productivity and development improvements. High elevations, cool temperatures, short growing seasons, steep slopes, and semiarid conditions limit the productivity of much of the more traditional agricultural zones. Gains in arable crop acreage are made at the expense of permanent pasture (Table 21–1), the only category of agricultural land use to decline between 1971 and 1981. This pressure results in

overgrazing on the remaining pastureland and encroachments on and *degradation of remaining forest* land.

Although the productivity changes in Turkish agriculture have kept pace with Turkey's rapid population growth (2.1% a year), the government's continued emphasis on commercial and industrial crops at the expense of gains in food and livestock production could result in serious future difficulties. A *critical challenge* in Turkey's development in the next decade is to extend the modernization and development that characterize its urban, industrial, and coastal districts to the still largely rural and traditional interior areas.

A tough, hard-working, disciplined, and largely homogeneous people, a strong military, and a growing agricultural and industrial base have made Turkey a *powerful state.* Yet this political and economic power is seldom applied outside its own borders. There are three reasons for this situation, all of them relating to issues never fully resolved by Atatürk's revolution. First, Western notions of *democracy* are imperfectly transplanted in Turkey. The nation was founded by a strong military leader, and the military perceives itself both as a guarantor of the ideals of the revolution and as a preserver of national unity. As a result, the army has intervened several times in the national government.

Second, while the state Atatürk created is secular, 98% of the population is Muslim. With 54% of the population still working in rural areas, the influence of the religious leadership remains strong, and its proper role in the national structure remains unresolved. Turkey is unlikely to remain unaffected by the current of fundamentalist revival now surging through the Middle East. This reassertion of a fundamentalist view of the state finds a sympathetic ear among conservative Turks.

Finally, Turkey's *ambivalent location* between east and west also causes uncertainty over the best foreign policy to follow. While Turkish leaders remember traditional quarrels with Russia, a vocal minority continues to agitate for closer ties with the USSR. Turkey's control over the entrance to the Black Sea makes it a valuable member of NATO, but in the event of war between NATO and the Warsaw Pact countries, Turkey would be sure to experience the military attentions of the USSR. Moreover, Turkey and Greece, another NATO member, are locked in dispute over Cyprus, control of the Aegean Sea, and offshore mineral- and oil-exploration rights. These circumstances add to Turkey's uncertainty about future directions.

Caught between Europe and the Middle East, between modernization and tradition, between freedom and authority, Turkey struggles to establish a *sense of identity* and purpose. When Turkey achieves agreement on a course of action, its regional effect can be enormous. Prospects for a consistently powerful regional role, however, are limited.

SMALL STATES OF THE EASTERN MEDITERRANEAN: UNIQUE ECONOMIES

Four countries on the eastern margin of the Mediterranean, together with one increasingly recognized de facto state organization (the Palestine Liberation Organization, or PLO), present *a different set of issues.* In these countries, agricultural and mineral resources, basic factors for many countries, are of secondary importance. They are superseded by questions of *human resources,* of *service functions,* and of various kinds of *inflows of support* for the nations concerned.

Jordan, Lebanon, Israel, Syria and Palestine are the states in this group. Syria is significantly larger than the other states (Table 21–4), but its occupation of parts of Lebanon in the 1980s, its similar problems and potentials, and its involvement in the affairs of other eastern Mediterranean states is so great that it would be unreasonable to exclude it. In this group, Palestine, or more specifically the Arab population of the former country of Palestine represented by the PLO, is unusual; it is a state without a defined national territory. Ironically, this condition mirrors the situation of Israel some years ago. Despite its uncertain status, it is important to include Palestine here because of the fundamental issues involved. If we accept the four smaller "countries" as viable political states and erect a national profile in each case, it is clear that there are some strong similarities, although individual differences are also important.

The eastern end of the Mediterranean has a reputation as a *strategic area* controlling important nodes of communication and trade. Routes linking Africa and Asia and the Persian Gulf and Europe pass through the area. Control of these routes has long been important. For a time Beirut was a vital air link between Europe and the East. Now, however, jet travel and the political crises of the late 1970s and early 1980s have encouraged travelers to route their flights through other intermediate points.

The *religious significance* of the area is remarkable. Here is the seedbed of the Jewish religion, out of which grew the various Christian denominations. The area is also significant to Muslims. While Mecca is the central holy place of Islam, Jerusalem, the site of the Prophet's ascension into heaven, is a major pilgrimage center. Medieval maps show Jerusalem as the center of the world with Asia, Africa, and Europe all focused toward it. Present power politics have greatly modified but not totally destroyed this view. From a religious and a geopolitical viewpoint, the area is still vital to the world at large.

Israel

Israel is a *major focal point.* It is unique among countries in that it became established territorially on the basis of two-thousand-year-old claims of the Jewish people to the area and as a result of the persecution of Jews in many countries of the world. The establishment of a nation in disputed land long occupied by indigenous peoples unsympathetic to the Israeli national cause was bound to provoke conflict. The small state of Israel, set up in 1948, was difficult to defend against surrounding Arab pressures. As a result of the 1967 war, Israel was able to "rationalize" its frontiers but at the territorial expense of Jordan and Egypt. The Palestinians, who had been dispossessed of part of their lands in the 1948 hostilities, found all their traditional territory in Israeli hands after the 1967 war. Not only did this war spark a new wave of *Palestinian refugees* into neighboring states, but subsequent hostilities, culminating in Israel's 1982 invasion of southern Lebanon, pushed refugees from the 1948 conflict into a new round of *forced migration.* Thus Israel's search for a secure and peaceful existence involves both permanent insecurity for Palestinians and possible counterproductive results for Israel itself.

Apart from the continued struggle to maintain and expand its territory, what kind of state is Israel? Statistics give some clues. In an area where per capita GNPs are generally low, Israel's is $5450, approximately four times that of its eastern Mediterranean neighbors. Despite a high rate of inflation, Israel is a comparatively *rich country* in a poor part of the world.

Some of the wealth comes from the industry of

TABLE 21–4 Land Use in Israel, Jordan, Lebanon, and Syria, 1971 and 1980

	Total Land Area (thousand hectares)	Arable		Permanent Crops		Permanent Pasture		Forest		Other	
		Thousand Hectares	Percent of Total Area	Thousand Hectares	Percent of Total Area	Thousand Hectares	Percent of Total Area	Thousand Hectares	Percent of Total Area	Thousand Hectares	Percent of Total Area
Israel											
1971	2,033	324	16	85	4	818	40	109	5	697	35
1980	2,033	325	16	88	4	818	40	116	6	686	34
Jordan											
1971	9,718	1,125	12	180	2	100	1	125	1	8,188	84
1980	9,718	1,190	12	190	2	100	1	125	1	8,113	84
Lebanon											
1971	1,023	240	24	85	8	10	1	95	9	593	58
1980	1,023	240	24	108	10	10	1	73	7	592	58
Syria											
1971	18,457	5,641	31	260	1	7,553	41	471	3	4,532	24
1980	18,457	5,282	29	454	2	8,378	45	466	3	3,877	21

SOURCE: Food and Agriculture Organization, *Production Yearbook* (Rome, 1982).

325

the people. The farming cooperatives of Israel are well known. The *Kibbutz* and the *Moshave* are two main types of agricultural production and marketing cooperatives that help Israel to maximize production from land that is fertile only in the narrow coastal plain. Efficient use of water, moved southward from the Sea of Galilee along the national water carrier, assists in the intense cultivation of citrus fruits, vegetables, and some grain crops (Figure 21–3). *High-quality farm exports* make a significant contribution to the economy, although only 7% of the population work on farms. Extremely efficient use of water is absolutely necessary for Israeli agriculture and urban life. Almost all of Israel's available water is currently being exploited, and desalinization of seawater is still too expensive to be a practical alternative. *One-third of Israel's area is unusable* (most of this comprising the Negev Desert); aridity makes much of the permanent pasture low in productivity. In order to increase agricultural yields, Israelis have had to become the most *efficient managers of water* in the world and to develop *innovative techniques* in irrigation, hothouse agriculture, and rainfall and run-off management.

Israel is basically an *urban country;* Tel Aviv along houses 20% of the country's population. *Industry* is an important part of the economy, and diamond cutting, textiles, the manufacture of woolen goods, and many other small industries provide a livelihood for the bulk of the population.

Another striking feature of the country is its *high level of education*. There are four main universities in Israel, and many immigrants are highly qualified academically and technically. Israel is one of the few countries that has difficulties because of the general high level of manpower resources. Moreover, in some fields the quality of Israel's technical and academic achievement is unsurpassed. Israel's work on hydrology, on horticulture, and in some electronic fields is particularly important. As a result, significant Israeli aid has been given to developing countries. With its high standards, technological sophistication, large, well-educated elite, and substantial European-derived population, Israel is more European in its values and orientation than it is Middle Eastern. Thus Israel is doubly at odds with its neighbors, for political conflicts are reinforced by a clash of cultural values that increases the difficulties of attaining mutually suitable compromises. This *conflict* is also mirrored within Israeli society, since Jews who migrated to Israel from other Middle Eastern countries often find themselves at odds with Jews whose origins are European.

Even with such rich and varied resources, it appears that Israel would not be a self-sufficient state were it not for the large *flow of capital* received from Jews around the world. The high burden of defense costs is one cause for this capital inflow, but the maintenance of current levels of living based on internal resources might be difficult even without the defense costs.

Lebanon

Lebanon, sharing the eastern Mediterranean coast with Israel, is also small both in area and population. Its *small size* in no way spares Lebanon from *cultural complexity*. Its society is highly fragmented

Irrigation is a necessary component of modern Middle Eastern agriculture. Here farm workers harvest a crop on a Kibbutz near Tel Aviv. (United Nations)

FIGURE 21–3 Small Eastern Mediterranean States. The small states along the eastern coast of the Mediterranean Sea are the scene of serious confrontation. The Arab-Israeli conflict continues. Part of the conflict revolves around territory occupied by Israel after wars with its Arab neighbors (West Bank, Gaza Strip, and Golan Heights). Palestinians displaced by the Arab-Israeli conflict present another source of confrontation. Lebanon, half Christian and half Muslim, has become embroiled in the conflict, and its existence as a nation is threatened.

along religious and ethnic lines. The political parties and armed militias that have proliferated during the last two decades usually are based on a Christian or Muslim community or sect. A combination of Lebanon's rugged mountain interior and strategic economic location attracted many of these groups

as refugees or traders. The very mountains that promise protection also place strict limits on the prospects for economic growth in the agricultural sector. As a result, Lebanese have for many years moved out from the poor agricultural lands of their country to become *merchants and entrepreneurs* in many parts of the world. The traders in many African capitals were until recently dominantly Lebanese; the Lebanese are the bankers of the Arab world and parts of Africa too. Large numbers of Lebanese moved to South America, and important Lebanese communities can be found in many American cities. Remittances from this overseas Lebanese population have supported many of the country's rural villages at living standards well above their usual subsistence level.

This overseas activity has always been balanced by the steady growth of Beirut as the financial center of the Arab world; as more and more traders returned to their homeland, this role has intensified. Beirut has also played a major *educational role* in the Middle East. At the American University of Beirut, 30% of the student population is composed of non-Lebanese Arabs. Because its curriculum has stressed a liberal arts orientation and because new universities are being founded elsewhere in the Middle East, American University and other Beirut institutions of higher learning have lost influence to schools stressing engineering, science, and technology. The mixture of *Christian and Muslim* that characterizes Lebanon, together with a considerable veneer of French culture from the colonial era, make Beirut the most *cosmopolitan city* of the Middle East. Yet there are deep social, cultural, political, and religious antagonisms that can flare into serious conflict.

Lebanon has no firm agricultural base, although bananas and citrus fruits grown in the coastal plain and deciduous fruits from the upland form the basis of an export trade with surrounding Arab states. *Industry* is only now beginning to expand. Metal goods, processed foods, textiles, and pharmaceuticals are leading sectors.

Lebanon was a major center for *Arab services.* Seventy percent of the Lebanese national income before 1975 was derived from the service sector, and 55% of the active population worked in service-related jobs. Moreover, until recently oil sheikhs were reluctant to invest in non-Arab countries. As a result, much of the capital that could not be employed in development projects at home was invested in Lebanese real estate. Thus, Lebanon was a *classic crossroads economy* whose lo-

327

cation resulted in a good standard of living (per capita GNP, $1070) and a steadily expanding economy as the Arab hinterland grew in wealth.

Lebanon's *status changed* in 1975. Tensions long masked by Lebanon's cosmopolitan veneer burst into the open. The pact that accompanied independence in 1943 guaranteed a dominant political and economic role to Christians, particularly the Maronites, the largest Christian group. In fact, the president of Lebanon always had to be a Maronite. Over time the numerical *balance between Christian and Muslim* populations changed, but the shift was not reflected in the political structure. Although jobs were also distributed on a community formula, the nation's wealth was not equitably shared. Moreover, Lebanon was host to a large Palestinian population. In 1975 *a civil war* erupted in which many were killed, and effective central government collapsed.

Lebanon's long-term *future remains cloudy.* Torn by conflicting internal pressures, it suffers from a major identity crisis. *Thoroughly Arab,* it is neither fully Christian nor Muslim, neither laissez-faire nor socialist, neither pro- nor anti-Palestinian. Until these problems are solved, Lebanon will be unable to realize its full potential.

Palestine

The most *volatile element* in the politics of the Middle East is the Palestinian Arab population. Creation of the state of Israel in 1948 resulted in the suppression of Palestinian nationalism and a dis-

persal of Palestinian Arabs among neighboring states. Many fled to the Gaza Strip, which fell under Egyptian control, to the West Bank (annexed to Jordan), or to Syria and Lebanon. Grouped into refugee centers, they subsisted on United Nations relief handouts. *Unabsorbed* into the structure and economy of their fellow Arabs, *unreconciled* to the loss of their homes and property, the Palestinian Arabs developed a distinct national consciousness. Neighboring governments have been unwilling to accept them as full members of their states. Only Lebanon has granted citizenship to limited numbers of Christian Palestinians. In many cases Palestinians have refused to be absorbed because they wish only to regain their ancestral land. Their numbers were swelled by additional *refugees* after the 1967 war, and the 1982 fighting in Lebanon further increased the refugee total (Table 21–5).

This population has never been accurately counted. Although the numerically largest *concentration of Palestinians* is found in Jordan, where they comprise nearly 40% of the total Jordanian population, it is only in the Gaza Strip and the West Bank that Palestinians constitute a distinct majority. Even on the West Bank, present Zionist settlement schemes call for the establishment of many new Jewish settlements.

Jordan

Jordan and the future of the Palestinian people are closely linked. Jordan's *population comprises three elements: Palestinians,* the pre-1948 settled *Arab*

TABLE 21–5 Total Estimated Palestinian Population, 1982

Country or Area	Total	Percent of Total Palestinian Population	Percent of Host Country Population
Egypt	50,000	1.0	0.1
Gaza Strip	400,000	8.3	99.0
Iraq	30,000	0.6	0.2
Israel	600,000	12.4	15.4
Jordan	1,300,000	26.8	39.4
Kuwait	300,000	6.2	21.4
Lebanon	700,000	14.4	21.9
Libya	25,000	0.5	0.8
Qatar	30,000	0.6	15.0
West Bank	800,000	16.5	97.0
Saudi Arabia	150,000	3.1	1.4
Syria	230,000	4.7	2.5
United Arab Emirates	34,000	0.7	3.4
United States	100,000	2.1	—
Other	100,000	2.1	—
Total	4,849,000	100	

SOURCE: *Boston Globe*, 11 August 1983.

population of the Transjordan, and the *Bedouin.* Originally created after World War I as a British mandated territory with a member of the Hashimite family as its ruler, Jordan possessed few agricultural and mineral resources and a sparse population of largely pastoral nomads. These nomads were loyal supporters of the Hashimite dynasty, and today approximately half of the Jordanian army is of Bedouin origin. After 1948 and until the 1967 war, Jordan occupied large sections of the West Bank and attempted to carry out programs of industrial and agricultural development throughout its territory. These programs have been hampered by Jordan's *limited resource base,* by significant land use problems associated with deforestation, overgrazing, and soil erosion, and by military conflicts with Israel and with supporters of the PLO.

In a sense, Jordan is *a homeland in search of a people.* Over time there has been a drift eastward across the Jordan River of Palestinian refugees and West Bank Arabs with Jordanian citizenship. The loyalty of this Palestinian population to Jordan is uncertain. Numerically they outnumber the Bedouin who dominate the government and army. Israel would prefer to see Jordan designated as the Palestinian homeland, since this would undercut PLO demands for a national territory in land now occupied by Israel. King Hussein, the Bedouin, and non-PLO Jordanians reject such a solution because they are certain to lose power as a result. The PLO opposes this proposal for both philosophical and pragmatic reasons. Unless it is linked in some fashion to the richer and better developed West Bank territories, an economically viable Jordanian state is hard to envisage.

Syria

Unlike its smaller neighbors, Syria is *endowed with resources.* These resources are primarily agricultural, but small-scale mineral and modest petroleum deposits exist. Although much of the country lies in the rain shadow of coastal mountain ranges, sufficient rainfall occurs in the central and northern areas to support *dry farming.* Here cereal grains and Mediterranean tree crops (olives, grapes) form the important cultigens. Eastward, rainfall becomes too small in amount and too variable in occurrence to make cultivation secure and successful.

Historically, the areas east of Homs and Haleb (Aleppo) were important zones of *nomadic animal husbandry.* They continue to be important producers of animal products for the urban market. Now organized on a cooperative basis, these pastoral districts are increasingly integrated into the cereal-based agriculture of the fertile semiarid zones along the Homs-Haleb axis. Efforts to intensify production and reduce the impact of drought are taking place through the use of *mechanization,* the integration of fodder rotation into grain-farming operations, and protection and restoration of rangeland productivity.

The major capital investments in agricultural intensification are concentrated on irrigation. *Irrigation* is an ancient technology in Syria. The shadoof (lever-operated bucket), qanat (underground water-collection tunnel), and noria or dulab (waterwheel) have long been used to raise water to agricultural fields. After World War II, major efforts were made to develop the irrigation possibilities of the *Orontes ('Asi) River* using modern technology. The gains from these development projects, although important, were minor when compared to Syria's total arable area. Attention then turned to the *Eurphrates Valley,* which represented the last major source of water available for irrigation. Work has begun on a giant dam project 20 miles (32 kilometers) south of Meskene. By creating an artificial lake 50 miles (80 kilometers) long, the new dam will provide water to irrigate 1.5 million acres (607,000 hectares). This project will more than double Syria's irrigated area and dramatically increase crop yields, as well as produce more cotton for export.

The resulting improvement of Syrian standards of living could prove considerable. The Euphrates Dam's *hydroelectric power* capacity of 12 million kilowatts will exceed the power potential of the Aswan High Dam in Egypt and should help compensate for Syria's lack of coal or oil energy resources. Construction of the dam, however, will make it necessary to displace more than 60,000 farmers from rich floodplain areas that will be inundated. Although plans call for resettling these displaced villagers, not all farmers are happy with the prospect of relocation. The *social costs* attendant upon such dislocation and social upheaval, together with possible unforeseen *ecological side effects* of the project, constitute the major hidden costs of this vast, technologically sophisticated scheme. The expectation is that local problems will be compensated by substantial gains at the national level and that the potential threat of conflict with Iraq over distribution of Euphrates River water can be avoided. There is little doubt that most Syrians consider such problems to be minor distractions on the road to progress and development.

Syria's major *regional significance* resides in its role as a frontline state confronting Israel politically and militarily. This role has involved Syria in several military confrontations—invariably losing ones—with Israel. It has also involved Syria in a protracted, and largely unsuccessful, intervention in Lebanon, as well as the loss of the *Golan Heights* area to Israeli annexation (a loss Syria does not accept). The high cost of the struggle with Israel includes a diversion of energy and resources from development. The struggle also partially masks conflict within Syria between orthodox Muslims and the exponents of Pan-Arab socialist nationalism.

No region of the world is more intimately associated in the popular mind with *oil* than the Middle East. This image is not unreasonable, because more than *one-quarter of the world's known oil reserves* are found in the region. Yet this wealth of "black gold" is unevenly distributed within the Middle East, and some nations, by accident of geology and of discovery, have much greater oil reserves than others. So abundant are the proven reserves and present output of such oil producers as Kuwait and Saudi Arabia that they deserve consideration in a separate category. The *oil industry dominates* the economy of these states and pours more money into national treasuries than the countries can spend, in the short run, on productive internal enterpises. With prosperity so closely tied to one resource, the oil states must grapple with a complex array of developmental prospects and problems.

These problems are greatest in the countries surrounding the Persian Gulf (Figure 22–1). The overwhelming *impact of oil* has dramatically *altered the economies* and *shaken the social systems* of the countries listed in Table 22–1. Even countries that produce limited amounts of oil, such as Oman, or that lack the resource entirely, such as the Yemens or several of the sheikhdoms that compose the United Arab Emirates, have become enmeshed in the oil economy of the region. Often their *citizens migrate* to neighboring countries in search of employment. This disruption has serious consequences for both labor-exporting and labor-absorbing societies. Countries such as Iran and Iraq that in the early 1970s possessed viable agrarian economies and whose economic structure mirrored the diversity of the integrated economies of the Mediterranean crescent were *destabilized* by social changes unleashed by an expanding oil economy.

Thus all countries in or near the Gulf have experienced change as a result of the exploitation of petroleum resources. This chapter examines these developments in *three contexts.* First, it explores

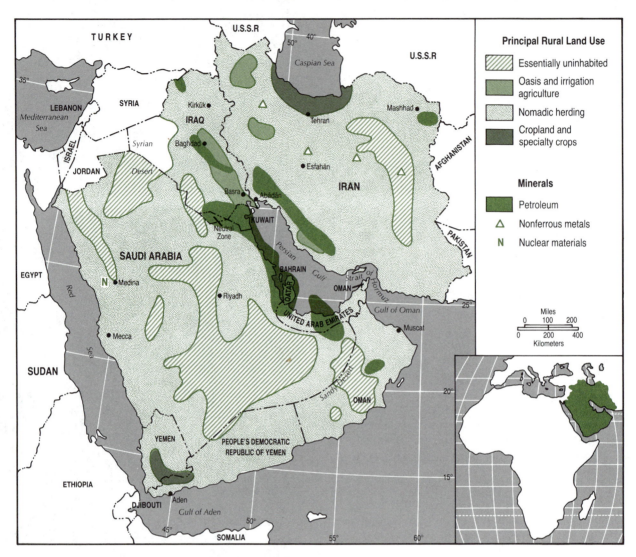

FIGURE 22–1 Land Use in the Persian Gulf States.
Aridity characterizes the Persian Gulf region, and much of the area is sparsely inhabited. The wealth derived from petroleum is being used, in part, to find ways to support the region's population in spite of the dryness.

the *regional contrasts* that characterize the region. Second, it reviews the uses to which the *petroleum-generated wealth* of the region has been put. Finally, it highlights the *impacts and changes* that have followed from petroleum-based development.

REGIONAL CONTRASTS

Although the states of the Gulf have a common experience through involvement in the oil economy, they also have other similarities and differ-

ences. The first similarity that they share is a *limited natural resource base*. Only hydrocarbons are available in abundance. Other minerals are nonexistent or found in small quantities. These minor deposits are not extensive enough to make mining profitable under present conditions. Only Iran possesses appreciable mineral reserves that a modern metallurgical industry can be based on. Because much of the Gulf region, particularly in its more mountainous districts, remains to be explored by modern methods, some caution is justified in any pessimistic assessment of the mineral resource potential of the region. But, for the present, only

TABLE 22–1 Gulf States: Population Density and Petroleum Production

Country	Population Density		Petroleum Production	
	Per Square Mile	Per Square Kilometer	Thousands of Barrels	Barrels Per Person
Bahrein	1,563	603	16,862	42
Iran	67	26	485,450	11
Iraq	86	33	346,750	24
Kuwait	233	90	342,151	214
Oman	12	5	120,654	121
Qatar	71	27	147,786	493
Saudi Arabia	13	5	3,512,697	338
United Arab Emirates (Abu Dhabi, Dubai, Sharjah)	43	17	558,998	399
Yemen, North	76	29	0	0
Yemen, South	16	6	0	0

SOURCES: Darwin O. Hemer, John E. Mason, and Gregory C. Hatch, "Oil and Gas Developments in the Middle East in 1981," *Bulletin of the American Association of Petroleum Geologists,* 66 (1982), 2011–33; and *Population Data Sheet* (Washington, D.C.: Population Reference Bureau, 1983).

petroleum and natural gas, often found in deposits of staggering size, are of great significance.

Second, *agricultural resources* are equally limited. Most of the Gulf states suffer from extremely *limited water supplies.* Water supplies are concentrated either in favorable highland areas where orographically induced rainfall occurs, as in Yemen or western Saudi Arabia, or in lowland oases where groundwater is close to the surface. Only in Iran and Iraq are substantial areas suitable for cul-

Petroleum is rapidly changing the face of the nations bordering the Persian Gulf. Oil from this region is shipped to the far corners of the world. Here in Saudi Arabia an oil refinery and storage depot stand in stark contrast to the desert inland. (UNATIONS from ARAMCO)

tivation. Iraq has both rain-fed agriculture in the north and irrigation potential in its arid south based on the Tigris and Euphrates rivers. Iran's mountains and Caspian Sea coast receive appreciable rainfall, although its central core is extremely arid. Yet in both Iran and Iraq, less than a quarter of the total land surface can be cultivated. Elsewhere slopes are too steep or rainfall too limited for non-irrigated agriculture. The *traditional qanat system* of tapping deep groundwater and of bringing it to the surface by gravity flow originated in Iran. It is an ingenious response to water shortage but is applicable only on alluvial fans. Population growth, agricultural intensification through irrigation, and urbanization all place *great pressure on local water supplies.* Overexploitation of groundwater, problems of waste disposal, economic limitations of desalting ocean water, and salt contamination of irrigated soils are problems that affect the region's limited water supplies.

A *third similarity* of most Gulf states is *low population densities* and *small total populations* (Table 22–1). There are exceptions. Iran and Iraq both have large populations, and some of the small states have very high population densities. Even in countries with low overall population densities, effective density is much higher since the population is concentrated in the small highly productive zones. Thus in Abu Dhabi approximately three-fourths of the emirate's total population is concentrated into one large urban area of the same name. The contrast between *small nodes of dense population* and high levels of economic activity and the

333

vast expanses of essentially uninhabited space characterizes the landscape of the Gulf.

Fourth, until oil development created new employment opportunities and sparked massive population movement to a few urban centers, much of this population was engaged in *traditional livelihood activities. Oasis agriculture* and *pastoral nomadism* constitute the two main traditional systems. *Urban traditions* generally are weak in oil states, and the educational and technological sophistication of the bulk of the population is limited. Skills lie almost entirely in the traditional sector of the economy and are unsuited to employment in many aspects of the oil industry. This *shortage of human skills* suitable to the modern sector of the economy has serious political consequences. It encourages the importation of skilled personnel from the industrialized countries as well as Indians, Pakistanis, Iranians, and Palestinians to fill specialized job niches. Dissatisfaction with these conditions has led most oil states to engage in vigorous *educational and job-training activities* in an effort to replace foreign experts with local personnel.

A fifth similarity is the overwhelming importance of *wealth derived from oil* in the local economy. The *size of oil deposits is staggering.* More than two-thirds of the known oil reserves outside North America, the USSR, and Eastern Europe are found around the Persian Gulf. Saudi Arabia alone may possess more oil reserves than any other nation; and it is already the world's third-largest producer after the United States and the USSR. Beginning with the discovery of oil in the Iranian hills at the head of the Gulf (1908), the pattern of location and development has spread southward through the coastal regions of the Arabian Peninsula. Not only have *new petroleum deposits* been located, but also knowledge of the areal extent and extractable quantity of oil in existing deposits has grown rapidly. Coupled with *improvements in the technology* of recovery and the traditional tendency to understate or otherwise obscure the reserve capacity, the *continued dominance* of the oil states in the world petroleum economy seems assured for the foreseeable future.

This dominance, however, is unlikely to be as great as it was during the 1970s. There are a number of reasons for this emerging trend. A combination of global *economic stagnation* and *energy conservation* has reduced demand significantly in the industrialized countries. *New petroleum deposits* have been located in other areas. *Political turmoil* in Iran and a prolonged *war* between Iran and Iraq damaged production, refining, and oil transportation facilities in both countries. Others have reduced output in an *effort to maintain high prices* as well as to conserve oil reserves as a base for possible petrochemical industrial development. Although these decreases in output are unlikely to continue indefinitely, it is also unlikely that *the Organization of Petroleum Exporting Countries* (OPEC), of which the Gulf states are critically important members, will regain its previous position of dominance in the oil industry.

This *period of dominance* allowed most oil-producing states to participate directly in oil operations. Conditions of the early years of development were reversed, and these countries gained more overt control over their own resources. In the past, *foreign companies,* often operating in a consortium such as Aramco (partly owned by several American oil firms), received concession rights to explore defined areas. Most revenues were generated from concession sales, royalties, and taxes paid on oil extracted and exported. Local governments had few skilled administrators; oil companies possessed a *monopoly* on extractive technology and on market distribution in the industrialized states. Moreover, the impoverished and politically weak states of the region lacked the capital to develop the oil resources.

Gradually, this *situation has changed.* As administrative and technical skills accumulated and as better understanding of the intricacies of oil economies developed, Middle Eastern countries demanded more favorable agreements with the oil companies. Tough new leaders appeared. Increased percentages of profits from the oil fields were returned to national treasuries, and greater pressure was placed on oil companies either to develop their concessions or relinquish them to someone who would. Many countries also *nationalized* at least a controlling interest in the firms that were developing their resources. This was made possible by a world market setting in which demand for oil outstripped supply. Nationalization and dramatic increases in oil prices took place in tandem. These unusual conditions are unlikely to recur in the near future.

A final characteristic of the region is the tremendous *impact of oil revenue* on the local society and economy. No state in the region has been able to isolate itself from oil wealth. The *stresses* placed on indigenous social and cultural systems have been great. This stress has led to controversy over the role *traditional values* should play in governing contemporary life. A wave of *fundamentalist Islamic revival* was responsible for much of the opposition to the shah of Iran. This fundamentalism stresses the relevance of traditional values and patterns of behavior in response to, and in opposition toward, the values of Western modernization. The more rapid and extreme the pressure for change, the more violent was the reaction from traditional centers of authority and belief. So violent was this response to secular modernization and nationalism in Iran that it swept away much of the new in its affirmation of traditional practices.

Another major consequence of oil development has been an *end to isolation* and an increase in the region's dependence on nonregional suppliers for basic necessities. Much of the food consumed in the Gulf is imported, and in many countries more than half the labor force is foreign. This *increase in dependence* is not equally shared by all countries, but it is present as an important disruptive force in all.

DEVELOPMENT AND THE PETROLEUM ECONOMY

The discovery and development of vast petroleum deposits have had a *drastic impact* upon the economies and sociopolitical systems of the oil states. A large petroleum output plus soaring oil prices have moved such former backwater states as Kuwait, Abu Dhabi, and Oman to international prominence. Formerly impoverished, the oil states now face the task of coping rationally with an embarrassment of riches. How to dispose of the *petrodollars* is a crucial issue. Although abundance has been the watchword of the recent past, the *long-term future* is by no means assured. Small states such as Oman have sufficient revenue for the present but must consider how to use the money wisely to prepare for the day when oil revenue is no longer available. The *need to diversify* an oil-based economy, to develop industrial processes not dependent on oil, and to invigorate an often sluggish agricultural sector before there is any long-term decline in oil revenues are obviously high priority items for the Gulf states.

There are *two approaches* that Gulf states have taken to deal with this issue. The first is to invest petroleum income within the nation in development projects that generate long-term income and employment, as well as to foster social services that improve material well-being in the short term.

The second is to invest capital that cannot be usefully absorbed internally in overseas enterprises.

Internal Investment of Petroleum Income

Large sums of petrodollars are currently expended to create *infrastructure improvements.* Ports such as Doha in Qatar have been built where none existed before. Investments in airports, highways, sewer systems, water systems, and pipelines have drastically changed the face of most oil states. *Housing projects* are also common. Oman is in the process of rebuilding its capital of Muscat. In constructing modern housing units for a growing urban population, Oman is destroying much traditional architecture, but to most citizens this change is seen as an inevitable and progressive aspect of modernization.

Substantial sums are also being invested in *agricultural improvements.* In Kuwait *desalinization* plants are now in operation whose output is used not only for drinking water but also for irrigation of vegetables to help feed the burgeoning population of Kuwait City. This type of agricultural operation is not economically feasible in the strict sense. Rather, it is justified by a desire for a reduced volume of food imports in a setting of nearly unlimited financial resources. In Saudi Arabia, where date palms provide the traditional staple crop, vigorous attempts have been made to diversify agricultural

A Kuwaiti fisherman stands on a quay, following a calling which shows little change over the centuries. Across the way are modern cars, the evidence of an ancient society revolutionized by oil revenues. (United Nations)

335

production. Encouragement of cereal and vegetable production promises to diversify the country's agricultural base. Production increases should result from such technological developments as the drilling of artesian wells to expand the agricultural area.

Few of these *agricultural development schemes* are practical in economic terms. Almost all of them result in considerable social disruption, and most require a level of skill that is beyond the experience of local participants. Frequently the technology employed contains backlashes that result in serious *land management problems.* Many of the irrigation schemes in Iraq, for example, have encountered serious *salinization* problems. Both continued government subsidies and the extensive involvement of expatriate labor and expertise have been required to keep such projects from collapse. Little effort is made to build on areas of local expertise. Especially neglected is the *pastoral sector* where considerable traditional experience still exists, and where historical use of low-productivity rangeland has been well managed. Today most Gulf states import their meat from industrialized countries such as Australia even though the population prefers to consume the local, now largely unattainable, sheep.

In Oman investment of oil revenues in agricultural development is especially crucial, since depletion of existing reserves early in the twenty-first century is anticipated, and prospects for additional discoveries are discouraging. Because agricultural land is limited by a low and unreliable rainfall and great expansion of farmholdings is not possible, Oman aims to achieve food self-sufficiency by improving the efficiency of existing agricultural systems, and by developing underground water resources (including the *qanat* system). Control of plant diseases, improvement of marketing and transport facilities, discovery of new and more productive plant strains, and the use of fertilizers and mechanized equipment where they meet the needs of local farming traditions are some of the projects now under way. Concentration on the agricultural sector is a *feature of development* in the oil states if a base is to be laid for economic viability in the future. Yet these traditional skill areas are the ones most likely to be abandoned in the face of competition from cheap synthetic products and low-cost food supplies imported from abroad.

Industrial development also is taking place. First priority is frequently given to the construction of cement factories to support the construction of industry. Plants based on *the refining of oil products,* such as fertilizer, chemicals, plastics, and gas liquefaction, are common. So too are consumer-oriented enterprises including soft-drink bottling plants, flour mills, fish-processing establishments, and textile-weaving firms. Often *craft traditions* form a focus for small-scale industrial development that caters to the tourist and export trade. The uniquely designed silver jewelry of Oman is one example. In many oil states, rug making, slippers, pottery, brass, copperware, and leather products such as hassocks represent a nexus for productive development as well as cultural continuity.

Few resources to support heavy industry are known. *Mineral exploration,* however, is in its infancy, and prospects exist for major discoveries. Reports circulate of major finds of gold, silver, nickel, and copper in mountainous western Saudi Arabia. If these reports are confirmed, Saudi Arabia may be able to translate its oil wealth into a diversified economic base that includes indigenous mineral-based industry.

Attempts to overcome *skilled manpower shortages* are also being made. Nearly every country has invested in modern universities and a rapidly developing primary and secondary school system. To staff these systems, numerous foreign teachers have been hired, often recruited from more educationally advanced countries such as Egypt. Large numbers of students, however, also study abroad under governmental contract. Upon their return they are funneled into decision-making positions at all levels of the government. It is their newly acquired expertise that underlies the rapid replacement of foreign personnel in the larger oil states of Iraq and Saudi Arabia and the tougher and more aggressive posture of governments dealing with the international oil companies.

Petrodollar Investment Overseas

Faced with an inability to absorb all of their oil revenue in internal investment projects, oil states seek *investment opportunities overseas.* Many of these transfers of funds represent the actions of wealthy private investors. Most frequently they reflect the actions of oil-state governments. Increasingly these transfers of capital have involved the purchase of property, banks, farmland, and industrial enterprises in the industrial countries. Governments have become active investors in the industrial infrastructure of the oil-consuming countries.

THE IMPACT OF OIL

While the Gulf's extraordinary economic importance from a global perspective is unlikely to endure, oil will continue for the next several decades to have *a profound impact.* This impact is most pronounced in three areas: (1) international political relationships; (2) internal social and economic conditions; and (3) internal political structures.

Changing Political Power Relationships

Oil wealth has *altered the balance of power* in several ways. Most of the oil states are now able to use their rapidly accumulating wealth to purchase arms with which to modernize their military forces. Possession of sophisticated weaponry made Iran the dominant political and military force in the Gulf during the last years of the shah's administration. This situation sparked demands for enhanced military preparedness among other states in the region. The Arab-Israeli conflict added an additional incentive to states in the region to achieve parity with real or perceived adversaries. The result has been *an arms race* in the Gulf in which the industrialized states of both the socialist and capitalist camps supply one or both sides of a potential conflict. Moreover, the region's dominance in OPEC gives it substantial force on the world political scene. Proposals to raise prices or threats to embargo oil shipments to supporters of Israel have sent tremors racing through the industrial world. The oil states are not insensitive to their *position of influence,* and there is every reason to believe that the oil states are not afraid to use their power. The *psychological satisfaction* derived from this change from an inferior to a preponderant political position explains much of the contemporary behavior of the oil states. They can no longer be ignored or dominated but must be dealt with as countries of great power and influence.

This position of influence gives the Gulf added *geopolitical significance.* The USSR and the United States compete for influence in this region. Because many of the European and Far Eastern allies of the United States depend on the oil of the Gulf for their economic well-being, any change in the internal politics of the region takes on added significance. The collapse of the shah's Iranian government, whose military forces were seen as a guarantee of regional security, raised fears of potential political and economic damage. Involvement by the USSR in Afghanistan is seen by the West as part of a Russian drive toward warm-water ports. It is also viewed as a threat to vital Western economic interests. This extraordinary level of interest and concern is a measure of the geopolitical significance of the Gulf in world affairs. The emergence of the *Gulf Cooperation Council* (Saudi Arabia, Kuwait, Oman, Qatar, the United Arab Emirates, and Bahrain) as a regional organization represents a local response to the need for a unified approach to common problems.

Changing Social Conditions

Developmental growth fueled by oil wealth has worked massive *changes in the societies* of the oil states. One important change has been *rapid urbanization.* Attracted by new job opportunities, social welfare programs, and a milieu of excitement and diversity, rural folk have flocked to the rapidly growing urban centers sparked by the oil boom. Doha, the capital of Qatar, leaped from insignificant village status to more than 80,000 inhabitants in the space of twenty years. Similar growth has taken place in the capitals of most oil states and

Much of North Africa and the Middle East is sparsely populated, since water is lacking. This satellite photograph shows an arid and sparsely inhabited part of Yemen in the southwest corner of the Arabian Peninsula. In the foreground is Wadi Hadramawi, a steepsided valley of an intermittent stream. In the background is part of the Gulf of Aden. (NASA)

sheikhdoms, Baghdad now exceeds 3.2 million. Tehran is a metropolis of more than 4.3 million, and Riyadh, Saudi Arabia, has passed the 600,000 mark. In some of the Gulf sheikhdoms, the majority of the population lives in the capital city: Kuwait City contains 750,000 inhabitants; two-thirds of Abu Dhabi's 450,000 population reside in the capital; Manama, Bahrain, is an urban center of 150,000, virtually all of that state's population. In these instances, city and state coincide. Second cities have also served as a regional migration focus in the large Gulf states. Among them are Mosul and Basra in Iraq, Esfahan, Tabriz, and Mashhad in Iran, and Mecca and Jidda in Saudi Arabia. Much of this growth has been accompanied by *shantytowns* and other temporary housing on the outskirts of the urban area. The result is a set of *sanitary and health problems* because of excessive crowding. These problems are gradually being removed as oil money is invested in improved housing and infrastructure.

Centers of economic growth and change, the expanding cities are also the scenes of *cultural conflict.* For rural migrants, adjustment to urban life is often difficult. Moreover, it is in the urban milieu that traditional Islamic and customary values are receiving their strongest challenge. Bombarded by a variety of exotic stimuli, ranging from the material possessions of the industrialized West to the clothes, values, movies, and behavior patterns of the foreign employees of oil and construction firms, the citizens of the oil states have been forced to adjust to a new social setting. In many cases the shift has involved moving physically and conceptually from the Middle Ages to the twenty-first century.

This *conflict of values* can produce serious social strains. Traditional leaders often react violently to apparent violations of cultural and social norms. In Iran much of the opposition to the shah was led by religious leaders who objected to the pace and direction of social change. When they gained power, one of their first actions was to make religious law the basis for social relations and for the legal system. The *resurgence of fundamentalist Islamic* approaches to social organization represents a profound challenge to the secular, modernist, Western-inspired governments of the region.

Equally profound are *differences in work habits and social priorities.* Western notions of the sanctity of time are seldom accepted by the new urban arrivals, and financial gains are often applied to social needs such as bride price or kin obligations rather than savings or job advancement. Tension between workers and foreign employees because of inadequate understanding of respective cultural values is often the result.

Great *income and status disparities* have also appeared as a result of the oil boom. Although some of the newly found wealth has filtered down to lower social echelons, much of the income gained is eroded by an inflationary spiral triggered by oil development and accelerated imports. Individuals close to the import trade, those serving as local representatives of foreign firms, and professionals of all sorts have benefited most from petroleum growth. The unskilled have been left behind. Some states such as Kuwait have instituted a progressive policy of free social services to improve living conditions, but others have been less farsighted. The gap between the ruling elite and the bulk of the population is often wide.

Changing Political Relationships

Social unrest stemming from culture change and income inequities shadows the political future of many of the oil states. Opposition to the central government of many of the oil states comes from a variety of sources. These *sources of tension* include ethnic and religious minorities within the state, traditional leaders alarmed at the changes taking place, rural people feeling neglected by a distant central government, and modernist forces wishing to accelerate the pace of change and channel its energy into a radical restructuring of society. In addition, many oil states have large numbers of foreign workers within their jurisdiction who feel no particular loyalty to the existing government. The result is a highly *unstable set of conditions* that could explode into conflict at any moment.

The *Kurds* are an example of an ethnic minority in Turkey, Iraq, and Iran that has long struggled for greater autonomy either within the structure of the existing states or through complete independence. In Iraq the Kurdish population of the north was for years engaged in armed rebellion against the Arab-dominated central administration. Although a desire for greater regional autonomy was the obstensible motive, increased access by Kurds to revenue from the northern oil fields in or near Kurdish territory was an underlying motivation.

Revolutionary Iran has many sympathizers in the nonorthodox religious minorities in the Gulf countries. Thus the implications of Iran's emergence as a regional power are numerous, even though they are very different from what they were during the shah's rule, and the prospects for confrontation over the political future of the Gulf region are very real.

Equally likely are governmental changes by *military coups.* Young officers in many of the oil states possess nationalistic and Pan-Arabic ideals. Impatient with the slow rate of economic progress, the pronounced social and economic inequalities in many oil states, and the corruption that frequently characterizes a boom economy, they are apt to use their position to seize power. Concern for excessive social permissiveness and employment of foreign nationals are also likely to spark unrest in the lower levels of the military hierarchy.

REGIONAL PROSPECTS

The petroleum economy is changing many features of the political, economic, and social fabric of the oil states. Prospects for considerable instability, exacerbated by the uncertainties of the Arab-Israeli conflict, are very real. The traditional life-styles of the oil states are struggling to adjust to the forces of change. The challenge facing the Gulf countries is how to harness their petroleum wealth to promote the enduring development of their economies. How well they can accommodate development and change without the drastic alteration of traditional values will determine the degree of stability and enduring change the region will experience.

FURTHER READINGS

ALLEN, J. A. *Libya: The Experience of Oil* (London: Croom Helm; Boulder, Colo.: Westview Press, 1981). A perceptive study of development without capital constraints, as well as its social and economic impacts.

BEAUMONT, PETER, GERALD H. BLAKE, and J. MALCOLM WAGSTAFF, *The Middle East: A Geographical Study* (London: Wiley, 1976). Systematic examination of problem areas blended with country case studies to make a most reliable overview of the contemporary Middle East.

CLARKE, J. I., and HOWARD BOWEN-JONES, *Change and Development in the Middle East: Essays in Honour of W. B. Fisher* (New York: Methuen, 1981). Contemporary change focusing especially on population dynamics and economic development.

COLE, DONALD P., *Nomads of the Nomads: The Al Murrah Bedouin of the Empty Quarter* (Chicago: Aldine, 1975). A cultural-ecological examination of a rapidly changing life-style that highlights the ambivalent aspects of cultural change.

COTTRELL, ALVIN J. (ed.), *The Persian Gulf States: A General Survey* (Baltimore: Johns Hopkins University Press, 1980). The issues of economic development examined in the light of the Gulf's culture, history, and art.

FARVAR, M. TAGHI, and JOHN P. MILTON (eds.), *The Careless Technology: Ecology and International Development* (Garden City, N.Y.: Natural History Press, 1972). A collection of articles on the problems resulting from the employment of alien, large-scale technology for development in Third World countries.

FERNEA, ELIZABETH WARNOCK, *Guests of the Sheik: An Ethnography of an Iraqi Village* (Garden City, N.Y.: Anchor Books/Doubleday, 1969). Organized around the seasonal cycle of village activities; a graphic and sensitive presentation of the life space of ordinary people, particularly women.

LUSTICK, IAN, *Arabs in the Jewish State: Israel's Control of a National Minority* (Austin: University of Texas Press, 1980). A perceptive study of the Palestinian minority in Israel and their relationship to the Israeli state.

PETERSON, J. E., *Yemen: The Search for a Modern State* (Baltimore: Johns Hopkins University Press, 1982). The impact of modernization on traditional life-styles and government policy are examined by a noted Middle Eastern authority.

WATERBURY, JOHN, *Hydro-Politics of the Nile Valley* (Syracuse, N.Y.: Syracuse University Press, 1979). Focuses on the political aspects of water resource development in the Nile Basin including Egypt's concern over control of the river's headwaters.

WILKENSON, J. C., *Water and Tribal Settlement in South-East Arabia: A Study of the Aflaj of Oman* (Oxford: Clarendon Press, 1977). A fascinating and satisfying multidisciplinary exploration of the fabric of economy and culture in a little-known part of the Middle East.

IX

Monsoon Asia

Clifton W. Pannell

23

Monsoon Asia: Overview

Three related but distinct regions make up *Monsoon Asia:* South Asia (Indian subcontinent), Southeast Asia, and East Asia (Figure 23–1).[1] More than one-half of the world's people live in Monsoon Asia, many in great poverty. In the valleys of the Indus and Huang He (Hwang Ho)[2] lie the remains of two of the world's important early *centers of civilization* and culture. Their cultural influences and traits have spread to other parts of Monsoon Asia where they were modified. The introduction of European habits and practices has added new dimensions to this large culture mosaic.

Recognition of South Asia, Southeast Asia, and East Asia as regions is based on a series of *cultural and geographical features.* The interaction of these different cultural features and environmental settings underlies the development processes in Monsoon Asia. South Asia is a physical unit, the Indian subcontinent. It is also a social unit with similar cultural origins and characteristics. East Asia has less physical unity, but many of the region's cultural characteristics originated in China. For example, Tibet and Inner Mongolia have special administrative identities within China; they also have close historical and ethnic ties with the Chinese. Southeast Asia, composed of many smaller and medium-sized states, has less cultural homogeneity than its two neighbors. In many ways it is a *buffer* between Chinese and Indian civilizations and has received significant contributions from both. The unity of Southeast Asia is based on its tropical and maritime character.

[1]Japan with its modern economy is not included in this treatment of Monsoon Asia. Yet its loss detracts little from the significance of the large region. In a region characterized by great contrasts, size, and large numbers, Japan is simply one of several medium-sized states both in population and area. Its remarkable economic growth of the last century, however, has distinguished it from other states in the region. Today, as a consequence, Japan serves as an excellent model of sustained progress and rapid development, highly visible to other Monsoon Asian states. Japan is covered in Chapter 14.

[2]See the footnote on page 376 for an explanation of the Chinese system of spelling Chinese names.

India and China are of special importance. Both countries have *gigantic populations,* and both are important industrial states. Both, however, are poor, and their per capita output of goods and services is very low. Both seek rapid economic growth, but by different systems. In each, successes and failures are evident.[3] Both are watched closely and sometimes emulated by other poor nations.

ENVIRONMENT: THE BASIS FOR HUMAN OCCUPANCY

The *distribution of population* in Monsoon Asia reflects an ordered relationship of people with particular physical environments. For example, the most dense population concentrations are in the great *alluvial valleys and floodplains* of the major rivers. The main clusters are located in the Indo-Gangetic Plain of India; in the Changjiang (Yangtze River) and Xi (Hsi) River basins and the northern plains of China; in the Tonkin delta of Vietnam; and along the Irrawaddy, Chao Phraya, and Mekong rivers in mainland Southeast Asia (see Figures 24–3, 25–2, and 26–5). Even where there are no great rivers, Asian peasants have clustered in the smaller alluvial basins and plains. The exceptions—peninsular India, the island of Java, and the Loess Plateau of north China—are explained by special circumstances and historical events. Several large arid areas—western and northwestern China and Mongolia—lie in Monsoon Asia. Sparse populations in these arid areas reflect an environment with limited potential for food production. Sedentary cultivation is difficult, and many inhabitants are pastoralists.

The point is that in Monsoon Asia people have gathered and proliferated most where the *agricultural potential* has been greatest. This concentration of population has meant that water has played a key role; and where water is available, agriculture generally has been oriented to *wet rice (paddy).* Although there are many different methods of growing wet rice and yields vary considerably, rice is the principal food crop in Monsoon Asia. Where environmental conditions permit, two crops a year are customary in China. In India, on the other hand, *double-cropping* is not widespread, even where the environment is suitable.

Population clustering reflects the degree of *accessibility* as much as the nature of the environment and the agricultural potential. Perhaps the best example of this principle is seen in the Yangtze Basin of central China. Historically, areas within the Yangtze Basin where navigation was possible prospered and grew in population. Away from navigable streams, the economy stagnated, and population grew slowly. More than a thousand years ago food grains were transported by water from the Yangtze region to the ancient capitals in the plains of northern China, a pattern that suggests a remarkable degree of economic and spatial integration for parts of China. Accessibility and the ability to move foodstuffs cheaply have been very important in determining where people could and did concentrate.

THE MONSOON AND THE RHYTHM OF SEASONS

Monsoon Asia is the southeastern quadrant of the Eurasian landmass. The term *monsoon,* believed to be of Arabian origin, describes a seasonal reversal in wind direction. In simplest form, this reversal results from the different heating and cooling rates of land and water in the summer and winter months, coupled with latitudinal shifts in major air masses. Thus during the summer in the northern hemisphere southwesterly (onshore) winds dominate, whereas during the winter northeasterly (offshore) winds are customary.

The Indian monsoon has *three distinct seasons,* and the most important is the June-September rainy season. The warm, moisture-laden winds blow from the southwest, strike the southern coast of Sri Lanka (Ceylon), usually in late May, and move north onto the Indian peninsula by June.

The rainy season in South Asia is followed by a comparatively mild, *dry autumn and winter season.* Although its duration varies by location, generally this period of cooler and drier weather extends from October to March. In March the *hot-weather season* begins, and by May it becomes almost unbearable with average daytime temperatures in many places above 100°F (38°C). Northern India, with its dry air and cloudless skies, is hottest, and relative humidity is very low. The landmass heats

[3]S. Swamy, ''Economic Growth in China and India, 1952–1970: A Comparative Appraisal,'' *Economic Development and Cultural Change,* 21 (1973), 1–83, and W. Mandelbaum, ''Modern Economic Growth in India and China: The Comparison Revisited, 1950–1980,'' *Economic Development and Cultural Change,* 31 (1982), 45–84. Mandelbaum and Swamy have developed analytical comparisons of the two states and compared a number of quantitative indexes of growth.

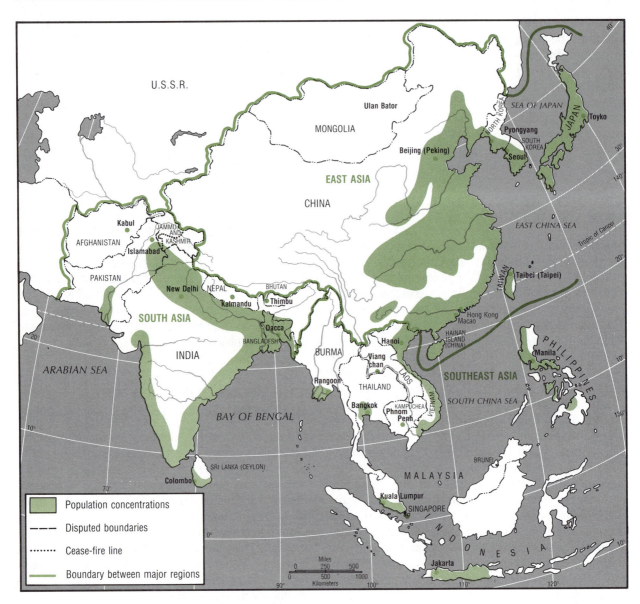

FIGURE 23–1 Monsoon Asia: National Political Units, Capital Cities, and Population Concentrations.

More than half the world's population is found in Monsoon Asia, and most of these people are concentrated in alluvial plains and river basins.

up, and a low-pressure system is created that begins to attract onshore wind, the precursor of the *summer, wet-season monsoon.* Usually southwesterly winds become well established quite suddenly in June and bring life-renewing precipitation. The rainy season begins anew, and the South Asian monsoon seasons have completed a full cycle. This southwest monsoon is divided into an Arabian Sea branch and a Bay of Bengal branch. Winds from the Arabian Sea strike the sharp ridge barrier of the Western Ghats, are forced upward,

and dump very heavy rains on the windward slopes of the Ghats. In the east, winds from the Bay of Bengal move rapidly northward and arrive at Calcutta in early June. These winds blow north as far as Khasi Hills, where a piling-up effect forces them northwest up the Ganges Plain. By the end of June the winds have reached Delhi, in early July, Lahore and Rawalpindi in Pakistan.

In *East Asia* the winter monsoon is characterized by cold and dry winds that sweep out of Siberia. In northern China average winter temperatures are often 5–10°F (3–6°C) cooler than at comparable latitudes and locations in Anglo-America. The winds first move eastward, then southward toward the equator. Some winds flow across the warm Sea of Japan and the East and South China seas, where great quantities of moisture are added to the air mass. When these winds strike land areas, heavy precipitation can result. More commonly, however, winters are dry and cool.

During the summer monsoon (May-September) in East and *Southeast Asia,* the winds move in great arcs toward a center of low pressure situated in central Asia. Moisture is brought overland, and the pattern of a summer maximum of rainfall described for South Asia is characteristic, with the monsoon first influencing the Indonesian Archipelago, then mainland Southeast Asia, and finally China.

The monsoons determine, in good part, the *agricultural calendar* of farmers throughout the region. Despite some use of irrigation, in South Asia it is the arrival of the southwest monsoon that brings the surge in farming operations and to which the peasants adjust many of their activities. During the extremely hot and dry season preceding the arrival of the summer rains, rural life has slowed. The earth is parched and cracked, and little can be done to prepare for sowing. With rain, human activity resumes at an intense pace. Soil preparation, planting, weeding, and the many tasks associated with the growing season must be compressed into a relatively brief time.

In East Asia, the monsoon pattern generally yields not only a wet-dry but also a warm-cool alternation of seasons. In winter, over much of China and especially that area north of the Qinling (Chin-ling) Mountains and Changjiang (Yangtze River), the Siberian high-pressure system controls climate. The result is the dominance of cold, dry air flowing east and south from the vicinity of Lake Baikal. Consequently, winters in north China are cold and dry and those in south China are cool and dry. Some winter cropping is possible, but generally the tempo of farming operations relaxes in the winter and practically ceases on the Loess Plateau and in Northeast China. Late spring (mid-April–mid-June) brings not only moisture but also amelioration of the low temperature. Farming activity picks up and the entire cycle of soil preparation and planting gets under way. Precipitation peaks during the summer months when it is needed most to nourish the rapidly growing crops under conditions of high average daily temperature and longer periods of sunlight essential for the process of photosynthesis.

THE CULTURAL BASIS

The cultural variety of Monsoon Asia is well expressed in the mosaic of ethnic and racial types and in the astonishing number of spoken and written languages. All three major gene pools—black, white, and yellow—have contributed to modern mankind in Monsoon Asia, although it is the Caucasoid (white) and Mongoloid (yellow) types that dominate contemporary racial patterns. More than *a thousand languages* are spoken in Monsoon Asia, and among these Mandarin Chinese, Bahasa Indonesian, Hindi, Bengali, and Chinese Wu are spoken by millions of people.

Religious affiliation is also varied. The dominant *Asian religions* are mystical in character and recognize many gods, but there are also the largely Christian Philippines (and formerly Vietnam) and Muslim states such as Indonesia, Malaysia, Bangladesh, and Pakistan. China and other Marxist states, including Vietnam, officially deny or discount religion. In some areas religious influences are declining with possible far-reaching effects as old traditions are cast aside.

Other *distinctive cultural features* are present in greater or less degree throughout Monsoon Asia. Architectural styles, agricultural systems and fertilization practices, and the organization of administrative systems have been used as a means to define specific regions. Political and administrative institutions and land-tenure relationships also provide good measures of similarity and differences among the major regions or cultural realms of Monsoon Asia.

PROCESSES OF DEVELOPMENT

The nature of the development process in different areas is also a distinguishing characteristic. A Chinese (Sinic) realm and an Indian realm are relatively simple to delineate and define by their *distinctive development strategies.* Southeast Asia is a shatter zone, fragmented into many political, cultural, and geographical subregions. The term *shatter zone* has been used commonly in political geography to describe regions of political and cultural instability and fragmentation. Normally such instability and fragmentation retard economic prog-

345

ress, but if shattering leads to solution of political turmoil, it may, over the long term, promote progress. For example, Bangladesh is more viable as a political unit than was culturally and physically divided Pakistan. Whether such fragmentation retards or accelerates economic growth and national development is hard to estimate. One method of examination is to compare the development performances of some of the various component states within the regions. Such a comparison may permit an evaluation of alternative routes to national development.[4]

Alternative approaches to development in Asia span the gulf between Communist economies and free-enterprise economies. Political approaches offer an equally broad set of alternatives from highly authoritarian systems of control to Western-style democracies. Some of the more obvious examples useful for evaluative comparisons of economic-

growth performance are India/China, South Korea/North Korea, and Burma/Thailand. Note that some economic growth goals may have been achieved at some sacrifice of personal and political freedom.

Political events since World War II have been unhappy for most of Monsoon Asia. Strife, war, famine, and hard times are common themes that run through the recent history of the region. These same themes occur in other developing areas, but such cataclysmic events as China's civil war, the end of colonialism in South and Southeast Asia, the Vietnam War, the internal ravaging of Cambodia, and the forging of Bangladesh are more common and frequent in Monsoon Asia. Moreover, such events in Monsoon Asia involve so many lives that they take on special significance.

Monsoon Asia has hope and promise as well as tragedy and misery. The whole range of success and failure in the experience of developing states exists, and most of the approaches can be studied. The panorama of the less developed world is here, and we can review and learn from the range of human conditions under situations of poverty and growth, tradition and modernity, technological backwardness and innovative change.

---

[4]Donald Fryer, *Emerging Southeast Asia, A Study in Growth and Stagnation,* 2nd ed. (New York: Wiley, Halsted Press, 1979). Fryer's geography, as the title implies, is focused on the theme of differential growth rates and performances. Monsoon Asia offers a number of excellent comparative situations.

24

South Asia: Past and Present

Known to the world as the *Indian subcontinent* or South Asia, this region today is composed of three large countries—India, Pakistan, and Bangladesh—and three smaller states—Sri Lanka (formerly Ceylon) and the Himalayan states of Nepal and Bhutan (Figure 24–1). To the northwest is Afghanistan, a transitional state with historic ties to both South Asia and the Middle East. In the adjacent Arabian Sea, Bay of Bengal, and Indian Ocean are numerous islands and island groups (of which the Laccadive, Andaman, and Nicobar islands belong to India) that also are part of South Asia.

India dominates the South Asian subcontinent, but several of the bordering countries are themselves large. Bangladesh and Pakistan rank among the ten most populous countries in the world, and Sri Lanka, Afghanistan, and Nepal are among the fifty largest countries in population.

The *creation of Pakistan* in 1947 and Bangladesh in 1971 resulted from special political, religious, and linguistic problems. The politically divided region of *Jammu and Kashmir,* whose status is still ambiguous, is another illustration of the political issues that have fragmented the subcontinent. Nepal and Sri Lanka are closely akin to India in cultural makeup, but each has managed to maintain autonomy and independence. The small Himalayan kingdom of Bhutan is a border state that shares some of the physical characteristics of Nepal. The origins and ethnic composition of Bhutan, however, are more closely connected with Tibet, and the cultural heritage is associated more with *Lamaistic Buddhism* than with the *Hindu* and *Muslim religions* more commonly associated with the subcontinent. Sikkim, formerly a semi-independent nation similar to Bhutan, was incorporated as an Indian state in 1975.

Included within these six countries is a considerable range in levels of development. All are poor by Western standards, but several have demonstrated substantial capacity to initiate new programs for development. Success to date, how-

FIGURE 24–1 Political Units, Principal Cities, and Major Rivers of South Asia. Although there are six other national political units in South Asia, India dominates the subcontinent as a consequence of its large population and area.

ever, has been difficult to measure. In most of these countries, especially Bangladesh, Pakistan, and India, problems related to recent political disputes and military conflict have inhibited growth.

An examination of the history and cultural origins of South Asia and of the contemporary situation can help us understand the nature of the de-

velopment processes under way in that region and the likelihood of their success or failure.

ORIGINS AND CULTURAL DIVERSITY

The early *origins of the population* are still in dispute, but it is well established that sedentary cultivators were farming the Indus floodplain more than five thousand years ago. Indeed, the term *India* is derived from the Sanskrit word for river used to describe the Indus. Although little is known of the ancient civilization that emerged in the Indus floodplain, it was focused on the relic cities of Mohenjo-Daro and Harappa (Figure 24–2). These cities were well planned and constructed and indicate the presence of an advanced civilization, the equal of ancient contemporary civilizations in lower Egypt and Mesopotamia. These early inhabitants are sometimes labeled *Dravidians,* referring to a highly mixed linguistic and cultural group that is found today mainly in southern India.

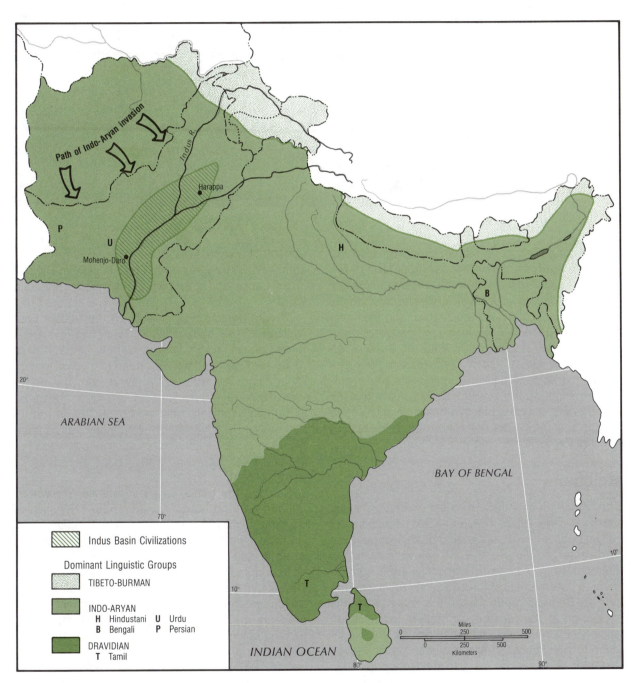

FIGURE 24–2 Early Civilizations and Major Languages of South Asia.
South Asia, although culturally and politically fragmented today, has its ancient roots in the Indus Basin civilizations. Today Hindi and related Indo-Aryan languages prevail throughout most of the subcontinent.

A cluster of houses on a hillside in the Katmandu Valley indicates building style and practices. In the distance may be seen the farm terraces that represent adjustment of agriculture to the rugged terrain of the Himalayas. (United Nations)

A second distinctive human group entered the subcontinent from the northwest mountains of Baluchistan and Afghanistan in successive migratory waves. These invaders were Caucasoid people, commonly described as *Indo-Aryans,* and they infiltrated the subcontinent over a long period of time. Although other invasions included the army of Alexander the Great, Islamic Moors, and Persians and Afghans, none had the significance of the earlier blending of the ancient Dravidians and early Indo-Aryans. It was the synthesis of these two groups that produced the *Hindu-based civilization* we today refer to as Indian.

Religions

The formation of India and Indian civilization is tied closely to the *Hindu religion.* A formalized set of religious beliefs with societal and political ramifications emerged very early. Local priests established themselves at the apex of the social strata and became a potent political force. So influential was the priestly Brahmin group *(caste)* that even today it is looked on as the capstone of modern Indian-Hindu society. Its members, whether rich or poor, continue to wield considerable power at the local level.

Other religious groups also developed in India. *Buddhism,* a spin-off of Hinduism, appeared early and enjoyed a flourishing but brief period of dominance. Buddhism spread throughout much of the

rest of Monsoon Asia, but its impact on the subcontinent was reduced. Today it is significant only in Sri Lanka and some of the northern mountain areas.

The introduction of *Islam* from the Middle East challenged the dominance of Hinduism about the tenth century A.D. It offered identity and status to those excluded and downtrodden in the well-established Hindu system. Islam came from the northwest, the traditional direction of Aryan invaders, and made itself felt most vigorously first in the Indus Basin. The invasion of Islam followed the main drainage systems and did not spread rapidly in peninsular India, a plateau with greater impediments to movement.

Perhaps the period of most intense *Muslim influence* was the Mogul period (1526–1739), which was the culmination of centuries of invasion by Islamic and Persian invaders. Resistance to the Islamic invasion was strong throughout the centuries, and India on the eve of British colonialism was highly fragmented and divided politically. The British entered as traders and commercial entrepreneurs in the eighteenth century but rapidly organized colonial territory and integrated the various sultanates and kingdoms, a process that continued to 1947 and the end of British colonial rule.

The impact of religion continues to shape the *political geography* of the subcontinent. Religious differences were responsible for the 1947 partitioning of the subcontinent into a Hindu component (India) and a Muslim component (Pakistan), from which East Pakistan broke off to form the state of Bangladesh in 1971. Although all three states have significant religious minorities, the *Hindu-Muslim split* has shaped and influenced political patterns and human activities more than any other. Other religious groups exist in relatively modest numbers, Jainists, Sikhs, Parsis, and Christians, but none has been effective in establishing a separate state based on their religion.[1]

Hindu religious beliefs proscribe killing animals. The cow is especially revered. Lacking systematic approaches to the breeding and slaughtering of cattle, a great proliferation of undernourished animals has resulted. Observers have criticized India's failure to use fully its bovine resources and have viewed the cattle population as essentially parasitic. Hindus, however, do use their cattle in several ways—dried cow dung for cooking fuel, milk as a source of protein, and the animals for draft purposes—and look upon them as a self-renewing, inexpensive resource. Whether or not this approach is a wise and efficient exploitation of a resource, it is an excellent example of the impact of culture on the development process. A negative aspect of the presence of 180 million cows and 60 million buffaloes is the overgrazing of common lands and the resulting damage to vegetation and soil, a serious environmental consequence.

The Hindu Caste System

Hindu society is a highly formalized order that determines virtually all facets of life for an individual. Birth determines a person's destiny, for despite recent legal proscriptions to the contrary, a man's occupation, diet, wife, and daily habits have been largely foreordained. In village India, the home of at least 70% of the population, those who perform the mean tasks are still "unclean" and thus *"untouchable."* The *caste system,* however, may be weakening in urban areas where the opportunity for social mobility exists.

It has been postulated that the lighter-skinned Aryans set themselves up as *Brahmins,* whereas darker-skinned Dravidians were forced into the lower castes. Such an ethnic-racial background to

the evolution of caste has never been clearly established. Moreover, there is some degree of caste mobility, and examples exist of castes with vigorous, wealthy members who have advanced in the social hierarchy.

The Human Mosaic

India is a human mosaic of immense variety and complexity. There is *no single India* any more than there is a representative Indian; rather, there are many different Indias. Differences in religion, ethnic groups, language, and caste illustrate the enormous diversity and present problems for development planners. Yet out of this diversity has emerged a common bond of allegiance to the Indian nation. The idea of nationhood represents the essential spirit of modern India.

Languages

Although four or five languages dominate in South Asia, there are *fifteen major regional languages* in India, for example, and several hundred dialects (Figure 24–2). The major regional languages are the basis in large part for the territorial division of India into states (Figure 24–1). The major languages are *Hindi* (India's national language) and its two derivatives, *Hindustani* and *Urdu* (the main language of Pakistan), as well as *Bengali* and *English.* Hindi and Bengali are Indo-European and derived from the Indo-Aryan Sanskrit. In the south, several Dravidian languages such as *Tamil* and *Malayan* are spoken, and they differ genetically from Indo-Aryan languages. The language difference, coupled with other distinctive traditions, is serious enough to give rise to separatist movements among the more than 100 million Dravidian speakers. The use of Urdu and Bengali in Pakistan and Bangladesh, respectively, provided one motive for political separation despite the common religion, Islam. English, despite its association with former colonialism, continues in use as a *lingua franca* and a semiofficial language of administration. Inasmuch as it belongs to no group, English serves as an important integrating element.

The Impact of Diversity on Territory

British colonialism played a powerful role in providing numerous *spatial linkages* such as railways to tie together the territory of the subcontinent into

---

[1] Sikhs have attempted to form a separate political territory, and there is some recognition of the existence of the Punjab state as a Sikh administrative subunit of India. It likely will remain part of the Indian Union.

operating administrative and economic regions. The withdrawal of the British in 1947 introduced a new formative phase in the political history of the subcontinent, a phase that is characterized by a high degree of domestic political instability. The political organization of the subcontinent expresses several cultural elements that have been able to assert themselves since British influence waned.

SOUTH ASIAN ENVIRONMENTS AS A RESOURCE BASE

South Asia, like China and the United States, spans a large area and contains *great physical variety*. Most of the subcontinent is tropical; only at higher elevations in the north and northwest is frost common. Consequently the main climatic variable is the amount of precipitation and its seasonality.

The landforms of the subcontinent are divided into three major regions (Figure 24–3):

1 The southern peninsular massif.
2 The interconnected drainage basins of the Indus, Ganges, and Brahmaputra Rivers.
3 The northern mountains composed of the Himalayas, Karakoram, Pamirs, Sulaiman, and Hindu Kush.

Considerable variety exists within each of these major units, but they stand apart as significant surface regions.

The Peninsular Massif

Peninsular India is an ancient *massif* composed of weathered crystalline rock. The peninsula is a plateau tilted up on the west to elevations exceeding 6000 feet (1829 meters) and sloping gently eastward. A narrow lowland fringes the coasts, but abrupt ridges *(the Ghats)* rise a short distance inland along both the eastern and western flanks. This coastal lowland is wider on the eastern flank where the main river basins broaden to form extensive alluvial plains. Drainage in the south is generally to the east, but north of Bombay the major rivers flow westward.

This peninsular massif contains most of the *mineral wealth* of India: copper, iron, gold, lead, manganese, and coal. Soils vary in fertility, but paucity of water is a more serious problem to agricultural production than is quality of soil.

The Northern Mountains

Several mountain ranges compose the northern margins of the subcontinent, and this *mountain wall* forms one of the most imposing physical features on earth. The height and magnitude of the mountain system exceed all others, and the geology of the mountain region indicates compressional forces and stress at work that are almost unparalleled. Fossil evidence suggests a fairly recent uplifting of gigantic rock waves that formed the Himalayas. The high ridges are aligned parallel to the margins of the plains farther south.

To the southwest of the Pamirs radiate two major ranges, the Sulaiman of eastern Afghanistan and western Pakistan and the Hindu Kush of Afghanistan. To the east are the Himalayas and the Karakoram, which form the wall that divides the Indian cultural realm from the Chinese. Other mountains have proved less of a barrier to movement and communications between the outside world and South Asia. India has long exchanged ideas and goods more with the Middle East and Southeast Asia than with East Asian civilization.

The Indo-Gangetic Plain

Of the three main physical divisions of South Asia, the contiguous plains of the Indus, Ganges, and Brahmaputra rivers are probably the most significant. Here is the *heartland* of ancient Hindu civilization, the core of India. The drainage systems considered together compose the largest continuous *alluvial plain* on earth, altogether more than 300,000 square miles (777,000 square kilometers). The plain is a structural depression formed at the same time as the Himalayas. The main drainage of the subcontinent has been oriented to this depression, and the resultant floodplain has been covered with vast quantities of alluvial material. Today the *Indo-Gangetic Plain* supports roughly half of the population of the subcontinent and is the political focal point for the three major states of South Asia. The selection of Delhi, a city located in a commanding position in the upper Ganges Basin, as the capital of India reflects well the Indian view of the special significance of this region.

Mineral and Land Resources

The *resource base* of South Asia, and India in particular, is impressive. Coal, iron ore, manganese, and chromite exist in large quantities, and smaller amounts of other valuable minerals are also pres-

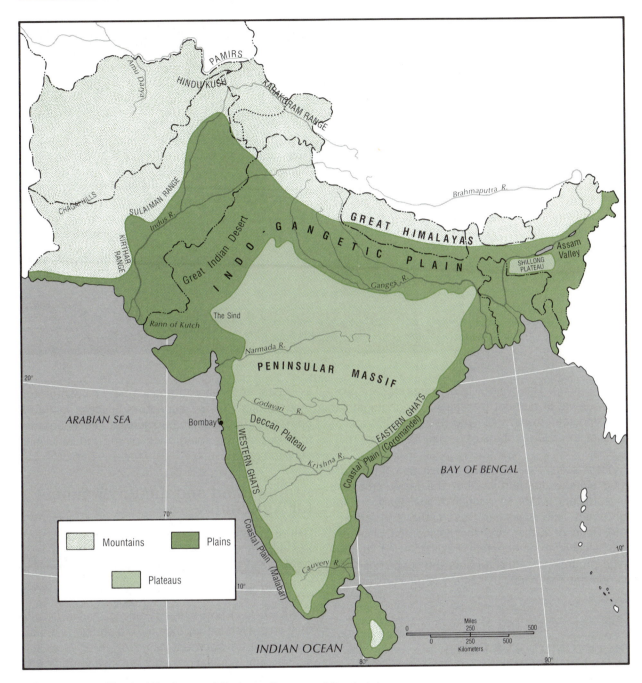

FIGURE 24–3 Physical Regions and Drainage Patterns of South Asia.
Three major physiographic regions compose the subcontinent: the northern and
western mountain wall, the southern plateau, and the intervening Indo-Gangetic
Plain. There are also narrow coastal plains on the eastern and western coasts.

ent. Perhaps the most serious mineral deficiency
that faces the subcontinent is the lack of petroleum. Future increases in food production may be
tied, in part, to increases in the production of pe-
troleum-derived chemical fertilizers. Recent offshore discoveries near Bombay have brightened In-

dia's prospects for future energy self-sufficiency.
Nevertheless, India imports more than 60% of its
petroleum, which represents a serious drain on its
scarce foreign exchange reserves.

The Indo-Gangetic Plain is one of the largest,
most *productive alluvial lowlands* found anywhere

Although much of Monsoon Asia is humid, part of the interior is dry. The Hindu Kush mountains in Afghanistan are barren and rocky. Shown here is the road from the city of Kabul to the Pakistan border. (United Nations)

on the globe. The availability of water and the partial renewal of nutrient materials through irrigation and floodwaters have allowed this great floodplain to sustain large numbers of people for several millennia.

Peninsular India contains a high percentage of land suitable for farming, but the major problem is sufficient water to nourish crops. Only in the dry west—the Rajasthan Desert, the Sind region, the Rann of Kutch, and the basins and plateaus of Baluchistan in western Pakistan—and in the high mountains of the north is the agricultural environment too harsh to sustain large numbers of people.

ECONOMIC SYSTEMS: AGRICULTURE

India and its South Asian neighbors are embarked on a course to promote national development and growth within the framework of a democratic political system. This fact in itself is significant, for many poor nations look to the experiences in India and China for development strategies. A salient feature of India's contemporary economy is the fact that most of the employed population work on farms. From their toil and production must come food for the large and growing population as well as capital necessary to finance new investment in other sectors of the economy. Success in development of the *agricultural sector is critical* both for food supply and for growth of the industrial and service sectors.

Arable Land and the Environment for Agriculture

Half of India's land is *arable,* a much higher proportion than in most other countries of Monsoon Asia (Table 24–1). About 338 million acres (137 million hectares) have been under cultivation in recent years.

Bangladesh possesses a ratio of arable land to total area even higher than India's. In this favorable agricultural environment has evolved an incredibly dense rural population, and the delta regions of the lower Ganges and Brahmaputra are as crowded as any place on earth.

Pakistan, by contrast, has a smaller percentage of land suitable for agriculture. The Indus floodplain and other areas where sufficient irrigation water is available are intensively cultivated and support a large population. Much of Pakistani territory, however, has limited agricultural potential because of aridity. Improvements in the irrigation systems may permit some expansion of the agricultural land.

TABLE 24–1   Ratios of Population to Land in Monsoon Asia

Region and Country	Number of People per Unit of Land		Number of People per Unit of Cultivated Land[a]	
	Per Square Mile	Per Square Kilometer	Per Square Mile	Per Square Kilometer
South Asia				
Afghanistan	60	23	487	188
Bangladesh	1,666	643	2,647	1,022
Bhutan	78	30	1,318	509
India	577	223	1,090	421
Nepal	269	104	2,209	853
Pakistan	291	112	1,160	448
Sri Lanka	608	235	1,839	710
East Asia				
China	271	105	2,608	1,007
Hong Kong	12,500	4,826	161,875	62,500
Korea (North)	398	154	2,170	838
Korea (South)	1,082	418	4,822	1,862
Mongolia	3	1	401	155
Taiwan	1,321	510	5,131	1,981
Southeast Asia				
Burma	142	55	958	370
Cambodia (Democratic Kampuchea)	87	34	526	203
Indonesia	201	78	2,018	779
Laos	41	16	1,108	428
Malaysia	115	44	886	342
Philippines	444	171	1,349	521
Singapore	12,500	4,826	80,937	31,250
Thailand	251	97	717	277
Vietnam	446	172	2,492	962
For Comparison				
United States	59	23	269	104
Japan	794	307	5,317	2,053

SOURCES: *Production Yearbook* (Rome: United Nations, 1980); and *Population Data Sheet* (Washington, D.C.: Population Reference Bureau, 1982).
[a]Includes cropped area and temporary pasturage. Does not include land in permanent pasture.

The *smaller states* of South Asia are even less favorably endowed with arable land than Pakistan. In most cases much of the surface terrain is too rugged, and significant agricultural development has taken place only in the level river valleys and basins.

The distribution of agricultural systems and crops in South Asia corresponds closely to variations in the environmental setting and particularly to water availability (Figure 24–4). In the dry west and northern mountains, herding is characteristic over much of the area. Herding is a land-extensive enterprise, however, and supports a very modest population. Most of the people are supported by subsistence agriculture in the irrigated valleys of the west and the mountain basins of the north. In the humid areas rice and jute are dominant. In the drier areas wheat, millet, corn, cotton, sorghum, and gram (a type of chick-pea) are characteristic. Estate agriculture (large-scale commercial production mainly for export) occupies some of the south-facing mountain slopes and nearby hills of northeast India. Tea in Sri Lanka and coconuts along India's Malabar Coast are also estate crops. These estate crops, a legacy of British rule, are grown in humid areas.

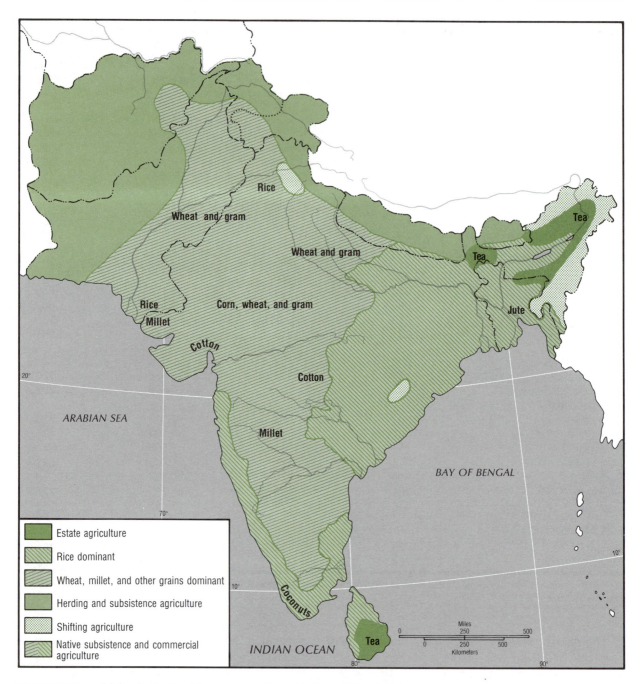

FIGURE 24–4 Major Agricultural Systems and Crop Regions of South Asia. Three main crop systems exist in the subcontinent, based largely on the availability of water. Rice, tea, and jute prevail in the most humid areas. Corn, cotton, wheat, sorghum, and millet are found in the drier interior. In the driest and highest areas of the west and north, where sedentary agriculture is marginal, herding and subsistence farming are common.

Traditional Indian Agriculture and the Village

Throughout India, agricultural life revolves around the local village. More than *500,000 villages* pepper India's landscape, and they contain more than 70% of the population. The economy, society, and politics of the village set the style and tempo of life. The village, as the primary production and social unit, holds the key to many insights about the character of modern India and the forces that oppose change and modernization.

Village life is largely *traditional and conservative.* In recent years, however, demonstrations and popular movements in the lower Ganges delta region illustrate that these peasants can be politically volatile. Where conditions of rural poverty have become intolerable, rebellion and violence have flared, and local leaders who espouse extreme leftist *Marxist* programs have emerged. Prominent among these was a group from the small village of Naxalbari. So radical were the solutions they advocated and so violent their actions that they and their followers have been labeled "*Naxalites,*" a political term that has come to stand for the most radical of Marxist-inspired rural revolutionaries. For most Indian peasants, however, life continues along patterns that are familiar and predictable.

For many centuries, a high degree of *autonomy and self-sufficiency* existed at the village level, and each village functioned largely as an independent operating and producing unit. Paralleling this economic independence was a social system in which occupation was a major indicator of caste and social rank. Inasmuch as most castes operated at the village level, this interaction yielded a strict and rigid social stratification. The few external contacts that existed generally were focused on neighboring villages, and intervillage linkages were formed along kinship lines of equivalent caste groupings. The commonplace picture of rural India as conservative, tradition-bound, and slow to change has been a reasonably accurate one. Improved communications networks and technical changes may gradually alter this picture.

Some change in villages occurred with the imposition of British colonialism over the country and the improved transport system associated with this colonial enterprise. Prior to British rule, India had been compartmentalized into a number of social, economic, and political units. Regional and local isolation began to break down with the improvement of the transport and communication network. As the villages were affected and integrated more into the national framework, powerful *social cleavages emerged* between villagers and the new elites in the urban centers. Such cleavages are characteristic of societies in transition from traditional to modern. New social and political institutions may be necessary to cope with the complexities of modernization and the consequences of national development.

INDIAN AGRICULTURAL DEVELOPMENT AND THE DILEMMA OF GROWTH

Two prominent factors have significance for the long-term alteration of the traditional Indian patterns of economy and society. These changes—rapid *population growth* and the *Green Revolution*—are interrelated and in some ways mutually reinforcing.

Boserupian Thesis

Enormous stress has been placed on the agricultural sector to increase food production continually. Although yields per unit area throughout most of India are still low by Western, Japanese, or Chinese standards (Figure 24–5), by and large production has increased and has managed to keep pace with the huge population growth. At least one economist, Ester Boserup, has suggested that *rapid population growth* is a positive factor because the intensified demand on the farming sector is the key that has opened the agricultural sector to reform and modernization.[2]

According to Boserup, without the stress of increased demand, the traditional agricultural system simply continues without basic alteration. Population growth, however, has forced the agricultural system to begin a process of transformation and modernization in step with the increased demand for agricultural products. *Boserup's thesis* is interesting if for no other reason than that it is contrary to the *neo-Malthusian* viewpoint of population growth (see Chapter 2). It should be pointed out, however, that Boserup's advocacy of population growth is qualified. She claims that the forced pace of rural modernization brought on by rapid population growth is a short-

---

[2]Ester Boserup, *The Conditions of Agricultural Growth* (Chicago: Aldine, 1968). See especially pp. 11–27.

is changed and the marginal productivity of such farm workers is improved, peasants will remain poor throughout much of India.[3]

India's Green Revolution

Although not a panacea for all the problems, the *Green Revolution* has played a significant role in the process of development under way in rural India. Perhaps in this revolution lies the greatest promise for India's rural modernization program.

In brief, the concept of the Green Revolution is a thorough-going process of *agricultural transformation* with the goal of maintaining agricultural production at a level well ahead of population increases. One aspect of the Green Revolution is the establishment of institutional supports to promote increased yields. These supports include irrigation projects; accessibility to fertilizers, insecticides, rodent controls, and herbicides; and improved cultivation techniques. Also included are education programs designed to stimulate the diffusion of new ideas and techniques in rural areas.

The introduction of new high-yielding varieties of grain, often called *miracle grains,* is the best-known aspect of the Green Revolution. A concerted effort was made in the 1950s and 1960s to improve the yielding qualities of rice, maize, and wheat. Plant geneticists eventually developed varieties that produced more seed and would respond better to the application of fertilizers. A particular characteristic bred into these miracle grains is that they are susceptible to more intensive cultivation; in essence, the *point of diminishing return* is delayed with increasing inputs. This feature of the new varieties is illustrated in Figure 24–6.

These miracle grains, including rice, are now more widely cultivated in India. Total grain production has increased substantially during the last two decades. The average yearly increase has been nearly 3%, a figure somewhat above the annual rate of population growth. Although the area under cultivation has expanded, much of this increase is attributable to improved yields per land unit. Yields per unit for all food grains, however, are still only half those of China; for rice the yield is less than one-third.

There are four reasons for the difference in yields between China and India:

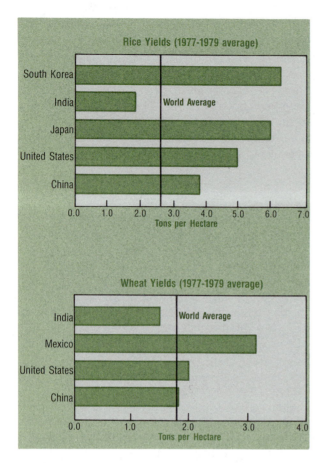

FIGURE 24–5 Grain Yields for Selected Countries. Although the expenditure of many man-hours of work per land unit is characteristic of Monsoon Asia, agriculture yields are often relatively low. For example, rice and wheat yields in India are significantly less than the world average. Japan and South Korea, however, have rice yields more than twice that of the world average.

term proposition. Once the transformation process is well under way, she predicts population growth will begin to taper off.

In India one of the problems in farm production is the *low level of productivity* of farm laborers. Rapid population growth means there is a crucial problem of labor absorption in rural areas with so many new hands to employ each year. The farm sector, as an employer of last resort, has traditionally absorbed this excess labor internally. In recent years such large quantities have been absorbed in some areas that the *marginal productivity* of the new laborers is zero. That is, if these workers were removed from the labor force, production would not be affected at all. Their contribution is, therefore, worth nothing. Unless this condition

[3]A. N. Agrawal, *Indian Economy: Problems of Development and Planning,* 8th ed. (New Delhi: Vikas, 1982).

1   China has a much higher proportion of land under irrigation (46% versus 28%).

2   China uses nearly two and one-half times as much fertilizer on 30% less cropland than India.

3   China's best land is generally used for grains, but some of India's best land, much of it irrigated, is used for other crops such as cotton and jute.

4   China double-crops its land at approximately twice the rate of India.

These differences indicate that India has a considerable potential for increasing grain production. If the use of miracle grains is expanded also, India's agriculture can be made much more productive.

Can India Feed Itself?

A program of *rural transformation* has been under way for some time in India. It could be argued that this transformation has both failed and succeeded. On the one hand, India's food production has to a large extent kept pace with the rapid population increase. Despite the frequent importation of foodstuffs, India has stood on the *threshold of food self-sufficiency*. Great gains are being made; unfortunately, these gains in production are offset in

Cattle are much revered in Hindu India. The animals are an important source of motive power in farming as well as manure for cooking and fertilizer purposes. There are approximately 240 million bovines (cattle and buffalo) in India. (C. W. Pannell)

part by the increased numbers of people. Yields per land unit remain low, which suggests that the agricultural transformation is still in its early stages. The sheer size of India's peasantry reminds us of the monumental task of transforming traditional agriculture.

On the other hand, in certain areas of India such as the Punjab and Haryana and scattered locations in Kerala, Karnataka, and other states, advances in agriculture have been rapid and impressive. In these places yields per unit—based on high rates of investment, *irrigation, double-cropping,* and *mechanization*—have raised farm incomes substantially and demonstrated the potential for progress and change in Indian agriculture. Available capital and a well organized and hard-working farm population are essential if such developments are to take place.

A negative feature of Indian farming is the unreliability of the *summer monsoon rains.* When these rains fail or are seriously diminished as in 1981 and 1982, farm production throughout much of the country is devastated and yields fall off sharply.

Yield per Land Unit (Output)

Miracle Grains

POINT OF DIMINISHING RETURN

Traditional Grains

Amount of Labor and Capital Applied per Land Unit (Inputs)

FIGURE 24–6 Hypothetical Difference in Yields and Point of Diminishing Return.
High-yielding or "miracle" grains offer substantially greater rewards for developing countries. As the graphs indicate, however, the increased yields result only from increased labor and capital inputs. Yields from the "miracle" grains are clearly much higher per unit of labor and capital input.

POPULATION AND POPULATION GROWTH

Discussing population and population growth in India and South Asia is never easy. Second among the world's countries in population, India is estimated to contain 730 million people. To this number roughly *15 million* new citizens are added each

The Indian government provides family-planning instruction and contraceptive devices in most cities and many villages. Much public education and many inducements are provided to encourage birth control. (United Nations/ILO)

year. Only China matches such huge annual increases in population. The question is how long such growth can be sustained.

India is not alone in rapid growth. South Asia as a whole has almost 1 billion people with an annual growth rate of 2.2%. Indeed, if the *high rates of growth* continue in India, Bangladesh, and Pakistan, South Asia soon may contain more people than East Asia.

It was noted that approximately 70% of India's population is directly *dependent on agriculture* for its livelihood. The agricultural sector must provide employment for huge numbers of new workers every year. At the same time, agriculture is attempting to modernize and become more efficient by substituting capital for labor to improve yields. A contradiction is obvious between the realities of India's current demographic situation and the nation's desire and capacity to initiate change in its traditional agricultural system.

Birth, Death, and Growth Rates in India

According to United Nations estimates, *life expectancy* for the average Indian is fifty years. The nation remains in stage 2 of the demographic transformation (Figure 24–7; also see Figure 2–3). Despite a considerable effort to control births, the annual birth rate is 3.6%. The death rate of 1.5% indicates that India has much to accomplish if the rate of growth is to be reduced.

Much public education and many inducements have been provided India's peasantry to practice various forms of *birth control*. Free transistor radios were given men who volunteered to have vasectomies, and animated cartoons depicted the disadvantages of too large a family. Government workers are provided special incentives for small families or penalized if they have more than two children. In 1975 and 1976 serious consideration

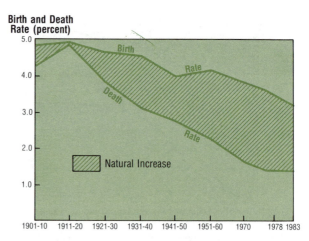

FIGURE 24–7 Indian Birth and Death Rate Trends, 1901–83.
Although the birth rate in India has been decreasing steadily in recent years, it has been offset by an equally rapidly declining death rate. Thus the net growth rate has been high, and India's annual population growth has been rapid.

was given to the use of legal methods for controlling family size. Political reaction to such stringent measures led to their abandonment. Today the official policy toward family planning is restrained. The problem appears to lie not so much in India's disapproval of birth control or neglect of *family planning* programs as in the sheer magnitude of the task of spreading effective family planning concepts and techniques to the more than 500,000 villages. Family planning techniques have been provided in most cities and many adjacent villages, but the main task lies in spreading the work throughout rural India.

Continuation of the high population-growth rate will give India a population of more than *800 million by 1990.* Based on what we know of India's cultural values and the esteem in which Indian peasants hold large families, it is unlikely that rapid reductions in birth and growth rates can be achieved in this century. Consequently, it seems likely that India's population will reach close to 1 billion by the turn of the century.

Population age and sex structure as evidenced in the *population pyramid* support the idea of continued growth (Figure 24–8). The familiar cone-shaped graph with the broad base parallels the pattern found in most poor countries (see Figure 4–11). There is a marked concentration of people in the younger age groups. These young people provide an ever-enlarging *potential for future growth* as more people enter the childbearing years. It is this pattern, perhaps as much as anything about India's demography, which suggests little or no letup in the rapid population growth in the near future.

Population Growth and India's Future

The picture of India's population growth is not attractive, but there are some hopeful signs. For one thing, the *pace of change* in today's developing countries is much greater than that for comparable demographic experiences in Western history. *New technologies* and techniques of communication, education, and public health appear continually, and change in values and traditions has been accelerated. Where the exigencies of the situation demand change, the pace of change may be rapid. If the information available from China is an accurate indicator of recent events there, it suggests that rapid and far-reaching demographic changes are possible even in large developing states. The

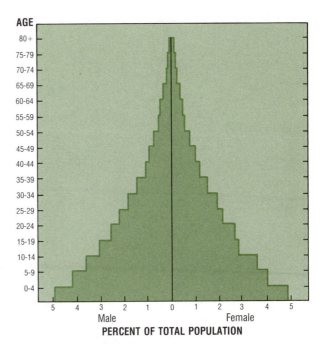

FIGURE 24–8 Population Pyramid for India. The population pyramid presented here is cone-shaped, with a wide base of very young people. The cone shape indicates a very youthful population with a great potential for future growth.

prospect that similar events will happen in India is another matter. Of all the impediments to rapid economic growth and development, population growth is one of the most serious.

Where the People Are

The population of the subcontinent is unevenly distributed (Figure 24–9). The middle and lower Ganges Basin, the lower Brahmaputra Basin, and the coastal littoral along the Eastern and Western Ghats are the most densely inhabited regions. This pattern appears to reflect the attraction of people to the most favorable agricultural environments, for here are concentrated dependable sources of water and productive alluvial soils. Such environments could be shaped and manipulated into productive farming regions that support a large number of people and sustain frequent croppings of intensive food grains such as wet rice.

The population of peninsular India is large, but it is less dense than the rice-growing alluvial lowlands of the great river floodplains and the coastal littoral. For one thing, much of the peninsula is dry and given over to such crops as wheat and millet.

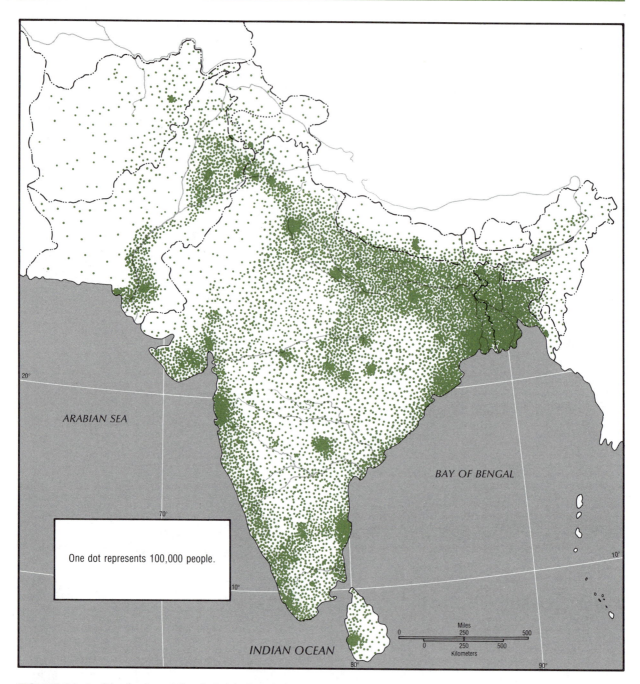

One dot represents 100,000 people.

ARABIAN SEA

BAY OF BENGAL

INDIAN OCEAN

Miles
250
0   500
0   250   500
Kilometers

FIGURE 24–9   Distribution of South Asia's Population.
South Asia has one of the greatest concentrations of people on earth. Yet, as the
map indicates, these people are not dispersed evenly throughout the subcontinent.
The most dense concentration is in the middle and lower Ganges Valley, especially
in the delta region, which is occupied largely by Bangladesh. Somewhat less dense
but important clusters are found throughout the peninsula and the Indus River floodplain.

Yields per land unit are lower than in the rice-growing areas; thus peninsular India cannot support as high a population density as the rice areas. In the drier desert and steppe regions of western India and Pakistan (Rajasthan and the Rann of Kutch) populations are sparse, a reflection of the marginal nature of these environments.

CENTRALIZED PLANNING AND ECONOMIC GROWTH IN INDIA

Since independence in 1947, India has developed its own form of *central planning* and government control to promote development. Such control has been expressed mainly in a series of five-year plans that have focused on the nationalization of key industries and services: iron and steel making, heavy machinery manufacturing, fertilization and chemical production, irrigation systems, and transportation and communication networks. In general, this planning was done to channel large investments to these enterprises. The important thing to remember, however, is that the plans are made by a national planning commission, and this commission has acted within a democratic context in which local, regional, and national interests are considered. Such procedures are not entirely different from those that operate in the American federal political system, although the United States has no single national commission charged with overall economic planning.

Since 1951 controversy has centered on *one principal issue:* the relative attention and investment provided manufacturing as contrasted with agriculture. An important subissue is to what extent the manufacturing sector is to be nationalized. After more than two decades of economic planning, it is clear that India has emphasized a *heavy industry approach.* Such a policy is similar to Soviet planning in the 1920s and 1930s and is aimed at making India self-sufficient in basic goods. This approach is termed *import-substitution* and, if successful, should reduce reliance on external suppliers.

Agricultural Investment Versus Industrial Development

Most Western planners have argued that India should focus more attention and investment on *modernization of agriculture.* Many Indians view an emphasis on agriculture, however, as extending the nation's dependence on foreign sources of basic equipment and industrial goods. It is, therefore, unappealing. Moreover, the Indians consider an agricultural emphasis will prolong India's political and military impotence in world affairs. This disagreement may have been a factor in the increased misunderstanding and mutual criticism that have characterized United States–Indian relations during the past three decades.

In recent years, rapid population growth and periodic droughts have shown the necessity for increasing agricultural production rapidly. Freedom from reliance on foreign manufacturers has been largely achieved, although India has sporadically looked toward external sources of foodstuffs to meet internal deficiencies. One consequence has been increased investments in those industries and services that support agriculture. An associated policy, aimed at providing more rural jobs in community development, self-help, and education-related projects, has received increasing attention. Over the years it has become obvious that creating industrial jobs in large cities will not suffice to provide employment for the additional millions who enter the labor force each year. The real crisis for India's development program continues to lie in the countryside. Here are more than 70% of the people, and some means must be found to satisfy not so much their long-term aspirations for education and a better life as their immediate needs for food, clothing, and housing. Indian planners and politicians are coming to realize that drastic solutions are required in rural India if the more sophisticated goods of industrial growth and national development are to be delivered.

The Industrial Economy

Industrial development and expansion in India evolved out of a tradition of *cottage manufacturing* associated with metal smelting and consumer goods. British plans for India in the nineteenth century neglected these native industries and concentrated on new industries that complemented manufacturing in Great Britain. British interests lay in introducing *labor-intensive industries,* such as textiles, or industries that relied on local commodities (jute processing, sugar refining, and leather tanning). The British hoped to sell India the more sophisticated manufactured goods. British colonial rule ensured that this policy operated.

World War I with its attendant shortages brought some expansion in industries such as *iron and steel,* cement, and paper. During World War II these developments intensified as India became a rear supply area for the China theater. Despite these improvements, India on the eve of independence had few machine-building and chemical industries.

British colonial policy concentrated industrial production in the three major port cities of Calcutta, Bombay, and Madras. The Indian government since independence has viewed this concentration as undesirable. A key feature of industrial planning has been to spread production centers throughout the nation while continuing to strive for the most economical location.

Planned Industrialization, 1947 to the Present

In line with the desire to achieve greater self-sufficiency, India has followed an *import-substitution strategy.* Planners have sought to develop industries to supply the entire range of producer and consumer goods except for highly specialized luxury goods. This range extends from the well-established textile, jute, and sugar-processing mills to locomotive, shipbuilding, and aircraft factories and

even includes computer manufacture and the processing of materials for atomic energy.

Production of a wide range of goods is possible by virtue of the domestic market's large size and the ability of producers to achieve economies of scale sufficient to keep prices low. India is reasonably well provided with resources and materials for industrial production.

Industrial Regions

Since independence a major effort has focused on distributing new industry throughout the nation. By locating industries closer to raw materials or markets, centers of production and growth may be spread more evenly.

Most prominent among the recently created industrial regions is the cluster of *heavy manufacturing* in the *Damodar Valley region* (Figure 24–10). Symbolized by the planned industrial city of Jamshedpur, this center of iron and steel production is largely *raw-material oriented.* Jamshedpur is close to the major iron ore deposits and coalfields of the subcontinent. Most of the recently established heavy-metal and associated chemical and machining industries are also located here. With each new

Jamshedpur, India's iron and steel center, is a planned industrial city and stands as a symbol of India's industrial growth. Iron and coal deposits are located nearby.

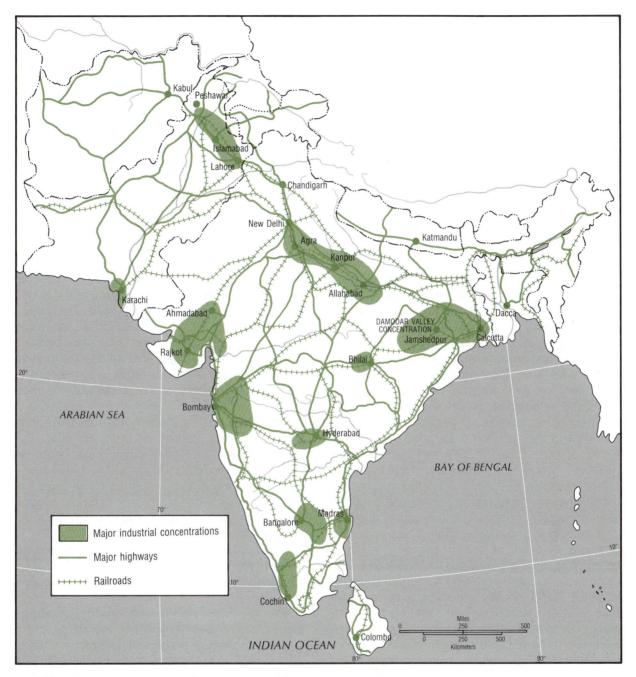

FIGURE 24–10 Surface Transportation and Industrial Concentrations in South Asia. Although poor, the subcontinent has a well-developed surface transportation network, in part inherited from British colonial investments. Several major industrial concentrations are also depicted.

project, the raw material and assembly factors that favored this concentration have been reinforced by economies of scale and agglomeration. The growth and concentration of Indian industry in the Damodar Valley have been impressive enough to make this region one of the major metal- and machinery-production centers not only in Asia but also in the world.

Some industries do not benefit through clustering. Many are oriented toward local or regional markets or are best located in association with agricultural or other raw materials that are dispersed more broadly. *Textile mills* are common in the cotton-producing region of Rajasthan, the paper mills and sugar refineries are located in peninsular India near the areas of heaviest sugarcane production. At the same time, older centers of industry continue to attract new establishments, for they possess the advantages of large markets, good accessibility and transportation linkages, and large and relatively skilled labor pools.

The Growth-Pole Theory

There has been a concentration of wealth in a few great cities, which presents a sharp contrast to the huge, impoverished, traditionally oriented rural hinterlands. India's desire to decentralize industry and encourage growth in many smaller cities corresponds with the *growth-pole development strategy*. This strategy aims to promote growth in many centers in an effort to spread economic activities and benefits throughout the state.

By clustering new industries in selected centers over a wide area, jobs are created in depressed areas and provide alternative opportunities for labor. Once established, these new industries spur other economic activities (multiplier effect). These centers and their economic activities grow as the forces of economies of scale and agglomeration become increasingly important. Moreover, by careful selection of specific industries in a center, the planner can exert some control and direction over resource development and economic growth.

Increased and diverse economic activity in turn becomes an *agent of cultural change.* Traditional behavior patterns are modified. In India's case, the caste system may be broken and opportunities for individual social mobility enhanced. Individual freedom of action is consequently improved. The effects of both economic and social change are then expected to be diffused throughout the surrounding countryside. By judicious use of the growth-pole theory, uneven concentrations of wealth and production are avoided or at least reduced. In addition, decentralization is thought by some to speed the pace of economic growth and development.

The growth-pole theory is controversial, and regional development experts have different opinions on how the theory should be used. For example, should poles be already-existing centers, or should they be new towns? How large should the growth-pole centers be? There are many such questions, and definitive answers are not yet available. What-

ever the merits or liabilities of such a strategy, it is difficult to argue against the strategy's political benefits. In a fragmented political situation such as India's, with its diverse local and regional interests, there are benefits for most if not all regions. The economic benefits balance regional interests and help to oil the national political machinery and provide government with a greater chance for success.

INDIA'S TRANSPORTATION NETWORKS

Railroads

India has possessed a *large and dense railway system* for almost a century. This system is the longest, most dense rail network in Asia, and it is one of the longest systems in the world. The usefulness of the rail system in promoting urban and economic growth is hindered, however, because it was designed to further colonial policies of stimulating exports and imports rather than internal development, and some of the routes are not interconnected.

Historically, the rail system has focused on the *four major urban centers* of Delhi, Bombay, Calcutta, and Madras. These cities were linked together by a skeletal framework of broad-gauge lines. Each city, in turn, was tied to its large *hinterland* by a network of lines of different gauges. Thus a fully integrated rail system did not exist, nor was there a means to stimulate growth at smaller centers in the interior. Such a varied rail system typified British colonial practice and variations in British administration and economic control.

Recently attempts have been to standardize gauges where feasible and not too costly, but a long-term project of *gauge standardization* is yet to be devised. After independence the system was divided into a series of regional railway networks under the direction of the government-owned Indian Railways. Despite problems of administration and operation, railways continue to be the most efficient and cheapest method for hauling goods and people through the nation.

The Highway System

In recent years attention has been given to highways as an alternative to rail transport. *A long-term plan* has been drawn up for construction of an *in-*

tegrated system of all-weather roads. These roads may provide an effective means of linking together the many areas of India and provide an effective agent for stimulating growth and economic activity in smaller cities.

Long-term highway development plans are designed to integrate most of the settled area of India through a network of main and secondary highways. The network is to reflect population densities. In the thickly settled agricultural regions a dense network of interlocking roads is to be built to place all villages within 2 miles (3.2 kilometers) of a secondary road and 5 miles (8 kilometers) of a main highway. In sparsely settled nonfarming areas, the network is to be thinner but still link the areas in a systematic manner.

In addition to new highway construction, recent efforts have concentrated on improving and upgrading existing roads. Important roads carrying much traffic will be widened to increase capacity. Other secondary roads will have their surface improved, and rural roads will be upgraded. The system may appear inadequate by West European standards, but compared to most other developing nations, and especially other large states, India has an extensive network of highways.

Air Transportation

Air transportation has developed rapidly in recent decades, and the Indian Government has promoted both an international air carrier (Air-India) and a domestic one (Indian Airlines). Most of the major cities and many intermediate cities are linked through the domestic carrier. International service is available through major gateway cities such as Delhi, Calcutta, Madras, and Bombay. Although service is not as frequent as in modern, wealthy countries, compared with most developing countries, air transportation in India is well developed.

Transportation and Development

The evolution and improvement of transportation networks are essential to economic development and political modernization. Without access to its territory and linkages that integrate regions, a state cannot function effectively or promote the kinds of programs necessary for proper economic growth. A good transportation system is imperative in a large developing state. For example, in China and the Soviet Union the expansion of transport networks has been critical to economic-growth policies and to the political-military goals of national strength and power.

India offers a somewhat contrasting pattern. A fairly dense transportation network was one of the consequences of British colonialism. It provided a broad base on which to develop a locational strategy of economic growth. Such a base permitted India to concentrate investment on other problems and simply improve the transport system that already existed. In the search to identify bottlenecks and critical problems in India's development path, it is unlikely that transportation will be seen as one of the key difficulties in the long struggle for economic growth and improved levels of living.

URBAN INDIA: CONTRASTING POINTS OF VIEW

Although perhaps only 22% of India's population resides in cities and towns, this 22% includes about 160 million people. India, by this standard, must rank as one of the world's leading urban states. These millions are contained in a variety of different-sized cities and towns. The prospects of future urban growth have given rise to several competing schools of thought and opinion about cities.

Some planners and scholars have suggested that India is neglecting to influence urban growth and thus can look forward to huge concentrations in a few great metropolitan areas: Calcutta, Delhi, Bombay, Madras, and Hyderabad. These scholars also suggest this concentration is bad, as income will be inequitably concentrated, and increasing urbanization will destroy the traditional social patterns focused on small, more personalized villages. They point to future dense crowding in huge supercities with attendant social breakdowns and note that political stress already is experienced in some of the large cities. Calcutta is often chosen as a poignant example.

Calcutta and Bombay

Greater Calcutta with its 9.1 million people is probably the most frequently cited example of everything wrong with big-city life and activities in a developing state. The poverty and disease of Calcutta, and the congestion in its streets, over which cars speed and sacred cows meander, have been chronicled so frequently that they need little repetition. Political rallies and urban violence also

Calcutta has been characterized as a city of poverty and disease. Beggars and street vendors line many of the side streets. (United Nations)

are no strangers to Calcutta's urban scene, and many sympathetic Westerners despair over the hopeless condition of India's largest urban center.

Bombay, by contrast, is sometimes viewed as an example of the positive aspects of urban India. The city is smaller (8.2 million people) and surrounded by a less densely populated and more prosperous agricultural hinterland. Bombay has always seemed one of India's most *advanced and progressive* places. Large and modern factories staffed by an industrious and relatively well-fed labor force create a large industrial output. These industries in turn support a thriving commerce and growing middle and upper classes of professional and entrepreneurial people. Cultural and educational institutions abound, and the city, with its international orientation and booming economy, is among India's most cosmopolitan centers. Yet Bombay has its problems too, as 40 to 50% of its population now reside in slums called *bustees,* and obtaining urban housing is a serious problem.

Both Calcutta and Bombay have good and bad points, and both have examples of growth and progress and of poverty and despair. Perhaps more important is that two views exist on the role of cities in India's future growth strategy. These two cities, like all of India's cities, will continue to grow.

Metropolis Versus Town

The alternative to metropolitan growth is dispersal of new economic activities in smaller cities and towns throughout the country *(growth poles).* Some planners claim that the greatest production efficiencies are achieved in the larger cities with good transport systems and skilled labor forces. Others assert that towns and cities in the 250,000 and smaller range are big enough to provide the *economies of scale* necessary for efficient production. These planners argue that India should direct more of its future economic growth to the smaller centers and spread the wealth more equitably. Such greater dispersal also appeals to local and regional political interests and satisfies the sectional interests of the Indian political system.

It is too early to predict exactly what will happen in the future, but already it seems that India's gigantic population will require the use and development of both styles of urban growth. The great metropolitan centers can be used intensively for the production of specific goods, especially those that are export oriented. The smaller cities and towns can continue to satisfy local needs and demands, both commercial and industrial, as they always have. To the extent that the economy modernizes, however, these local and regional production centers will have to become more sophisticated as they are drawn increasingly into the national and even the international economy.

PAKISTAN

What began in 1947 as a political problem of partitioning British India's *Muslims and Hindus* into two nations has had serious economic consequences. The political divisions and economic troubles associated with the partitioning have continued, and violence has erupted periodically. Most such outbreaks have involved Indian and Pakistani forces fighting over control of Kashmir. The latest outbreak, however, involved the eastern territory of Pakistan, a Bengali-speaking region. Unfortunately for Pakistan, the linguistic and *sectional differences* between Urdu- and Bengali-speaking Muslims overcame whatever affinities based on religion existed previously. The result was that East Pakistan, with help from the Indian Army, established itself as an independent state. Pakistan was left as a single territorial unit focused on the Indus River floodplain.

Although the reasons for the *political fragmentation* of Pakistan into two independent states are many, the basic difficulty centered around the position of one group, the West Pakistanis. The *Urdu-speaking West Pakistanis* dominated the politics and military affairs of the nation, although their popular support was a minority position. The numerically dominant *Bengali-speaking East Pakistanis* were to some extent exploited and, as they saw it, oppressed. Following the free elections in late 1970 in which the East Pakistani Awami League Party won a majority in the Pakistan National Assembly, disagreements broke out and fighting followed. The state ultimately fragmented into two political entities: Pakistan (the western sector) and Bangladesh (the eastern). Such fragmentation is another example of the diversity and variety of the Indian subcontinent and has far-reaching political, social, and economic consequences for a large number of people.

The New Pakistan

The Pakistan that emerged from the political division is considerably different from the joint state that existed before 1971. The differences are political as well as economic. First, the *new Pakistani state* is patterned politically along British lines, to

which is added an interesting blend of local religious feeling and socialist doctrine. The economy has become increasingly *socialistic* with less freedom for the free-enterprise sectors of industry, commerce, and finance. Such a socialist pattern, to some extent, may reflect Pakistan's attraction to China and its policies of state control throughout the economic system.

Although separation from Bangladesh has led to the presumption of greater political and social coherence in Pakistan, other problems associated with a large area and a heterogeneous population have become more apparent. The large, arid Baluchistan region west of the Indus floodplain, with its various tribal interests and political groupings, has become an area of major *domestic difficulty*. In the northwest, tribal elements that look to Afghanistan or Iran for political support and cultural appreciation create other problems. Pakistan somehow must knit together these diverse ethnic threads into a strong social fabric if the progress it so much desires is to be achieved. The recent Soviet invasion of Afghanistan led to a large influx of Afghans into Pakistan. Pakistan has found itself embroiled in this conflict against the Soviets. This situation has further complicated its relations with India.

Jammu and Kashmir

No problem in contemporary Pakistan is of greater magnitude than the territorial dispute with India over *Jammu and Kashmir* (Figure 24–1). The cease-fire line drawn in the 1950s has become an effective boundary, and India occupies the most productive and densely populated parts of the region. The resolution of the territorial claims has not yet been found. In today's political climate the stalemate over Jammu and Kashmir is not likely to change. Since division into two nations, Pakistan's position has been weakened in economic and military strength vis-à-vis India. India emerged from the Pakistan-Bangladesh dispute as the dominant political and military force on the subcontinent, a position that is unlikely to be challenged in the near future.

Growth and Progress

Pakistan, like its neighbors, is *poor,* and its past political problems have aggravated its condition of impoverishment. Most of the population is concentrated in the *watershed of the Indus,* and two main productive regions stand out. One of these regions is focused on *Karachi*, the early capital and a large port and industrial center located at the mouth of the Indus. The upper Indus region, focused on Lahore and the new capital, Islamabad, is the modern political center and region of greatest contemporary planning concern.

The economy of Pakistan continues to be largely *agrarian*. Even where irrigation is commonly practiced, as in the Indus Basin, weather patterns often play a key role in determining agricultural output and thus the size of the gross national product. Severe flooding in 1972, for example, seriously reduced crop production and the rate of economic growth.

Agriculture is based on *three main crops:* wheat, rice, and cotton. Canesugar is also important, as are pulses from which cooking oils are extracted. Government objectives, as implemented by various rural development programs, are to increase agricultural production through increased applications of chemical fertilizers, better seed strains, and improved farming techniques. Agricultural imports and exports have generally balanced, although recently there has been a modest export surplus.

Industry accounts for a smaller percentage of the GNP than does agriculture, although it is growing more rapidly. A broad range of industrial establishments exists, but the *socialization of industry* may alter the past patterns of growth by reducing levels of entrepreneurial investment. Traditionally, cotton textiles have been the most important foreign-exchange earner, and a high rate of investment in these activities continues.

The years since separation from Bangladesh have witnessed more rapid economic growth, yet Pakistan continues as an *impoverished country* with a traditional, agricultural economic system. The large area and population provide many advantages, and industrial production can be geared to exploit the advantages of scale economies. With help from the Chinese and other external sources of investment capital, along with the end of the East Pakistan problem, Pakistan is able to focus more energy on its economy than was possible in past years.

BANGLADESH

The creation of a new country is never simple or easy. The conditions of South Asian politics and society make it especially difficult to create the permanent loyalties and symbols so necessary to the evolution of successful statehood. Events of recent years have been especially frustrating for the citizenry of the former territory of East Pakistan. It was against a background of political, social, and linguistic discrimination that Bangladesh emerged as *a new nation-state*. India's military backing was important in forging the new state, but it was the desire for self-determination that led the former eastern sector of Pakistan to wrench itself free from the confinement and pressure its coreligionists in West Pakistan had imposed on it.

The Delta Environment

Bangladesh is an unusual place. Its tropical *delta environment,* limited territory, and large number of people, together with language differences, present a host of problems. These features, and the failure of the former Pakistan government to recognize and cope with them, largely explain the desire of Bangladesh to seek its own destiny as an independent state. Bangladesh, with nearly 97 million people occupying 55,598 square miles (143,999 square kilometers), is one of the most *crowded* places on earth. Envision crowding 97 million people into the state of Illinois or Georgia for a reasonably close approximation of the population density.

Most of the land area of Bangladesh is low-lying and encompasses the intricate network of streams and distributaries that form the mouths of the Ganges River and its confluent, the Brahmaputra. Much of the area is composed of delta lands and gives a common water character to the country. The extent of this delta environment is large and almost without parallel elsewhere.

It is this delta environment that has permitted so many people to live on so little land. Almost two-thirds of the land area is farmed, and a *productive agricultural system* has evolved based largely on the cultivation of *paddy rice.* Rice is not the only crop produced, but is by far the most important. A variety of other food crops and some specialized cash crops such as jute are also produced in large quantity, but rice is the main provision of the great mass of people.

The rhythm of alternating wet and dry seasons already described for India also governs agricultural life in Bangladesh; 80–90 inches (203–229 centimeters) of rain concentrated in the summer months provide much of the water necessary for rice cultivation. A finely developed system of *irri-*

Many of Bangladesh's people lost all their material possessions in the battles of the early 1970s that led to Bangladesh independence. Millions of people fled the fighting for refugee camps in India. (United Nations)

gation supplements this rainwater and provides the controls for *sustained high yields* necessary to feed so many people. Unfortunately, tropical storms that originate over the adjacent warm Bay of Bengal disrupt life by causing periodic floods in many coastal areas. Destruction and loss of life are great as a consequence of the extreme crowding on the delta. It is a sad irony that so much productive agricultural land is vulnerable to the vagaries of natural catastrophe.

Prospects for Economic Growth

Bangladesh entered nationhood in 1971 after considerable suffering and natural disaster. The war of independence caused great damage to the infrastructure. Bridges and transportation systems were wrecked, ports were clogged and not functioning, and power facilities were damaged. A great storm had caused enormous loss of human and animal life, and agricultural production had declined sharply. The United Nations sponsored relief efforts, and these were supplemented through additional donations from other foreign national and private agencies. Bangladesh was not without friends and sympathizers, and these external suppliers saw the nation through the immediate crisis.

The immediate goal of restoring production was set out in a one-year plan. This plan was designed only to restore the country to its previous, preindependence levels of production. In 1973 a *five-year development plan* was adopted to promote future growth. An annual goal of 5.5% was set for production increases, primarily in the agricultural sector. This goal was unrealistic in view of the many natural and human problems. The principal difficulty was the high degree of reliance on foreign capital. During its first five-year development

371

plan, Bangladesh procured about 60% of its investment capital externally. It is a poor country and is not a particularly good credit risk. Nevertheless, it has found some capital assistance from sympathetic foreign sources and may rely on these sources at least for a short time. Unfortunately, the capital has frequently been used to purchase foodstuffs, the most critical need, with little left for development projects.

The *second five-year plan,* launched in 1980, has placed more stress on *industrial production* and has sought to encourage investment from the private sector. Laws have been enacted to encourage foreign private investors to invest their capital in Bangladesh, and several large joint projects, such as a major urea fertilizer plant, are being developed. Industrial production in such items as sugar, caustic soda, diesel engines, motor vehicles, and television sets has risen sharply in recent years. Nevertheless, agriculture continues to be the mainstay of the economy, and the agricultural sector accounted for approximately 55% of the gross domestic product (GDP) in 1981. *Rice and wheat* production, the two most important food grains, have increased steadily in recent years.

Two major problems beset Bangladesh's economy: (1) the persisting unfavorable balance of trade and (2) the poor fiscal position of the country. A substantial national and foreign debt exists, and the annual service on this debt amounts to more than $1 billion. If foreign aid continues in sufficient amount to help Bangladesh with its trade, fiscal, and food problems, the outlook is reasonable for continued economic growth. Population growth, based on the very high annual birth rate (4.9%), unfortunately undermines economic progress, and there is little indication of any success in reducing the annual birth or population growth rate.

SRI LANKA

Of all the South Asian states, Sri Lanka (Ceylon until 1972) is perhaps the most modern and certain of the direction in which its society and economy are moving. Although poor, it has initiated a massive effort to cut the birth rate and to promote improvements in the agricultural economy. Ceylonese agriculture, with its *large plantation sector,* has long been commercialized. The plantation sector, however, has been plagued with inefficiencies, and certain aspects of the agricultural economy have stagnated in recent years. Such

agricultural stagnation has perpetuated low yields and intensive manual inputs. This economic bottleneck must be overcome before real growth can be achieved.

Land and Environment

Sri Lanka is a pear-shaped island state of 25,332 square miles (65,610 square kilometers) that lies 12 miles (19 kilometers) southeast of India. The island is a *tropical land,* although the hot and humid climate is moderated at higher elevations in the interior mountains. The climate is shaped primarily by a seasonal pattern of rainfall in which the southeast monsoon dominates between May and August and brings heavy rains to the southwest quarter of the island. From November to early spring, the wind direction is reversed, and the eastern flank receives most of the rain. Frost is unknown, and average monthly temperatures vary only slightly from the coldest to the hottest month.

About 44% of the island is forested, and much of today's plantation agricultural complex was formerly forest land. *Mineral deposits* of graphite, quartz, feldspars, and gemstones are commercially significant, and exploration for additional minerals continues.

Technological Backlash

Sri Lanka, like many other developing nations, has applied some *technological innovations.* Occasionally these innovations can lead to unforeseen problems. The case of *malaria* and the insecticide *DDT* is an example. During the 1950s and the 1960s, the use of DDT was widespread on the island, and the incidence of malaria was drastically reduced. In 1963 there were only seventeen reported cases of malaria in Sri Lanka. Lowland humid areas previously little used because of malaria mosquitoes were settled, cleared, and put to agricultural use. Within a short time, a sizable population lived and worked in these lowland areas. When the damaging effects of DDT became widely recognized, its use was severely curtailed. Without an effective means of mosquito control, the number of reported malaria cases rapidly increased. In 1974 more than 315,000 cases of malaria were detected. Sri Lanka is now faced with a hard choice among three options: abandoning these new croplands, using more DDT to control the increasingly resistant mosquitoes and accepting other possible damage to human health (contaminated water supply) and to the environment, or funding other more costly control measures.

Sri Lanka is not alone in facing problems of *technological backlash.* Similar DDT-malaria problems exist throughout the world's humid tropics and especially in Monsoon Asia. In India, for example, the number of malaria cases increased from 1.3 million in 1972 to 2.5 million in 1974. Egypt, with its Aswan High Dam, has another type of technological backlash. The Aswan Dam was constructed to control the flow and floods of the Nile River and to use the impounded water to improve and expand irrigated agriculture. The dam, however, reduces the amount of fertile silt deposited downstream, necessitating increased use of chemical fertilizers. The sardine fishing industry in the Nile Delta has been adversely affected because the organic matter in the river has decreased greatly. Also, the incidence of disease and parasites, particularly the debilitating schistosomiasis, has increased because of the environmental changes wrought by the change in the river flow.

Contemporary Patterns of Development

Sri Lanka is confronted with a serious *economic dilemma.* Conflicts between the interests of consumers and the long-term goals of economic development have led to inadequate funding to satisfy both the demands of consumers and the goals of centrally planned economic development. Another problem has been the serious inflationary spiral that has affected all of the world's economies. Export prices of tea, rubber, and coconuts, the principal cash crops, have boomed; but these have been offset by equally large increases in the cost of imports: rice, sugar, flour, and crude oil. The net effect has been to impede the rate of capital formation and thus restrict the rate of investment to promote economic growth.

Social problems also continue to plague the island. Of the 15.6 million inhabitants, 70% (known as *Sinhalese*) are of Indo-Aryan origin and descended from the ancient settlers. About 20% are more recently arrived South Indian Dravidians (*Tamil* speakers). The Sinhalese are Buddhists, and the Tamils are Hindus; thus the division is not only ethnic and linguistic but also religious. Violence between the Sinhalese and Tamils has been characteristic for several centuries. More recently, newly arrived Tamil plantation workers have created additional stress, and the government has worked to

repatriate some of these people to India. In July 1983 more than two hundred persons were reported killed in a week of violent clashes between the two groups.

Sri Lanka's government has followed relatively enlightened economic and social policies and energetically promoted population control and economic growth. Political and social problems and unrest have been confronted and in most cases resolved. The economic problems of this small island state with a *limited resource base,* however, are serious and place Sri Lanka in a vulnerable position.

Another small island Asian state (Taiwan) has been in a similar situation, and Sri Lanka may draw some inspiration from the remarkable economic growth and progress of Taiwan. The lesson learned is that such an achievement requires *disciplined effort* from the people as well as *investment capital.* If these two criteria are met, Sri Lanka may have the same kind of success as Taiwan. In that case, the characteristics of small-island accessibility, a high level of administrative control, and an integrated economic system become advantages. These advantages may be developed to the point that they outweigh the liabilities of a limited resource base.

AFGHANISTAN

Afghanistan is not climatically within Monsoon Asia. Its *location is transitional* between South Asia and central Asia and has long figured prominently in the political and military history of the Indian subcontinent. Since the time of Alexander the Great, the strategic location of Afghanistan with its command of the mountain passes that link eastern, southern, southwestern, and central Asia has attracted conquerors and rulers. The recent history of the country reflects the continuation of this competition between British and Russian interests in the nineteenth and early twentieth centuries and more recently among the Soviets, Indians, Pakistanis, and Chinese.

The Physical Basis for Livelihood

Most of Afghanistan is *high and enclosed,* similar to the intermontane basin region of the western United States (Figure 24–11). With the exception of the Kabul River and its tributaries that flow east into the Indus system, all other drainage in Afghanistan is internal (no outlet to the sea).

373

FIGURE 24-11 Afghanistan: Physical and Agricultural Regions. Afghanistan exhibits many Middle Eastern features such as aridity and Muslim religion. Yet Afghanistan's contact with the outside world has been through South Asia. Afghanistan's population is conservative and regional loyalties are often more important than allegiance to the nation as a whole.

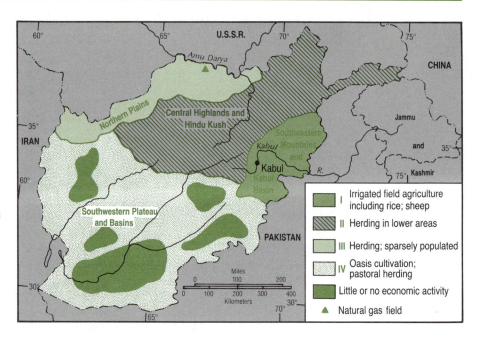

The dominant landforms of Afghanistan are the Hindu Kush Mountains, which separate the plains of northern Afghanistan along the Amu Darya (Oxus) from the rest of the nation. Despite this division, the country has managed to maintain a strong political coherence. South of the Hindu Kush are a great desert plateau and basin region in the southwest and south and a large alluvial basin centered on Kabul in the east. The Kabul Basin is the most densely settled part of the country and forms the heartland of Afghanistan.

Most of Afghanistan is dry, and precipitation averages 20 inches (51 centimeters) or less. This precipitation is concentrated in the winter and early spring, and much of it falls as snow. The seasonal distribution is the reverse of the monsoonal pattern found farther south and east. Daily temperature variations and seasonal ranges are great. These temperature ranges further illustrate the nonmonsoonal nature of the climate.

Isolation and a rugged land surface have impeded the introduction of technical innovations and the speed of economic change and growth. The difficulties of surface transportation and the sparseness of the *road network* illustrate this deficiency. There are no railways in Afghanistan, and the major roads serve only the main urban centers. Large sections of the country have no modern means of transportation, although in recent years a great deal of investment has been made in transport improvement. The impoverishment and extremely low level of economic development are not surprising in view of the impediments to internal movement of goods, ideas, and people.

Afghan Society

Afghanistan comprises a series of different *tribal groupings* and is dominated by the most numerous of these, the Pashto-speaking Pathans, who are concentrated in the Kabul Basin. Of the nearly 14.2 million Afghans, probably 90% are of Caucasian stock, and the remaining 10% are Mongoloid. Although strong dialectal differences exist among the major tribal groups, a *common religion (Islam)* provides a cultural cement that is very powerful in holding this traditional society together.

One of the most striking features of Afghanistan is its demography. The country possesses one of the highest annual birth and death rates in the world, 4.8 percent and 2.3 percent, respectively. Such a pattern suggests that Afghanistan is still in the *initial stage of the demographic transformation*. When coupled with a per capita GNP of $170, Afghanistan emerges as one of the poorest and least developed states on earth.

The Changing Economy

Subsistence farming of grains, fruit, and a few specialty crops, along with animal herding, are the traditional mainstays of the Afghan economy. More than 80% of the population is employed in these activities. In the Kabul Basin, the most productive region in Afghanistan, irrigation is commonly practiced with an intensive agricultural system that includes *paddy rice.* Recent agricultural innovations introduced from elsewhere, such as new seed strains and better techniques of cultivation, are expected to increase agricultural productivity in Afghanistan and meet the future demands of a growing population. Although Afghanistan possesses reserves of several important *minerals,* only natural gas production is significantly developed. The Soviet Union has assisted with the production of natural gas and purchases most of the gas.

Economic Development and the Five-Year Plans

Since 1957, Afghanistan has followed a series of government-formulated *five-year plans.* These plans were designed originally to bring the country into an age of economic development. Initially the major focus was on *capital improvements:* roads, bridges, irrigation systems, and power production and transmission. Beginning with the third five-year plan in 1968, however, more attention has been focused on social and educational problems. Funds recently have been funneled into improving literacy levels and into the commodity-producing sectors of the economy. Annual economic growth rates of 4% or better are sought.

These five-year plans have relied heavily on *foreign financing,* most of which has come from the Soviet Union, the United States, and West Germany. These countries and also the World Bank have assisted in arranging some of the foreign trade and special external purchases. Afghanistan also has helped pay for some of its modernization program by exports of cotton, wool, rugs, fruit, and natural gas. It seems likely, however, that the country must continue to rely on outside aid to finance economic development for some years to come.

Soviet Invasion and Recent Political Problems

Afghanistan is an extremely poor, landlocked buffer state that has only recently begun the push toward modernization. The process of modernization has come under increasing stress in recent years in part as a result of *Soviet pressure.*

Afghanistan, a monarchy until 1973, has undergone severe political upheaval in the last decade. A republic was proclaimed in 1973 after a coup d'etat, but the republic was superseded in 1979 by the Soviet-backed communist democratic republic of President Babrak Karmal. In order to ensure the tenure of the Karmal government, the *Soviet Union invaded Afghanistan* in 1979 and has fought a bloody war against Afghan insurgents. The insurgents are supported in neighboring Pakistan and China and also have received some assistance from other foreign governments. The Soviets, who anticipated a quick resolution of the fighting when they initially invaded, found themselves facing an increasingly serious dilemma of having fighting troops in a foreign country and yet being unable to stabililize the country into a *client state.*

25

China: Origin and Development of Civilization

As noted in Chapter 23 two of mankind's oldest and richest *civilizations* originated in Monsoon Asia, one of them in the Indus Basin. Farther north and east, in the drainage area of another great river, the *Huang He*[1] (Hwang Ho) or Yellow, there gradually developed the other great culture of Monsoon Asia, the Chinese. Although little is known of the earliest origins of the Chinese, and the written record dates back only 3700 years, enough archeological evidence has been found to suggest that the *Chinese civilization* emerged largely independently. In north China, a society evolved out of late Stone Age culture and developed techniques of sedentary agriculture; writing; formal religious, social, and political systems; and distinctive technological and architectural practices.

NORTH CHINA CULTURE HEARTH

Loess Land and the Physical Environment

Much of north China is covered by a mantle of fine, windblown, dustlike soil material, *loess,* called *huang-tu* (yellow earth) by the Chinese (Figure 25–1). Loess, common not only in China but also in places in Anglo-America and Europe, is usually associated with past glacial activity. So extensive has been the deposition of loess in north China that the material is as much as 500 feet (152 meters) deep in places. Although loess is of eolian

---

[1]Huang He is an example of the system *(pinyin)* now commonly in use in China to render Chinese terms and names in English. This system follows standard rules and is being substituted for the older forms of romanization (e.g., Hwang Ho) such as the Wade-Giles system. In order to minimize problems, we introduce the *pinyin* and follow it with the Wade-Giles or other conventional spelling in parentheses, unless the *pinyin* spelling is the same as the conventional one; examples are Beijing (Peking) and Xi'an (Sian). Certain names that are well known to Westerners through established usage will continue to be used here, although the *pinyin* romanization will be given at least once; examples are Yangtze River, Canton, and Manchuria.

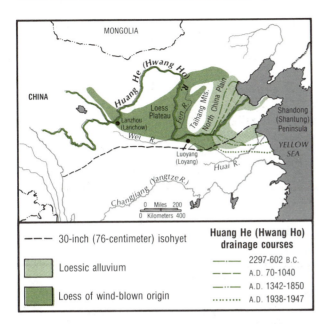

FIGURE 25–1 The Huang He and Other Major Rivers of China.
Although not China's largest river, the Huang He, with its main tributaries such as the Wei, has been the focus of much of China's ancient history. The map indicates the areas of loessial soil as well as the major shifts in the bed of the river.

(wind-borne) origin, erosion has taken place, and much of the material has been carried off the Loess Plateau by the Huang He and redeposited as *alluvium* in the North China Plain.

Climate in north China is characterized by dry, cold winters and hot summers. Sandstorms are common in spring, and dry winds of Siberian origin sweep down from the north and west during the cooler months. On the Loess Plateau, annual precipitation is between 10 and 20 inches (25 and 51 centimeters). On the North China Plain, precipitation is greater, between 16 and 30 inches (41 and 76 centimeters). South of the 30-inch isohyet (equal rainfall line), loessial materials dissipate and are not present in the Changjiang (Yangtze River) Basin. Throughout north China, *rain is concentrated in summer;* and although this concentration is good for cropping, the *evapotranspiration* is high enough to render the Loess Plateau and much of the North China Plain marginal for many crops unless irrigation is used.

It is interesting that such a *marginal area* is the site of the *birth of Chinese agriculture.* Yet, as Ho has noted, it was on the Loess Plateau and not on the North China Plain that Chinese agriculture first

developed.[2] His explanation is that the loessial soils were fertile, porous, and easy to work. The early Chinese with their primitive wooden digging tools could operate much more effectively on such soils than on the dense, more compact alluvial soils of the low-lying floodplain farther east.

The Huang He and Loess

The arid and semiarid environments of north and northwest China for a long time have been closely intertwined with the *evolution of human activities* and cultural practices. Within these environments several features have been important for the Chinese and their technological and economic systems. Chinese civilization developed in and around the drainage basin of the Huang He. Although *irrigation* was not a vital part of the earliest agricultural settings in north China, the Huang He has affected the Chinese in many ways. The difficulty of controlling and using this river has been a long chapter in Chinese history. The chronicle has not always been happy, and the phrase "China's sorrow" long ago gained currency as a means of describing the devastation brought on by this powerful and unusual river.

The *Huang He* rises among the swamps and lakes found in the eastern part of the Tibetan Plateau. The river initially flows east and north and after 1000 miles (1609 kilometers) reaches the vicinity of the modern city of Lanzhou (Lanchou). The river, with its steep gradients, has cut its way down to the western Loess Plateau, a drop of some 10,000 feet (3048 meters). Once on the plateau, where its gradient and velocity are suddenly reduced, the river broadens and meanders. There the Huang is broad and shallow. In the arid climate considerable amounts of its volume are lost to irrigation and evaporation, and the river is a sluggish and undramatic stream. At the top of the Great Loop, the river abruptly shifts direction and begins to flow south. As it changes direction its character also begins to change. Several factors account for this change:

1 The river begins to flow through poorly consolidated and easily eroded loess lands.
2 Numerous tributary streams, some of which are large, join the main channel; they too flow through the easily eroded loess and

[2] P'ing-ti Ho, "The Loess and the Origin of Chinese Agriculture," *American Historical Review,* 75 (1969), 31.

377

Loess is an unconsolidated, friable soil that covers large areas of Shaanxi (Shensi) Province in north China. This eroded hillside lies on the Loess Plateau of southern Shaanxi Province near the city of Xi'an. (C. W. Pannell)

contribute enormous quantities of material to the main stream.

3 As the river flows south, it descends more rapidly and its capacity to carry silt increases.

4 There is more rainfall, and more water enters the Huang and brings quantities of silt through the tributary network.

Together, these factors mean the stream is much larger and carries a *heavy silt burden.* Conditions are further complicated by the seasonality of rainfall, concentrated in the summer months of July and August. Such a concentration of rainfall results in great changes in the volume of flow of the river. In fact, the Chinese claim the Huang in a flood in the nineteenth century flowed at a discharge rate 250 times greater than its lowest flow, a record for the world's major rivers. Finally, the absence of heavy vegetative cover and the nearness of a number of important tributaries combine to increase the speed with which rainwater is funneled into the main stream, a speed that is analogous to water flows in desert areas.

All of these features combine to produce *floods of devastating proportions.* At the south end of the Great Loop, the river flows almost due east and within a short distance drops sharply off the Loess Plateau and enters the North China Plain near the city of Luoyang (Loyang). The *mechanics of stream*

flow operate as follows: high silt load, sudden reduction of gradient and velocity and thus loss of capacity to carry the high load, and seasonal concentration of precipitation. The results long have been of profound consequence to the great number of Chinese living in the North China Plain, for the annual floods have become legendary, and Chinese history is filled with accounts of the mighty river.

Most rivers flood, but the nature of the floods on the Huang is spectacular. The character of the stream, coupled with *ecological abuse* of the loess lands and the progressive degradation of local vegetation, has led to increased flooding over time. Moreover, the river is continually building up its bed with the dumping of its heavy sediment load, and the *stream bed has become elevated* throughout much of its path across the North China Plain. The Chinese have responded by building *levees,* but the levees must continue to be heightened as the river bed is elevated. However, it is impossible to build a dike high enough to protect against all floods. Such a vicious spiral of progressively higher levees has resulted in elevating the bed high above the surrounding floodplain. Consequently, when major floods have occurred, the river has broken out of its bed and inundated enormous areas of the North China Plain and has been terribly difficult to rechannel.

Since 602 B.C. the Huang He has changed its course fifteen times and flowed to the sea through *different channels.* Some of these channels have entered the sea at distances as much as 500 miles (805 kilometers) south of the present mouth (Figure 25–1). At one time the flow was actually channeled south into the Huai He and ultimately into the Yangtze. Such an event might be analogous to seeing the Mississippi River flood and alter its channel to the degree that it actually flowed to the Gulf of Mexico through the Rio Grande.

Flooding of the Huang He is of *monumental proportions.* When the river flows out of its channel, the results are catastrophic. For example, in 1938 the Chinese Nationalist government destroyed the dikes on the south side in an effort to slow the advance of the invading Japanese Army. More than 6 million acres (2.4 million hectares) were flooded and caused the relocation of an estimated 6 million refugees. As many as 500,000 people may have died as a result of this induced flood. Nine years passed before the river was rechanneled to resume its flow north of the Shandong (Shantung) Peninsula.

The Li River in Guangxi Province flows through a region underlain by limestone in China's karst region. The limestone towers create beautiful landscapes, but they also create difficult problems for farmers. (Eastfoto)

DIFFUSION OUT OF THE NORTH CHINA HEARTH

The Huang He and the Loess Plateau have meant many things to China and its people. Among them, one theme is common: The Chinese have long *struggled with an environment* over which they exercised little control and direction. The ecological conditions that confronted the Chinese as they developed sedentary agricultural practices in north and northwestern China were unusual and probably not dependable. It may be that the Chinese, through their various activities of cultivation and forest removal, rendered this environment even more intractable and uncertain. Whatever the causes, the *Chinese began to migrate south* away from the ecological rigors of the Loess Plateau and the North China Plain. In so doing, they were confronted with a substantially different environmental setting that offered *new opportunities* for agriculture and more dependable physical conditions for human sustenance. Migration south also extended the influence of cultural practices, as tribal people who inhabited south China were acculturated and absorbed by the more advanced Chinese.

Landscapes and Environments of South China

Southeastern China possesses a more intricate and *complex surface geography* than the broad plains and plateaus of the north. The main rivers and their tributary networks dissect the uplands into a system of river valleys separated by rugged and sometimes high ridges. With the exception of the Yangtze and Xi (Hsi) drainage basins, there are *no extensive low-lying floodplains.* The consequences of the diverse and rugged landscape are restricted movement and communication, *isolation* of local areas, and *sectionalism.*

The Chinese adapted easily to the alluvial lowlands and floodplains and began *specialized rice growing.* Native peoples who were not absorbed were forced into the rugged uplands. So rough was some of the terrain that these native peoples isolated themselves and preserved some degree of *ethnic and cultural independence* and integrity. In some cases ethnic separateness has continued to the present. Where tribal people remained in the lowlands, they were converted to Chinese in thought, speech, and practice, a process known as *sinicization.*

Thus it happened that the environments of south China, too, played a role in shaping the historical and social patterns that evolved. On the one hand, the Chinese were drawn to the environment, for it proved more tractable and dependable and could support more people through the mechanism of water control and rice cultivation. Sustained food production permitted the sustenance of large numbers of people and provided a *safety valve area* when conditions became bad in the traditional north China hearth. On the other hand, the

379

ruggedness and diversity of landscapes in the south promoted *political fragmentation* and regional interests. China, like the United States, has old and powerfully established sectional interests. Even under a Communist government, with its emphasis on centralization, sectionalism has not disappeared.

The Outer "Non-Chinese" Regions

Several distinct areas are located outside what is commonly referred to as China proper. These are Tibet, Xinjiang (Sinkiang), Nei Monggol (Inner Mongolia), and Dongbei (Manchuria) (Figure 25–2). These regions, with the exception of the Manchurian Plain, all share some *common physical characteristics*. They are for the most part *high and dry,* with great extremes of temperature between winter and summer. Much of their mountain and

basin area is characterized by internal drainage, salt marshes, and desert, and the environment has never supported a dense, productive agricultural system. Traditionally peopled by *non-Chinese,* these areas have long been divided politically and culturally between China and neighboring countries and people. Most *livelihood* is derived from herding, oasis cultivation, and subsistence farming. Today these non-Chinese regions, except for Outer Mongolia, have once again been attached to China, but their features of physical geography and human population serve to distinguish them from China proper.

The Three Chinas

Based on broad environmental and ecological factors, at least *three distinct Chinas* are discerned. To some extent these may be identified with hu-

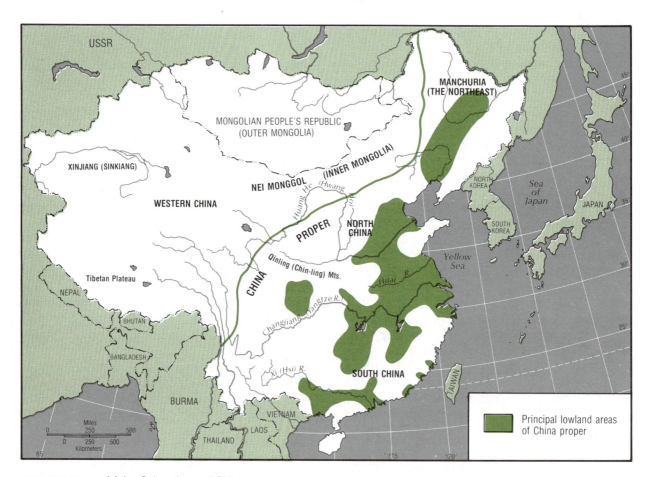

FIGURE 25–2 Major Subregions of China.
On this map are indicated the major subregions of China and the principal lowlands. The heavy line separates roughly the eastern third of the country, which is occupied predominantly by Han Chinese and thus is viewed as distinct from outlying non-Han regions.

380

man populations, but occupancy styles and activities are a better way to describe them. In brief, the three Chinas exist as *large ecological regions.* The first is *western China*—the dry, high China with its highly specialized, adaptive livelihood styles closely attuned to the physical conditions of environment. Its people are Turkic, Tibetan, and Mongol. The other two parts of China lie in the east, and these two may be divided between north and south. *The south* begins south of the Qinling (Chin-Ling) Mountains–Huai River line and includes the main agricultural lands of the Yangtze and Xi (Hsi) drainage systems. Its climate is mild and wet, much like the climate of the southeastern United States. Its agriculture is mainly wet rice. Although included in China proper and typically inhabited by Han Chinese, it entered the Chinese orbit during the late formative period. Its dialects and human population are diverse enough to indicate its more recent entry into the great Chinese cultural system.

Finally, there is *north China,* the source area out of which China was born and grew. North China includes the middle and lower Huang He drainage basin, to which should be added Manchuria. This is a cold, dry region in winter, but the summers are hot and humid and support intensive Chinese-style agriculture. The people are largely Mandarin speakers; and although there are ethnic infusions of Manchu, Mongol, and other tribal elements, it was here that the proto-Chinese came to be.

Based on the features of environment and oc-cupancy, these three broad regions persist to distinguish the main areal parts of the Chinese state even in the face of strong central-government efforts to counter this diversity. Environment and geographical features as much as politics continue to shape the face of modern China.

SOCIAL AND POLITICAL UNITY

One of the most interesting and unusual facts about China is its *permanence and long political tenure.* Despite changes in dynasties, occasional disruption by outside groups, and short terms of competing factions, China has managed to persist and to maintain control over essentially the same territory for some two thousand years. For Europe, it would be as though the Roman Empire had remained virtually intact to the present, and its institutions of government had, with certain modifications, survived.

The Administrative Structure

One reason for China's long-term cohesiveness has been a powerful and *effective central political apparatus* that prevented fragmentation. The traditional approach to governance was based on a large and *centralized bureaucracy* that operated with surprising efficiency on a modest budget. The constant circulation of bureaucrats from one post to another reduced the growth of powerful local

Southern and central China are primarily areas of wet rice farming. Rice paddies dotted with houses shaded by trees are a common rural scene. This photo shows members of the Muyu People's Commune in Hupeh Province weeding their rice fields. (Eastfoto)

interests. Through these and related policies, the central administration was able to maintain a remarkable degree of control even in the absence of modern means of communication and transportation. When times were bad, or if a dynasty was overthrown, a new dynasty emerged and the political system persisted and remained effective. In the absence of an integrating economic system, the centralized bureaucracy provided some of *the glue* that held the diverse parts of the empire together.

The Chinese Language

The people of China spoke *a variety of different dialects,* many mutually unintelligible. In contrast, written Chinese based on *ideographs* (picture symbols that represent ideas, objects, and actions) was used in all parts of China. It was the language of government and provided the central means of communication among the various dialect groups. Here, then, was another aspect of Chinese culture that promoted integration of its territories.

That the written language was known only to a small fraction of the total population was not a great disadvantage. The *literate group*—scholars, government functionaries, and the wealthy landed gentry—formed the leadership group at the local level. They were the decision makers, and it was this group with whom the central government communicated and to whom it addressed calls for taxes, laborers, and military conscripts. Both the central-government apparatus and the common written language became national symbols for the Chinese and important components of the emerging cultural system of the Chinese state.

Confucian Social Order

Another key integrating element was the *Confucian tradition.* Confucius and his many succeeding disciples and interpreters established for China a social order whose total impact has been pervasive for more than two thousand years. Confucius outlined a set of *social and behavioral patterns* for ancient Chinese society. These views were transmitted through an important segment of China's rich literature, and over time became embedded in the social system that ordered most, if not all, relationships among individuals, families, and the political system. In general, the Confucian precepts emphasized *stability.* Because any fundamental change in the existing order of social, economic,

and political relations was excluded from these precepts, Confucianism often has been viewed as rigid and conservative.

It is not surprising that *Confucius has been blamed* frequently in recent years for disallowing innovation and technical change and preventing China's economic modernization. Yet such criticism oversimplifies the nature of traditional China. Confucianism was but one of many underpinnings of traditional Chinese society. Nevertheless, in the last few years Communist leaders in China have waged a vigorous campaign to exorcise the ghost of Confucius from the new China in their drive to shift attitudes on such matters as how many children a family should have or what the nature of social stratification should be. Communist planners search and *strive for a new order* as they seek economic modernization and growth, and Confucian ideas have no place in their revolutionary society. Consequently, it seems likely the idea of Confucius as a great man of Chinese history and letters will continue to be challenged.

THE IMPACT OF THE WEST

In the early part of the nineteenth century, new influences entered China along its coasts and rivers to challenge the traditional government system. The British Navy, for example, forced China to accept the superiority of Western power in the *Opium War* of 1839–42. *The West* had progressed greatly in the seventeenth and eighteenth centuries, largely because of new techniques and innovations associated with the Industrial Revolution. China had progressed little toward these new technologies and developed few modern scientific forms of inquiry. European powers forced China to *open its doors* to the outside world and to begin the long drive toward modernization.

In the middle of the nineteenth century, there occurred one of the great Chinese political upheavals, the *Taiping Rebellion,* a partial result of the new challenge from the West. The Taiping Rebellion was followed at the turn of the century by the *Boxer Rebellion.* Finally, the old dynastic order could stand no longer, and fledgling revolutionary forces succeeded in establishing the Chinese Republic.

Through these years, China was increasingly set upon by people who were not susceptible to acculturation and who introduced modern weapons and industrial technologies superior to indigenous techniques. For the first time, China was confronted with *an external force* that asserted and promoted its own racial and cultural superiority and could support its claim with power, money, and technology.

CHINA'S REVOLUTION AND THE RISE OF CHINESE COMMUNISM

Among the ideas and innovations introduced into China in the nineteenth and twentieth centuries were the concepts and views of *Western democracy and Marxism.* Following the overthrow of the Qing Dynasty in 1911 and a brief period of near anarchy, the Chinese Republic was eventually established under the guidance of the revolutionary patriot *Sun Yat-sen.* His party, the Chinese Nationalists, announced the ''Three Principles of the People'' by which there was to be reform in politics, society, and the economy. This reform looked toward the enfranchisement of individuals and their improved status and economic well-being within the state.

Upon Sun's death in 1925, *Chiang Kai-shek* took over the *Nationalist party* and government, and except for brief periods, he ruled the Nationalists and their exiles in Taiwan until his death in 1975. Chiang had a more conservative political outlook than Sun, and he soon aligned himself with the urban commercial interests and large landholders in the rural areas.

A number of young Chinese had been sent to study in Europe and the United States. Among the ideas they were exposed to was *Marxism,* a political philosophy of great interest as a result of the Communist revolutionary movements in Europe and Russia. With the success of the Russian Revolution and the establishment of the Soviet Union, a base of support existed, and Marxist ideas gained currency in China as a competing means to achieve the goal of reforming and modernizing China.

The various Marxist groups in China fought bitterly among themselves. A major conflict existed as to whether the Communist Party should *focus on converting the urban workers* as the main strategy for revolution or whether a *rural strategy* that focused on the peasantry would be more appropriate. Among those who supported the rural-peasant line was a young librarian in Peking. His name was Mao Zedong *(Mao Tse-tung).*

After a short period of unsuccessful competition with the Nationalists, the Communist urban strategy failed, and the Communists increasingly turned to the peasantry for support. In the 1930s, however, Chiang Kai-shek attacked the rural Communists and drove them out of south and central China to a remote location in the Loess Plateau. The retreat of the Communists from the south to the isolated Great Loop of the Huang He is now an epic in the lexicon of Chinese revolutionary literature, *the Long March.*

Driven into hiding, the Communists regrouped. With the Japanese invasion in 1937, the National-

Chinese citizens line up in front of the mausoleum of the late chairman Mao Tsetung. The tomb is located in Tian-An-Men Square in the center of Beijing. (C. W. Pannell)

ists were preoccupied with fighting the Japanese and no longer were able to prosecute vigorously their campaign against the Communists. In a short time, the Nationalists managed to lose the political and military initiative. By war's end in 1945, the Communists had formidable military and political arms, and the Chinese Communist Party offered itself as a new and energetic answer to the plights of the poor, landless peasant. The issue was joined. The Communists increasingly established their power in north China, and the Nationalists no longer seemed able to compete for popular support. The results are well known; Chiang Kai-shek and the remnants of his army retreated to Taiwan, and Mao Tse-tung and his followers established a *new, revolutionary government.* These actions were the culmination of events that began early in the nineteenth century when the West seriously began to challenge the traditional Chinese system. The ascendancy of a Communist government in 1949 was but another change in a century of turmoil for China. The Communists, now firmly in power, are carrying on the processes of change as rapidly as possible.

26

China and Its Neighbors: Changing Societies and Economies

After 1949, with the ascendancy of the Communist government in China, the main objective of the new political order was to make over China completely. The new government sought to modernize China's economy and society as its political system had been made over during the years of revolution. Since 1949 radical new economic and social programs and policies have been promulgated. These revolutionary policies and the events associated with them form a modern phase of China's development.

This chapter also includes brief surveys of some of China's East Asian neighbors: Hong Kong, Taiwan, and the two Koreas. Hong Kong, Taiwan, and South Korea have free-market economies and offer interesting comparisons to the centrally directed economic systems of China and North Korea.

CHINA'S POPULATION

China as long ago as the year A.D. 2 is believed to have had a population of 60–70 million. Although the population may have fluctuated by more than 20 million during specific periods, ten centuries later it was still around 60 million. Such a demographic history suggests that in early China *a crude balance* was achieved between the number of people and the ability of the agricultural system to support them. Without some innovation or technical change, the population did not grow.

Population Growth

By 1400 the population had begun to increase, and by 1751 it had reached 207 million. A century later China's population had grown to about 420 million. Better techniques of farming, technical innovations, and greater internal movement of large amounts of food grains led to substantial growth in the population.

The outbreak of the *Taiping Rebellion* in 1850 began a *century of turmoil, war, and demographic change.* The consequences of this change were a high death rate and a decrease in the rate of population growth. After the Communists assumed power, they reported a 1949 population of 542 million people, an increase of about 25% over the preceding century. Since 1949 population growth has been great. Although recent figures are rather crude, 1982 official Chinese census figures indicated that 1,008,000,000 *Chinese* lived on the mainland. (The 1983 estimated population is 1,023,300,000.) In addition in 1982 there were 24 million Chinese residing in Taiwan, Hong Kong, and Macao.

Communist Policies

A most significant act of the Communist government when it took over in 1949 was to institute a comprehensive program of state-assisted *public health*. This program led to a rapid decline in the death rate. A high birth rate also continued with a resulting increase in annual population growth that began to subside only after 1960. *Family planning* has been increasingly stressed in recent years, and much effort has been expended in changing social attitudes about the value of female children. This change in attitude is necessary in order to persuade Chinese families not to keep having children until they have one or more sons. In 1979 it was reported that China had implemented a policy to promote *one-child families.* This policy employs economic incentives and advantages for families with one child and penalizes women who give birth to more than two children. Although it is too early to assess the significance of this policy, it indicates just how serious the Chinese government is about restricting future population growth.

At present it is estimated that the annual birth rate has dropped to 2.3%. This rate is a significant decline from the 4.5–5.0% of traditional times. Offsetting the decline in birth rates is a continued decline in death rates to an estimated 0.8% (Figure 26–1). Thus the annual growth of 1.5%, if accurate, indicates that China is midway in its course of *demographic transformation* (see Figure 2–3).

Additional statistics suggest that although China continues to be a relatively poor state in terms of per capita income, the capacity exists to reduce birth rates rapidly in step with the demographic transition from traditional patterns to modern. Recent visitors have reported statistics that indicate drastic drops in the birth rate in urban precincts in

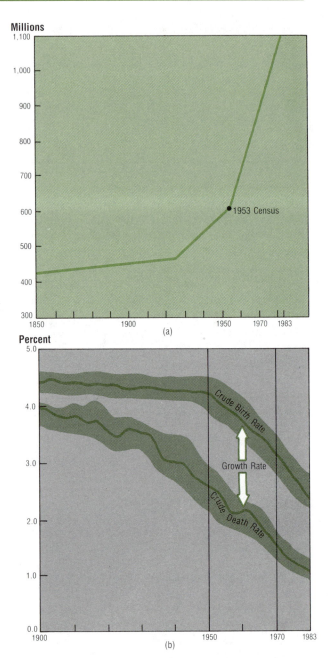

FIGURE 26–1 Chinese Population Growth and Trends in Birth and Death Rates.
(a) Estimated Chinese population growth, 1850–1983.
(b) Trends in birth and death rates, 1900–1983. The rapid population growth is explained by the fact that the death rate has dropped more rapidly than the birth rate.
SOURCES: ''China: Population in the People's Republic,'' *Population Bulletin*, 27 (1971), and Population Reference Bureau.

Shanghai and in Sichuan Province. If these statistics reflect national patterns, China has made great strides toward getting its population growth into balance with its resources and production, an achievement unprecedented for such a populous country.

Most claims about China's birth and death rates are difficult to evaluate inasmuch as China is so large, and family planning programs implemented in a large, modern city such as Shanghai may not have diffused to many rural areas. Nevertheless, enough information on rural areas has been made available to suggest that the family-planning policies and programs are now being extended to rural areas. The *paramedical teams,* for example, function in part to disseminate knowledge and instruments of birth control. For this reason among others, some authorities have suggested that the Chinese birth rate has actually dropped more sharply and rapidly than indicated above. Some have even gone so far as to suggest the Chinese birth rate, based on recent trends, will drop below the rate of birth in the United States within the next decade. If such a trend is real and continues, the Chinese have indeed demonstrated the powerful role of central-government direction in producing *rapid demographic results.*

Achieving a more moderate growth rate has not come easily. *Traditionalists* believed that large families were good, the natural order of things in Confucian terms. As in agrarian societies throughout the world, children, and especially sons, were seen as economic assets to look after their elders in their declining years. Although the Communist government was skeptical about birth control initially, it appears that after 1960 *family planning* was viewed with increasing favor as necessary to rapid economic progress. Except for the period of the Cultural Revolution (1966–71), it seems that China has vigorously promoted family planning. *The "pill"* is the most commonly used method of birth control, although a variety of other practices are also employed, including sterilization, especially of females, and abortion.

The extensive public health and welfare system is an effective agent in promoting birth control, and the *"barefoot doctor"* (paramedic) is heavily involved in spreading the word about family planning into remote areas. With the improvements in medicine and greatly reduced death rates, parents can expect their children to reach adulthood. The idea that children alone will provide support for their aged parents is also outmoded, for the state now cares for people too old to work who have no descendants to look after them.

Other measures may be even more influential in affecting birth rates in the countryside. Especially critical are the *"work point"* and family production systems whereby total income of rural work teams and families is apportioned according to the total work points or total production compiled by the production units. The two systems may contradict each other. For example, the more children in a family, the smaller each individual's share of the family total will be if adult work points are used. Total production, however, may increase if child labor is used in a family production unit. On the other hand, the allocation of private gardening plots to each family increases with the number of members in the family to a maximum of four. Families with two children thus receive the same amount of land as those with six children. This system is used as a kind of *negative tax.* Harsher measures are also employed; for example, young people are discouraged from marrying before a certain age (twenty-four for women, twenty-nine for men). In urban areas young married couples may be denied housing before they reach the required age.

The success of these programs and policies indicates that China has made remarkable progress in altering the value system of its citizens. A new ideal of a small family with one child has been proclaimed. If we can believe the available information, this ideal is being accepted although not always happily. There have been numerous examples where women have suffered and been criticized heavily if their "one child" turns out to be a girl. Such a *change in values* appears to place China in opposition to the established Marxist position on population growth. The *Marxist position* accepts any population growth as good in that the labor potential is increased. Such a position, however, seems to conflict with policies of social enlightenment such as the emancipation of women from their traditional roles. Chinese Communists may perhaps justify family planning on the basis of its role in gaining *greater status for women.* Also, it may well be that the Chinese are not too concerned with the link between Marxist theory and Chinese practice as applied to their goals of modernization. Pragmatism and success may be more important.

Where the People Are

In China *people are concentrated* where they have been for many centuries. Today more than 900 million people reside within the eighteen provinces of China's traditional core area. One large area, Manchuria, has been populated within the last century. *Han Chinese* recently have also migrated in large numbers to the region of the Great Loop of the Huang He. If a line were drawn connecting Kunming in southern Yunnan (Yunan) Province with Quiqihar (Tsitsihar) in western Manchuria, more than 90% of China's people would be found east of that line. Although the eastern part of China contains less than one-third of the total area (Figure 26–2), it also contains most of the well-watered, *productive agricultural land.* Thus there exists in China a *correlation between arable land and population.* With the dense settlement of the central part of the Manchurian Plain in this century, the last major region of the Chinese state able to support a dense population was occupied. Future development of agricultural lands will depend on expensive and difficult irrigation and reclamation projects if population densities similar to the core region are to be supported.

West of the Kunming-Quiqihar (Tsitsihar) line, rainfall decreases to the point that settled agriculture is marginal. Perhaps for this reason most of the region has traditionally been the domain of *pastoralists* and *specialized oasis farmers.* Although two-thirds of China's territory is included, only about 70–80 million people live here, and many are members of China's minorities. These relatively empty areas of China would appear to invite migrants from crowded eastern China. The nature of the environment with its paucity of rainfall, however, is alien to the traditional methods of intensive farming common in eastern China. Perhaps new technologies, such as mechanized agriculture and improved irrigation techniques, will open much more of this region to denser settlement, but it seems unlikely that Han people will live in western China in the concentrations so common in the more humid east.

CHINA'S CHANGING SOCIAL PATTERNS

Changes in *China's society* have been dramatic in recent years. In part, these are the changes expected in any *society transforming* itself from traditional to modern, from rural to urban, and from agricultural to industrial. To these changes has been added Marxism, which seeks to transform the *basic value system* rapidly and thereby speed up modernization. The special Chinese brand of communism, *Maoism,* has contributed a distinc-

tive flavor and demands radical approaches and great speed in the development drive.

The Traditional Pattern

Society in *traditional China* focused for the most part on family relationships at the village level. For more than two thousand years, a *family-oriented* social system existed based on ethical norms and values outlined by *Confucius* as early as 436 B.C. In this system, order was based on the primacy of the head of the household, and his desires ruled over the younger generation who lived in the same household. Male heirs, who could carry on the family name and line, dominated the system. Women were important for the work they performed but had little role in formal decision making and no proprietary rights. They were excluded from a meaningful role in affairs of the family and the village.

Beyond the family, there were established common surname associations or *clans.* These clans frequently acted to satisfy larger mutual interests and would temporarily side with other clans when some joint purpose was served. In addition to clans, other voluntary associations and groups emerged. Often these were focused around a religious objective or some common commercial or special-activity interest.

Such was traditional society. It centered on the *village,* and most of the relationships that grew up were village oriented and based on *kinship ties.* To a considerable extent, villages were *self-sufficient.* There was a tie between the economic system with its self-sufficiency and the *inward-looking* social system. It was the political system as much as anything that linked the village with the rest of China. Social links with the outside did exist, but for most peasants, these links were not well established and were rarely used.

Society in the New China

Social revolution is as much a part of the Maoist remolding of China as change in the political and economic systems. In fact, all three are related components of a broader process of development. The success of Mao's revolution was based at the village level. Social reform and land redistribution in the village followed wherever the Communists were able to establish themselves.

Among the most important consequences of the new social revolution was the *shift in family*

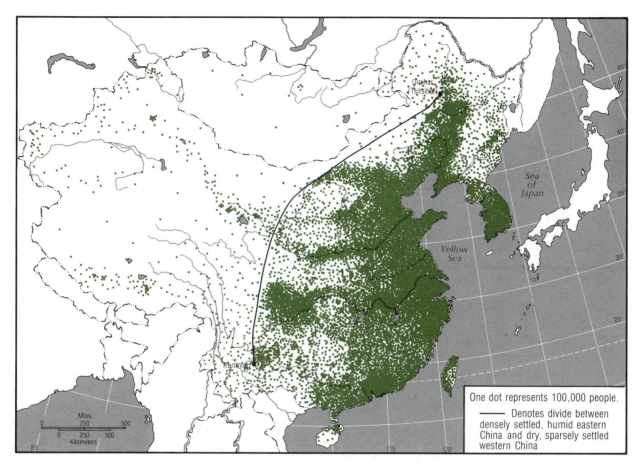

FIGURE 26–2 Distribution of Population in China and Rimland States. China's population is heavily clustered east of a line connecting Kunming and Qiqihar (Tsitsihar). Here reside approximately 95% of the Chinese people, a reflection of geographical features of better accessibility and improved opportunities for farming.

dominance. Authority has shifted from the traditional head, the eldest male member of the family, to a younger man, usually the oldest son, who is also the most productive member of the family. It is a reversal of the established Confucian norm and an indicator that social changes in China are fundamental and far-reaching.

Another abrupt change from established practice was the creation of *new roles and rights for women.* Women have been freed from their traditional subservience and lower status and have been encouraged to spend time away from home, often as salaried members of the labor force. Men at the same time have been encouraged to view their wives as equals and to share in the household chores. Men also have had to recognize the right of their wives to join in political activities and associations that promote the group interests of women. With this new set of relationships, a basic

alteration in the established Confucian order has been set in motion.

THE AGRICULTURAL ECONOMY

China's contribution to the development process may be its emphasis on *rural areas as economic growth centers.* The government believes that the village has played a key role in revolutionizing China in less than a generation. It is this accomplishment, based on the special approach to and reform of traditional village China, that is believed to be a unique contribution of the Chinese to the theory and process of development and modernization.

The economy of China has always been rooted in and *dependent on agriculture.* Even after three decades of Communist rule, China remains an ag-

389

ricultural country with approximately 80% of its people engaged in farming and related activities. Despite the fact that farming continues to dominate employment activities, the nature of the rural economy has changed greatly. *Collectivization* of land and organization of rural activities by *communes* have altered many traditional rural patterns and especially the uneven distribution of landholdings. At the same time, some of the old cropping practices and farming techniques linger.

Traditional Agriculture and the High-Level Equilibrium Trap

One traditional means whereby China increased agricultural production was through *expanding its cropland.* This process reached its zenith in the south with its rice-growing environment, as the frontiers of Chinese settlement expanded over the centuries. With the settlement of Manchuria in this century, no more attractive agricultural frontiers remain. The primary path to greater agricultural output in the *future must come from increased per-unit yields.*

Other traditional techniques of increasing crop output in China were simple: add more labor and night soil (decomposed human excreta) and intensify *multiple-cropping.* Unfortunately, increased crop output led to more people. A greater population in turn reinforced the cycle by providing more laborers and more human waste. Gains in production were countered by increased population, and many people existed at a level of subsistence. This system developed a kind of *self-reinforcing chain reaction* whose result could only be a descending spiral of poverty for most people. There were no stimulants for technical breakthroughs or investments in ideas or products that might have led to technological innovation. Innovation in turn might have lessened the traditional demand for labor and reduced the incentive for large families. The population continued to expand in step with expanded commodity output. Per capita wealth and demand remained stagnant and may even have decreased after the sixteenth century. It was this situation, described as a *high-level equilibrium trap,*[1] that resulted in increasing poverty and created such stress in the traditional Chinese economy and society. It has been argued that this stress was the main precondition to the demands for the reform and modernization of China.

Peasant life in traditional China was frequently difficult, and even the village elites and main landlords were hardly wealthy by Western standards. Several levels of peasants existed: landlords, middle peasants who farmed their own land, and poor peasants. The poor peasants frequently hired themselves out, or otherwise became indentured or tenanted, and thus associated with one of the wealthier peasants. The landholding relationship and pressure for land became more urgent with increasing population, for the effect of more sons was to fragment more and more the existing cropland and to reduce continuously the amount of farmland available to most peasant families.

Such a situation, when coupled with the vagaries of *natural catastrophes* of drought, flood, earthquake, or locust plagues, yields an unpleasant picture of traditional peasant life. The picture was especially bleak in north China where the natural events were more erratic and less subject to control. Floods and drought would lead to food shortages, and many families would flee to the south in search of a better situation. Landlords would squeeze poor families harder, and often the difficulties would lead to local violence. In some cases such violence may have been widespread enough to crystallize into larger movements leading to political change.[2]

Modernization of Agriculture

In addition to socializing agriculture through state ownership of land and the formation of rural communes, Chinese planners in recent years have attempted to raise crop yields and production. Many approaches have been followed: land reclamation, better land management, improved irrigation, increased dry cropping, and market incentives. Perhaps the most promising of these new approaches are *innovations* in the form of chemical fertilizers, hybrid grains, insecticides, and improved irrigation systems. All of these techniques are forms of substituting *capital for labor* in the agricultural sector, a practice common in the West but only recently applied intensively in China.

In the early 1950s the new government was fortunate to have several years of good weather, and agricultural output increased. The *"Great Leap Forward"* program of 1958, however, coincided with several years of poor weather, and crop produc-

---

[1] Mark Elvin, *The Pattern of the Chinese Past* (Stanford, Calif.: Stanford University Press, 1973), pp. 203–19.

[2] A chronicle of such events on the eve of Communist takeover appears in William Hinton, *Fanshen: A Documentary of Revolution in a Chinese Village* (New York: Vintage Books, 1966).

tion fell. Since then, except for some uncertainty during the years of the *"Cultural Revolution,"* output of food grains is believed to have increased gradually in pace with the growing population.

Despite increased crop production, China has frequently purchased grain (wheat usually) from several Western states. The major suppliers have been Australia, Canada, and the United States. Such sales suggest that China may not produce enough food to support its huge population. At the same time, however, China exports a considerable amount of high-cost foodstuffs (vegetables, pork, condiments, and specialty foods) to Hong Kong, Southeast Asia, and Chinese communities all over the world. China may well be a *net exporter of foodstuffs,* despite the large grain purchases. China has claimed self-sufficiency in food production since 1971. Grain production figures for 1981 are provided in Table 26–1.

Food exports notwithstanding, China's foreign grain purchases suggest a *weakness in agriculture* that makes modernization and increased productivity of the system ever more pressing. One major problem is that little good land is available for agricultural expansion, at least under present technologies. For example, the United States, with only one-quarter the Chinese population, has 153 million acres (62 million hectares) more arable land. With only dry or marshy lands available for future expansion, China faces an increasingly difficult situation with its growing population. Statistics on rural land use show that cropland has actually decreased to about 245 million acres (99 million hectares) in recent years. Yet China depends on the agricultural economy to help form capital in order to finance development projects. It is imperative that per-unit yields be increased rapidly if agriculture is to continue financing development and living levels are to be raised simultaneously.

An associated and equally urgent problem is that as the economy modernizes, it will increasingly urbanize, and China's cities are commonly found on the best agricultural land. As cities grow, good agricultural land is lost to urban land uses. Here is a built-in conflict, wherein modernization and economic development programs are likely to result in the loss of a significant amount of the best agricultural land.

Crop Regions and Crop Types

Food *grains,* including rice, are the leading crops produced in China. As a general rule, *rice* is grown wherever environmental conditions permit. In the

TABLE 26–1 China: Output of Selected Agricultural and Industrial Products, 1981

Agricultural		Industrial	
Product	Metric Tons (millions)	Product	Metric Tons (millions)
All grains	325	Coal	620
Rice	143	Crude oil	101
Wheat	58	Crude	
Soybeans	9.2	steel	36
		Cement	84
		Chemical	
		fertilizer	12.4

SOURCE: *Beijing Review,* 25 (1982).

southeast, where frost is uncommon, *double-cropping* of rice is the usual pattern. Rice tends to produce the largest quantity of edible grain on a given unit of land, provided there is plenty of labor, nutrient material, and water. Thus the alluvial valleys and basins of the southeast and central China where rice can be grown are among the most productive and densely inhabited parts of China (Figure 26–3).

Dry cereals, wheat, millet, and kaoliang (a type of grain sorghum) tend to dominate foodstuffs in the drier, cooler north, although cash crops such as cotton are also common. In the south the more common cash crops are tea, fruits, sugarcane, and tung. The most striking pattern of Chinese agriculture is the transition from wet rice in the south to dry cereals north of the Qinling (Chin-ling) Mountains–Huai River divide.

The Commune

Probably the most important economic unit in modern China is the *commune.* Although at one time the term "commune" was used to refer to urban neighborhoods and districts, it is in the rural areas that communes have functioned best as operating units. Today's commune is large in area (an average of 60 square miles, or 155 square kilometers) and usually composed of the territory and fields of a number of small hamlets and villages. The commune is organized hierarchically and made up of several *brigades,* which in turn are composed of many *teams.* In addition, a central office conducts administrative affairs and maintains ties with other political, social, and economic units.

Communes first appeared in China in 1958, about a decade after the ascendancy of the Communists. The formation was the achievement of a

391

FIGURE 26–3 Agricultural Regions of China.
China is a large, middle-latitude land with a variety of environments and, as a result, many different cropping systems. The most significant regions are the humid rice-growing south and center; the cooler north and northeast, where coarse grains and wheat are grown; and the dry west, with its oasis and herding complex.

stage of communism outlined by the late Chairman Mao Zedong (Mao Tse-tung) to *promote modernization* through one of the great movements that have swept revolutionary China. In this case, the *"Great Leap Forward"* was designed to push economic growth to a level of output equal to that of the United Kingdom within a decade. To accomplish this goal, it was necessary to rely on small-scale *rural industries* as well as the production of *large factories.* The communes were organized to support the new efforts to modernize, but they built on an established pattern of village production and social, political, and economic *self-sufficiency.* Not only were these communes to grow their own food, but also they were encouraged to build local industries to produce clothing, tools, and other necessities—local production for local consumption. At the same time any surplus could

be sold and shipped to other regions. Return goods would complete the flow and suggest the modern adaptation of a well-established Chinese pattern of spatial organization and interchange.

Individual families are assigned to specific labor teams. In general, all adult members and some older children of a farm family perform labor and are awarded *work points* for their labor. The work points form the basis for the share the family will receive from the total earned by the commune after expenses and taxes have been paid. In addition, the family works a small private plot of land assigned it by the commune and usually owns a couple of pigs and a few ducks or chickens.

The *private plots,* as in the Soviet Union, are very important. About 5% of China's arable land is composed of such plots, and their output of fresh vegetables is probably close to 50% of total national production. Of equal importance, however, is the support they provide for swine production, a primary source of animal protein in the Chinese diet and of organic fertilizer for crops. As the Chinese discovered during the "Great Leap Forward," a decline in swine production brought about through curtailment of the family-operated private plots had far-reaching consequences for agricultural output. The private plots were restored to individual families after a brief but disastrous interruption.

Production Responsibility System

Since the late 1970s a new rural production program has been introduced: the family *production responsibility system.* Under this system farm families are allocated a certain amount of land to farm; ownership of the land remains with the commune. The family contracts a production quota with the production team and commune, and the family can keep whatever is produced beyond the quota after expenses and taxes are paid. The surplus can be used by the family or sold in private markets. The system fixes much of the farm decision making in the farm family. It is expected that the system will increase production, improve efficiency, and effect better crop selection.

How much better off a family in one of China's rural communes is today than a poor peasant family of fifty years ago is not easy to evaluate. Most members of every family must still *work long and hard.* Today, however, there is generally *enough food* for everyone, even if the government must import it. A basic educational opportunity is provided everyone in free *schools,* and a *health* service of national dimensions provides reasonable if modest care to all at very low cost. Along with these benefits goes a certain amount of political education, and the state expects everyone to comply with its directives without question.

INDUSTRIALIZATION AND ECONOMIC GROWTH

Location and size have been fortunate for China. One of the important consequences of China's great size and particular location is a *rich and diverse resource base,* although access to these resources has sometimes been difficult. The rich resource base has helped promote industrialization and economic growth. It includes not only a *wide*

variety of environments for the production of food and cash crops, some of which (cotton, tobacco, soybeans, and tung nuts) are closely associated with industrial processing, but also substantial reserves of *fossil fuels* (coal and petroleum) and *minerals* (iron ore, tin, manganese, copper, and molybdenum) (Table 26–1). These resources serve the state by assuring *a degree of economic self-sufficiency* that requires less dependence on external sources. China, therefore, does not need to expend scarce capital on imported raw materials. Indeed, foreign exchange can be gained by the sale of some of these resources. Recently, for example, China has financed the purchase of sophisticated machinery and technology from Japan through the sale of crude oil. However, imports are beginning to be financed more through long-term credit arrangements.

Accessibility, by contrast, is more difficult to assess, for in much of China the landscape is rough, and transportation has traditionally been slow and expensive. Water transportation links China increasingly with the outside world. In this century new emphasis has been placed on other forms of surface and air transportation, as China has sought to tie together the far-flung parts of its vast territory.

Contemporary Patterns of Industrialization

After seizing power at midcentury, the Communist leaders and planners established a basic *goal of industrializing* the state as rapidly as possible. It appears that this policy and its implementation were based on the *Soviet model* of investing most in *heavy industry* (producer goods) to the relative neglect of *consumer goods* and agriculture. The Soviet Union initially provided a great deal of technical assistance in constructing new plants and provided, for a price, much of the equipment and machinery necessary.

At the beginning, things went well. Good weather provided large agricultural yields, and the *Soviets cooperated* and helped complete a number of important industrial and construction projects. Strains, however, soon began to appear between the two states, and in the late 1950s the Soviets withdrew their assistance. Soviet engineers abandoned some projects in the middle of construction. It was at that point that the Chinese changed direction in central planning and established their own path for industrialization and economic growth.

393

The "Great Leap Forward," "Walking on Two Legs," and the "Four Modernizations"

At the time of the *"Great Leap Forward"* in 1958, it was difficult to gauge the full nature of the new changes under way in industry. Subsequent events have permitted a better understanding of the new policies for economic growth. An established part of Communist policy was to *develop the interior* and eradicate the "evil" influence of the *treaty ports and colonial cities* as centers of industry and innovation. This new policy, according to the planners, would serve to develop backward areas, *take industry to the raw materials,* reduce transport costs, and provide better protection for the centers of production. A conflict existed, for it was recognized that the treaty ports as great industrial centers had good transportation access, were well supplied with skilled labor, already had established industrial and power facilities, and were low-cost centers for industrial production. All of the amenities and advantages of large cities already existed. As the Chinese planners knew, these assets are very difficult and expensive to create and build anew.

During the exuberance of the "Great Leap Forward," a new experiment was undertaken to short-cut the process of industrialization. This experiment focused on local small units of industrial production; the *"backyard"* iron *furnace* was one example. The Communist planners sought to reduce dependence on the great industrial centers by promoting *local industries* designed to produce goods required at the local level. In part, this new *policy failed.* Products from local industry were often inferior or useless and in some cases were not cheap. Apparently, the principle of local production for local consumption survived, however, for in the late 1960s it reappeared in a new guise, "Walking on Two Legs."

"Walking on Two Legs" refers to the use of both *small and large* units of production, the *traditional and modern sectors,* and the *rural and urban* locations. It might best be described as a *policy of dual industrialization.* Moreover, this policy has been strengthened by increasing local industrial production and economic self-sufficiency at the commune level. The lesson learned is that *small-scale industrial units have limited capacity* and can produce only certain goods in limited quantities.

Most recently China has moved to a policy referred to as the *"Four Modernizations."* Here the policy is to modernize four broad sectors—*agriculture, industry, science,* and *national defense*—by the end of the century. The frequently stated goal is for China to become a modern, socialist industrial state by the year 2000, ready to enter a new stage of economic growth and able to compete with major world powers such as the United States and the Soviet Union. In carrying out this modernization, China has indicated a willingness to borrow and use Western technology and scientific knowledge.

Industry, Cities, and Transportation

Despite the aim of China's new planner to shift the centers of production to the interior, the *well-established cities* such as Shanghai, Tianjin (Tientsin), Shenyang, Guangzhou (Canton), Wuhan, and Beijing (Peking) continue to dominate as the great industrial cities. A few *new centers* have been developed. The Inner Mongolian city of Baotou (Paotou), with its new steel mills, is a frequently cited example. Many other older cities in the interior, such as Urumqi (Urumchi), Chengdu (Chengtu), and Changsha, also have been given new industries. Nevertheless, the major centers of industry continue to be the cities of the coast, the Yangtze ports, and the great Manchurian cities, a testimony to the validity of their initial selection.

This situation is neither startling nor illogical if one examines a transportation map of China (Figure 26–4). These *great industrial cities* for the most part are those best served by rail and sea and are the most accessible. By virtue of their large populations and established industries, they are the centers of the greatest consumption of goods, fuel, and raw materials. Such elements and activities have a tendency to feed each other, to reinforce the growth already well established.

One plan to counter the influence of the treaty ports was to build *new rail links* throughout the interior. The Chinese have built and continue to build an impressive if not dense rail network throughout the interior. The new growth centers, however, will be hard pressed to rival the advantages of international trade linkages and sea transport shared by the former treaty ports.

China's industrial and transportation geography may be usefully compared with India's. India possesses a dense network of railroads focused on the major industrial and commercial centers, all of which were inherited from British colonial days.

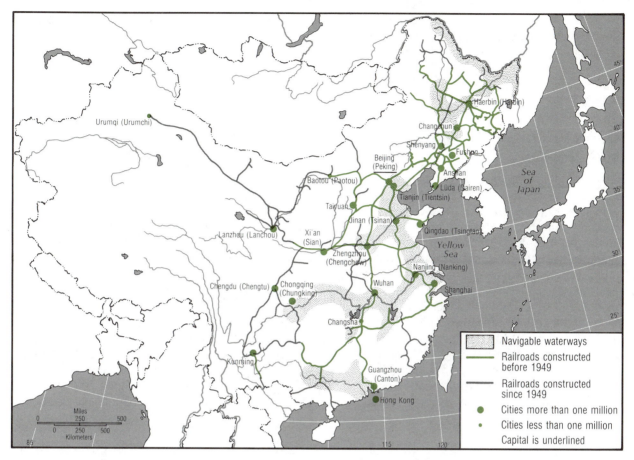

FIGURE 26–4 Railways and Waterways of China.
China has a modest network of railways, although many new lines have been constructed in recent years. The rail system is supplemented by an extensive network of rivers and canals, which are especially important in central China.

China, by contrast, contains *two patterns:* the indigenous interior urban and water transport network, which has been supplemented with a modest but expanding rail system, and the coastal and northeastern cities (treaty ports) and transport arteries. The main industrial areas are associated with the second, although in recent years the two have been increasingly integrated.

Both China and India have used preexisting networks of transportation and cities in their recent development strategies. Both countries have developed new industrial centers, and it is hard to know whether Baotou is more important to China's economy than Jamshedpur is to India's. Both countries seek to develop new areas in the interior and to redistribute wealth away from the coastal areas that are associated with a colonial past. Neither, however, is likely to follow too closely such

a locational policy if it impedes the rate of economic growth and progress.

Regional Inequality: A Continuing Problem

One of the most serious problems facing China is *regional inequality* between interior and coastal regions. China's interior, especially away from the Yangtze River, has remained impoverished, while Manchuria and coastal areas with better accessibility and rapid industrial development have become more prosperous. This trend has led to increasing income inequalities in different parts of China with sharp regional differences between the lower Yangtze delta and interior locations such as the Loess Plateau. Policy statements have emphasized the need to reduce such inequalities and to im-

395

prove the regional distribution of industrial production. How seriously such statements can be taken is uncertain, but the ideological commitments and policy statements indicate the Chinese are serious about reducing the regional inequalities in their large country.

Oil: A Promising Development Opportunity

Oil is perhaps the best example of an industry that has been developed very rapidly under Communist guidance. Before 1949 less than 3 million tons of oil had been produced over a half century of consumption. After the establishment of the People's Republic in 1949, the Soviets provided both equipment and personnel for oil production. By 1955 annual crude oil production totaled nearly 1 million tons. By 1970 production had jumped to 20 million tons, and estimates of production in 1981 were 101 million tons. Much of the great new production was associated with the development of the Daqing (Taching) oil field in Manchuria. It was on the basis of production from this field that China achieved *self-sufficiency* in petroleum production and initiated oil exports to Japan.

Estimates based on proven and probable *reserves* suggest that China has between 1 billion and 2 billion tons of oil in the ground. Future developments, especially those offshore, will require very large investments in equipment and very sophisticated equipment, neither of which China possesses at present. The future of Chinese oil production, then, may be tied closely to political events, for the Chinese may need assistance from other countries such as Japan, the United States, or other Western states. The potential for greater resource exploitation, coupled with the opportunity to capitalize on that resource for capital formation, would suggest the Chinese will consider carefully such factors in determining their future relations with foreign powers.

Future Trends

Today, China produces, in *modest quantities,* a full array of industrial and scientific goods: automobiles, airplanes, locomotives, computers, and atomic bombs. Iron and steel, cement, and electric production exceed the level of the United Kingdom. Fossil fuel production is large and climbing rapidly, based on new petroleum discoveries. Total industrial production, however, is small by United States or Soviet standards. When divided among China's great population, the *per capita output is very small.*

China's economy by many standards remains *agrarian* in nature. In terms of the Rostow model outlined in Chapter 4, China would appear to be about to enter the stage of "economic takeoff." It is conceivable that the recent shift in foreign policy and the efforts to buy new technologies will accelerate economic development. Both Japan and the Soviet Union industrialized quickly. Special conditions and traditions, however, when coupled with the huge population, suggest that China's economic development will proceed in a somewhat different manner and at a more modest pace.

CHINA'S RIMLAND

Surrounding China on its eastern and northern flanks are several nations or colonies that have been linked to China either directly or through ancient *cultural and historical ties* (Figure 26–5). In Korea physical isolation and a degree of political autonomy have promoted a separate and distinctive culture. In other cases Chinese territory and Chinese-speaking populations are involved (Hong Kong, Macao, and Taiwan). The Chinese government considers these three areas integral parts of the Chinese state.

Mongolia, too, is considered Chinese territory but of a somewhat different nature. Its status, in the Chinese view, is analogous to Tibet, Xinjiang (Sinkiang), and Nei Monggol (Inner Mongolia). Although its inhabitants are recognized as non-Chinese, their history and destiny are so interlinked with China as to make them proper subjects for political integration with the People's Republic. Were it not for Soviet opposition and power, the Chinese probably would have forced the incorporation of the People's Republic of Mongolia into their state.

Hong Kong

At the mouth of the Xi (Hsi) River estuary, just within the tropics along China's southeast coast, is the *British Crown Colony* of Hong Kong. Hong Kong comprises 404 square miles (1046 square kilometers) of territory on which live 5.2 million people. Its main resource is the hardworking population whose efforts have brought prosperity to the colony. Traditional functions of Hong Kong were

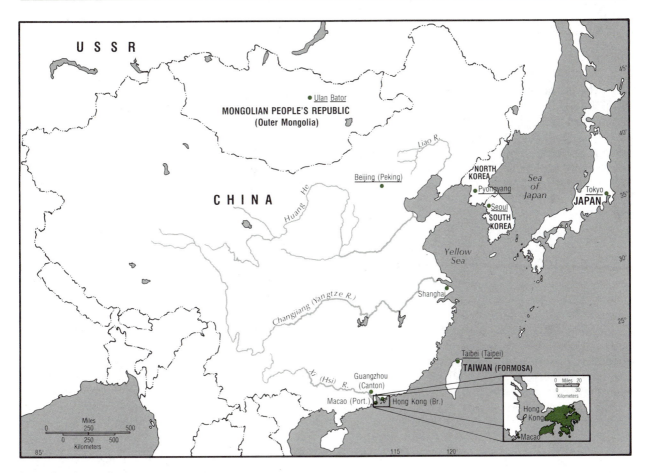

FIGURE 26–5 Main Political Units of East Asia.
China dominates East Asia not only because of its size but also in terms of political geography.

trade and commerce. From its beginning, the British colony was an *entrepôt* and trading center in south China, and it served this function for a century. Hong Kong's *harbor* is deep and spacious. The development of steam-driven ocean vessels with increasingly deep drafts gave Hong Kong an advantage over neighboring Macao. Although financed and controlled by Britain, Hong Kong functioned as an appendage of China's foreign commercial and trade structure and traditionally accounted for a substantial part of China's foreign trade.

After the ascendancy of the Communist government in mainland China, Hong Kong's future appeared bleak. Military defense of the colony against China is not feasible. Somehow the colony has survived, and indeed prospered, although there have been periods of great uncertainty and instability.

With the interruption in its traditional functions of trade and transshipment in 1949, Hong Kong of necessity turned to new enterprises. By 1970 more than 40% of the labor force worked in *manufacturing.* Textiles and garments accounted for nearly 45% of this employment and for 60% of the colony's exports. Other important industries are plastics (toys, recreational goods, and sundries), wigs, electronic components, and machine tools.

Banking, insurance, and other forms of *commerce* continue as important activities and make up the other major employment field. Hong Kong continues its role as an entrepôt, and approximately $4 billion worth of goods are transshipped through the colony. This *transshipment trade,* consisting of manufactures such as textiles, pharmaceuticals, watches, and foodstuffs, raw materials, and fuels, is increasing rapidly. About one-fifth of it goes to Japan. Singapore, the United States, Indonesia, and Taiwan are among other important destinations of reexports. *Tourism* is also

397

extremely important to the economic health of the colony. More than 100,000 tourists arrive in Hong Kong each month. The earnings derived from the tourist trade and from other invisible sources of foreign exchange, such as banking and insurance, help offset an unfavorable balance of trade that amounted to $11 billion in 1981.

One of the revealing facts about Hong Kong's character is that about half of its citizens were born elsewhere, mostly in China. Some 98.5% of them are described as Chinese, based on their language and place of origin. At the end of World War II, Hong Kong's population was about 600,000. By 1960 this number had swollen to 3.13 million. At present, the colony's population is 5.2 million. *Immigration* has accounted for this remarkable growth in such a short period, and one must look to events in China for an explanation. China's recent political activities and history, such as the "Great Leap Forward" and the "Cultural Revolution," have apparently spurred a migration. The mainland government at times has allowed some of its citizens to leave, and others have left China illegally. More than 400,000 immigrants have entered Hong Kong since 1977.

With such a large *population* and *modest land area,* Hong Kong is really a city-state and is one of the most crowded places on earth. The colony's population density is 12,900 per square mile (approximately 5000 per square kilometer). Despite the dense population, 90% of the land area is either in rugged uplands and scrub vegetation or under cultivation. Only 10% of the area is urbanized, that part of the colony in which its inhabitants live and work. For example, in one urban district there are about 430,000 people per square mile (166,000 per square kilometer), one of the most dense urban concentrations on earth.

The future of Hong Kong's political *status is in doubt.* The Chinese view Hong Kong as Chinese territory, although they appear to recognize the present lease arrangement as good until 1997. The British argue that the lease pertains to the New Territories and part of Kowloon, but they claim the island of Hong Kong and a few square miles of Kowloon Peninsula are British territory in perpetuity. Unless some arrangement is agreed on that provides protection for British and other foreign investment, the local economy will suffer, and the enormous contribution Hong Kong makes to China through its purchase of Chinese goods will be jeopardized. It has been estimated that 40% of China's

foreign exchange earnings are gained through Hong Kong.

Taiwan

Recent political events involving China have shocked Taiwan. By 1979 only twenty-three countries continued to recognize Taiwan, and it was displaced as the representative of China to the United Nations in 1971. Perhaps the most shocking blow was when Taiwan's long-time ally, the United States, withdrew diplomatic recognition in 1979 in favor of the People's Republic of China. Although Taiwan's *economic growth* and progress in recent years have been impressive, its political *future is clouded.* The island state contains three major dialect groups and a variety of political factions. All are related in one way or another to the island's past as a frontier territory of China, and all have opinions on China's pressing desire to reintegrate Taiwan into China.

China has had ties with Taiwan for more than a thousand years and, since the sixteenth century A.D., has sent a steady stream of immigrant farmers and adventurers, first to the P'enghu (Pescadore) Islands and then on to Taiwan. At various times this stream has been interrupted, and the island has been occupied by the Dutch, Spanish, French, and Japanese. Despite these non-Chinese occupations, ever since Chinese farmers began to arrive and cultivate the productive alluvial western coastal plains and basins, the island has been Chinese in ethnic and linguistic composition.

The *eastern two-thirds* of the island is high and rough. Except for a narrow valley and a small alluvial basin in the northeast, little suitable land exists for intensive farming. The *western one-third* of the island is composed of a series of alluvial plains and basins separated by broad cobble-strewn river beds. A dense network of roads and rail lines covers these plains and basins, and the densely inhabited parts of the island are well integrated spatially.

Taiwan lies astride the tropic of Cancer 100 miles (161 kilometers) east of the China mainland. Its climate is *monsoonal,* with most precipitation occurring in the summer. In the northeast quadrant, however, the pattern is reversed, with heavy winter rains from the prevailing winter northeasterlies. Summer and autumn typhoons are an annual threat and often bring great devastation and flooding. Frost is unknown at lower elevations.

The combination of climate and alluvial lowlands has resulted in a *productive paddy-rice agricultural environment* like that of southeastern China. Other crops include sugarcane, tropical fruit, tea, and a rich variety of vegetables. *Intensive cropping* of the alluvial area has given Taiwan one of the world's most productive agricultural systems.

In 1885 the Chinese Qing Imperial government made Taiwan a province, two centuries after it had been absorbed by the empire. The new governor set about to modernize and develop the island as a showpiece province. Ten years later, after the Sino-Japanese War of 1895, the Japanese took over the island as a colony and accelerated the pace of investment and growth. The combined efforts of late Qing China and Japan to modernize Taiwan through the construction of roads, dams, irrigation systems, bridges, a railway system, new cities, and ports laid the stage for a *transformation* of the traditional economic system.

The fifty years of Japanese colonialism also brought improved standards of public health, a decline in the death rate, and rapid population growth. At the beginning of this century, there were about 2 million people on the island. By 1920 population had climbed to 3.65 million. In 1949, when the Nationalist Chinese government was exiled on Taiwan, there were more than 6 million inhabitants. To these were added more than 1 million soldiers, functionaries, and refugees from the mainland. Rapid population growth has continued, and now there are 18.9 million inhabitants, an incredible number of people for an island of 13,892 square miles (35,980 square kilometers), roughly the combined size of Maryland and Delaware.

A significant consequence of *Nationalist rule* of Taiwan was *land reform*. A comprehensive and profound land reform program was instituted in the 1950s giving land to those who tilled it. A program of government-supported, low-interest loans provided financial support. *Land redistribution* was accomplished, and output has climbed steadily since that time (Table 26–2).

Despite Taiwan's very high ratio of people to arable land, the island, with its *double-cropping* of rice and abundant vegetable, grain, and fruit production, remains basically self-sufficient in food. During the last decade it has exported sugar, rice, mushrooms, pineapples, and pork. The relatively high cost of these items in relation to imports of other less expensive food grains has permitted Taiwan to remain close to agricultural self-sufficiency.

TABLE 26–2 Taiwan Agricultural Production

Year	All Agricultural Products (1976 = 100)	Rice (1976 = 100)
1962	59.1	77.9
1965	70.6	86.6
1967	77.5	92.8
1971	86.9	85.3
1975	90.9	91.9
1977	104.0	87.5
1981	105.8	

SOURCE: *Industry of Free China,* 58 (August 1982), 58–59.

Taiwan has *industrialized* with help from United States foreign aid contributions, which were both large and well conceived during the 1950s and 1960s. So successful was the United States aid program that no new projects were funded after 1965. Growth was both rapid and self-sustained, as indicated by the dramatic annual increases in per capita GNP—from $260 in 1969 to $500 in 1975 to more than $2300 in 1983. Associated with the remarkable economic progress has been a shift in the structure of the economy to an urban-oriented commercial and industrial economic system.

Korea

In many ways Korea might have become part of China. Through much of its history, it has existed as a loose political appendage of the Middle Kingdom and owes China a profound *cultural debt*. Much of what is considered traditionally Korean began as Chinese: religion, art forms, architecture, part of the written language, political and social institutions, agricultural practices, and land-tenure relationships. Despite the many cultural threads and practices from China that have entered the Korean fabric, somehow an independent and separate entity has emerged. Modern Koreans, with their own distinctive culture, langage, and history, see their country as an individual East Asian civilization. They have worked and suffered to establish this *cultural integrity* and are proud of it. Despite the political division of the nation into North and South Korea after World War II, Koreans have maintained a *sense of nationhood*. In recent years the two countries have discussed reunification, but no concrete actions have been initiated.

Korea's location as a peninsula jutting off the Asian mainland toward Japan has made it a sensi-

A shirt factory in Taiwan's duty-free zone. Materials are imported free of tariffs, processed by low-cost labor, and then exported to world markets. (United Nations)

tive and *strategic bridge*. It has been so used by various warring groups for centuries. Yet traditionally it has been under China's protection and influence. In 1905, after the Russo-Japanese War, Korea became a Japanese protectorate. Five years later it was annexed and became a part of the Japanese Empire. Although this *colonial rule* lasted only until 1945, its impact was far-reaching, for Japanese investment and development projects initiated the transformation of a hitherto traditional economy and society. Without such Japanese contributions as roads, railways, ports, mines, factories, and irrigation systems, today's growth and progress would likely not have been possible. Nevertheless, Japanese colonial rule left a legacy of distrust and dislike that continues to sour relations between the nations.

Korea's 84,565 square miles (219,023 square kilometers), roughly the size of Minnesota, lie in the middle latitudes, and *climatic patterns* largely reflect the peninsula's location and position. Summers are warm and moist; winters are dry and cold, a reflection of the dominant air mass, the Siberian high, that controls much of the winter climate of eastern Asia. In North Korea, especially at higher elevations, winters are very cold and harsh. Farther south, the influence of the adjacent water bodies moderates the temperature and provides a better opportunity for moisture accumulation in the air. The southern part of the peninsula thus has a much greater potential for agricultural production (Figure 26–6). Its less severe winter, longer growing season, and higher average daily temperatures

make a cool-season second crop possible, and a dense agricultural population is supported.

Landforms reinforce the agricultural pattern, for the largest alluvial plains are in the south. Most of the peninsula is mountainous, focused on a north-south-trending spine, with peaks rising to 9000 feet (2743 meters). Only about one-fifth of North Korea is suitable for cultivation. Although this rugged relief is a serious obstacle to intensive agriculture, the mountains offer other types of resources and economic opportunity. Hydroelectric power production, timber, and mining are major activities in the mountainous regions, and Korea is well endowed with coal, iron ore, tungsten, graphite, magnesite, and a variety of other minerals. Most of the *mineral wealth* is concentrated in the north and to some extent has offset the greater agricultural potential of the south.

During the colonial rule (1905–45), the Japanese, in making their investments and constructing cities, factories, and transport lines, focused on the north. Over time much of the *industrial output* of the state became concentrated in the northern half of the peninsula. The south, with its more intensive wet-rice agriculture, supported twice as many people and was looked on as a *breadbasket*. Thus developed the apparent geographic complementarity of the two halves of the country, an *industrial north* and an *agricultural south*.

Such a regional characterization has never been entirely fair or correct, and large cities in the south, such as Seoul, Pusan, and Taegu, long have had important industrial functions. The north,

41.3 million)—and the *Korean War,* which hardened this division, new economic patterns have emerged. Each half has pursued policies to round out its economic structure in order to achieve greater *self-sufficiency* and reduce reliance on external sources of supply. To some extent, both have succeeded in promoting economic growth and in providing more of the manufactured goods necessary for their own economies. It has been suggested that over time the move toward self-sufficiency by both north and south will lead to an acceptance of the political division as permanent, inasmuch as the two halves no longer need each other.

Both North and South Korea have prospered in recent years, and both countries are increasingly looking toward the *conversion of their economies* to industrial, service-oriented systems from the traditional, agrarian patterns. Growth rates in per capita GNP have been impressive in recent years. South Korea especially has had good success in exporting textiles, apparel, footwear, and electronics to consumer-oriented economies such as Japan and the United States. *Heavy industry* is a new component of South Korea's industrial base.

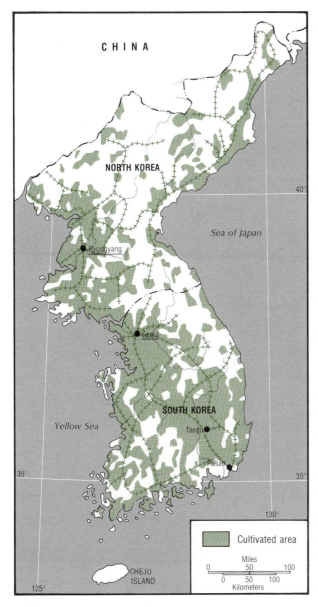

FIGURE 26–6 Cultivated Area of Korea. The Korean peninsula is divided politically between north and south. The southern part has considerably more cultivated land area than does the north.

although a focal point for much Japanese energy in building an industrial base, has been traditionally an agricultural land also. Recently the North Korean government has claimed further progress in agricultural investment, great improvements in farm production, and the successful socializing of agriculture.

Since the *division of the country* into two halves—a Soviet-oriented north (population 19.2 million) and a Western-oriented south (population

Seoul, South Korea, has nearly 9 million inhabitants and is growing at a rapid rate. (United Nations)

401

In keeping with the plans for economic growth, both countries have invested heavily in new roads, railways, mines, irrigation systems, power sources, and ports. For example, South Korea has constructed a four-lane expressway linking Seoul with Pusan. *Cities are growing* at prodigious rates, and the capitals, Pyongyang in the north and Seoul in the south, are among the most impressive. Indeed, Seoul with 9 million inhabitants is one of the world's largest cities. The rate of economic expansion in both the industrial and agricultural sectors, coupled with the resource base and the industrious, educated, and disciplined labor force, suggests a happier economic future for both Koreas.

27

Southeast Asia: Islands and Mainland

Prior to the nineteenth-century European occupation of most of Southeast Asia, little was known in the West about this part of the world. Traditionally the region has been a *zone of contact* and interaction between *India* on the west, from which it drew much of its early culture, and *China* to the north. For many centuries the Chinese were interested in Southeast Asia and had at various times extended their culture and influence to what is now known as Vietnam. In the last 150 years, many Chinese have migrated southward, and several Asian countries today have large and significant Chinese minorities. During the nineteenth century and the first half of the twentieth, the *age of colonialism* for Southeast Asia, many divisive and heterogeneous forces fragmented the region. World War II and invasion by Japan brought a new political awakening and changes in traditional values to this corner of Asia.

Perhaps most important, World War II ushered in a new age and signaled the decline of European colonialism in Asia. The war exposed the weaknesses of European colonial control over most of Southeast Asia. After the war, the British, French, and Dutch colonies emerged as independent nations. This *transfer of power* was not always easy. Often local groups fought among themselves for political power, and the polarization between Marxist-oriented movements and other approaches to government has been especially bitter. Political democracy has been challenged everywhere and has been under serious stress even in places as modern and "Western" as Singapore and the Philippines. In southern Vietnam, democracy was never really taken seriously, and nowhere in the Southeast Asian region does it thrive.

Today, with the emergence of Asian *nationalism,* the traditional view of the region as a cultural interface between China and India is declining. The future is likely to focus more and more on indigenous views and patterns that demonstrate national pride rather than on those features that have been borrowed from India or China. On the other hand,

China and India as military and political powers are geopolitical realities that all Southeast Asian states recognize and must consider in designing their own futures.

GEOGRAPHICAL OUTLINES AND PATTERNS

Despite the idea that Southeast Asia is a distinctive region, within the region there exist both *cultural and physical contrasts* and diversity. Perhaps most obvious is the division between the mainland—Burma, Thailand, and Indochina (Laos, Kampuchea [Cambodia], and Vietnam)—and insular and archipelagic Southeast Asia—the Philippines, Indonesia, Brunei, Singapore, and Malaysia. These two parts are separated by shallow waters that lie over the Sunda Shelf (Figure 27–1).

Southeast Asia extends more than 3000 miles (4828 kilometers) from east to west (Burma to New Guinea). If the western half of New Guinea (West Irian) is included, Southeast Asia contains about 1.72 million square miles (4.45 million square kilometers), including 807,000 square miles (2.09 million square kilometers) on the mainland and 915,000 square miles (2.37 million square kilometers) of insular territory. Although this region lies near the equator, it stretches to almost 30°N in northern Burma, and a considerable part extends as far as 20°N.

Areal Organization

One of the most conspicuous features of human activity in *mainland Southeast Asia* is the formation of *national corelands* around the major river basins: in Burma, the Irrawaddy; in Thailand, the Chao

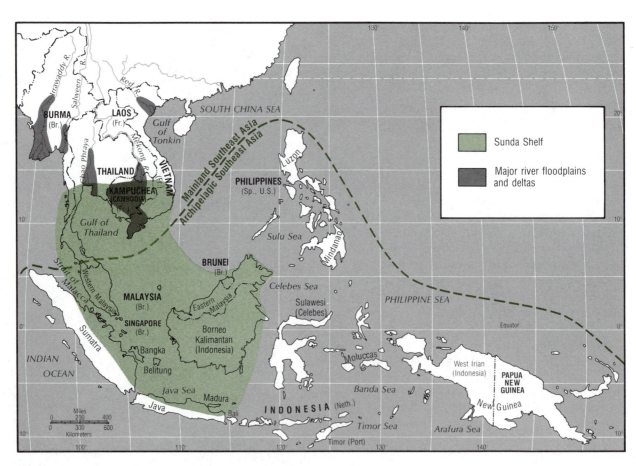

FIGURE 27–1 Mainland and Archipelagic Realms of Southeast Asia. Southeast Asia may be divided between the mainland and an archipelagic realm to which has been added the Malay Peninsula. On this map national political units are identified along with their former colonial associations.

404

Human settlement is largely water oriented in mainland Southeast Asia. Along the rivers are fertile alluvial soils irrigated with river water. The rivers also provide transport routes. (United Nations)

Phraya; in Laos, Kampuchea, southern Vietnam, the Mekong; and in northern Vietnam, the Red River. It is these great rivers that provide the soil-enriching floodwaters, the water supplies for *irrigation* systems, and the primary corridors of *transportation* and access to the adjacent broad alluvial floodplains. The large river basins contain the most *productive agricultural environments* and the most dense concentrations of people (see Figure 23–1).

By contrast, *archipelagic Southeast Asia* shows no single clear pattern of areal organization and formation of national territories. In the archipelagos of Southeast Asia, the role of European colonial policies in shaping the patterns of development and political evolution was especially prominent. To a great extent, it was the activities of several European powers—the Spanish in the Philippines, the Dutch in Indonesia, and the British in Malaya and northern Borneo—that gave rise to the political boundaries that form the outlines of the modern states.

Marine Location and Accessibility

Among the most significant features of the geographical setting of Southeast Asia are its *maritime orientation* and location. It is composed of a series of islands and peninsulas, and no country, with the exception of Laos, is without a shoreline and adequate anchorage for seagoing ships. Moreover, the Strait of Malacca, separating Sumatra from the Malay Peninsula, is one of the great shipping corridors, and most maritime traffic between Europe and eastern Asia passes through it. *Accessibility* ranks high among the assets of Southeast Asia, and the seas that separate the various countries and islands are much more a tie than a barrier.

This feature of maritime accessibility can best be illustrated through the example of *Singapore*, the region's most important *entrepôt*. British colonials created this great city from a tropical island and mangrove swamp in the mid-nineteenth century. Its location, carefully selected by Sir Thomas Stamford Raffles at the southeastern end of the Malacca Strait, gives Singapore control over access to and through this part of the world that is the basis of its past growth and likely prosperous future. Like Europe, Southeast Asia has long benefited as much from its location and accessibility as from the resources of its physical environment. The difference is that the use of this access has developed more slowly than in Europe.

405

Mainland Southeast Asia

For most of mainland Southeast Asia, the dominant *physical features* are the rugged cordilleras that splay out from the Himalayas to the north and arc to the south. These mountains are underlain by an ancient crystalline mass of stable granite material. This mass and its subterranean extension, the Sunda Shelf, geologically tie together much of Southeast Asia's islands, peninsulas, and mainland as a physical unit. The north-south-trending mountains of mainland Southeast Asia, although physically related to the taller Himalayas to the north, have been heavily weathered and rounded in the tropical rainy climate. They are much lower than their counterparts in southern Tibet. (Few are higher than 10,000 feet, or 3048 meters.) The ranges parallel one another and separate the major river basins that form the corelands of the five countries of mainland Southeast Asia. These main ranges are, from east to west, the Annamite Chain of Vietnam, the Shan Highlands of western Thailand and eastern Burma that extend the length of the Malay Peninsula, and the Arakan Yoma of western Burma. Active seismic forces such as faulting are not part of the physical landscape directly associated with the mainland highlands and the Sunda Shelf but are common along the archipelagic southern flank of the shelf and from the Philippines to New Guinea.

Although accounting for less than half the total land area of Southeast Asia, the mainland component probably contains the greatest *cultural fragmentation.* For example, the groups that inhabit the highlands are made up of a number of diverse tribes with cultural orientations and linguistic systems different from those of their more advanced, less isolated, lowland cousins. Most of these highland people, by virtue of the territory they occupy, have found themselves politically contained in one or another state whose political, social, and religious ideals they may not share. Long ago, the stage was set for conflicts between the politically dominant majorities that inhabit the major river basins and the groups that occupy interior highland locations. This problem is serious in almost every country of mainland Southeast Asia.

Archipelagic Southeast Asia

Contrasted with the stability and regularity of the north-south-trending axes of mountains and river valleys of mainland Southeast Asia is the younger *active belt of volcanism* associated with many of Southeast Asia's islands. A string of volcanic islands stretch from Sumatra and Java east to Celebes and the Moluccas and north to the Philippines. Not only is this area one of the most geologically active regions on earth, but it is also a highly diverse land surface, a good reflection of the newer processes of landscape formation. It is part of the circum-Pacific belt of volcanism known as *the Pacific Ring of Fire.* Moreover, the influence of volcanic materials on local soils has been profound enough to affect fertility levels and cropping patterns and add to the complexity of the region's physical makeup.

The islands of Southeast Asia and the Malay Peninsula have been called the "Malay realm" because of the ethnic and linguistic affinities of the *dominant Malay people.* However, the presence of large numbers of Chinese and other significant *minorities* makes this label rather inappropriate. Indigenous peoples of ancient Australoid or negroid stock are also present in certain locations; in New Guinea the native people are predominantly negroid. Moreover, the Filipinos are largely Christians, while most other Malays are Muslim, and religious differences inhibit understanding and interchange between these otherwise closely related peoples.

TROPICAL ENVIRONMENTS AS SPECIALIZED ECOSYSTEMS

Despite a diversity in landforms, the shared marine orientation and tropical climate give Southeast Asia a common set of environmental characteristics. These commonalities promote similar economic patterns and provide the basis for a unified approach to economic growth and development.

Effects of Equatorial and Tropical Location

Within 5–6° of the equator, high average humidity and temperatures are common. Singapore (2°N), with little seasonal variation in rainfall or temperature, is a *classic equatorial climatic* station. Similar climatic regimes are found throughout much of the Indonesian and Malaysian archipelagos, although locally landforms and prevailing winds may alter the typical pattern somewhat.

Away from the equator, seasonal rainfall produces *distinct wet and dry seasons.* In general, these seasons follow the alternations of the monsoon wind system. During the northern hemisphere summer, *the monsoon* takes the form of a southerly wind from April through September, and during winter, a northerly wind from October through March. Although most rainfall is concentrated in the summer months, surface relief creates some variations. Examples of unusual and unexpected patterns are found in the autumn maximum of rainfall along Vietnam's central coast and the summer dry areas in the rain shadow of Burma's southeastern mountains.

A Fragile Environment

The uniformly high temperatures and heavy rainfall of the tropical environment produce a *fragile ecological condition.* Such a condition demands careful management if use of the environment is not to result in serious and often nearly irreversible damage to the natural resource base. Tropical rain forests, the original vegetative cover of much of Southeast Asia, are characteristic of most of the humid tropics of the world.

The *rain forest* appears luxuriant and prolific, and often the uninformed assume that a rich soil supports this thick forest cover. In fact, the soil is not rich but has been heavily leached of plant nutrients. The rich forest growth is supported in large part by the sparse accumulation and rapid decomposition of its own litter on the floor of the forest. The presence of the forest is based on the *rapid recycling* of the nutrients. Thus the growing forest in large part produces its own food supply. The process involves a great number of species of trees, plants, and vines. It is an ecological system extremely sensitive to any alteration.

One *exception* is the fertile *alluvial soils* of the floodplains and deltas where plant nutrients are periodically added to the soil by flooding. A second exception is where *volcanic action* has produced soil of great fertility. The areas of these fertile soils are densely settled and provide the basis of livelihood for most of Southeast Asia's peoples.

One product that is especially prominent among the resources associated with the tropical environment is timber. *Tropical hardwoods* include such valuable species as teak, ebony, and mahogany, which are exported for furniture, veneer, and plywood. In addition to saw timber are trees such as rattan palm used for carvings or household accessories.

Imperata Grasses: Challenge of the Tropics

One of the most serious problems in Southeast Asia's rain forests has been the replacement of the forest by tough, fire-resistant, fibrous grasses of the genus *Imperata.* These grasses are known lo-

The tropical rain forest of Malaysia covers over 75% of the land. Lumber and forest products rank third in the country's list of exports. Research programs have been started to diversify the forest industry by introducing pine and other fast-growing trees from which paper and pulp can be manufactured to meet growing internal needs. (United Nations)

cally by a variety of names such as cogonales, cogon, or just plain elephant grass.

Where a large area of the forest has been cleared and improperly maintained, or where small plots are continuously cultivated, these *grasses invade the fields* and once established are difficult to eradicate. *Imperata* grasses cannot be controlled by indigenous farming methods. Heavy equipment and machinery are required for *Imperata* removal—equipment the people do not have and cannot get. In Southeast Asia where the population growth rate is high and agriculture an important means of livelihood, the loss of land to *Imperata* is a serious problem. *Imperata* grasses have taken over large areas in the Philippines, Thailand, and Indochina.

Interestingly, *Imperata* has not been a significant problem where land is farmed strictly in the traditional shifting (migratory) manner and where demographic stability is maintained. The small for-est clearings are cultivated for only a few years and then abandoned. Forests around the edges of the clearings gradually encroach on the clearings and shade out the *Imperata.*

Shifting Cultivation

Over large areas of the humid tropics, including Southeast Asia, *shifting agriculture* is widespread (Figure 27–2). It is a *low-yield system,* and the use of hand tools is characteristic. The cycle of clearing, burning, cultivating, and fallowing enables the soil to store up plant nutrients between cropping periods. The nutrients are then released over a short period when the land is cropped. In some cases the forest is allowed to grow back to maturity, but more commonly the *slash-and-burn* phase is repeated after a dozen years or so. The objective is to use the land only until yields decline and to

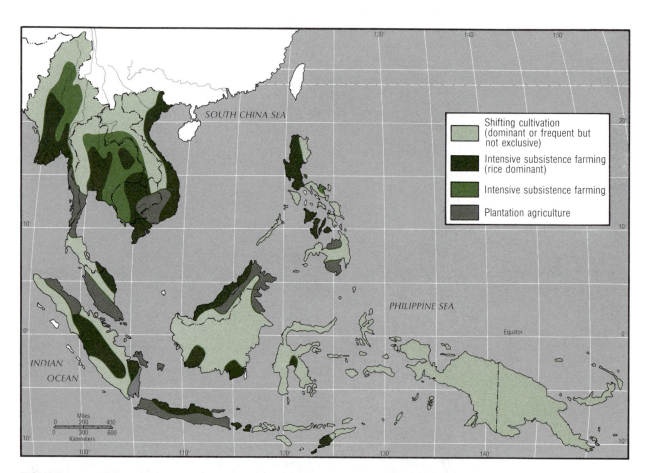

FIGURE 27–2 Major Agricultural Systems of Southeast Asia.
The three main agricultural systems are identified by location and approximate areal extent. Although intensive rice cultivation supports a high percentage of the population, shifting cultivation and plantation agriculture occupy the majority of the cultivated land.

408

stop cultivation before any serious soil destruction takes place.

Shifting cultivation is an adaptive, though perhaps primitive, method to re-create natural conditions. A great number of *different crops,* vines, cover crops, tubers, and pulses are intertilled and replicate in part the natural characteristics of the tropical ecosystem. The diversity of the shifting cultivator's cultigens adds strength to the crop system by providing a more balanced diet and a more assured food supply, as well as protecting the soil. The ability to grow things quickly and get the surface covered protects the soil from rain and sun and retards the process of leaching. All of these techniques indicate that where populations are sparse and the land is given sufficient time to lie fallow, *shifting cultivation is conserving* rather than destroying. The system can support a sparse population indefinitely. If the population grows or if too-frequent cropping takes place, the end result may be degradation of the tropical rain forest and a rapid decline of soil nutrients. If the balance is disturbed, large areas may be lost to imperata grasses. The land may then be useless both to shifting cultivators and to other sedentary users who might desire the land for more intensive exploitation.

Paddy-Rice Agriculture

Within Southeast Asia the most common form of agriculture is the cultivation of *paddy rice* (Figure 27–2). In all but one country (Malaysia) paddy rice accounts for more than half the agricultural area; moreover, it supports most of the population as the primary food grain. Thus there is a remarkable areal association between the low-lying alluvial river basins and dense concentrations of people. Rice has long been a major export commodity of Burma, Thailand, Kampuchea, and Vietnam. However, recent unrest has disrupted agricultural production, and traditional exports have declined.

Paddy rice is *grown in several ways* in Southeast Asia; all depend on the presence of water to sustain plant growth. The main difference is the source and control of water. In some cases, as in the Irrawaddy Delta of Burma and parts of the Philippines, rainwater is the source. More common is the use of supplementary water. In the floodplain of the Chao Phraya in Thailand and along the lower Mekong in Cambodia and Vietnam, the irrigation of wet rice depends on the annual flooding of the rivers. In other locations, as in Java, water is provided and controlled through irrigation systems. Irrigation comprises the most sophisticated type of wet-rice cultivation and also gives the highest yield per land unit.

Commercial Agriculture

Estate or *plantation agriculture* in Southeast Asia involves the extensive use of capital and is commercialized. Because of its association with money and international trade, this form of agriculture is a dy-

Wet rice is the principal crop of Southeast Asia agriculture. Rice is usually cultivated and harvested by hand. (United Nations/FAO/ H. Null)

namic *force for modernization* and economic progress and plays a special role in the development of the region.

A number of different crops have been utilized in estate cropping, and some have succeeded better than others. Moreover, there is considerable variation in where particular crops are grown. In Java and the Philippines, *sugar* has dominated, but until recently special marketing arrangements and guaranteed prices for certain quantities resulted in stagnation. In contrast, *rubber* trees, concentrated mainly in Malaysia and Sumatra and grown on estates and on small holdings, have been very successful and account for more than half of the area under estate crops in Southeast Asia. Other important cash crops include *coconuts, coffee,* and recently vast plantings of *oil palm.* All of these crops are remarkably well suited to the local environments, although the rubber tree, imported from the New World by way of England in the late nineteenth century, is probably the most adaptable to a greater variety of conditions.

Costs of cultivation, harvesting, processing, and transport to market, along with market price, affect the success of commercial crop cultivation. Since rubber can be cultivated and harvested cheaply and is relatively easy to process and transport, natural-rubber producers have managed to remain competitive with synthetic producers. Other crops such as coconuts or oil palm may require more sophisticated forms of processing; often only an estate or firm large enough to produce in great quantity can be financially successful with these crops.

The foreign-owned estate with its greater capitalization has distinct advantages. Yet, as Donald Fryer has pointed out, the foreign-owned estate, despite its many advantages and its potential success as a money-maker and development device in Southeast Asia, has a *political problem.*[1] The bulk of Southeast Asia's population resents these vestiges of the colonial period, and many of the voices of Asian nationalism strongly disapprove of the continuation of what they view as a form of Western European economic colonialism and exploitation.

Immigrant Chinese also proved adept at growing cash crops. With their special closely knit social organization, the Chinese were able to meet labor and marketing requirements within their ethnic group and thus possessed special advantages over all other groups. *Native smallholders* were unable to develop the marketing and transport linkages so vital to Chinese success. Thus cash cropping of several varieties existed; some of it was estate based, and some was focused on smallholders.

Cash cropping accounts for about half the estimated cultivated area in Southeast Asia (20 million acres, or 8 million hectares). The contributions of these crops are much greater than is obvious from the area or the number of people employed. Not only does the estate-crop sector earn capital through the sale of crops, but the operations fundamental to large estate cropping are also powerful elements in economic modernization. Modern estate cropping can promote economic growth and change through its role as a model for modern techniques of cultivation, processing, transportation, and marketing.

Alternative points of view regarding estate cropping are not so optimistic.[2] Some authorities assert that estates, especially when controlled by foreign investors, impede rather than promote economic growth. Their argument is that a *dual economy* is created, and that it can lead to friction within the producing area and a maldistribution of wealth. Moreover, estate production is characteristically limited to one or two crops that are sold to only a small number of nations.

MINERAL RESOURCES

Southeast Asia possesses some *important minerals.* Among these are metallic minerals and fossil fuels. Several minerals exist in large concentrations. *Tin* is probably the best known, and the major tin-producing belt, located in granitic ore and placer deposits of the Malay Peninsula and adjacent islands, is among the most productive and accessible in the world (Figure 27–3). Other important minerals found in large quantities include *iron ore* (the Philippines and the Malay Peninsula), *manganese* (the Philippines and Indonesia), and *tungsten* (Burma and Thailand). Significant minerals in lesser quantities include gold, bauxite, copper, zinc, and chromium. Tin leads all minerals in the value of its production, and this mineral has played an important role in the development and economic growth of the Malay Peninsula and the Indonesian islands of Bangka and Belitung.

[1]Donald Fryer, *Emerging Southeast Asia, A Study in Growth and Stagnation,* 2nd ed. (New York: Wiley, 1979), pp. 66–78.

[2]Paul Harrison, *Inside the Third World,* 2nd ed. (London: Penguin, 1981).

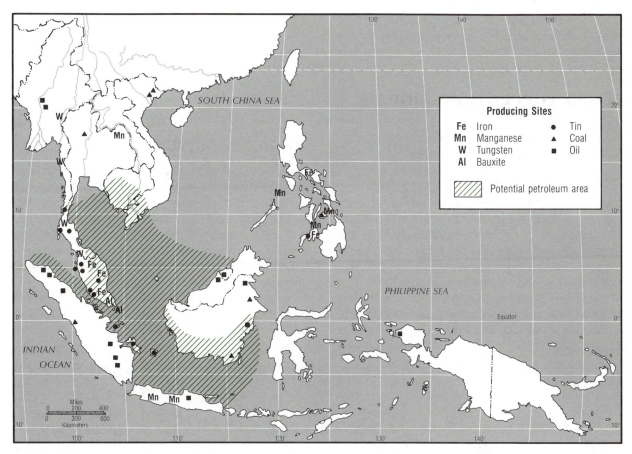

FIGURE 27–3 Mineral Production of Southeast Asia.
Although several important minerals are found in Southeast Asia, tin and oil are easily the most important. The major producer of tin is Malaysia, followed by Indonesia and Thailand. Indonesia is the major producer of oil and ranks as one of the world's major exporters.

Of greater importance than the metallic wealth are the extensive fossil fuel resources, especially oil. *Coal* is found in exploitable quantities in several states (the Philippines, Indonesia, Thailand, and Vietnam), but only the deposits of Tonkin are ample enough for the requirements of industrialization. *Petroleum,* by contrast, exists in great quantities in the Indonesian archipelago and along the northern coast of Borneo. The existence of this resource is particularly fortunate, for Indonesia is among the poorest states of the region and greatly needs the foreign exchange from the sale of crude oil and natural gas. Indonesia exports more than $10 billion worth of petroleum products yearly, much of it going to Japan.

The Sultanate of Brunei, on the coast of north Borneo, is small and rich in oil. Its status is analogous to the small oil sheikhdoms of the Persian Gulf, with a small population and a large resource of oil. It stayed out of the federation of Malaysia in order not to have its wealth diluted by joining a larger state.

Since 1970 almost every state in Southeast Asia has been involved in *the search for oil.* On the mainland several large sedimentary basins appear attractive as potential sources, but the major hope seems to lie offshore. The Sunda Shelf underlies the Gulf of Thailand, the Java Sea, and much of the South China Sea with an average depth of about 150 feet (46 meters). Offshore drilling is relatively simple in such shallow water, and the future may hold considerable promise for this particular resource. Although there is the question of political control and ownership of the offshore deposits, these deposits may be a means of generating capital and assisting economic development.

411

If offshore oil exploration shows large reserves, faster rates of economic growth throughout the region may be anticipated.

RAPID POPULATION GROWTH

When compared with certain parts of Monsoon Asia, Southeast Asia is *not densely populated.* The exceptions—parts of the islands of Java and Luzon; the lower delta regions of the Irrawaddy, the Chao Phraya, the Mekong, and the Red rivers; and the city-state of Singapore—achieve densities common along the lower Yangtze and Ganges rivers (Figure 27–4). All the areas noted are the core regions of their respective countries. The total population of the region is estimated at slightly more than 375 million. With annual growth rates of 2.2%, many Southeast Asian countries will double their populations in about thirty-two years and will confront the *demographic dilemma* that faces mod-

ern India and China. Perhaps the major implication of the current situation is that Southeast Asia has a great opportunity for economic growth and development, yet this opportunity must be viewed in the context of rapid and potentially dangerous population growth.

REGIONAL DISUNITY

Although Southeast Asia has extensive mineral resources, a potentially productive agricultural environment, and a modest population density, problems do exist. These problems—economic, political, and social—are serious enough to raise fundamental questions about the ability of the region to achieve many of the goals of economic growth and national development.

In some ways the contemporary map of Southeast Asia resembles the map of Europe. The region is *fragmented* into a number of independent

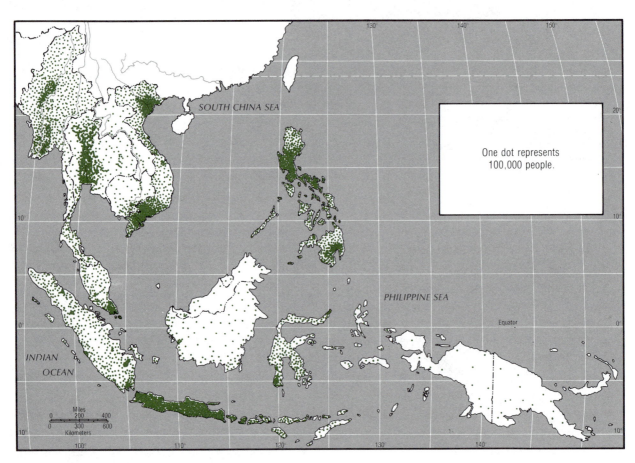

FIGURE 27–4 Population Distribution in Southeast Asia.
Southeast Asia has a large and rapidly growing population that is dispersed unevenly throughout the region.

states, most of which are relatively small in area if not in population. *Political instability* and *conflict* have been commonplace and represent a serious challenge to the viability of sovereign states throughout the region.

Insurgency in Southeast Asia

Among the most serious problems that face Southeast Asia is the question of the *political legitimacy* of various groups and governments. World War II and the end of European colonialism were followed by the rise to power of popular *native governments* that in turn ushered in a new chapter in the history of self-government in the region. These local governments represent several different political perspectives.

Of those political movements available in Southeast Asia, the *Communist Party* was among the best organized. The stress on sound and effective organization and discipline among the party members was obviously a strength. In opposition, most former colonizing powers, who did not want to see Communist governments in control of former colonies, worked vigorously to install political regimes acceptable to the West as they prepared to grant independence to their colonies. The challenge of well-organized insurgency movements inspired and supported by Marxists then became a serious threat to the political survival of several states and the stability of the entire region.

Virtually every country in Southeast Asia has been affected by *insurgent political-military efforts.* Malaysia had a serious insurgency problem in the 1950s, but sharp reactions coupled with improved social conditions and popular support of the government appear to have curbed this movement. In Indonesia, the Communist Party was legalized and in the early 1960s was the third largest in the world. A bloody coup d'etat resulted in the party being outlawed, a policy that continues today.

In the late 1940s the Philippines had a violent Communist insurgency, the Huk movement, that was crushed but has been sporadically rekindled. In addition, a Muslim-inspired separatist movement has plagued the southern island of Mindanao. Civil rights were suspended by President Ferdinand Marcos for a number of years, and the Philippines today is potentially explosive politically. Chronic unemployment, a poor distribution of income, rapid population growth, and stagnant economic growth suggest conditions that increasingly are threatening long-term political stability and prospects for economic growth.

Indochina has experienced more military trauma over a Communist or non-Communist government than any other part of the region. After full-scale wars that involved both France and the United States with the Vietnamese insurgents, the enlarged Vietnamese state has sought to establish its hegemony over both Laos and Kampuchea and thus control the entire Indochinese peninsula. Border clashes between Vietnamese and Chinese forces and between Vietnamese in Kampuchea and neighboring Thais reflect the desire of China and Thailand to curb Vietnamese strength and dominance in the peninsula.

Until *political stability* is attained, there is likely to be little, if any, real opportunity for economic growth and progress. Only in those states that have achieved at least a credible level of peace and stability and an outward-looking government (Malaysia, Singapore, and Thailand) has economic progress been significant. It may be that Indonesia, after a decade of domestic strife and revolutions, should also be considered one of Southeast Asia's progressive states, but it is still too early to gauge accurately how significant the economic growth in Indonesia has been and how successful the development program is.

Efforts to Forge Regional Unity

Counterpoised to the problems and difficulties raised by insurgent movements, revolution, and military actions have been the efforts to forge *regional unity* in Southeast Asia. The formation of the *Southeast Asia Treaty Organization (SEATO)* in 1954 was primarily a political-military arrangement aimed at containing Asian communism. The Vietnam War rendered SEATO ineffective and resulted in its dissolution.

The *Association of Southeast Asian Nations,* or *ASEAN,* is a political-economic organization that was formed in 1967. Its aim is to accelerate the pace of regional development and growth through cooperative trade and shared projects. ASEAN is specifically interested in promoting economic growth through regional integration and cooperation. Thailand, the Philippines, and Malaysia previously belonged to the Association of Southeast Asia, and Indonesia and Singapore joined them in forming ASEAN. Brunei joined in 1984.

Another important agency supporting the goal of economic growth in the region is the *Asian Devel-*

opment Bank, which is headquartered in Manila. Similar in concept and operation to the World Bank and other regional financial institutions that promote development, this bank is financed through deposits of wealthy states and lends money for specific projects among the participating states primarily in Monsoon Asia. Projects range from irrigation and agricultural growth schemes to major transportation and industrial investments.

Cultural Pluralism

Contemporary Southeast Asia exhibits diverse political and ethnic qualities. To a large extent, society and polity are closely related, and many of the region's political problems stem from the strains of multiethnic societies. Cultural pluralism is evident in economic organizations, for there exists a close association between specific occupations and particular ethnic groups. Certain ethnic groups are heavily concentrated in urban locations, whereas others are focused in rural areas or around mining sites. Pluralism and separatism rather than integration have characterized the economies, societies, and politics of most states in the region.

In most of Southeast Asia there are large Chinese and Indian minorities (Figure 27–5). The sizes of these minorities vary considerably from country to country. In some cases, as in Malaysia, the Chinese comprise about 35% of the population and the Indians another 10%. In the Philippines and Indonesia, however, the Chinese represent only 2–3% of the total population, and the number of Indians is too small to be of any consequence. Some Chinese immigrants have been integrated and absorbed into the native population. They have been urged or forced to assume local identities, to use the local language, and to take local names. In Thailand integration has been reasonably successful. Chinese have been present in large numbers

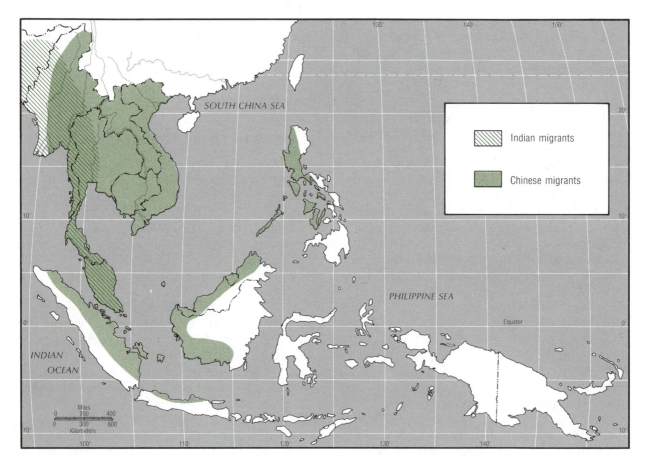

FIGURE 27–5 Nonnative Ethnic Groups of Southeast Asia.
Two main nonnative ethnic groups are found in Southeast Asia—Indians and Chinese.
The areas of principal concentrations are shown.

414

for a long time. Much intermarriage has taken place, with resulting cultural assimilation. Yet a range of problems inherent in a multiethnic society exists, and no approach has proved entirely satisfactory in coping with the difficulties involved in dealing with several different groups. *Racial and ethnic divisions persist* and occasionally result in violent political outbursts and activities.

The Chinese in Southeast Asia

The Chinese have been involved in Southeast Asia for many centuries. The history of Vietnam, for example, is in part a history of Chinese attempts to extend their influence southward. Yet it was in the nineteenth century, in part associated with British colonialism, that extensive *Chinese migration* to the region took place. The British were interested in developing a tractable and industrious labor force to exploit tin deposits in the Malay Peninsula and to assist in agricultural development programs in northern Borneo. Inasmuch as the British held the native Malays in low regard for purposes of organized labor, they sought workers from the outside and encouraged the immigration of Chinese from South China.

Over the years, large numbers of Chinese emigrated to Malaya and other parts of Southeast Asia. Ties between China and the migrants in Southeast Asia were fundamentally social and cultural but were rapidly extended to trade, production, and shipping. Chinese in Malaya quickly acquired land and began to produce *cash crops* such as coconuts, sugar, spices, and rubber. They also established *small tin mines* with modest capital requirements. Their social system provided them

with the linkages to assure a constant supply of cheap labor and ready access to markets, both local and international.

Over time, the Chinese carved out a large niche in the economies not just of Malaya but in all states in Southeast Asia. While not often involved in primary production, they concentrated in cities and *specialized in trade, shipping, and finance.* In some cases they gained control of the rice-milling industry, a key aspect of most of the agricultural economies of Southeast Asia. In other cases they dominated virtually all forms of *local commerce,* partly because of their ability to play the role of middleman in linking the colonial administrations of the European powers with the native populace. The European colonials were unwilling to perform these functions, and the native population had no acquaintance with the particular crafts of commercial entrepreneurship. The Chinese filled a vacuum. In the process they established themselves in a position of privilege in the various Southeast Asian societies.

The Chinese gradually enlarged their traditional economic system until their position was too powerful to be controlled by external forces. Moreover, the more Chinese there were in a concentration, the more vigorously they *resisted acculturation and integration,* and the customs and traditions of old China have been better preserved among the Southeast Asian communities than in China itself.[3] It is this *cultural distinctiveness,* coupled with a *privileged economic position,* that has

[3]Victor Purcell, *The Chinese in Southeast Asia,* 2nd ed. (London: Oxford University Press, 1966), pp. 30–39, presents a good summary of Chinese society and customs in Southeast Asia.

Throughout Southeast Asia the Chinese form an important ethnic group. Many Chinese are concentrated in the cities, where they are shop owners and traders. Shown here is a street in Bangkok, Thailand. (United Nations)

created enormous resentment against the Chinese in Southeast Asia. Their clannishness has made them an easy and convenient target for the frustration and anger of impoverished masses of Southeast Asians and their political spokesmen.

THE PRIMATE CITY: SOUTHEAST ASIAN DILEMMA

Nowhere in Southeast Asia are the problems of a fragmented society, a dual economy, and an uncertain political system so conspicuous as in the great *primate cities* of the region. In these cities are found the largest concentrations of wealth juxtaposed with the most dense clusters of poverty. Here the main commercial, political, administrative, educational, and medical institutions are located "cheek by jowl" with the greatest squatter concentrations and the most vocal and often radical politicians, the most modern stores and factories, and the largest number of itinerant peddlers and vendors. Indeed the term "primate" was coined to describe the single city in a country that had a disproportionately large concentration of the administrative, economic, educational, and service activities and wealth of the country along with a large percentage of the total urbanized population.

Such a phenomenon is not peculiar to Southeast Asia. In Southeast Asia, however, it occurs in several countries. The great capitals—Bangkok, Rangoon, Jakarta, and Manila—are examples (Figure 27–6). These are the largest cities of their countries, as well as the greatest ports and commercial trade centers. Each contains the major international airport and accounts for a major share of its nation's industrial output. Here, too, are found the best universities, hospitals, research centers, newspapers, and other cultural institutions. In some cases, as in Thailand or Burma, there are few cities with more than 100,000 inhabitants.

Primate Cities and Modernization

Primate cities in Southeast Asia *dominate the political and economic life* of their respective countries. Yet it is not at all clear if the consequences of this dominance are positive or negative. The division between city and countryside has raised a serious question about the productive benefits of these cities. It has been suggested that this *urban-rural division* is also a division between a modernizing economy and society and a traditional economy and folk society. Some authorities argue that the condition of primate cities dominating the national scene and economy is temporary and self-adjusting and requires no special corrective strategy. Others argue that primate cities live off the rest of the country and are bad and that strategies designed to interrupt their growth and deconcentrate wealth and benefits are needed.

If an *economic disparity* between the primate city and the rest of the nation exists, the stage is set for a more serious consequence, the emergence of two distinct and antagonistic political camps: one urban-oriented, modern, and sometimes foreign-associated and the other rural-based, traditional, and native.

STAGNATION, TRANSFORMATION, AND GROWTH

A *variety of approaches to development* have been attempted in Southeast Asia. These range from Communist to socialist to free enterprise. No one approach has succeeded, and it is possible to evaluate the achievements of different approaches in specific countries. Such an evaluation may permit some conclusions on the relative merits of these development strategies in the context of the Southeast Asian region.

Burma

Despite the fact that Burma is a large state and has a resource base more than sufficient to support its present population, its growth record has been poor. Its per capita output of goods and services actually declined between 1939 and 1969, a situation that speaks poorly of Burma's particular approach to development. The approach, however, is unique, a sort of *authoritarian isolationism* and locally derived socialism.

In 1973, after years of stagnation and perhaps even declining economic well-being, the military government softened its policies slightly. In addition to promulgating a new constitution, the government relaxed its efforts at isolation and began to encourage more foreign contacts. Since then things have begun to improve in Burma, and the record of economic growth has been especially impressive. The gross national product during the 1970s and early 1980s has increased substantially. The agricultural sector has shown impressive

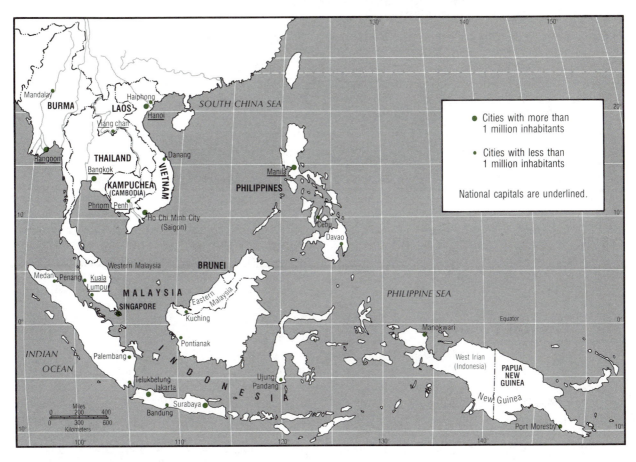

FIGURE 27–6 Major Southeast Asian Cities.
Many of Southeast Asia's large cities are primate cities, which grew rapidly during the former colonial era. Good examples are Rangoon, Manila, and Bangkok.

growth. This high-level performance in agriculture reflects the increased use of *Green Revolution* seeds, fertilizers, and insecticides.

Moreover, since Burma is turning more to an external and *open development policy,* English is once again being taught in the schools, an indication of a more worldly outlook.

Indonesia

After gaining independence from Dutch colonialism in 1949, Indonesia emerged as an important voice among the nonaligned states of the world. Led by the charismatic but troubled Sukarno, the island republic in the 1950s and early 1960s attempted to portray itself as a leader among those states that were following a new form of *independent socialism.* In Indonesia, the political rhetoric and the ambitions of the leaders far outdistanced economic performance and the ability of the state to support

its claim that Indonesia was the best model for other less developed areas. This situation led to severe social and political stress that finally resulted in a 1965 coup d'etat and the overthrow of Sukarno.

Since 1966 the military regime has set its major sights on repairing the economic neglect and damage of the Sukarno regime. One aspect of this policy was to seek assistance from outside sources such as Japan, the United States and the World Bank. The performance of the *military government* has not been dramatic, nor has the economic growth rate been spectacular, but there is a new, sober emphasis on solid achievements in economic planning and growth. Completion of *Repelita I* (the first five-year development plan) in 1974 was one such achievement in which annual economic growth averaged about 7%. The goals of *Repelita II* (the second five-year plan) were more ambitious and focused on faster rates of growth in the agri-

417

cultural sector. *Repelita III* (1979–84), the third five-year plan, looked to growth in earnings from petroleum sales to finance development.

Indonesia is *still a poor country.* It does show, however, that economic conditions can change and improve even where the situation is unstable and the prospects gloomy. Indonesia justifiably may point with pride to its recent record of growth in gross national product. One uncertain aspect of Indonesia's achievement is the limited extent of distribution of the fruits of the new growth among the masses of peasants. Unless some improvements are made available to this large group, the outlook for future political and social stability is poor. Another serious problem is the large and rapidly growing population whose numbers must be fed and employed adequately to prevent social and political unrest.

The Philippines: Progress from the Green Revolution

Although political events and developments in the Philippines during recent years have involved Muslim insurgency and cause for concern about the true nature of Philippine democracy, some sectors of the Philippine economy show considerable vigor and promise. One of the most successful examples is found in the *agricultural sector* as seen in the experience of certain rural areas in growing newly developed strains of high-yielding rice. In addition, other *innovations* include increased use of insecticides, chemical fertilizers, and mechanical threshing machines; better systems and methods of irrigation; and more efficient use of labor gangs to provide adequate labor during periods of peak demand.

Such new and improved cropping practices, especially when coupled with the *Green Revolution* seed strains developed at the International Rice Research Institute in Los Baños, have resulted in dramatic increases in yield, farm output, and farm income in some rural areas.[4] For example, in a four-year period coinciding with the introduction of new rice strains (1965–69), yields nearly doubled; such gains are becoming more common throughout the Philippines as the new rice strains and supporting technologies are being diffused. Yet without a concomitant reduction in the rate of population growth, the Philippines may discover in another generation or two that the technical break-

through in agricultural production does not lead to per capita gains in income with long-term improvement in levels of living. If the *time cushion* provided by the new agricultural breakthrough is not used wisely, the Philippines may fall into the same *high-level equilibrium trap* as did China centuries earlier.

Malaysia and Singapore

Probably the *two richest countries* in Southeast Asia are the progressive neighbors Singapore and Malaysia. Upon independence in 1963, the two were politically integrated, but the *Chinese-Malay dichotomy* was severe enough to disrupt the political alliance. Despite political separation, both the city-state of Singapore and the Federation of Malaysia continue to thrive economically and stand as solid examples of prosperity in Southeast Asia. Despite the partially *authoritarian* nature of politics in the two countries, both have vigorous and *growing private sectors* in their economies. These dynamic *free-market forces,* coupled with an outward-looking attitude, may account for their steady economic growth. Certain important aspects of Singapore's economy, however, have more state control than exists in Malaysia. For example, roughly half of the population of the city-state is now housed in public projects; the private sector has not been as successful in housing as in other forms of enterprise.

Singapore, with its per capita GNP of roughly $5200, stands next to Japan as the most prosperous place in Monsoon Asia. Despite the physical constraints and problems of its small area, the city-state has made good use of its location and *traditional economic functions* of trade, shipping, and service. Building on this base, it has recently turned to sophisticated *manufacturing* in which the skills and industry of its talented and disciplined labor force are employed. Cameras, watches, and electronic components are some of the products. Oil refining, which depends on Indonesian and Persian Gulf crude, is also a major industry for Singapore. Singapore has succeeded in many ways. The island republic is a "have" among "have-nots," an *island of prosperity* in a larger sea of poverty. Singapore demonstrates, as does Japan, that even under adverse conditions Asians can achieve living levels comparable to those of Europeans, and they can do it in a relatively short time.

Malaysia's growth rate has been more modest than Singapore's; nevertheless, the federation has progressed steadily in recent years. Annual eco-

---

[4]Robert Huke, "San Bartolome and the Green Revolution," *Economic Geography,* 50 (1974), 47–58.

nomic growth rates of 4–8% have been maintained consistently. Malaysia has been especially fortunate in the exploitation of its *natural and agricultural resources.* Tin, rubber, timber, coconuts, and more recently palm oil are the main products. It is difficult, however, to predict whether high demand levels for these commodities will continue.

Perhaps the most impressive thing about Malaysia's progress is that the division in Malaysian society between Malay natives and Chinese has not stifled or interrupted the growth. Despite their *ethnic and political differences,* the two groups have managed to coexist and cooperate to promote their common well-being. If this lesson can be learned and applied throughout the region, the prospects for rapid economic growth seem likely.

Vietnam

The *1973 Paris Peace Agreement* that ended direct United States participation in the fighting in Vietnam brought a short-term peace to what, until the 1954 Geneva accords, was Indochina. A year later fighting erupted again, and many observers were astonished at the speed of the collapse of South Vietnam. Since 1954 the bitterness among the Vietnamese people with respect to which group would guide the political destiny of Vietnam has sharpened, and the *Vietnam War* of the 1960s and 1970s was a painful expression of this bitterness.

The dispute between the two Vietnams in part stemmed from the desire of the northern half to incorporate the southern half and forge *one unified Vietnam.* The two halves complement each other economically, for the north is more industrial and is a food-deficient region. The south traditionally is a rice surplus area and of value to balance the shortages. There was, however, an argument for separating the two halves, for the Annamite Mountains, Indochina's most prominent physical feature, touch the sea midway down the coast very near the former demilitarized zone. Historically, the north and south have rarely been politically integrated.

The South

The traditional rice surpluses of *southern Vietnam* disappeared during the war of the 1960s. The problem of peasant insurgency and rural insecurity became so great that much productive land was simply abandoned or little cultivated. The war had many other consequences for the *predominantly agrarian economy* of southern Vietnam. Some of these consequences accelerated the pace of economic growth: for example, United States aid, among other things, provided training programs for Vietnamese civilians, made huge investments in capital improvements (roads, harbors, cities, power grids, and irrigation systems), and drew large numbers of people into urban centers to contribute to the industrial and commercial sectors of the economy. The abrupt withdrawal of United States soldiers and civilians in 1971–72 and the cutbacks in United States military and economic aid created problems of unemployment, inflation, and instability in the economy. These problems continued and, coupled with insecurity in rural areas, created an atmosphere of uncertainty about South Vietnam's future. This uncertainty may have played a part in the quick demise of the government in early 1975 and its replacement by a Marxist government.

The potential for economic growth is great. Most promising is the enormous *potential for increased agricultural production.* Restoring abandoned land to cultivation and opening new land in the Mekong Delta marshlands could promote greater output. Increased yields on lands already cropped through improved irrigation, new seed strains, and more chemical fertilizer and insecticides are equally promising.

The North

Northern Vietnam's heartland is focused on the Red River drainage basin and the Tonkin Delta. In many ways this region, with its dense population and intensive *double-cropping* rice complex, resembles Chinese occupancy patterns found farther north. The northern Vietnamese, however, have not been able to replicate the agricultural achievements and yields of the Chinese, and northern Vietnam has long suffered from *food shortages.* These shortages have, to some extent, been offset by the presence of *metallic minerals and coal.* Based on these resources, the former French colonial regime initiated a number of *industrial development projects.* The Communist government has expanded these industries. During the 1960s both the Soviet Union and China contributed heavily to the north and countered United States aid in the south. The war and especially the heavy bombing, however, left much of the industrial complex in ruins.

Since the war Vietnam has embarked on a crash program to rebuild the industrial establishment in the north. Agricultural deficits have continued, and

419

in 1981 a 4-million-ton deficit in rice was reported. The Vietnamese continue to depend heavily on the Soviets for direct assistance.

In early 1979 *a new conflict* erupted in Indochina. China staged a limited, three-week invasion of northern border areas of Vietnam. China justified this action on the basis of Vietnam's mistreatment of Chinese living in Vietnam, many of whom were driven out of that country. Another factor was Vietnam's earlier invasion of Kampuchea, an ally of China. Negotiations followed, but much bitterness remains between the two countries.

Although it is difficult to predict the political direction in Indochina, Vietnam appears set on establishing its dominance throughout the peninsula. In attempting to establish control Vietnam likely will encounter serious antagonism from China.

Implications for Future Growth

There is *no single path* to rapid economic growth and social change in Southeast Asia. Despite the absence of a unitary approach to modernization, some lessons may be learned from past experience. Those states that have become preoccupied with their own problems or ambitions and have little or no cooperation with a community of states have performed poorly. Only after reshaping their policies have their performances improved. Burma and Indonesia are good examples. The potentials for growth and progress in Southeast Asia are enormous. Moderate policies that emphasize the goals of growth and take advantage of international assistance appear to be the most successful. The shift in Indonesian policies and political and economic behavior after Sukarno's ouster suggest the desirability of a moderate policy. Moreover, the turnabout in Indonesia is an excellent indication of the effect that political policies have on the economy and how quickly things can change for the better—a hopeful outlook for many other less developed, impoverished areas of the world.

FURTHER READINGS

BENNETT, GORDON, *Huadong, the Story of a Chinese People's Commune* (Boulder, Colo.: Westview Press, 1978). A well-illustrated and clear account of life in a commune in southeastern China near Hong Kong. This report, developed as source material on China for the China Council, is appropriate for both teachers and introductory students.

CHANG, JEN-HU, ''The Agricultural Potential of the Humid Tropics,'' *Geographical Review,* 58 (1968), 335–61. Chang's thesis is that under present technologies the humid tropics may be producing near their capacity, an argument based on the climate characteristics of the region and associated photosynthetic properties of food grain.

CENTRAL INTELLIGENCE AGENCY, *People's Republic of China Atlas* (Washington, D.C.: Government Printing Office, 1971). Excellent regional textual and cartographic coverage of China: population distribution, physical geography, transportation, agriculture, ethnic patterns, minerals, industries, and administrative structure.

CONGRESS OF THE UNITED STATES, JOINT ECONOMIC COMMITTEE, *China After the Four Modernizations* (Washington, D.C.: Government Printing Office, 1982). An excellent collection of recent papers that covers economic policy, manufacturing and extractive industries, population and labor utilization, agriculture, and foreign economic relations.

DUTT, ASHOK (ed.), *India: Resources, Potentialities, and Planning* (Dubuque, Iowa: Kendall-Hunt, 1972). A good, straightforward survey of India and its resources for development from the perspective of a geographer.

DUTT, ASHOK, et al., *India in Maps* (Dubuque, Iowa: Kendall-Hunt, 1976). A companion to Dutt's 1972 work that provides text and maps of India's physical, cultural, economic, and political geography. Base maps for student use are included.

FAR EASTERN ECONOMIC REVIEW, *Asia Yearbook* (Hong Kong: Far Eastern Economic Review, annual). Up-to-date coverage of political, economic, and social patterns in all countries of the region.

FISHER, CHARLES A., *South-East Asia* (London: Metheun, 1966). Still a superior geography of Southeast Asia.

FRYER, DONALD W., *Emerging South-East Asia: A Study in Growth and Stagnation,* 2nd ed. (New York: Wiley, 1979). A work oriented toward economic topics that focuses on the contrast between the stagnant and progressive economic systems of the region.

GINSBURG, NORTON (ed.), *The Pattern of Asia* (Englewood Cliffs, N.J.: Prentice-Hall, 1958). Although out of date, still one of the best regional texts on the geography of Asia.

HARRISON, PAUL, *Inside the Third World,* 2nd ed. (London: Penguin, 1981). A readable account and interpretation of the roots of poverty, some of which focuses on the humid tropics.

MYRDAL, GUNNAR, *Asian Drama: An Inquiry Into the Poverty of Nations* (New York: Twentieth Century Fund, 1968). A famous Swedish economist's examination of planning problems in economic development in East and Southeast Asia.

NOBLE, ALLEN, and ASHOK DUTT (eds.), *India: Cultural Patterns and Processes* (Boulder, Colo.: Westview Press, 1982). A series of essays that focus on the diversity and unity of India's culture with emphasis on spatial distribution of cultural form.

PANNELL, CLIFTON W. (ed.), *East Asia, Geographical and Historical Approaches to Foreign Area Studies* (Dubuque, Iowa: Kendall-Hunt, 1983). In this recent volume geographers and historians analyze cultural differences and development in East Asia.

PANNELL, CLIFTON W., and LAURENCE J. C. MA, *China, the Geography of Development and Modernization* (New York: Halsted Press, 1983). An up-to-date geography of China's efforts to develop in the context of large area, population, and resource base.

SCHWARTZBERG, JOSEPH E., *A Historical Atlas of South Asia* (Chicago: University of Chicago Press, 1978). A definitive historical atlas of India, a major contribution to geographic knowledge on South Asia.

SPATE, OSCAR H. K., and A. T. A. LEARMONTH, *India and Pakistan, A General and Regional Geography,* 3rd ed. (London: Metheun, 1967). The closest thing to a standard geography of the subcontinent.

SPENCER, JOSEPH, and WILLIAM THOMAS, *Asia, East by South, A Cultural Geography,* 2nd ed. (New York: Wiley, 1971). A major work on Monsoon Asia with a cultural approach.

GLOSSARY

agricultural calendar The chronological sequence of farming operations throughout the year including the period of land preparation, sowing, cultivation, and harvesting different crops.

Agricultural Revolution A period beginning 7000–10,000 years ago characterized by domestication of plants and animals and development of farming.

Ainu The proto-Caucasian population of ancient northern Japan. The Ainu are still found in small numbers on the island of Hokkaido.

alluvium Material that has been transported and deposited by water, often very fertile.

apartheid The policy of the South African government of strict white-nonwhite segregation.

Appalachia That part of the Appalachian Highlands of the United States designated as a poverty area.

arable Land that can be cultivated profitably.

Association of Southeast Asian Nations (ASEAN) A political-economic organization formed in 1967 to promote cooperation among member nations.

autarky Self-sufficiency, or the policy of economic independence.

Autonomous Region The highest-level political unit in China by which ethnic minorities are theoretically allowed some local self-rule and preservation of customs.

Aztec One of the four high civilizations of pre-Columbian Latin America centered on the area around present-day Mexico City.

bauxite The ore from which aluminum is obtained.

Bedouin An Arab who lives by nomadic herding in the deserts of North Africa and the Middle East.

Benelux The nations of Belgium, the Netherlands, and Luxembourg considered as a group.

Boers Literally Dutch farmers, specifically the Dutch living in the Republic of South Africa.

boreal forest The large coniferous forest extending across Canada and part of northern United States.

borderlands Specifically the region in southwestern United States in which Hispanic influence is particularly strong. More generally, "borderlands" refers to any sparsely settled and poorly controlled area along international boundaries.

Boserupian thesis The theory developed by Ester Boserup that rapid population growth can speed economic development.

Boxer Rebellion   1899–1900 political uprising promoted by a Chinese secret society; directed, in part, against the penetration of China by foreign interests and foreign missionaries.

Brahmin   The highest group in the hierarchy of the Hindu caste system comprising basically Aryans who are teachers and religious leaders.

buffer state   A small relatively weak country between two large nations; the buffer state serves to reduce conflict between the larger nations.

bustee   Spontaneous settlement or shantytown composed of huts constructed of nonpermanent material on land owned by someone else. Calcutta has had a particularly serious problem with such shantytowns.

by-product   A secondary or incidental material obtained in the mining or manufacturing process; for example, the primary mineral being mined may be lead, but silver is also recovered out of the lead ore, or in refining petroleum for gasoline, heavy hydrocarbons are also obtained that can be used in the chemical industry.

Campo cerrado   A vegetative type found in the Central West region of Brazil and covering 75% of the region, used principally for cattle grazing (mixed grassland and woodland).

Campo limpo   A grassland vegetative type found in the Central West region of Brazil; the agricultural potential of campo limpo land is considered very low.

Campo sujo   A grassland with scattered woods found in the Central West region of Brazil; the agricultural potential of campo sujo land is considered low.

capital   Any form of wealth that can be used to produce more wealth; resources.

capital goods   Items produced that can be used to create wealth or other goods. For example, a home washing machine is a consumer good, but the machines used to manufacture the washing machine are capital goods.

cardinal temperatures   The temperature range for a specific plant within which it can grow; the upper and lower temperature limits for plant growth.

Caribbean Common Market (Caricom)   A grouping of British and former British colonies in and around the Caribbean Sea formed to reduce trade barriers between member states.

carrying capacity   The maximum number of animals or people an area can support; population-

carrying capacity depends on many variables such as nutrition level, level of living, and trade.

caste   A rigid system of social stratification based upon occupation with a person's position passed on by inheritance; derived from the Hindu culture.

Central American Common Market (CACM)   A group of five Central American nations (Guatemala, El Salvador, Honduras, Nicaragua, and Costa Rica) that strive for economic integration.

centrally planned economies   Nations with a productive system controlled by the national government and based on communistic and socialistic principles.

Cerradão   A scrub woodland vegetative type found in the Central West of Brazil; agricultural potential of cerradão land is considered low.

Chibcha   One of the four high civilizations of pre-Columbian Latin America centered on the area around present-day Bogotá, Colombia.

Ch'ing (Qing) Dynasty   The last imperial dynasty of China, A.D. 1644–1911; founded by the Manchus.

circular causation   A development theory based on the idea of an upward or downward spiraling effect.

city-state   A sovereign country consisting of a dominant urban unit and surrounding tributary areas.

climax vegetation   The concept that if an area is without interference by mankind for a long enough period, an assemblage of plants will eventually dominate that reflects climatic, soil, and moisture conditions.

coke   Coal that has undergone partial combustion and that is almost pure carbon; used for smelting iron ore.

Cold War   The 1945 to mid-1960s period of hostility, just short of open warfare, between the Soviet Union and the United States and its allies.

collectivization   The process of forming collective or communal farms, especially in Communist countries; nationalization of private landholdings.

command economy   A centrally controlled and planned livelihood system as, for example, the socialist and Communist systems.

commercialism   The system of interchange of goods; the activities that lead to sale and exchange of products.

common market   A customs union among a group of nations with no or reduced tariff walls

424

among members and a uniform tariff system with the outside world.

(The) Common Market The European Economic Community (EEC), the European Common Market.

commune The basic socioeconomic unit of present-day China; the Chinese rural collective community.

comparative advantage The idea that if there are several alternative uses of resources in a region, the most advantageous use tends to be selected.

consumer goods Commodities produced for use by an individual or family such as a car, radio, and clothing; designed for consumption; do not assist in creating additional wealth.

conurbation A network of urban centers that have grown together.

coreland The area of dense population that forms the urban-industrial heart of a nation; also the cultural, economic, political center of a nation or group of people.

cottage (household) industry A system of manufacture in which raw materials are processed in the worker's home.

Council of Mutual Economic Assistance (CMEA or Comecon) The organization formed in 1949 to regulate economic relations between the Soviet Union and Eastern Europe.

crop rotation The practice of using the same parcel of land for a succession of different cultigens in order to maintain or improve yields.

cultigen A culturally improved plant; a domesticated plant.

cultural determinism The theory that a person's range of action is limited largely by the society within which he or she lives; for example, food preference, desirable occupation, and laws.

cultural landscape The mankind-modified environment including fields, houses, highways, planted forests, and mines, as well as weeds, pollution, and garbage dumps.

cultural norm A standard of conduct sanctioned by a society.

cultural pluralism The presence of two or more groups within the same area who follow different ways of life.

Cultural Revolution The upheaval in China during the 1960s when old cultural patterns were condemned and new Maoist patterns were strongly enforced.

culture The ways of life of a population that are transmitted from one generation to another.

culture complex A group of culture traits that are employed together as, for example, how clothing is made and distributed to consumers.

culture hearth The source area for particular traits and complexes.

culture realm The area within which the population possesses similar traits and complexes as, for example, the Chinese realm or Western society.

culture trait A single element or characteristic of a group's culture as, for example, dress style.

Dark Ages The period, about A.D. 500–1000, during which Europe was in political, economic, and cultural decline.

debt peonage A system formerly used in Latin America to hold the labor force in bondage; debts accumulated by one generation were passed on to the next.

demographic transformation A theory, based on Western European experience, of the relationship between birth and death rates and urbanization and industrialization.

developed Used in this book to identify nations or regions that have a high level of economic production per person; *rich, advanced,* and *modern* are synonymous terms.

Development A process of continuous change for the better. Economic development is the progressive improvement in livelihood of an area, country, or people. Development is usually measured on a per capita basis for comparative purposes.

diffusion The spread of an idea or material object over space.

double-cropping The raising of two crops in succession in the same field during one growing season.

Dravidian One of the earliest inhabitants of India; dark-skinned Caucasoids of peninsular India. Also a family of languages that includes Tamil, Kannada, and Telugu.

dual economy The existence of two separate economic systems within a region; common to the poor world where one system is geared for local needs and another for the export market.

economies of scale The decrease in production costs brought about by high-volume production; mass production techniques are commonly used.

425

ecosystem The assemblage of plants and animals in a particular environment and their interdependence.

ejido A form of land tenure in Mexico by which land is given to a farming community. The community may allocate parcels of land to individuals. Title to the land, however, remains with the community.

entrepôt A center, usually a port, where goods are collected for redistribution; often a transshipment point.

entrepreneur The "organization man"; a person who organizes economic activities.

environment That which surrounds; the setting, both natural and cultural, within which a group lives.

environmental determinism A general geographical theory, now largely discredited, according to which the physical environment controls, directs, or influences what mankind does; variations in the physical environment should therefore be associated with different levels of economic well-being.

environmental maintenance The care of physical surroundings that is necessary for sustained productivity.

erosion The picking up and transporting of earth materials by moving water and other natural agents.

estáncia A Spanish term used in Latin America to describe a large rural landholding usually devoted to stock raising, especially horses and cattle; similar to a ranch.

estate The term characteristically used in parts of Africa and Asia for a large export-oriented agricultural enterprise usually owned by a foreign company and using local labor; a plantation.

ethnic group A group of people who share a common and distinctive culture.

ethnic religion A religion associated with a particular group and area such as Shintoism, Taoism, or Hinduism.

European Coal and Steel Community (ECSC) A common-market-type organization set up after World War II to facilitate movement of coal and iron ore among the Benelux nations, France, and West Germany.

European Economic Community (EEC) The common market structure consisting of nine full members with a goal of economic integration.

European explosion The spread of Western culture traits and complexes worldwide.

European Free Trade Association (EFTA) A group of seven European nations (Austria, Denmark, Norway, Portugal, Sweden, Switzerland, and United Kingdom) who joined together in 1959 to reduce tariffs among member nations. The transfer of Denmark and the United Kingdom to the European Common Market brought an end to the EFTA.

European Parliament An elected body of representatives chosen from the various European Economic Community nations whose function is to exercise democratic control over the executive and administrative units of the Common Market.

evapotranspiration The combined loss of water from direct evaporation and water transpired by plants.

exploitive agriculture Farming with the idea of maximizing short-term income with no consideration of long-term effects.

extensive agriculture A farming system characterized by small inputs of labor or capital per land unit.

factory A manufacturing unit based on quantity production, a distinct division of labor, and use of inanimate power-driven machinery.

fazenda A Portuguese term used in Brazil to describe a large rural landholding; these units may be devoted to crop or animal production or to both.

feed grains Cereals such as maize (corn), oats, and sorghum used as provender for poultry and livestock.

ferroalloy A mineral such as manganese mixed with iron to make steel; ferroalloys give steel a variety of properties such as rust resistance or strength.

field rotation The movement of cropping from one parcel of land to another; a practice common in tropical areas of low population density.

fiord (fjord) Glacially eroded valley that subsequently has been invaded by the sea.

Flemish Belgians who speak a Dutch dialect and occupy northern Belgium; one of the nation's two principal national groups.

flow resources Renewable resources such as trees, grass, rivers, and animals.

food grains Cereals such as wheat, rye, rice, maize (corn) used for human consumption.

form utility The change in shape, constitution, or character of a material to increase its usefulness and hence its value as, for example, smelting iron ore or processing cloth into clothing.

fossil fuels Organic energy sources formed in past geologic time such as coal, petroleum, and natural gas.

Four Modernizations Development policy and program that China's leaders designed in the late 1970s to promote the country's modernization to the end of the twentieth century. Four broad sectors for development are included: agriculture, industry, science, and national defense.

free trade The unrestricted movement of goods among independent nations.

fund resources Nonrenewable resources such as minerals.

General Agreement on Tariffs and Trade (GATT) A multinational agreement of the mid-1960s by which signatory nations agreed to reduce tariffs and, in general, to improve international trade.

geography A branch of knowledge concerned with the study of how and why things are distributed over the earth. Four traditional emphases of geography are: (1) the study of distributions, (2) the study of the relationship between people and their environment, (3) the study of regions, and (4) the study of the physical earth.

Ghana Empire A West African trading state that flourished from the ninth to the thirteenth century but dates from earlier times; the modern nation of Ghana has taken its name from this empire.

Golden horseshoe An industrial district in Canada extending from Toronto to the western end of Lake Ontario to Hamilton.

Gosplan The Soviet Union state agency charged with making national development plans; usually has five-year goals.

great circle The shortest distance between two points on the surface of the earth. Ships and planes usually use a great circle route whenever possible, especially if a great distance is to be covered. Any line that divides the earth into equal halves is a great circle.

Great Leap Forward The 1958–1961 attempt in China to socialize agriculture and increase production both in farming and industry.

Green Revolution The use of new high-yielding hybrid plants, mainly rice, corn, and wheat, to increase food supplies along with development of the infrastructure necessary for greater production and better distribution to the consumer.

gross national product (GNP) The total value of all goods produced and services provided during a single year.

growing season The period during the year that crops can be grown without artificial heat.

growth-pole theory The concentration of development efforts on selected sites, usually urban, with the expectation that the improvements made will spread outward from the site.

Gulf Cooperation Council A regional organization to deal with common problems in a unified manner. The Council consists of Bahrain, Kuwait, Oman, Qater, Saudi Arabia, and the United Arab Emirates, countries with large oil reserves and small populations, situated along the Persian Gulf.

habitat An environment; the place where an organism lives.

hacienda Spanish term used in Latin America for large rural landholding usually devoted to crop and animal production; these units formerly had a high degree of internal self-sufficiency and operated under the patron system.

Han people The name used by the Chinese to refer to themselves to distinguish "cultured" Chinese from others; most of China proper is populated by the Han.

hardwood Angiosperm trees, commonly called broadleaf, and usually deciduous.

heavy industry A term applied to manufactures that use large amounts of raw materials such as coal, iron ore, and sand and that have relatively low value per unit weight.

high-level equilibrium trap The condition of increasing population growth tied to increasing production so that levels of living remain constant; this situation apparently existed in traditional China.

hinterland The tributary area of a port or city; the size and productivity of the hinterland are usually reflected in size and wealth of the port or city.

Hispanic A person of Spanish origin who retains some aspects of his or her cultural heritage living in the United States. The Hispanic element is a governmentally recognized minority. Hispanics are the fastest growing minority in the United States.

humus Partially decayed organic material that is an important constituent of soils; improves water-holding capacity, provides some plant nutrients, and makes the soil easier to cultivate.

hybridization The selective crossing of different varieties or species of plants or animals to produce offspring possessing certain desired characteristics.

Imperata A Southeast Asian type of grass that invades cleared areas and is difficult to control; elephant grass.

import substitution The policy to encourage local production of needed goods rather than importing them; subsidies, loans, and protective tariff regulations are often the means of assuring local production.

Inca One of the four high civilizations of pre-Columbian Latin America centered on the city of Cuzco, Peru, and extending in the Andes of South America from southern Colombia to central Chile.

Indo-Aryan One of the earliest inhabitants in India who came from the northwest and spoke a Sanskrit language; light-skinned Caucasoids of northern India.

Industrial Revolution The period of rapid technological change and innovation that began in England in the mid-eighteenth century and subsequently spread worldwide. The principal attribute of the revolution has been the development of cheap and massive amounts of controlled inanimate energy.

inertia Used in this book in the sense of continuation in an area of an economic activity, especially manufacturing, after the original factors favoring the location have ceased.

infrastructure The services and supporting activities necessary for a commercial economy to function, such as roads, banking, schools, hospitals, and government.

intensive agriculture A farming system characterized by large inputs of labor or capital per land unit.

intercropping The raising of two or more crops in the same field at the same time.

Iron Curtain A popular expression for the boundary between Eastern and Western Europe signifying the difficulty of movement of goods and people into Eastern Europe.

isohyet A line along which all points have the same precipitation value; usually designated in average yearly amount.

isotherm A line along which all points have the same temperature; usually designated in average monthly or yearly values.

jihad A holy war, traditionally declared by the spiritual and secular leader (caliph) of Islam, against unbelievers.

job out Subcontracting by a manufacturer to a specialized firm or person for production of a specific item or component.

jute A hard fiber used in bagging and cordage produced primarily in Bangladesh and adjacent parts of India; obtained from a species of *Corchorus.*

Kemal Atatürk The founder of modern Turkey after World War I and the demise of the Ottoman Empire. Atatürk strove to create a westernized, secular state.

kibbutz An Israeli collective farming community.

kogai A Japanese term meaning environmental disruption such as air and water pollution.

kolkhoz A Soviet collective farm. Members of the *kolkhoz* lease the land from the government.

Koryo Dynasty The dynasty lasting from A.D. 918 to 1392 during which Korean culture matured.

kraft process A method of processing resinous woodpulp that yields a strong fibrous paper used for wrapping paper and cardboard. Development of the process allowed the use of pine and other coniferous trees previously of little utility.

La Plata Group A subgroup of the Latin American Free Trade Association comprising Argentina, Brazil, Paraguay, and Uruguay; oriented toward cooperation in development of the watershed of the Río de la Plata.

latifundia Literally large landholdings; the control of most of the land by a small percentage of landowners.

428

Latin America That part of the New World south of the United States comprising a cultural region largely, but not totally, of former Spanish or Portuguese colonies.

Latin American Free Trade Association (LAFTA) A loosely organized multinational grouping of Latin American nations that have joined together for economic integration.

least-cost point The location where costs of assembling materials, their processing, and shipment to market are lowest.

legumes Plants of the family Leguminosae, especially those used agriculturally, such as beans, peas, alfalfa, and soybeans. Bacteria on the underground nodules of these plants have the ability to remove nitrogen from the air and fix it in a form usable by plants.

level of living The actual material well-being of a person or family as measured by diet, housing, and clothing.

life-style The mode or manner of behavior of an individual or group; the way of life as evidenced in dress, material possessions, and diet.

light industry A term applied to manufactures that use small amounts of raw materials and employ small or light machines.

lingua franca An auxiliary language used among peoples of different speech; commonly used for trading and political purposes.

linkage The connection or interdependence among industries either horizontally (over area) or vertically (at different stages of production); examples of vertical linkages are wheat from the farmer to miller to baker to store to customer.

local relief The difference in elevation between the highest and lowest points of an area.

loess Deposits of wind-transported, fine-grained material that are usually easily tilled and quite fertile.

machine tool Any power-operated instrument used to shape, form, or cut materials as, for example, a drill press or lathe.

Malthusian doctrine A theory advanced by Thomas Malthus that human populations tend to increase more rapidly than the means to care for the population.

manioc (cassava) A woody plant of the genus *Manihot* having tuberous roots that are used extensively in tropical areas as a food source; source of tapioca.

Maoism The philosophy and behavior patterns extolled by the Chinese Communist leader Mao Zedong (Mao Tse-tung).

marginal land An area in which the production costs are almost equal to income and little or no savings are possible.

market economies Nations with a productive system based upon capitalistic principles.

Marshall Plan The system of economic and technological aid given by the United States to Western Europe after World War II.

Mata de primeira classe A vegetative type found in the Central West region of Brazil and indicative of excellent land; first-class forest.

Mata da segunda classe A vegetative type found in the Central West region of Brazil and indicative of good agricultural land; second-class forest.

materialism Devotion to tangible objects as opposed to spiritual needs and thoughts.

Maya One of the four high civilizations of pre-Columbian Latin America situated in southern Mexico and northern Central America.

megalopolis Originally the continuous urban zone between Boston and Washington, D.C.; now used to describe any region where urban areas have coalesced to form a single massive urban zone.

Meiji period The period, dating from 1867, during which the Japanese society and economy were transformed from feudal to modern.

mercantilism The philosophy by which most colonizing nations controlled the economic activities of their colonies; in essence, the colony existed for the benefit of the mother country.

mestizo A person of mixed European and Indian blood in Latin America.

Mexicanization The majority ownership of a corporation operating in Mexico by Mexicans. By means of tax policy and legislation, the Mexican government has attempted to control foreign investment and create greater local participation in these large companies. Many other nations have similar policies.

Mezzogiorno Southern Italy; one of the poorest regions of Western Europe.

Middle Ages The period in European history from about A.D. 500 to 1350; the first part of the Middle Ages is also called the Dark Ages.

millet (dura) A widely cultivated cereal grass,

particularly in parts of Africa and Asia, used as both a feed and food grain.

mineral deposit   A naturally occurring concentration of one or more earth materials.

minifundia   Literally, small landholding; most of the landowners combined control only a small percentage of the total area in farms.

miracle grains   The new hybrid varieties of maize, wheat, and rice that produce a greater quantity of useful calories than traditional varieties.

mixed forest   Woods composed of both deciduous and evergreen trees; a transitional woodland.

Mogul Period (1526–1739)   Period in which Muslim power was consolidated and reached its zenith and controlled most of the Indian subcontinent (1707).

monsoon   The seasonal reversal in surface wind direction over the southeast quarter of Asia.

mulatto   A person of mixed European and African blood in Latin America.

multiplier effect   The idea that for each new worker employed in an industry, other jobs are created to support and service the workers.

Muslim   One who surrenders to the will of God (Allah) as revealed by the Prophet Muhammad; a follower of Islam (submission).

nation-state   A political grouping of people who occupy a definite area and who share a common set of beliefs and values.

nationalism   The emotional attachment of an individual or group to a country or region.

nationalization   The expropriation of an enterprise by the government and the operation of the enterprise by the government.

Naxalite   A radical, Marxist, rural revolutionary in India.

Neolithic period   The later part of the Old World Stone Age; characterized by well-developed stone implements and some food raising.

New World   The American continents of North and South America.

night soil   Human excrement used as a fertilizer, a practice common in parts of Monsoon Asia and Europe.

North Atlantic Treaty Organization (NATO)   A multinational military alliance of Western nations founded in 1949.

obstacle   Anything that inhibits a person from achieving a want or a desire or a goal.

Old World   The continents of Europe, Asia, and Africa.

Operation Bootstraps   The government of Puerto Rico's post–World War II development plan emphasizing manufacturing, agricultural reform, and tourism.

Opium War   The 1839–42 British-initiated war designed to protect British commercial interests in China. The pretext was the refusal of the British to cooperate with the Chinese in halting opium traffic; the Chinese then sought to restrict British commercial activities.

ore   A mineral deposit that is economically profitable to mine; the material mined.

Organization for African Unity   A group of African states concerned with the problems of political and economic relations among African nations and between Africa and other parts of the world.

Organization for Economic Cooperation and Development   Originally established to assist the Marshall Plan fund distribution; in 1961 reorganized and now strives for expanded international trade and poor world development.

Organization of Petroleum Exporting Countries (OPEC)   A thirteen-nation group of oil-producing countries that controls 85% of all the petroleum entering international trade. These nations have joined together to regulate oil production and prices; a valorization scheme.

Ottoman Empire   A Turkish sultanate that controlled a large area from southern Eastern Europe through the Middle East and into North Africa; the empire survived for about six centuries until it collapsed after World War I.

paddy (padi) rice   A term used in Monsoon Asia referring to the rice plant; also to rice grown in flooded fields.

Paleolithic Age   The earlier part of the Old World Stone Age during which time the human race lived by hunting, fishing, and gathering.

Palestine Liberation Organization (PLO)   An umbrella organization created by the Arab states in 1964 to control and coordinate Palestinian opposition to Israel. Originally subservient to the wishes of the Arab states, the PLO is now dominated by the independent, nationalist ideology of El-Fatah, the largest and most moderate of the Palestinian resistance groups.

pampas A Spanish term for grasslands; also used as a proper noun to identify the most important region of Argentina.

patron system The economic and social relationships between the large landowner in Latin America and his workers; a paternalistic relationship.

pattern The distributional arrangement of a phenomenon as, for example, the distribution of population, climate, or cropping.

permafrost Permanently frozen ground.

petrochemical industries Manufacturing that uses raw materials derived from petroleum or coal; examples of such industries are fertilizer, synthetic rubber and fiber, medicines, and plastics.

photoperiod Length of daylight or the active period of photosynthesis.

place utility The change in location of a product usually to increase its value as, for example, the movement of oil from the Arctic to the eastern United States.

placer deposits A concentration of one or more minerals in alluvial materials.

plantation A large agricultural producing unit emphasizing one or two crops that are sold and having distinct labor and management groups.

plural society *See* cultural pluralism.

polder A tract of land reclaimed from the sea and protected by dikes; about 40% of the Netherlands is polders.

population explosion The rapid natural increase in human numbers within a short time period, generally within the last hundred years.

population pressure The strain or demands a population places on an area's resources; the ratio of number of animals or people to carrying capacity.

poverty Material deprivation of such severity as to affect biologic and social well-being. The lack of income or its equivalent necessary to provide an adequate level of living.

primary activities Economic pursuits involving production of natural or culturally improved resources, such as agriculture, livestock raising, forestry, fishing, and mining.

primary processing Manufacturing that uses the products of primary activities such as wheat, iron ore, fish, and trees; the products of primary processing may be sold to consumers (e.g., fish) or used as a raw material for further processing (e.g., steel bars).

primate city An urban center more than twice the size of the next largest city; a primate city has a high proportion of its nation's economic activity. Such cities are most obvious in the poor world.

principle of the use of best resources first If accessible, high-quality resources will yield a greater economic return than resources of lower quality; therefore, high-quality resources tend to be used first.

producer goods A product used to create income, such as a tractor, newspaper press, or processing machine.

productive capacity The total amount of resources that can be marshaled in an area under a given level of technology. Although it is a useful concept, productive capacity cannot be accurately measured.

proven reserves The amount of a given mineral known to exist that is economically feasible to mine.

pulse Certain cultivated leguminous plants used for food, especially peas and beans.

push-pull migration A theory used to explain the movement of people from rural areas to urban centers; the migrants are forced out of one area by limited opportunity and attracted to cities by perceived advantages.

qanat (kanat) An underground tunnel, sometimes several miles long, tapping a water source; the water flows through the tunnel by gravity.

region A portion of the earth that has some internal feature of cohesion or uniformity as, for example, the trade area of a city or an area of similar climate.

regional specialization The division of production among areas; each area produces those goods or provides those services in which it has some advantage and trades for goods that can be produced more cheaply elsewhere.

Renaissance The reawakening of arts, letters, and learning in Europe from the fourteenth to sixteenth centuries; the transition period between medieval and modern Europe.

Repelita Indonesian five-year plan for modernization and economic development of the country. For example, Repelita III (1979–84) is the third five-year plan.

431

resource Anything a person can use to satisfy a need.

Sahel The semiarid grassland area along the southern margin of the Sahara Desert in western and central Africa.

salinization The accumulation of salts in the upper part of the soil often rendering the land useless. Salinization commonly occurs in moisture-deficient areas when insufficient or salt-laden water is used for irrigation.

Sandinistas The term applied to Nicaraguan rebels who overthrew the Somoza regime in 1979 and who formed the post-Somoza government of Nicaragua. The Sandinistas took their name from an early Nicaraguan revolutionary.

secondary activities The processing of material to add form utility; manufacturing.

secondary processing Manufacturing that uses the products of primary processing, such as flour and steel bars, to produce a commodity that is more valuable, such as bread and automobiles.

sedentary (settled) agriculture A farming system based on continual cropping or use of the same fields.

Sertão A Portuguese term used in Brazil to refer to the frontier spirit and life; the backlands.

Shang Dynasty The period from about 1765 to 1123 B.C. in China, during which most of the cultural attributes of modern China were established.

shantytown An urban area of low socioeconomic characteristics common in most cities in the poor world. Shantytown dwellers are often squatters. Shantytowns differ from slums in the sense that slums are in the process of decay whereas shantytowns may be in the process of improvement.

shatter belt A politically unstable region, especially Eastern Europe before Soviet domination.

shifting agriculture A farming system based on periodic change of cultivated area, land rotation.

Sinitic Chinese or Chinese related.

softwood Gymnosperm trees, commonly called needle-leafed and usually evergreen.

Songhai Empire A west African trading state that flourished in the fifteenth and sixteenth centuries and centered along the middle Niger River.

Soviet Socialist Republic (SSR) The highest-level political-territorial division of the USSR, in which major ethnic groups are permitted some semblance of self-identity.

sovkhoz A Soviet state farm operated by the state with paid employees.

spatial organization The structures and linkages of human activities in an area.

split technology The presence of two forms of production. One form is characterized by modern methods of production and the other by traditional production practices. Industrial split technology is common in Japan.

stages of economic growth A theory developed by Walt Rostow in which five stages of economic organization are recognized: traditional society, preconditions for takeoff, the takeoff, drive to maturity, and high mass consumption.

standard of living The material well-being judged to be adequate by a society or societal subgroup as measured by diet, housing, and clothing.

Standard Metropolitan Statistical Area (SMSA) A governmental designation in the United States for a central city of at least 50,000 inhabitants and the surrounding counties.

steppe A large grassland region in southwestern Soviet Union. Any area of short grass; also an arid climate more moist than a desert but still water-deficient.

subsistence agriculture A farming system in which the farmer and his family consume most or all of the production; noncommercial.

suspended culture A society whose behavioral patterns undergo little or no change.

sustained yield forestry Harvesting timber only at its annual growth rate.

taconite A low-grade iron ore formerly considered worthless but now, because of technological advances, an important resource.

taiga The large coniferous forest extending across northern Soviet Union.

Taiping Rebellion One of China's major uprisings (1850–64) that proved almost fatal to the Manchu dynasty. The rebellion arose out of local conflicts, but quickly grew and spread throughout most of the country. Embracing vague principles of equality and religion, the rebellion failed, but it demonstrated the weaknesses of Manchu control.

tariff   The system of duties or customs imposed by a nation on imports and exports.

technocratic theory   A theory that asserts technology increases at a rate greater than population.

Tectonic plate   The theory that continental portions of the earth are divided into sections that ''float'' on a denser mass. The boundaries between plates are zones of instability in which volcanism, earthquakes, and mountain building occur.

Tennessee Valley Authority (TVA)   A United States regional development commission charged with the planning and execution of development projects in the Tennessee River Valley.

terra roxa   A latosol found in Brazil and other tropical areas that is especially suitable for coffee cultivation.

tertiary activities   Economic pursuits in which a service is performed, such as retailing, wholesaling, government, teaching, medicine, repair, and recreation.

Third World   The countries aligned with neither Communist nor Western nations; generally these are poor nations.

threshold   The number of customers needed in order to support a particular activity or business.

time cushion   A period made available by technological advances to bring population growth under control before population food demands become greater than the supply; a Neo-Malthusian concept. Technological advances include the Industrial Revolution, and more recently, the Green Revolution.

Tokaido Megalopolis   A large multi-nuclei urbanized region extending from Tokyo to Osaka in Japan.

Tokugawa period   The period (1615–1867) during which the focus of power in Japan shifted to the Kanto Plain area and many of the modern Japanese characteristics were firmly fixed in the culture.

Trans-Amazon Highway   An ambitious project to construct an east-west highway the length of the Brazilian Amazon; part of a larger project to develop the Amazon Basin.

transhumance   The practice of moving animals seasonally between summer alpine and winter lowland pastures.

Treaty of Rome   The agreement signed in 1957 by the Benelux nations, France, West Germany, and Italy that created the European Economic Community.

tribalism   Allegiance primarily to the local group and continued observation of the group's customs and life style.

tsetse fly   A member of the genus *Glossina*, common in parts of Africa, that transmits a number of diseases harmful to humans and domestic animals.

tubers   Food plants with edible roots or other subterranean parts such as the white potato, sweet potato, manioc, and yams.

tundra   Originally a vast treeless plain in the Soviet Union; now the treeless area polarward of the limit of forest and equatorward of polar ice cap; also the climate of this type region.

ujamaa   A Swahili word meaning ''familyhood'' that expresses a feeling of community and of cooperative activity. The word is used by the Tanzanian government to indicate a commitment to rapid economic development according to principles of socialism and communal solidarity.

underdeveloped   Used in this book to identify the nations or regions that have a low level of economic production per person; *poor, undeveloped, emerging, backward, less developed,* and *developing* have also been used to designate such areas.

underemployment   The incomplete use of labor either because a person works only part-time or seasonally or because individuals are employed inefficiently as, for example, the use of five people to perform a task that two could do, or a person working below his or her skill level.

universalizing religions   Religions considered by their adherents as appropriate for all mankind. Proselytizing is common.

untouchables   The lowest group in the hierarchy of the Hindu caste system comprising basically Dravidian people who have the most menial occupations.

urbanism   The social-cultural aspects of city living; the way of life of those who live in cities.

urbanization   The agglomeration of people, the process of becoming city dwellers.

U-turn migration   The movement from large urban centers to the suburbs and middle-sized cities, principally in the United States, Japan, and Western Europe.

valorization   The attempt to control a commodity's price by a nation or a group of nations that produces the item.

value added   The residual value of a product or service after all production costs or other charges have been subtracted.

virgin and idle land program   The attempt by the USSR to expand agriculture to the east into Siberia and central Asia.

Walking on Two Legs   The Chinese development approach that stresses both traditional and modern production methods, city and rural areas, and large and small production units.

Walloons   French-speaking Belgians, who are concentrated in southern Belgium; one of the nation's two principal national groups.

White Australia Policy   The attempt to exclude nonwhites from migrating permanently to Australia and to encourage whites, especially British, to settle in Australia. Officially this policy was termed the Restricted Immigration Policy.

wood pulp   Ground wood used for newsprint and book paper; usually softwood.

xerophytes   Plants that are able to withstand moisture deficiency; vegetation common to desert areas.

Yayoi   The culture group located in Kyushu, from which many Japanese traits and complexes evolved.

zaibatsu   A large Japanese financial enterprise similar to a conglomerate in the United States but generally more integrated horizontally and vertically.

zero population growth   The maintenance of a stable population in which births and in-migration are balanced by deaths and out-migration.

Zulu   A large African tribe of relatively high cultural attainment located along the southeastern coast.

APPENDIX

From	To	Factor
Acres	Hectares	Multiply acres by 0.4047
Centimeters	Inches	Multiply centimeters by 0.3937
Cubic feet	Cubic meters	Multiply cubic feet by 0.0283
Cubic meters	Cubic feet	Multiply cubic meters by 35.3145
Feet	Meters	Multiply feet by 0.3048
Gallons (U.S.)	Liters	Multiply gallons by 3.7853
Hectares	Acres	Multiply hectares by 2.4710
Inches	Centimeters	Multiply inches by 2.54
Kilograms	Pounds	Multiply kilograms by 2.2046
Kilometers	Miles	Multiply kilometers by 0.6214
Liters	Gallons	Multiply liters by 0.2642
Meters	Feet	Multiply meters by 3.2808
Miles	Kilometers	Multiply miles by 1.6093
Square kilometers	Square miles	Multiply square kilometers by 0.3861
Square miles	Square kilometers	Multiply square miles by 2.59

Selected National Statistics

Nation	Population (millions)	Area Square Miles (thousands)	Area Square Kilometers (thousands)	Birth Rate (%)	Death Rate (%)	Infant Mortality (less than age 1) (%)
Anglo-America						
Canada	24.9	3831	9922	1.5	0.7	1.1
United States	234.2	3679	9529	1.6	0.9	1.1
Western Europe						
Austria	7.6	32	84	1.3	1.2	1.3
Belgium	9.9	12	31	1.3	1.1	1.2
Denmark	5.1	17	43	1.0	1.1	0.8
Finland	4.8	130	337	1.3	0.9	0.8
France	54.6	211	546	1.5	1.0	1.0
Germany (West)	61.5	96	249	1.0	1.2	1.2
Greece	9.9	51	132	1.5	0.9	1.8
Iceland	0.2	40	103	1.9	0.7	0.8
Ireland	3.5	27	70	2.1	0.9	1.1
Italy	56.3	116	301	1.1	1.0	1.4
Luxembourg	0.4	1	3	1.2	1.1	1.1
Netherlands	14.4	16	41	1.3	0.8	0.8
Norway	4.1	125	324	1.3	1.0	0.8
Portugal	9.9	34	88	1.6	1.0	2.6
Spain	38.4	195	505	1.4	0.8	1.0
Sweden	8.3	174	405	1.1	1.1	0.7
Switzerland	6.5	16	41	1.2	0.9	0.9
United Kingdom	56.0	94	244	1.3	1.2	1.2
Eastern Europe						
Albania	2.9	11	29	2.7	0.7	4.7
Bulgaria	8.9	43	111	1.4	1.1	2.0
Czechoslovakia	15.4	49	128	1.6	1.2	1.7
Germany (East)	16.7	42	108	1.4	1.4	1.2
Hungary	10.7	36	93	1.3	1.4	2.1
Poland	36.6	121	313	1.9	0.9	2.1
Romania	22.7	92	238	1.8	1.0	2.9
Yugoslavia	22.8	99	256	1.7	0.9	3.1
Soviet Union	272.0	8600	22275	1.9	1.0	3.3
Japan	119.2	146	378	1.3	0.6	0.7
Australia	15.3	2968	7687	1.6	0.7	1.0
New Zealand	3.2	104	269	1.6	0.8	1.2
Latin America						
Argentina	29.1	1068	2766	2.4	0.9	4.1
Bahamas	0.2	5	14	2.0	0.5	3.2
Barbados	0.3	0.2	1	1.7	0.8	2.3
Belize	0.2	9	23	3.2	0.8	6.3
Bolivia	5.9	424	1098	4.3	1.6	13.0
Brazil	131.3	3286	8512	3.1	0.8	7.6
Chile	11.5	292	757	2.2	0.7	3.3
Colombia	27.7	440	1139	2.8	0.8	5.6
Costa Rica	2.4	20	51	2.9	0.4	1.9
Cuba	9.8	44	115	1.4	0.6	1.9
Dominica	0.1	0.3	1	2.1	0.5	2.0

Population Under 15 Years Old (%)	Life Expectancy at Birth	Annual Growth Rate (%)	Per Capita GNP (dollars)	Daily Food Supply (calories)	Annual Per Capita Energy Consumption (gallons of oil)	Percent of Labor Force in Agriculture
23	74	0.8	11,230	3374	2690	5
23	74	0.7	12,530	3576	2469	2
19	72	0.1	10,250	3535	1041	9
19	72	0.1	11,980	3583	1348	3
20	74	− 0.1	12,790	3418	1195	7
23	73	0.4	10,380	3100	1251	12
22	74	0.5	12,130	3434	995	9
17	72	− 0.2	13,520	3381	1325	4
22	73	0.6	4,540	3400	568	38
26	76	1.2	12,550	2950	640	9
31	73	1.2	5,350	3541	763	19
21	73	0.1	6,830	3428	687	11
17	72	0.1	13,900	3320	1195	8
21	74	0.4	11,140	3338	1348	6
21	75	0.3	13,800	3175	2383	8
25	71	0.6	2,534	3076	299	25
25	73	0.7	5,770	3149	564	15
19	75	0.0	14,500	3221	1700	5
19	75	0.3	17,150	3485	1027	5
20	73	0.1	8,950	3336	1127	2
36	69	2.0	840	2730	221	61
22	72	0.3	4,413	3611	1080	38
24	70	0.4	5,610	3340	1365	11
19	72	0.0	7,286	3641	1743	10
22	70	0.0	4,188	3521	814	16
25	71	1.0	4,187	3656	1160	31
26	71	0.8	2,546	3444	962	33
24	70	0.8	2,789	3445	488	31
25	70	0.8	4,701	3460	1224	23
23	76	0.7	10,330	2949	852	13
25	73	0.9	11,190	3428	1394	6
29	73	0.8	7,580	3345	978	9
28	70	1.5	2,560	3347	407	13
44	69	1.6	3,632	2299	1457	—
28	71	0.9	3,500	3172	195	17
49	61	2.4	1,110	2504	120	29
44	50	2.7	601	1974	94	50
38	63	2.3	2,214	2562	212	40
32	66	1.6	2,560	2656	239	20
38	63	2.0	1,334	2364	188	27
37	70	2.5	1,476	2550	168	30
29	72	0.8	1,410	2720	230	24
—	58	1.6	750	2081	41	—

(continued)

Selected National Statistics, *continued*

Nation	Population (millions)	Area Square Miles (thousands)	Area Square Kilometers (thousands)	Birth Rate (%)	Death Rate (%)	Infant Mortality (less than age 1) (%)
Dominican Republic	6.2	19	49	3.5	0.9	6.7
Ecuador	8.5	109	284	4.2	1.0	8.2
El Salvador	5.0	8	21	3.5	0.8	5.3
Grenada	0.1	0.1	0.3	2.5	0.7	1.5
Guatemala	7.9	42	109	4.2	1.0	6.6
Guyana	0.8	83	215	2.8	0.7	4.3
Haiti	5.7	11	28	4.1	1.4	11.3
Honduras	4.1	43	112	4.5	1.0	8.7
Jamaica	2.3	4	11	2.7	0.6	2.8
Mexico	75.7	762	1973	3.2	0.6	5.5
Nicaragua	2.8	50	130	4.7	1.1	8.9
Panama	2.1	30	78	2.9	0.6	3.4
Paraguay	3.5	157	407	3.5	0.8	4.6
Peru	19.2	496	1285	3.7	1.1	8.7
Puerto Rico	3.4	3	9	2.3	0.6	1.9
St. Lucia	0.1	0.2	1	2.8	0.7	3.0
St. Vincent	0.1	0.2	1	2.7	0.6	6.0
Surinam	0.4	63	163	2.8	0.8	3.6
Trinidad	1.2	2	5	2.5	0.6	2.6
Uruguay	3.0	68	176	1.8	1.0	3.4
Venezuela	18.0	352	912	3.3	0.5	4.1
Subsaharan Africa						
Angola	7.6	481	1247	4.7	2.2	15.3
Benin	3.8	43	113	4.9	1.9	15.3
Botswana	0.9	232	600	5.1	1.7	8.2
Burundi	4.5	11	28	4.6	2.2	12.1
Camaroon	9.1	184	475	4.4	1.8	10.8
Cape Verde	0.4	2	4	2.8	0.9	8.1
Central African Republic	2.5	241	623	4.5	2.0	14.7
Chad	4.7	496	1284	4.5	2.3	14.7
Comoros	0.4	1	3	4.6	1.7	9.2
Congo	1.7	132	342	4.4	1.8	12.8
Djibouti	0.3	9	23	4.9	2.3	7.3
Equatorial Guinea	0.3	11	28	4.2	1.8	14.2
Ethiopia	31.3	472	1222	4.8	2.3	14.6
Gabon	0.7	103	268	3.5	2.0	11.6
Gambia	0.6	4	11	4.9	2.8	19.7
Ghana	13.9	92	239	4.8	1.6	10.2
Guinea	5.4	95	246	4.6	2.1	16.4
Guinea-Bissau	0.8	14	36	4.0	2.1	14.7
Ivory Coast	8.9	124	321	4.7	1.8	12.6
Kenya	18.6	225	583	5.4	1.3	8.6
Lesotho	1.4	12	30	4.1	1.3	11.4
Liberia	2.1	43	111	4.7	1.5	15.3
Madagascar	9.5	227	587	4.6	1.6	7.0
Malawi	6.8	46	118	5.1	1.8	17.1
Mali	7.3	479	1240	4.7	2.1	15.3
Mauritania	1.8	398	1031	5.0	2.2	14.2
Mauritius	1.0	1	3	2.6	0.7	3.5

Population Under 15 Years Old (%)	Life Expectancy at Birth	Annual Growth Rate (%)	Per Capita GNP (dollars)	Daily Food Supply (calories)	Annual Per Capita Energy Consumption (gallons of oil)	Percent of Labor Force in Agriculture
43	61	2.6	1,338	2094	103	50
45	60	3.1	1,220	2104	130	52
46	62	2.7	590	2051	70	51
—	71	1.8	872	2431	42	—
44	59	3.2	1,159	2156	50	56
38	70	2.0	723	2431	214	22
44	52	2.7	297	2100	13	74
47	58	3.5	591	2015	50	63
38	71	2.1	1,182	2660	278	22
44	66	2.6	2,250	2654	334	37
48	56	3.6	874	2446	91	40
39	70	2.3	1,908	2341	189	34
42	65	2.7	1,557	2824	50	50
42	58	2.6	1,122	2274	147	38
31	73	1.6	3,010	2460	718	4
50	67	2.2	850	2207	73	—
—	67	2.1	520	2284	40	—
49	68	2.0	2,950	2450	426	22
31	70	1.9	5,267	2694	1007	16
27	70	0.9	2,820	3036	255	11
42	67	2.8	4,170	2435	611	19
44	42	2.5	790	2133	42	60
46	47	3.0	326	2249	14	46
50	50	3.4	902	2100	—	81
39	42	2.4	235	2254	3	84
42	47	2.7	793	2069	30	83
34	61	1.8	320	2346	24	57
43	43	2.5	328	2242	11	88
42	40	2.1	140	1762	5	85
44	47	2.9	313	2180	18	64
44	47	2.6	1,108	2284	43	35
—	46	2.7	476	—	145	—
42	47	2.5	175	—	19	75
43	40	2.5	142	1754	4	80
34	45	1.5	3,909	2403	256	77
45	42	2.1	348	2281	17	78
47	50	3.2	402	1983	53	54
44	45	2.5	298	1943	17	82
39	42	1.9	185	2344	12	83
45	47	2.9	1,174	2517	47	79
50	55	4.1	432	2032	36	78
40	52	2.9	538	2245	—	87
48	54	3.2	536	2404	90	71
44	47	3.0	332	2486	19	87
48	47	3.3	200	2066	14	86
46	43	2.6	185	2117	6	88
46	43	2.8	484	1976	37	85
34	65	1.9	1,342	2576	81	28

(continued)

Selected National Statistics, *continued*

Nation	Population (millions)	Area		Birth Rate (%)	Death Rate (%)	Infant Mortality (less than age 1) (%)
		Square Miles (thousands)	Square Kilometers (thousands)			
Mozambique	13.1	302	782	4.5	1.8	11.4
Namibia	1.1	318	823	4.3	1.4	11.9
Niger	6.1	489	1267	5.1	2.2	14.5
Nigeria	84.2	357	925	5.0	1.7	13.4
Réunion	0.5	1	3	2.3	0.6	1.6
Rwanda	5.6	10	26	4.9	1.8	10.6
São Tomé	0.1	0.4	1	3.9	1.0	7.2
Senegal	6.1	76	197	4.8	2.2	14.6
Seychelles	0.1	0.2	1	2.9	0.7	24.0
Sierra Leone	3.8	28	72	4.5	1.9	20.6
Somalia	5.3	246	638	4.6	2.1	14.6
South Africa	30.2	471	1220	3.6	1.0	9.5
Sudan	20.6	968	2506	4.7	1.7	12.3
Swaziland	0.6	7	17	4.9	1.6	13.4
Tanzania	20.5	365	945	4.6	1.4	10.2
Togo	2.8	22	56	4.8	1.8	10.8
Uganda	13.8	91	236	4.6	1.5	9.6
Upper Volta	6.8	106	274	4.8	2.2	21.0
Zaire	31.8	906	2345	4.6	1.8	11.1
Zambia	6.2	291	753	4.8	1.6	10.5
Zimbabwe	8.4	151	391	4.7	1.3	7.3
North Africa and The Middle East						
Algeria	20.7	920	2382	4.6	1.4	11.6
Bahrain	0.4	0.3	1	3.7	0.9	5.3
Cyprus	0.7	4	10	2.0	0.9	1.8
Egypt	45.9	387	1002	4.3	1.2	10.2
Iran	42.5	636	1648	4.3	1.3	10.6
Iraq	14.5	168	435	4.6	1.2	7.7
Israel	4.1	8	21	2.4	0.7	1.5
Jordan	3.6	38	98	4.7	1.1	6.8
Kuwait	1.6	7	17	3.7	0.4	3.1
Lebanon	2.6	4	10	3.0	0.9	4.1
Libya	3.3	679	1760	4.7	1.3	9.9
Morocco	22.9	275	713	4.4	1.3	3.1
Oman	1.0	82	212	4.8	1.7	12.7
Qatar	0.3	4	10	3.1	0.9	5.3
Saudi Arabia	10.4	830	2150	4.4	1.3	11.2
Syria	9.7	71	185	4.6	0.8	6.1
Tunisia	6.8	63	164	3.5	1.0	9.8
Turkey	49.2	301	781	3.1	1.0	12.1
United Arab Emirates	1.4	32	84	3.1	0.7	5.3
Yemen, North	5.7	75	195	4.8	2.1	16.0
Yemen, South	2.1	129	334	4.8	1.9	14.4
South Asia						
Afghanistan	14.2	250	648	4.8	2.3	20.5
Bangladesh	96.5	56	144	4.9	1.8	13.5
Bhutan	1.4	18	47	4.1	1.9	14.9
India	730.0	1237	3204	3.6	1.5	12.2

Population Under 15 Years Old (%)	Life Expectancy at Birth	Annual Growth Rate (%)	Per Capita GNP (dollars)	Daily Food Supply (calories)	Annual Per Capita Energy Consumption (gallons of oil)	Percent of Labor Force in Agriculture
44	47	2.7	240	1906	28	67
44	52	2.9	1,410	2218	—	49
47	43	2.9	336	2139	10	91
48	49	3.3	873	1951	17	55
32	65	1.7	3,830	2649	123	29
47	47	3.1	250	2264	6	91
—	—	2.9	382	2060	25	—
45	43	2.6	499	2261	53	76
40	65	2.2	1,797	—	114	—
44	47	2.6	319	2150	18	60
44	43	2.6	282	2033	16	84
42	61	2.6	2,509	2780	696	30
44	48	3.0	380	2184	28	78
46	47	3.3	844	2281	—	74
46	52	3.2	299	2063	11	83
47	47	3.0	391	2069	23	68
45	54	3.1	356	2110	8	83
45	43	2.6	237	1875	6	83
45	47	2.8	225	2271	21	75
47	50	3.2	586	2002	172	68
48	53	3.4	815	2576	158	60
47	56	3.2	2,129	2372	134	32
39	66	2.9	7,490	—	2399	—
25	72	1.1	3,759	3047	300	35
39	56	3.1	654	2760	113	50
45	55	3.1	2,160	3138	243	40
46	56	3.4	3,020	2134	138	43
33	72	1.7	5,450	3141	728	7
48	61	3.6	1,623	2107	110	21
46	70	3.4	25,850	—	1269	2
37	66	2.1	1,070	2495	217	12
47	57	3.4	8,560	2985	472	20
46	57	3.1	869	2534	63	53
45	49	3.1	5,924	—	139	62
32	58	2.2	27,790	—	5045	—
44	54	3.2	12,720	2624	311	62
48	65	3.8	1,569	2684	194	32
40	59	2.5	1,417	2674	124	35
38	62	2.1	1,511	2905	161	54
31	63	2.3	25,660	—	2831	—
45	43	2.7	459	2192	15	76
46	45	2.9	512	1945	109	47
45	40	2.5	170	2695	18	79
45	47	3.1	144	2100	8	74
42	44	2.3	100	2028	—	93
39	50	2.1	253	2021	48	71

(continued)

Selected National Statistics, *continued*

Nation	Population (millions)	Area Square Miles (thousands)	Area Square Kilometers (thousands)	Birth Rate (%)	Death Rate (%)	Infant Mortality (less than age 1) (%)
Maldives	0.2	0.1	0.3	4.3	1.3	12.0
Nepal	15.8	54	141	4.4	2.1	14.9
Pakistan	95.7	320	829	4.3	1.5	12.4
Sri Lanka	15.6	25	66	2.8	0.6	3.7
East Asia						
China	1023.3	3692	9561	2.3	0.8	4.4
Hong Kong	5.2	0.4	1	1.7	0.5	1.0
Korea, North	19.2	47	121	3.2	0.8	3.4
Korea, South	41.3	38	98	2.5	0.8	3.4
Macao	0.3	0.006	0.02	2.6	0.8	7.8
Mongolia	1.8	604	1565	3.7	0.8	5.4
Taiwan	18.9	14	36	2.3	0.5	0.9
Southeast Asia						
Brunei	0.2	2	5	3.1	0.4	1.8
Burma	37.9	261	676	3.8	1.4	9.9
Indonesia	155.6	753	1950	3.2	1.5	9.2
Kampuchea	6.0	70	181	3.8	1.9	20.1
Laos	3.6	91	237	4.4	2.0	12.8
Malaysia	15.0	128	333	3.1	0.7	3.1
Philippines	52.8	116	300	3.4	0.8	5.4
Singapore	2.5	0.2	1	1.7	0.5	1.2
Thailand	50.8	198	513	2.6	0.7	5.4
Vietnam	57.0	127	329	3.7	0.9	9.9

SOURCES: *Population Data Sheet* (Washington, D.C.: Population Reference Bureau, 1983); *World Development Report, 1981* (Washington, D.C.: World Bank, 1981); supplemented by other international statistical sources.

Population Under 15 Years Old (%)	Life Expectancy at Birth	Annual Growth Rate (%)	Per Capita GNP (dollars)	Daily Food Supply (calories)	Annual Per Capita Energy Consumption (gallons of oil)	Percent of Labor Force in Agriculture
45	47	3.0	391	1797	—	—
42	44	2.3	156	2002	3	93
45	51	2.8	349	2281	44	57
35	66	2.2	302	2126	28	54
32	69	1.5	304	2453	167	71
25	76	1.2	5,460	2883	480	3
39	64	2.4	1,130	2837	569	50
33	64	1.7	1,720	2728	328	36
34	—	1.8	2,020	1987	76	—
42	64	2.9	780	2523	333	56
32	72	1.8	2,360	2805	440	37
35	66	2.7	11,890	2959	2839	—
41	54	2.4	183	2286	14	67
38	49	1.7	519	2272	47	59
35	37	1.9	—	1926	6	74
42	45	2.4	120	2082	20	76
39	64	2.4	1,817	2610	153	51
41	62	2.7	789	2189	71	47
26	71	1.2	5,220	3074	1242	2
40	61	2.0	769	1929	75	77
40	54	2.8	200	1801	28	71

INDEX